Temperate Fruit Crops: Theory to Practical

About the Editors

Dr. S.N. Ghosh (Dr. Satyanarayan Ghosh) started his career as Junior Horticulturist (Assistant Professor) in All India Co-coordinated Research Project in Cashew, where he worked for a period of almost more than 7-years after which he joined as Associate Professor and later as Professor of Horticulture on 27.07.1998. During his service as Junior Horticulturist, most significant achievement was release of a cashew variety-JHARGRAM-1 as well as development of package and practices for cultivation of cashew in West Bengal. Another significant achievement during his carrier was popularization of Mosambi-sweet orange in West Bengal through development and dissemination of location-specific production technology. For his significant contribution in Citrus, "Dr. Shyam Singh Best Scientist Award on Citrus Research -2009-2010", was conferred upon Dr. Ghosh by National Research Centre for Citrus(ICAR, Govt. of India), Nagpur, India. Success story of his last 20 years of concentrated efforts is introduction of grapes in West Bengal. He has developed technology for higher production of sweet table grapes in West Bengal. Besides, he has developed number of technologies for cultivation of Aonla, Pomegranate, Sapota, Ber, Bael, Custard apple, Jackfruit, Water apple, Wood apple, Mango, Guava, Karonda and Fig in red and laterite zone of West Bengal.

Dr. Ghosh shouldered additional responsibilities in many administrative position like Officer-in-Charge of All India Coordinated Research Project on Cashew (BCKV Centre); Associate Director of Research of the Regional Research Station, Jhargram, West Bengal; Head, Department of Fruits and Orchard Management, B.C.K.V.; Dean, Faculty of Horticulture, Bidhan Chandra Krishi Viswavidyalaya, West Bengal; Associate Dean, College of Agriculture, Susunia, Bankura, West Bengal. He had organized one National and one International Symposium on Minor Fruits in 2006 and 2011 respectively. *International Society of Minor Fruits, Medicinal & Aromatic Plants* (ISMFM&AP) formed under dynamic leadership of Dr. Ghosh and now he is President of the Society. The society (ISMFM&AP) Published a journal-International Journal of Minor Fruits, Medicinal and Aromatic Plants and Dr. Ghosh is associated with the journal as Editor-in-Chief (web site-http://www.ijmfmap.in) since its inception in 2015.

Dr. Ghosh has successfully guided 9 Ph.D. students and 10 M.Sc. students and handled 12 Research Projects as Principal Investigator funded by ICAR, State Government and Private Organizations. He has published more than 200 research papers in different reputed journals at national and international level. He has visited a number of countries like Morocco, China, Malaysia, Portugal, USA, Azerbaijan, Sri Lanka and Bangladesh and presented his research papers in International conference, seminar/ symposium.

Dr. Ghosh has published 5 nos. edited books on fruit crops viz., i. Tropical and Sub-tropical Fruit Crops: Crop Improvement and varietal wealth (Part-I and Part-II); ii. Temperate Fruit Crop Breeding: Domestication to Cultivar Development (Part-I and Part-II); iii. Breeding of underutilized fruit crops (Part-I and Part-II); iv. Underutilized Fruit Crops: Importance and Cultivation (Part-I and Part-II) and Minor Fruits: Nutraceutical Importance and Cultivation (Part-I, II and III).

Dr Amit Kumar presently is working as Assistant Professor in the Division of Fruit Science, Faculty of Horticulture, Sher-e-Kashmir University of Agricultural Sciences and Technology-Kashmir, Shalimar, Srinagar. He has worked on different temperate fruit crops viz. apple, pear, peach, nectarine, almond, walnut, strawberry and olive. At present, he is also associated with the improvement of temperate fruits on various researchable aspects viz. standardization of production and post-harvest technologies for High Density Planting of newly introduced varieties, raising of quality nursery planting material of temperate fruit crops, establishment of 'Field Gene bank' of different temperate fruits and rootstocks.

He has been associated with two externally funded and six university funded projects as Principal Investigator/Co-investigator. Guided four M.Sc. and one Ph.D. student. Published 68 research papers in National and International journals, 17 book chapters, 14 popular articles, 20 oral/poster presentations in International and national Conference/Congress/ Symposium.

He has received Award for Excellence in Research in 4th Faculty Branding Awards, Kolkata (West Bengal) in 2016. He is the recipient of Young Scientist Associate Award of Society for Recent Development in Agriculture, Meerut in 2016 and Young Scientist Award of Agricultural Technology Development Society, Ghaziabad in 2017.

Dr. Munib Ur Rehman working as Assistant Professor, Division of Fruit Science, Faculty of Horticulture, Sher-e-Kashmir University of Agricultural Sciences and Technology-Shalimar, Srinagar, since 2009. He has been associated with three externally aided projects as Co-PI and also been associated with research group working on crop improvement, pesticide compatibility studies and standardization of production protocols for temperate fruit and nut crops. He has guided two M.Sc. students besides post graduate teaching. He has published more than 25 research publications in different reputed journals of national and international level and other 10 review articles, book chapters and extension bulletins.

Dr R.R. Sharma, After post graduation from IARI, New Delhi, he joined as Scientist there in 1997. At present, he is working as Principal Scientist at the Division of Food Science & Postharvest Technology, IARI, New Delhi.

Dr. Sharma has made significant contributions in the field of production and post harvest management of fruits. He has developed six varieties of mango such as Pusa Arunima, Pusa Surya, Pusa Lalima, Pusa Pitamber, Pusa Pratibha, Pusa Shresth, which all are regular bearer and bear attractive coloured fruits, and have good shelf life. These varieties have become very popular among the farmers in different parts of India and abroad. He has recommended Sweet Charlie, a Floridian strawberry for commercial cultivation in India which has attracted strawberry growers in our country. He standardized the micropropagation protocol for grape and has provided solution for the problems of fruit cracking and unfruitfulness in lemon.

Dr. Sharma reported the occurrence of fruit pitting, a new disorder in mango and albinism, fruit malformation and buttoning in strawberry for the first time. He has standardized the shrink wrapping technology for apple and kiwifruit, for extending shelf-life and maintaining quality. He has also standardized fruit bagging technology for apple, guava and pomegranate, which has become quite popular among the farmers of Himachal Pradesh and Maharashtra

Dr. Sharma is a prolific writer and has the distinction of publishing 130 articles in journals of International and national repute, 10 books, 30 book chapters, 10 training manuals and 170 popular articles. His total citations are 2470 with an h-index of 22 and i10-index of 55.

He has served as International Mango Registrar for 4 years (1994-2004). He is the recipient of several honours and awards such as Rajiv Gandhi award of Govt. of India (2014), Education award, Govt. of India (2010) and Dr Rajinder Prasad Award of ICAR (2005 and 2009), Dr R.N. Award of IARI (2004 and 2009) and *Himachal Shri* award of Govt of Himachal Pradesh. He is Fellow of National Academy of Agricultural Sciences (2017), National Academy of Biological Science (2016, The Horticultural Society of India (2013) and the Bioved Research Society (2009). At present, he is member, Editorial Board of 10 International journals including prestigious journals such as Scientia Horticulturae (The Netherlands), International Journal of Fruit Science (UK), American Journal of Plant Science (US), Journal of Food Processing and Technology (US), and Stewart Postharvest Reviews (UK), and Agricultural Sciences (US). He is the Chief Editor of International Journal of Food Processing and Postharvest Technology, India.

Temperate Fruit Crops: Theory to Practical

Dr. S.N. Ghosh
Faculty of Horticulture, Bidhan Chandra Krishi Viswavidyalaya
Mohanpur, Nadia, West Bengal, India- 741252

Dr. Amit Kumar
Division of Fruit Science, Faculty of Horticulture
Sher-e-Kashmir University of Agricultural Sciences and Technology-Kashmir
Shalimar, Srinagar, Jammu and Kashmir-190 025, India

Dr. Munib Ur Rehman
Division of Fruit Science, Faculty of Horticulture
Sher-e-Kashmir University of Agricultural Sciences and Technology- Kashmir
Shalimar, Srinagar, Jammu and Kashmir-190 025, India

Dr. R.R. Sharma
Division of Food Science and Postharvest Technology
ICAR-Indian Agricultural Research Institute, New Delhi- 110 012, India

nipa

NEW INDIA PUBLISHING AGENCY
New Delhi – 110 034

NEW INDIA PUBLISHING AGENCY
101, Vikas Surya Plaza, CU Block, LSC Market
Pitam Pura, New Delhi 110 034, India
Phone: + 91 (11) 27 34 17 17 Fax: + 91 (11) 27 34 16 16
Email: info@nipabooks.com
Web: www.nipabooks.com

Feedback at feedbacks@nipabooks.com

ISBN: 978-93-87973-15-2

Composed and Designed by NIPA

Preface

Fruits are grown extensively under varying climatic conditions ranging from temperate to tropical. Fruits are regarded as protective food as they are rich source of essential minerals, vitamins and antioxidants, which protect us from several fatal diseases. In the world scenario, temperate fruit crops dominate over tropical or subtropical fruits whereas in India, temperate fruits are grown largely in west Himalayan ranges. Temperate zones for fruit culture cannot be defined exactly by lines of latitude or longitude but are usually regarded as including those areas where frost in winter occurs, even though rarely. Temperate fruits are valued for their attractiveness, taste and nutritional quality. Besides, they are also an important source of foreign exchange. The cultivation of temperate fruits is the mainstay of the economy of hill farmers of the country. However, their national productivity is far below the international level as their production is encountered with several complex problems, which if not addressed in time, may result in complete crop failure.

Hence, there is an urgent need to address the causes of low productivity so as to meet the increasing demand, especially in view of no further significant area expansion and burgeoning population pressure. One of the major factors behind the low productivity is use of low productivity cultivars, which is very often seen in many areas, lack of quality planting materials, poor adoption of improved technologies of cultivation such as provision for adequate pollination, high density planting, insufficient infrastructure for input supply etc. Furthermore, to aggravate this situation, different biotic and abiotic stresses like drought, low snow fall, insect-pests and diseases are ever there to challenge the temperate fruit cultivation. Therefore, it is necessary to have a sound knowledge of temperate fruit cultivation so as to achieve the goal of higher productivity with superior quality, particularly in changing and challenging environment.

Thus, keeping in view, the challenges faced by the temperate fruit industry and a uniform syllabus approved by the ICAR for undergraduate and PG students, we planned to write an edited book entitled '*Temperate Fruit Crops: Theory to Practical*'. The book is an effort to summarize the current research information on temperate fruits in a comprehensive treatise. Authors attempted to provide the information and physiology behind current cultural practices as well as future trends.

The book will have 15 chapters on important temperate fruits such as apple, pear, apricot, peach, plum, cherry, strawberry, kiwi, almond, olive, hazelnut, pistachio nut, pecan nut, quince and walnut. Besides dealing with principles and cultivation practices of 15 temperate fruits, this book contains information on all difficulties faced by a farmer with practical application on all cultivation aspects. Efforts have been made to provide the maximum information about the nutritive and medicinal value, origin and distribution, area of cultivation, genus and different species, cytogenetic, botanical description including flowering and fruiting, physiology of flowering, fruit set and development, abiotic factors limiting production, varieties, suitable climate and soil, different methods of propagation including micropropagation, all aspects of cultivation, insect-pest, diseases and their control, physiological disorder and their control measures, harvesting, yield, post harvest management and storage, value addition, economics of cultivation, trade and marketing, organic cultivation, effect of climate change and future research thrust of individual crop in such a way that the information may be easily transmitted among the scientists, researchers, students, policy makers and entrepreneur. We are of the strong opinion that this book would certainly help the students in writing any UG or PG level examination. In addition, this book will also serve the purpose of fruit growers due to its well fabricated content.

This book could not have come in its present form without the helps received from various colleagues and students in different stages during the preparation of the manuscript and for which they are profusely thanked.

We acknowledge our family members who were ever there to extend unconditional support during the preparation of this book and we thank all of them from the bottom of our hearts. We thank the group of outstanding contributors who made this project possible. Each is recognized as an authority in a particular research area and enthusiastically contributed his or her knowledge to making this book a fine source of information.

We extend our gratefulness to New India Publishing Agency (NIPA), 101, Vikas Surya Plaza, CU Block, LSC Market, Pitam Pura, New Delhi-88, India for their keen interest in publishing this important book and providing generous help in bringing this book in befitting manner.

Dated: April, 2018

S.N. Ghosh
Amit Kumar
Munib Ur Rehman
R.R. Sharma

Contents

PART-I

List of Contributors

Akash Sharma
Division of Fruit Science, Sher-e-Kashmir University of Agricultural Sciences and Technology-Jammu, Jammu & Kashmir, India

Amit Kumar
Division of Fruit Science, Sher-e-Kashmir University of Agricultural Sciences and Technology-Kashmir, Shalimar, Srinagar, Jammu &Kashmir-190 025, India
Email: khokherak@rediffmail.com

Anirudh Thakur
Department of Fruit Science Punjab Agricultural University, Ludhiana, Punjab-141 004, India
Email: anirudhthakur@pau.edu

Babita
Department of Fruit Science, Dr. Y S Parmar University of Horticulture and Forestry Nauni -Solan Himachal Pradesh- 173230, India

D.P. Sharma
Department of Fruit Science, College of Horticulture, Dr. Y S Parmar University of Horticulture and Forestry, Nauni-Solan, Himachal Pradesh-173230, India
Email: dptabo@gmail.com

Irfan Ahmad Bisati
Division of Fruit Science, Faculty of Horticulture, Sher-e-Kashmir University of Agricultural Sciences and Technology-Kashmir, Shalimar, Srinagar-191121, Jammu & Kashmir, India

Ejaz Ahmad Parray
Division of Fruit Science, Faculty of Horticulture, Sher-e-Kashmir University of Agricultural Sciences and Technology-Kashmir, Shalimar, Srinagar-191121, Jammu & Kashmir, India

G.H. Rather
Division of Fruit Sciences, Faculty of Horticulture, Sher-e-Kashmir University of Agricultural Sciences and Technology-Kashmir, Shalimar, Srinagar-191121, Jammu & Kashmir, India
E-mail: drghassan@rediffmail.com

J. S. Chandel
Department of Fruit Science, Dr. Y. S. Parmar University of Horticulture and Forestry Nauni, Solan, Himachal Pradesh-173230, India
Email: chandeljs@yahoo.co.in

Jahangir Ahmad Baba
Division of Fruit Sciences, Faculty of Horticulture, Sher-e-Kashmir University of Agricultural Sciences and Technology-Kashmir, Shalimar, Srinagar-191121, Jammu & Kashmir, India

Kousar-Javaid
Division of Fruit Science, Faculty of Horticulture, Sher-e-Kashmir University of Agricultural Sciences and Technology-Kashmir, Shalimar, Srinagar-191121, Jammu & Kashmir, India

M. Maqbool Mir
Division of Fruit Sciences, Faculty of Horticulture, Sher-e-Kashmir University of Agricultural Sciences and Technology-Kashmir, Shalimar, Srinagar-191121, Jammu & Kashmir, India

M.U. Rehman
Division of Fruit Sciences, Faculty of Horticulture, Sher-e-Kashmir University of Agricultural Sciences and Technology-Kashmir, Shalimar, Srinagar-191121, Jammu & Kashmir, India
Email:munib11.mr@gmail.com

Mahital Jamwal
Division of Fruit Science, Sher-e-Kashmir University of Agricultural Sciences and Technology-Jammu, Jammu, Jammu & Kashmir, India

N. Sharma
Department of Fruit Science, College of Horticulture, Dr. Y S Parmar University of Horticulture and Forestry, Nauni-Solan, Himachal Pradesh-173230, India

Nimisha Sharma
Division of Fruits and Horticulture Technology, Indian Agricultural Research Institute, New Delhi-110012, India

Nirmal Sharma
Division of Fruit Science, Sher-e-Kashmir University of Agricultural Sciences and Technology Jammu, Jammu, Jammu & Kashmir, India
Email: nirmalsharma.77@rediffmail.com

Parshant Bakshi
Division of Fruit Science, Faculty of Agriculture, Sher-e Kashmir university of Agricultural *Sciences and Technology of Jammu. Chatha-180009, India*
Email:bakshi_parshant@rediffmail.com

Pramod Verma
Department of Fruit Science, Dr. Y. S. Parmar University of Horticulture and Forestry, Nauni Solan, Himachal Pradesh-173230, India

R.R. Sharma
Division of Food Science and Postharvest Technology, ICAR-Indian Agricultural Research Institute, New Delhi- 110 012, India, *E-mail: rrs_fht@rediffmail.com*

Rehana-Javaid
Division of Fruit Science, Faculty of Horticulture, Sher-e-Kashmir University of Agricultural Sciences and Technology-Kashmir, Shalimar, Srinagar-191121, Jammu & Kashmir, India

Rimpika
Department of Fruit Science, College of Horticulture, Dr. Y S Parmar University of Horticulture and Forestry, Nauni-Solan, Himachal Pradesh-173230, India

Shabbir Hussain
Division of Fruit Science, Sher-e-Kashmir University of Agricultural Sciences and Technology Kashmir, Shalimar, Srinagar, Jammu and Kashmir-190 025, India

Sayan Sau
Department of Fruits and Orchard Management, Faculty of Horticulture
Bidhan Chandra Krishi Viswavidyalaya, Mohanpur, Nadia, West Bengal-741252, India

S. Lal
Central Institute of Temperate Horticulture, Old Airport, P.O., Rangreth, Srinagar-190007 Jammu & Kashmir, India
Email: shivcith@gmail.com

Shafat Ahmad Banday
Division of Fruit Science, Faculty of Horticulture, Sher-e-Kashmir University of Agricultural Sciences and Technology-Kashmir, Shalimar, Srinagar-191121, Jammu & Kashmir, India

Shahid Qayoom
Division of Fruit Science, Faculty of Horticulture, Sher-e-Kashmir University of Agricultural Sciences and Technology-Kashmir, Shalimar, Srinagar-191121, Jammu & Kashmir, India

Tashi Angmo
Division of Fruit Science, Faculty of Horticulture, Sher-e-Kashmir University of Agricultural Sciences and Technology-Kashmir, Shalimar, Srinagar-191121, Jammu & Kashmir India

Umar Iqbal
Division of Fruit Sciences, Faculty of Horticulture, Sher-e-Kashmir University of Agricultural Sciences and Technology-Kashmir, Shalimar, Srinagar-191121, Jammu & Kashmir, India

V.K. Wali
Division of Fruit Science, Sher-e-Kashmir University of Agricultural Sciences and Technology Jammu, Jammu, Jammu &Kashmir, India

Vishal S Rana
Department of Fruit Science, Dr. Y.S. Parmar University of Horticulture and Forestry Nauni-Solan Himachal Pradesh-173230, India
Email: drvishal_uhf@rediffmail.com

Yasmeen-Gull
Department of Physics, Government Degree College, Pulwama-192301, Jammu &Kashmir
India, Email: yasmunib@gmail.com

Zubair Ahmad Tak
NICRA, KVK-Malangpora, Sher-e-Kashmir University of Agricultural Sciences and Technology Kashmir, Shalimar, Srinagar-191121, Jammu & Kashmir, India

9

Pear (*Pyrus* sp.)

Gh. Hassan Rather, Rehana-Javid, M. M. Mir, Umar Iqbal Irfan Ahmad Bisati, Ejaz Ahmad Parray and Tashi Angmo

1. Introduction

Pear is among the oldest of the world's fruit crops, figuring in both Bible and the tales of Homer. They are by far the most important of the deciduous tree fruits, are widely grown in temperate and increasingly, in tropical regions, and figure prominently in world trade. The fruit is primarily grown for the fresh fruit market, which is much more remunerative than that for processing.

Pear (*Pyrus communis*, Rosaceaea family), is a deciduous fruit, grown under a wide variety of climatic regimes, ranging from cold dry temperature hilly conditions to warm humid subtropical conditions. It is grown in many places of India like Jammu and Kashmir, Himachal Pradesh, Karnataka, Tamil Nadu, Kerala, Utter Pradesh. Sandy pear plants were brought to Amritsar by travellers from China who settled at Harsha Chinna village near Amritsar about 2,500 years ago. The Kashmir valley represents one of the major Pear (*Pyrus communis*.L.) growing areas of the world. Pear was introduced in India from China. The time at which Pear was introduced to Kashmir is not precisely known, however Yukteshwar Kumar writes in his book "A History of Sino-Indian Relations" that peach and pear came to India during the early years of the Eastern Han period (25 AD to 220 AD). This pre historic fruit has been under cultivation both in Europe and Asia for about 3000 years (Hedrick *et al.*, 1921). Between 16th and 18th centuries, many European writers travelling Kashmir noted fantastic royal gardens of the Mughals, mentioning the abundance of peach, pear, plum and cherry trees, with fountains spouting cold, icy water. Pear (*Pyrus communis*. L), originating from the German word 'peer' meaning fruit, is a fascinating fruit steeped in rich history. This fruit is known as 'tang' in Kashmiri language. Low chill requiring pears are being cultivated in plains of

North India. Cultivars requiring high chilling hours (900-1000) are cultivated at higher hills of Jammu and Kashmir, Himachal Pradesh and Uttarakhand. Majority of the Pear production is actually limited to Kashmir however, low chilling cultivars grow in subtropical regions.

1.1. Nutritional and medicinal values

The pear fruits available in the market belong either to *P. communis* or *P. pyrifolia*. Pear fruits are excellent source of vitamin C, less allergenic than many other fruits and its juice is sometimes used as the first juice introduced to infants (Vadivel and Janardhanam, 2005). Pears are consumed fresh and canned.

The pear is a nutritious fruit having enough essential nutrients. Pear predominantly contains carbohydrates, organic acids, vitamins, minerals, pectic substances, aroma constituents, tannins and enzymes (Table 1). Although no single plant can provide with adequate level of all nutrients required by human being, yet the pear fruit contains many essential nutrients like carbohydrates, proteins, ash and moisture, it also provides with some essential elements like calcium, phosphorus, sodium and magnesium moderate enough to meet with the recommended daily allowances. Moreover, the fruit is rich in amino acids (Table 2), of which 26 known amino acids are found in fruit seeds and 17 in pulp. Nine of these are considered as essential while eight are non-essential amino acids (Mahammad *et al.*, 2010).

Table 1: Composition of edible portion (pulp) of important types of pears grown in India

Components	Pear Bagugosha	Pear Kashmiri Nakh	Pear Country	Pear Kieffer
Moisture(g/100g)	86.5	83.6	86.0	86.3
Protein (g/100g)	0.4	0.2	0.2	0.2
Fat (g/100g)	0.1	0.3	0.1	0.2
Minerals(g/100g)	0.3	0.4	0.3	0.2
Fibre (g/100g)	2.1	0.6	1.0	1.4
OtherCarbohydrates (g/100g)	10.6	14.9	12.4	11.7
Calcium (mg/100g)	20	20	6	10
Phosphorous (mg/100g)10	20	20	10	10
Iron (mg/100g)	1.5	1.0	1.0	0.4
Vit A(IU/100g)	0	0	14	9
Vit B_1 (IU/100g)	-	-	0.02	0.03
Vit B2 (IU/100g)	-	-	0.03	0.02
Nicotinic acid mg/100g	0.2	0	0.2	0
Vit C(mg/100g)	1	3	0	7

(*Source*: Anon, 1969)

Table 2: Composition of Amino acids in pears seed and pulp. (mg/100g protein)

Amino acids	Seeds	Pulp
Protien	26.00 ± 0.01	14.38 ± 0.07
Leucine	6.86	4.98
Isoleucine	3.69	3.18
Methionine	1.05	0.72
Phenylalanine	3.76	3.13
Lysine	4.38	3.28
Therionine	2.20	2.12
Tyrosine	3.04	0.34
Valine	3.86	3.02
Aspartic	8.56	5.79
Glutamic	11.73	6.58
Serine	3.14	1.77
Glycine	3.67	2.81
Alanine	4.45	2.71
Histidine	3.26	2.87
Arginine	4.81	3.09
Cystine	1.68	0.56

(*Source*: Mahammad *et al.*, 2010

Pear, has been used in traditional medicine as to reduce excess weight, to cure gout, removes cholesterol, lowers blood pressure, normalizes heart pulse, helps in regeneration of tissues and improves immunity in infants, perfectly suitable for Diabetics, cures diarrhoea and Colitis. Also the bark of pear tree acts as Diuretic and Analgesic.

Pear is considered as weight loss diet due the presence of vitamins particularly Folic acid which is necessary for cellular construction & whose deficiency leads to cancer like that of gall bladder. It removes excess Cholesterol from the body and normalises the heart pulse. As it contains Calcium, it is good for growth of infants, help them in fighting infection, leads to elaboration of muscular mass and also eliminates ammonia from the organism's body which is due to the presence of amino acid Arginine. Boiled pears are good to cure gastric disorders. It helps in bone formation and keeps the nervous system in a good state. It is also important in expulsion of uric acid from the body thereby protects the organism from Gout.

2. Area and Production

In India Area and production of pear during 2014-15 was 44 thousand ha and 308 thousand tonnes respectively (Horticultural Statistics at a Glance, 2015).

3. History & distribution

The advantage with pear is that it can be cultivated in both temperate and subtropical regions of India, although it is more acclimatized to temperate zone. There are different accounts on the origin of Pear from North or South America to continent of Australia. The earliest written record of cultivated Asian pear groves in Japan is in the manuscript of the Emperor Jito in AD 693 (Kajiura, 1994). The European pear (*P. communis*) is grown in five major regions: Europe, North America, South America, South Africa, and Oceania, while production of Asian pear (*P. pyrifolia*) is concentrated in Asia. The Occidental Group are native to Eastern Europe and South Western Asia while as the Oriental Group is native to Eastern Asia including China, Japan and Manchuria. Vavilov (1951) proposed three centres of origin for cultivated pears i.e., Chinese centre, central Asiatic centre and near eastern centre. Zeyen and Zhukovsky (1975) proposed fourth centre for pear i.e., Siberian centre. Chinese centre comprising the region of north and central China, Japan and Korea. *P. pyrifolia, P. ussuriensis, P. betulifolia* and *P. calleryana* originated in this tract. Central Asiatic centre includes Western Tian-Shan, Pamir Alai, Uzbekistan, Tajkistan, Afghanistan and North West India. *P. communis, P. salicifolia, P. regeli* and *P. pashia* are indigenous species of this centre. Near Eastern centre comprising Asia Minor and Caucasus mountains. *P. syriaca* and *P. caucasus* species are found in this centre. Siberian centre species are *P. communis, P. cordata, P. nivalis* and *P. salvifolia.*

P. pyrifolia (Nakia Sand pear and Nakh in Kashmiri) is the most widely distributed of the Oriental Pears, extending over the entire central and Southern regions of China as well as Formosa and Japan. The pear (*P. pyrifolia* Burm. F) is originated in China, where its culture dates to 2500-3000 years. It may have been introduced into India from China during the time of lord Kanishka (120-170 A.D.), who settled some Chinese hostages around village Harsa Chhina (Amritsar district) by them. Its cultivation appears to have spread to other parts of Northern India from this place, but this is only a possibility.

The main Pear producing Countries of the world are China, Italy, USA, Russia, Germany, Japan, Spain, France, Turkey and Argentina. Successful cultivation of many low chilling cultivars such as Patharnakh, Le Conte, Keiffer and Chinese Sandy Pear have been reported even from the hot places of northern India (Jawanda, 1981). These low chilling cultivars are also cultivated in the hills of South India (Hayes, 1957).

Kashmir has the unique distinction of possessing an unbroken historical record from ancient times to the present day. Huen Tsiang, a Chinese pilgrim (AD 631) who travelled and stayed in the valley for 2 years mentions Pears as being cultivated in profusion. Pear is grown in six districts of Kashmir Budgam, Anantnag, Baramulla, Srinagar, Kupwara and Pulwama, Budgam being the largest producer.

4. Taxonomy and cytogenetics

The name pear is derived from Latin, *pera* or *pira*, with some variants like in French as *poire,* in German as *peer,* and in Greece as *acras* as wild type and *apios* as cultivated pear. The place name 'Perry' and 'Pharisoulopol' can indicate the historical presence of pear trees. The term "pyriform" is sometimes used to describe something which is "pear-shaped". Pear is classified into Magnoliophyta division, Class Magnoliopsida and Order Rosales. It is a member of the Rosaceae family and the Pyrus genus. Pyrus consists of 60 species, of which 22 are primary and the rest non primary including botanical varieties, sub species and interspecific hybrids (Westwood, 1976). It belongs to Equisetopsida C. Agardh class of vascular plants, Magnolidae Novak ex Takht subclass, characterized by plants that have ribbed leaves and flowers. Belonging to the Rosales Bercht. & J. Presl order, and Rosaceae Juss family, with hermaphrodite flowers, polypetalae and perigynics stamens. The pear, of *Pyrus* L., gender is a fruit of big importance for the agriculture of moderate latitude countries, being cultivated on a large scale in China, Western Europe, and the United States (Dzhangaliev *et al.*, 2003; Fideghelli, 2007).

The Maloideae subfamily, where the Pyrus gender belongs, has a basic chromosome number as x=17. Most of the accepted theories suggest it as an allotetraploid or allopolyploid from the cross between two primitive forms of Rosaceae family, Prunoidae with x=8 and Spiraeoideae with x=9. This theory was based on the observation of a predominance of univalent (unpaired chromosomes) and not from multivalent chromosomes during meiosis. Subsequently, isozyme studies supported this theory (Weeden and Lamb, 1987). Most cultivated pears are diploid (2n=34), but there are few polyploidy cultivars of *P. communis* and *Pyrus x bretschneiderii*. According to Weeden and Lamb (1987), the speciation of Pyrus occurred without a change in chromosome number. The number of species varies greatly (20-75 species) according to the interpretation of each author (Erhardt *et al.*, 2002). There are 23 wild species cataloged, all native to Europe, temperate Asia, and northern mountainous regions of Africa (Westwood, 1982).

Species diversity is concentrated in western Eurasia to eastern Asia and especially in China, but several species are mentioned by many authors, without a consensus, which hampers an organization, as many are hybrids between species, and in some cases, different regions use different names for the same cultivars (Masseron, 1992).

4.1. Taxonomic Classification of Pear

Kingdom: Plantae

Division: Magnoliophyta

Class: Magnoliopsida

Order: Rosales

Family: Rosaceae

Subfamily: Spiraeoideae

Tribe: Pyreae

Genus: *Pyrus*

Species: About 30 species

4.2. Classification of Pear

On the basis of origin and commercial fruit production:

1. European pear (*Pyrus communis* L.)
2. Japanese pear (*P. pyrifolia* Burm.)
3. Chinese pear (*P. bretschneideri* Rehd. and *P. ussuriensis* Maxim)

Furthermore, the genus is also divided into two native groups, i.e. Occidental and Oriental pears

Important species of Pear

Pyrus species are classified into two groups based upon original geographical distribution. European and West Asian Species (The Occidental group): These are native to Eastern Europe and south Western Europe including South Afghanistan, Transcaucasia (Azerbaijan Georgia and Armenia) and Asia Minor (Non-European Portion of Asia– Turkey, Greece and Europe). Their fruits are having persistent calyx and small fleshy pedicels and usually are pyriform. The principal species are *Pyrus communis*, Linn. and *Pyrus nivalis*, Jacq. (Snow pear). Besides, there are many species that have horticultural significance, presented in Table 3.

4.2.1. European pear (*Pyrus communis* L.)

This is the major pear of commerce. It probably originated from the Caucasian countries and northern Persia, or possibly, the original centre of distribution was in Cashmere and the north-western Himalayas, where there also pear forests. Fruits are variable in shape, usually pyriform, sometimes round- conic, turbinate or occasionally round-oblate, green, yellow, red, russet, or combination of these, flesh with few or many grit cells at maturity. Major cultivars includes: Williams' Bon Chrétien (Bartlett), Beurré D'Anjou (D'Anjou), Beurré Bosc (Kaiser Alexander /Bosc/Kaiser), Doyenné du Comice (Comice).

4.2.2. French Snow pear (*Pyrus nivalis* Jacq.)

It is commonly known as the "snow pear", is grows in south-east Europe to western Asia. Fruit can be eaten raw or cooked; they have a mild sour taste. It is a very hardy able to withstand a small supply of water or very high or low temperatures. Fruits small, roundish pyriform to globose, late ripening, acidic, becomes sweet when ripe. Distribution is mainly in Austria and wild in France.

4.2.3. East and Northern Asian Species (The Oriental Group)

These are native to Eastern Asia, including China, Japan and Manchuria. Fruits are characterized by deciduous calyx and non-fleshy pedicles, and they are usually apple shaped. The principal species are *Pyrus pyrifolia and Pyrus bretschneideri* however *Pyrus ussuriensis, Pyrus serrulata, Pyrus betulifolia, Pyrus calleryana and Pyrus phaeocarpa* also come under this group.

4.2.4. The Sand Pear (*P. pyrifolia* Burm.F. Nak.)

It is native to China, Japan, and Korea and known by many names, including: Asian pear, Chinese pear, Korean pear, Japanese pear, Taiwan pear, and sand pear. Fruits are heavily russetted, flesh with grit cells, and have little aroma refreshing, sweet taste, and crisp texture. Important cultivars include: 'Chojuro' 'Kosui' (the most important cultivar in Japan), 'Hosui', 'Imamuraaki' 'Nijisseiki', and 'Patharnakh' etc.

4.2.5. Chinese white pear (*Pyrus bretschneideri* Rehd.)

The Chinese White pear is a species of pear native to northern China. Fruits are yellowish or sometimes brown, apple shaped or pyriform (Verma., 2015).

Asian pears are grown in high hills to mid-hills and some cultivars such as 'Patharnakh' are grown in the sub-tropical plains of Punjab, Uttar Pradesh Rajasthan, Tamilnadu and North-Eastern States(Verma *et al.*, 2010).

Table 3: Useful horticultural traits carried by native pear species

Scientific name	Useful characteristics
European	
P. communis L.	Resistance to fire blight, white spot, pear scab, pear decline, pear psylla; cold hardy.
P. nivalis Jacq.	Resistance to pear psylla ; cold hardly.
P.cordaa Desv.	Resistance to codling moth; adaptation to warm winters, high pH, drought.
Circum-Mediterranean	
P. amygadaliformis Vill	Resistance to codling moth; adaptation to warm winters, high pH, drought and clay soils.
P. elaeagrifolia Pall.	Resistance to codling moth; cold hardly.
P. syriaca Boiss	Adaptation to drought.
P. longipes Coss. & Dur.	Cold hardy; adoption to wet ad dry soils.
P. gharbiana Trab.	Adaptation to low pH.
Mid-Asian	
P. glabra Boiss.	Adaptation to low pH
*P.*salicifolia Pall.	Adaptation to drought.
P. regelii Rehd.	Resistance to pear psylla
P. pashia Buch.-Ham ex D. Don.	Adaptation to warm winters, low pH, clay soils.
East Asian	
P. pyrifolia (Burn) Nak	Resistance to pear scab.
P. pseudopashia T.T. Yu	Resistance to codling moth; winter hardy.
P.ussuriensis Maxim	Resistance to fire blight, bear psylla; winter hardy.
P. calleryana Decne	Resistance to fire blight, Fabraea leaf spot, pear psyalla, codling moth; adaptation to warm winters, low pH, wet, dry and clay soils.
P. betulifolia Bunge.	Resistance to pear psylla, codling moth; adaptation to warm winters, low pH, wet, dry and clay soils.
P. fauriei C.K. Scheid.	Resistance to fire blight, pear psylla, codling moth; adaptation to warm and cold winters, low pH, wet and clay soils.
P. hondoensis Kik. & Nak	Resistance to fire blight.
P. dimorphophylla Mak.	Resistance to powdery mildew, pear psyalla, codling moth; adaptation to low pH and wet soils.
P. kawakamii Hayata	Resistance to codling moth; adaptation to warm winters.

(*Source:* Hancock, 2008)

5. Botanical description

Pear is a tall, vigorous, spreading perennial deciduous plant attaining a height of 10 to 13 m. Its leaves are oval and more or less round with the leathery texture, At times the leaves at the base are shaped like a wedge. The flowers are almost 2.5 cm (1 inch) in width and normally white in colour. The flowers of the apple and **pear** trees look alike, with the exception that, at the base, the 5 styles separate. Pear trees begin flowering at the age of 5-8 years mostly on spurs and terminal shoots. Floral biology and flowering in pear has been presented in Table 4. Fruit of pear is a globose or pyriform pome with persistent or deciduous calyx, flesh is usually with grit cells. Walls of the locules are cartilaginous with black or nearly black seeds.

Pears are classified into three groups according to the number of carpels and fruit size: small fruits that have two carpels known as Asian pears; large fruits with five carpels as European pear and fruits with three to four carpels are hybrids of fruits mentioned above. Asian pears have a crisp texture, with sweet and juicy pulp. The pulp is characterized by having 'stone cells' which are sclerenchyma cells that differ from fibre because of their very elongated structure. These stone cells also offer a sandy texture to the fruit. While as, the European pear has a buttery and juicy texture, with characteristic flavour and aroma (Layne and Quamme, 1975).

Table 4: Floral biology and flowering of Pear

Flower	Hermaphrodite, white. Rarely pinkish, 20-30 stamens
Inflorescence	Corymbs
Flower bud differentiation	Takes place in July; FBD takes 13-14 days; flower duration range from 21-29 days
Anther dehiscence	On the day of anthesis, mostly in afternoon; 12.00 and 4.00 hr.
Emasculation	At the balloon stage of development
Pollination	By insect (Honey bees more effective in fruit setting)
Pollen collection	1-2 days before anthesis
Anthesis	Peak period is between 10.00 and 1.00 hr
Flower position	Bears mostly on spurs and terminal shoots

5.1. Flowering and pollination

Flower initiation depends upon the levels of organic and inorganic storage compounds and endogenous growth regulators notably cytokinin and gibberllins (Banno *et al*., 1985).

Flowers of pear are self incompatible and fertilization and fruit set require cross-pollination with pollen from an appropriate pollenizer cultivar. For cross-pollination of pear, insects are required to carry pollen between blossoms. Large, symmetrical pears are produced when full pollination occurs. Fruit from poorly pollinated flowers are likely to be shed within a short period of time, but those that do develop will be malformed and small. Early fertilization is also desirable to allow the development of mature fruit. Although pollen foragers find the flowers appealing, attracting bees to pear flowers can be difficult as the sugar concentration of the nectar is low. Also, pears bloom early in the spring when pollinator activity can be low. These pollenizer cultivars should make up 10-15% of trees in the orchard, and be spaced evenly between production trees.

Honey bees are the pollinators of choice. The standard recommendation is to place 2-5 strong colonies per hectare when 10- 20% of flowers are in bloom. To maximize pollination in pears (whose flowers can be less attractive than weeds are to foraging insects). Managing competing blooms is an important concern when managing pollination. If encouraging wild pollinators, or establishing an alternative pollinator such as the blue orchard bee, is of interest to the orchardist, alternate forage is required for these bees to complete their life cycle. One option for orchardists is to plant forage between tree rows which will flower after the crop bloom period. However, growers should also be aware that alternative forage may attract bees to orchards off -bloom. This can result in bee kills for neighbouring beekeepers if the grower uses insecticides. (Delaplane and Mayer. 2000).

5.2. Genetics of Gametophytic Self Incompatibility (GSI) present in pear

GSI in pear (Rosaceae) is determined by multiple alleles at a single locus, the S locus. Compatible pollinations occur when the S allele carried by the male gametophyte is different for either of the alleles carried by the diploid sporophyte (style). Incompatible pollinations occur when the S allele carried by the male gametophyte is the same as either of the alleles carried by the diploid sporophyte (style).

5.2.1. Managing Incompatibility

- Compatible pollinizer cultivars must be present in the orchard (Table 5).
- Honeybees are usually required to effect movement of pollen between and among compatible cultivars.
- Bloom timing between compatible cultivars must overlap sufficiently.
- Orchard layout must facilitate pollen flow. Honeybees typically fly up and down rows, not across.

Table 5: Pear pollinizer compatibility chart

Main Variety	Pollinizers
Corella	Ya Li, Tsu Li
Lemon Bergamot	Josephine
Josephine	Packhams
Howell	Packhams
Abate Fetel	Williams, Beurre Bosc
Packhams	Josephine, Howell, Nijisseki
Williams	Buerre d'Anjou, Buerre Bosc
Comice	Buerre d'Anjou, Williams, Buerre Bosc
Buerre Bosc	Williams, Comice
Winter Nelis	Williams, Buerre Bosc

Source: Verma, 2015

5.3. Fruit set and fruit development

Generally fruit set is not a problem in commercial pear cultivars, being as high as 73.2% in cross pollination and ranges from 17.3% to 32.6% in open pollination (Dhaliwal *et al.,*1982). Among the various growth regulators, the role of gibberellins is definitely more important in setting of pear fruits. Gibberellin treatment enhanced the proportion of parthenocarpic fruits, especially in cross pollinated trees and, interestingly, GA_{4+7} showed more pronounced effect on self-pollinated trees and GA_3 on cross pollinated trees (Yuda *et al.*, 1981). Besides Gibberellin fruit set has also been augmented by the application of SADH, CCC, 2,4,5-TP and Boron sprays. Length, diameter and weight of pear fruits increase from fruit set to maturity in a sigmoidal pattern. The rate of growth of fruit is determined by cultivar, soil moisture, heat stress, fruit set, thinning and growth regulator sprays. Sugar content increases throughout the growth period. Total phenolics are high upto 20 days after fruit set gradual decline occurs subsequently.

5.4. Physiology of flowering, fruit set and development

Flowering, fruit set, and development in pear parallels with other temperate zone fruits, higher temperatures during the vegetative period and the lack of sufficient chilling pose problems not observed under most conditions in the temperate regions. Retardation or stoppage of growth favours flower induction particularly in pome fruits, that form terminal flower buds, but growth rate may be excessive under high temperatures. Marginal chilling may results in erratic bloom, with flowers opening over an extended period of time. This leads to problems in pollination, in timing of pesticide and thinning treatments, and variable size and maturity of the fruits on the same tree. The flower bud formation in fruit plants of the temperate climatic zone like pear is a very complicated biological phenomenon. However, some of the most characteristic moments of the bud transition from the vegetative into the reproductive state and their further development still not been sufficiently clarified. The existence of versatile relations is presumed, between the genetic control, the hormonal balance and the presence of sufficient amount of assimilates in the plant as a whole and more precisely in the forming flower buds. The development of flower buds is related to the characteristics of the fruit tree species and cultivars, ecological conditions and agricultural practices. Quality of the reproductive organs depends on the factors and conditions for flower bud formation, which in turn influences the quantity and quality of fruit production. The initiation and development of flower buds in pear can be successfully regulated by means of scientifically well-founded agri-technical practices such as pruning, fertilization, irrigation and treatment with growth regulators.

5.5. Physiology of fruit ripening

Kondo and Seto (2004) demonstrated the changes of important acids that are involved in fruit ripening process. Changes of jasmonic acid (JA) during ripening in 'La France' pear fruit (*Pyrus communis*), and interactions between JA and abscisic acid (ABA) were investigated. Although JA increased rapidly after harvest, i.e. 165 days after full bloom (DAFB), it did not coincide with the changes of internal ethylene concentrations. The increase of JA after harvest may have been caused by water loss from the fruit after harvest. The result suggested that JA may influence endogenous ABA synthesis separately from ethylene. The result also demonstrated that ABA may influence endogenous JA synthesis via ethylene.

Late pear cultivars such as Passe-Crassane (PC) require a long chilling treatment before they are capable of ripening. Early cultivars such as Old-Home (OH) have no cold prerequisite. The regulation of 1-aminocyclopropane-1-carboxylic acid synthase (ACS) genes was studied by El-Sharkawy *et al.* (2004) in OH, PC and in OH x PC hybrids in order to determine the role of this gene family in the cold requirement. Of the seven Pc-ACS cDNAs isolated, four (Pc-ACS1a/b and Pc-ACS2a/b) showed differential expression associated with the cold requirement. Pc-ACS1a transcripts accumulated throughout the cold treatment and, with Pc-ACS2a, during ripening of cold-dependent cultivars. Pc-ACS1b and Pc-ACS2b were detected only during ripening of cold-independent genotypes. Furthermore, Pc-ACS2a transcript accumulation was negatively regulated by ethylene, whereas Pc-ACS2b was positively regulated by the hormone. Pc-ACS3, 4 and 5 transcript accumulation was similar in all genotypes. Genetic analyses of OH, PC, and 22 OH x PC progenies demonstrated that late, cold-dependent cultivars were homozygous for Pc-ACS1a and 2a whereas early, cold-independent cultivars were heterozygous for Pc-ACS1(a/b) and homozygous for Pc-ACS2b.

Lelievre *et al.* (1997) briefly demonstrated the ethylene bio-synthesis for normal ripening of Passe-Crassane pears. ACC synthase and ACC oxidase cDNA probes were isolated to evaluate the role of chilling on ethylene biosynthetic gene expression in pear. A 3-months chilling treatment at 0 degrees C strongly stimulated ACC oxidase activity, and to a lesser extent ACC synthase activity. The levels of mRNAs hybridizing to ACC synthase and ACC oxidase probes also increased dramatically. Fruit stored at 18 degrees C immediately after harvest did not exhibit any of these changes, whilst fruit that had been previously chilled exhibited a burst of ethylene production associated with high activity of ACC oxidase and ACC synthase upon rewarming. ACC oxidase mRNA strongly accumulated in rewarmed fruits, while ACC synthase mRNA level decreased. The chilling-induced accumulation of ACC synthase and ACC oxidase transcripts was strongly reduced when ethylene action was blocked during chilling with 1-methylcyclopropene (1-MCP). Upon rewarming, ACC synthase and ACC oxidase transcripts rapidly disappeared in 1-MCP-treated fruits. A 5-week treatment of non-chilled fruits with the ethylene analogue propylene led to increased expression of ACC oxidase and to ripening. However, ethylene synthesis, ACC synthase activity and ACC synthase mRNAs remained at very low levels. These data indicated that ACC synthase gene expression is regulated by ethylene only during or after chilling treatment, while ACC oxidase gene expression can be induced separately by either chilling or ethylene.

6. Propagation

6.1. By seeds

Seedlings do not produce identical replicates of maternal or paternal parents. Each seedling is unique. Nonetheless, seedling rootstock production is quite practical: it is quick, inexpensive, easy to perform and produces a large number of plants. Another benefit is that the trees produced have rooting systems that grow downward and are symmetrical. To produce a seedling, seeds are removed from mature fruit of the desired rootstock, stratified and planted.

Seeds of pear, Kainth and Shiara and extracted from fully mature fruits for raising seedlings. The seeds need stratification (5-7°C) for proper germination. The seeds are kept in the stratifying medium till the uppermost layer of seeds pushes the sand upward and protrudes the radical. The germinating seeds are sown in the beds of directly in the nursery rows immediately at a distance of 10-15 cm in rows 15 cm apart.

6.2. By hardwood cuttings

Hardwood cuttings are typically collected from the current year's shoots during the fall and immersed in auxin (either higher levels for a brief dip or lower levels for an entire day). Hardwood cuttings of 25 cm long and 1.5 cm in diameter, each with 8 buds of one-year-old shoots from mature pear tree taken in February and dipped in 4000 p.p.m. IBA for 10 seconds resulted in best rooting percentage (30.00 - 32.75%) and number of roots/cutting (18.55-22.03) (El-Shazly and El-Sabrout, 1994).

The effects of IBA and NAA on the rooting of quince cv. Portugal cuttings were investigated by Pio *et al.* (2006) Half of the woody cuttings (20 cm in length, without leaves) from the middle portion of the branches was immerged in IBA solution and the other half in NAA, each at concentrations of 0, 1.000, 2.000, 3.000 and 4.000 mg/l, for 5 seconds. After they were put in plastic sacks (0.65 d m^3 capacity) filled with soil +sand (as substrate) at 2:1, v/v, they were placed in a greenhouse with 50% brightness. After 75 days, the rooting and sprouting percentages, number of sprouts and leaves, and sprout length were determined. IBA was superior to NAA and the 2.300 mg IBA/l was the superior.

6.3. By softwood cuttings

Softwood cuttings are harvested during early summer, immersed in high levels of auxin, and placed in a greenhouse with occasional misting to be hardened and ready for the nursery by fall. Both hard- and soft- wood cuttings require a humid storage environment for root formation. If the shoots are harvested early, they typically grow in the nursery for one year before reaching an adequate size for transplantation.

6.4. Grafting

Pear plants are commonly propagated by T-budding during April-September or tongue grafting done during December –January. In case of interstock used, double shield budding is suitable method. Whip and tongue grafting in January-February is followed in order to reduce juvenility.

The similarity or differences of peroxidase isozymes in rootstocks and scions may influence their graft compatibility. Gulen *et al.* (2002) conducted a stydy to identify peroxidase isozymes that may be used as markers to predict compatibility between pear (*Pyrus communis* L.) and various quince (*Cydonia oblonga* Mill.) clones. 'Bartlett' (BT) and 'Beurre Hardy' (BH) pear cultivars are known to form incompatible and compatible grafts, respectively, with quince rootstocks. The two pear scion cultivars were budded on 'quince A' (QA), 'quince BA-29', and 15 selected quince clones from Turkey. Bark and cambial tissues were taken from nonbudded rootstocks and scions, and 4 cm above and below the graft union for peroxidase isozyme analysis performed by starch gel electrophoresis. Isoperoxidase analyses were also performed on samples from the graft unions collected 12 months after grafting. Many isozyme bands were observed commonly in the two scions; however, one anodal peroxidase A was detected in BH (compatible scion) but not in BT (incompatible scion) samples. This isoperoxidase was also detected in QA, Quince BA-29, and nine of the Turkish quince clones. Another isoperoxidase, band B, was detected in BH but not in BT or any of the rootstocks. However, the compatible (BH/QA) and moderately compatible (BT/BA-29) graft union tissues contained bands A and B whereas incompatible graft union tissues (BT/QA) lacked both. Graft union samples involving BT and five Turkish quince clones (705, 609-2, 702, 804, and 806) had both 'A' and 'B' isoperoxidases while one or both of these bands were absent in nonbudded graft partners. Field observations of 3.5 year-old grafts of BT and Turkish quince clones revealed that the vegetative growth (vigour) of

BT scion was significantly greater, when grafted on these five clones, than that in graft combinations with other clones. It was suggest that matching of isoperoxidase 'A' in quince rootstocks and BH pear scion may be associated with a compatible graft combination. Additionally, presence of isoperoxidases 'A' and 'B' in the graft union tissues may be used as an indicator to predict a compatible graft between BT and quince rootstocks.

Another method of forecasting pear-quince graft incompatibility, a phenomenon of early tree senescence, was advocated by Musacchi *et al.* (2000) by detecting presence of polyphenols in pear-quince grafted plants. Assays of bark and phloem samples excised from two-year-old potted pear trees for polyphenol detection by HPLC and TLC showed that (+)-catechin and (-)-epicatechin increased in the quince-incompatible cvs. 'Beurre Bosc' and 'Bartlett'. Both metabolites were found in year 2 after grafting and before the appearance of visible incompatibility symptoms.

6.5. Budding

Gautam (1996) noted highest success (87%) in chip budding in pear when operation was made on 16th September. Grafting success and plant growth with chip budding were superior to tongue grafting.

6.6. Micropropagation

Micropropagation produces a clonal replicate of a tree using tissue culture. Micropropagation is a very effective and quick propagation method, but is not currently used in pear production due to the high labour and equipment costs. This method has been used by researchers, and it is expected to become common commercial practice in the future.

Micropropagation utilizes seedling tops, young shoots or suckers collected from a mature pear tree. The plant material, always containing an axillary bud, is sterilized and placed on a mixture of nutrient enriched sugar agar and plant hormones. This mixture receives long-day length lighting to produce new shoots. After the new shoots are rooted with high concentrations of IBA, they are exposed to a couple days of darkness followed by short-day length lighting to allow for proper growth. Shoots can be grown continuously and sliced to produce multiple clones of the same pear tree.

Micropropagation techniques were developed by Nadosy (1997) for 5 important American-bred hybrid pear rootstocks (OHF (Old Home x Farmingdale hybrids) 34, 51, 69, 87, 230), BA-29 and Sydo quinces, some pear seedlings and 2 important commercial pear cultivars (Clapp's Favourite and Bartlett [Williams' Bon Chretien]). Cultures of shoot tip explants taken in December from plants raised in a protected environment exhibited 70-100% sterility, regardless of the disinfectant used, whereas explants taken from outdoor plants exhibited higher levels of infection (0-10% sterility). Material from OHF 51, BA-29 and pear seedlings could be transferred from the multiplication medium to the root induction medium after 3 weeks. The lower concentration of major elements in WPM medium was more suitable for OHF rootstocks than the higher concentrations in MS media. Low concentrations of auxins (0.05 mg IAA/litre) and BA (0.5 mg/litre) were more effective for shoot multiplication than higher concentrations. OHF 87 displayed excellent proliferation and elongation. Rate of multiplication was high in OHF 230, BA-29 and in most wild pear seedlings, and was poor in OHF 51 and quinces. Rooting was generally better on a medium containing 200 mg IBA and 15 g sugar/litre than on a medium containing half-strength MS salts, 4 mg thiamine, 100 mg meso-inositol, 5 mg IBA and 15 g sugar/litre. Acclimatization was easiest if begun in March to take full advantage of summer conditions. Micropropagation of the commercial pear cultivars proved to be very difficult.

Micropropagation of six OH x F (Old Home x Farmingdale) pear rootstocks was examined by Mehri *et al.* (2004). Explants of genetic resources were initiated from branches of field-grown trees. Among the macroelements and organics tested (Cheng, Lepoivre, MS and Viseur media), Cheng medium was the most effective for OH x F 230, 217 and 69 and the best combination of BA and auxin (NAA or IBA) was 1.0 and 0.3 mg/litre, respectively. For 230 and 69, NAA promoted shoot formation better than IBA. The best micropropagation medium for OH x F 40, 34 and 87 was Lepoivre with 1.0 mg BA/litre and 0.1 mg IBA/litre. For all the accessions tested, the regeneration was achieved in both Cheng and Lepoivre media while MS and Viseur media seemed to be less efficient. OH x F 230, 87 and 69 responded as well as all the media. Viseur medium with 0.01 mg NAA/litre and 1.0 or 1.5 mg BA/litre produced, for all the clones, a lower proliferation rate and a lower number of shoots per explant but greater shoot length, suitable to make roots.

Erig *et al.* (2004) worked to determine the auxin type and concentration to promote in vitro rooting of quince (*Cydonia oblonga* cv. MC) as rootstock for pear (Pyrus spp.), and to evaluate the survival of rooted microcuttings during acclimatization under ex vitro conditions. The treatments consisted of 3 auxin types (IBA, NAA and IAA) at 5 concentrations (0, 5, 10, 15 and 20 micro M). The microcuttings were cultivated for 7 days in 1/2 Murashige and Skoog's medium containing myo-inositol (100 mg litre^{-1}), sucrose (30 g litre-1), agar (6 g litre^{-1}) and auxin. The microcuttings were then transferred to a new medium without auxin. Rooted microcuttings originating from several treatments were acclimatized under ex vitro conditions. The auxins (IBA, NAA, and IAA) had the same effect on rooting percentage, and rooting was greatest when the auxins were applied at 10 micro M. NAA resulted in the greatest mean root length. Callus formation increased with auxin concentration, being more pronounced with IBA and NAA. Plant survival of 65.12% was obtained after 30 days of acclimatization.

The stem segments with axillary buds of BA-29 (*Cydonia oblonga* Mill.) were used by Wang *et al.* (2010) as experimental explants to study the different disinfection methods and the effects of hormone on adventitious bud occurrence and multiplication. The results indicated that: (1) During primary culture process, the medium of 1/2MS+6-BA 1 mg.L-1+IBA 0.1 mg.L-1+GA 0.2 mg.L-1 was the best for induction and multiplication, and the explants sterilized in 0.1% HgCl2 for 5 min got the best result with the lower contaminate rate of below 20% and the highest survival rate of above 70%. (2) The optimal multiplication medium in subculture was MS+6-BA 1 mg.L-1+IBA 0.4 mg.L-1+GA 0.05 mg.L-1, the multiplication rate was above 3.8, and the new bud has strong growth potential. (3)1/2MS+NAA 0.5 mg.L-1+IBA 0.5 mg.L-1 was the best medium for rooting, with a rooting rate of 81% and 7.6 roots on average. After acclimatization for one week, the survival rate was up to 80% on 30 days after transplanted into sterilized medium.

6.7. Rootstocks

6.7.1. Rootstocks for European pears

Clonal quince rootstocks, which can induce different degrees of dwarfing relevant to high-density planting, are used in areas with mild winters, especially in Western Europe, but are generally sensitive to fire blight and to lime-induced chlorosis. The cropping per unit of tree size varies with the scion cultivar (Carrera and Ortiz, 1984). The different attributes of pear rootstock is given in table 6 and 7.

6.7.1.1. Quince C ('EMC'): It was selected and produced in virus-free form at East Malling. The rootstock is dwarfing as a result of the heavy and precocious cropping that it induces.

6.7.1.2. Quince A: It is also from East Malling (EMA), and is semi-dwarfing. It induces high productivity in relation to tree size but does not induce such precocious cropping as in 'Quince C'.

6.7.1.3. EMH (QR 193-16): This is a much more recent quince rootstock introduced by East Malling in 2001. The rootstock is more dwarfing than 'Quince A' but usually rather more invigorating than 'Quince C', while consistently inducing larger fruits of pear scion cultivars than either of these.

6.7.1.4. Adams 332: The rootstock is from Belgium, and is intermediate between 'Quince C' and 'Quince A' in its effect on scion vigour. It induces precocious cropping.

6.7.1.5. Cydo: It is from France, and is of similar vigour to 'Quince A' and is particularly recommended for 'Commic'.

6.7.1.6. BA 29: The Rootstock is from France, and is rather more invigorating than "Quince A' and is widely used on poorer soils. It has also been recommended for inherently weak-growing and precocious cultivars, e.g. Passe Crassane (Carrera and Ortiz, 1984).

6.7.1.7. OHF (Old Home x Farmingdale): Clones 51 and 333 are more invigorating than 'BA 29', but are resistant both to low temperature and fire blight.

6.7.1.8. *Pyrus communis*: Seedlings especially of the main commercial cultivars such as 'D' Anjou' and 'Bartlett', are still widely used although trees on these can be excessively vigorous and they are susceptible to fire blight.

6.7.1.9. Pyrodwarf (Rhenus 1) : A breeding programme for obtaining *Pyrus communis* rootstocks suitable for modern pear production has been undertaken since 1980 at Geisenheim Research Institute, Germany. Out of >800 seedlings obtained from the cross Old Home x Bonne Luise d'Avranches, about 10 clones with low to medium vigour have shown potential. The first dwarfing clone, named Pyrodwarf, results in low vigour, high precocity and yield efficiency, uniform fruit size, good anchorage and winter cold hardiness, no sucker development, and no lime-induced chlorosis on high pH soils, and shows good compatibility with all pear varieties. The variety name will be Pyrodwarf and the patent name, Rhenus 1 (Jacob, 1996).

6.7.2. Rootstocks for Asian pears

6.7.2.1. *Pyrus pyrifolia:* This seedling rootstock is generally used in the region. Japanese pear trees generally have their vigour controlled by being trained on a horizontal trellis and therefore developing a dwarf rootstock has not been a priority.

6.7.2.2. *Pyrus calleryana:* Strains of this rootstock being tolerant of flooding have been selected in Japan.

Table 6: Rootstock attributes

Property	Rootstocks
Dwarfing	Quince-C, Quince-A, Pyrodwarf, Pyro-233, O x F-97, OH X F 51, Oregon 211, Oregon 249, OH x F 34, OH x F 69, OH x F 87, OH x F 230, OH x F 233
Cold hardy	OH x F series
High temperature tolerant	Oregon-211 and 249, Oregon-260, 261 and 264
Drought tolerant	Oregon 211 X 249, Oregon 260, 261 and 264
High soil moisture tolerant	Oregon 211 X 249, Oregon 260, 261 and 264
Low soil pH tolerant	Oregon series
Resistant to powdery mildew	Oregon 211 and 249
Resistant to crown and root rot	Oregon Series X OH x F series
Resistant to crown gall	Oregon series and OH x F series
Resistant to canker/gummosis	OH x F series, Oregon 260, 261 & 264
Nematode resistant	Oregon series

(*Source*: Verma, 2015)

Table 7: Important characters of commercially used rootstocks

Rootstock species	Place	Scion Vigour	Black End	Pear Decline	Fire Blight	Woolly Pear Aphid	Cold Hardiness	Soil Susceptibility
P. pashia	N. India, Southern China	vigrous	R	MR	MS	S	S	Susceptible to lime induced chlorosis
P. communis	America, New Zealand, Australia, India & S. Africa	-do-	R	MR	S	S	R	Tolerates a wide range of soil texture and soil moisture
P. Nivalis (Perry pear)	Europe	-do-	R	MS	S	-	R	-do-
P. calleryana	USA, China, Australia	-do-	R	MR	R	R	S	-do-
P. betulifolia	China, N. Italy & USA	Very vigrous	R	R	R	R	MS	-do-
P. ussuriensis	Asia, Plains of N. America	vigrous	MS	S	R	R	MR	Sensitive to excessive moisture
P. pyrifolia	Japan, China, Korea	-do-	S	S	MR	S	MR	Tolerates a wide range of soil texture and soil moisture

(R: Resistant, MR; Moderately resistant, S: Susceptible, MS: Moderately Susceptible)*Source:* Campbell, 2003

7. Cultivars

Important European cultivars of pear are presented in Table 8.

Table 8: European pear cultivars

Sl.No	Cultivar	Origin	Important characters
1.	Bartlett	England	Widely adapted to soil, climatic conditions, fruit medium large, ovate-pyriform, turning yellow on ripening, flesh white, melting, firm, fine and juicy, flavour mildly subacid and fine aroma. Anjou, Bosc, Comice and Seckel are good pollinizers for Bartlett.
2.	Max Red Bartlett	England	Bud mutant of Bartlette, Productive, fruit skin colour is dark cranberry red, shoots and leaves have reddish tings.
3.	Anjou	France	Tree vigorous, regular producer, fairly resistant to cold and fire blight. Fruit large, bright green, greenish yellow at ripening, flesh fine with high desert quality, excellent keeping quality. Bartlett, Bosc, Comice and Nijiseiki are good pollinizers for Anjou
4.	Bosc	Belgium	Early winter pear, excellent fruit quality, vigorous, precocious, productive and cold hardy. Fruit medium large, oblong ovate, yellow skin with brown russet.Bartlett, Anjou, Comice and Nijiseiki are good pollinizers for Bosc.
5.	Comice	France	Tree upright, open and inconsistent in bearing. Fruit large, flesh of outstanding texture and flavour, late cultivar. Bartlett, Bosc, Anjou and Nijiseiki are good pollinators.
6.	Conference	England	Fruits medium to large, long-pyriform in shape, green smooth skin with uneven russet. It is late maturing with long storage life. Can produce parthenocarpic fruits and is heavy and regular bearer, but susceptible to fire blight.
7.	Flemish Beauty	Belgium	Fruit large, obovate, smooth and attractive. Flesh pure white, very juicy free from grit cells, high keeping quality, susceptible to blight.
8.	Hardy (Beurre Hardy)	France	Used as interstock on quince rootstocks, fruit of good size and shape with golden-brown skin speckled with pink, flesh white, juicy and fair to good flavour, good keeping quality.
9.	Eldorado	California	Winter type, late season cultivar, moderately vigorous and productive, fruit large, pyriform with narrow neck, moderately firm.
10.	Packham's Triumph	Australia	Tree upright, regular in growth, good yielder but susceptible to blight, stony pit virus and cold. Flesh fine, juicy, free of grit cells, rich and pleasing flavour.

Contd.

11.	Magness	USA	Fire blight resistant, vigorous, upright, spreading and male sterile. Fruit medium size, oval partially covered with russet. Flesh soft, very juicy, free of grit cells.
12.	Moonglow	USA	Fire blight resistant, attractive fruit, soft, flavour mild and sub-acid.
13.	Victoria	-	fruits medium to big in size, colour greenish yellow with reddish patches, pulp very soft, juicy and sweet.
14.	Winter nellis	Belgium	Fit for plain and hilly areas, fruits medium sized, roundish obovate and narrow near stalk end and yellow in colour.

(*Source:* Mitra *et al.*, 1991)

Important Asian Types and Hybrids of pear are presented in Table 9.

Table 9: Asian Types and Hybrids

SI. No.	Cultivar	Important characters
1.	Baggugosha	Important variety of Kashmir Valley, tree vigorous and irregular in bearing. Fruit small, greenish yellow with tapering stem end. Flesh sweet and gritty, avg. yield 100 kg/tree.
2.	Kieffer	Cross between French pear and Oriental pear, tree well adapted to different geographical, climatic and soil conditions, moderately resistant to fire blight, fruit medium large, brownish , gritty and hard.
3.	Le Conte	Low chilling cultivar, tree medium growing, spreading, bear quite heavily. Fruit small, round, yellowish green. Flesh white, juicy acidic sweet, susceptible to blight.
4.	Gola	Prolific bearing, roundish fruit, medium, excellent keeping quality.
5.	Patharnakh	Low chilling cultivar, grown in submountaneous region of India. Plant can withstand high temperature, hot winds, water logging and drought conditions. Tree very vigorous, heavy yielder. Fruit medium-sized, round with prominent dots.
6.	Nijisseiki	Commercial cultivar of Japan, medium in vigour, spreading and heavy yielder. Fruits medium in size, round crispy and juicy, skin colour yellow, flesh colour cream.
7.	Gold Nijisseiki	Irradiation induced mutant of 'Nijisseiki', resistant to black spot.
8.	Nitaka	Product of early breeding programme, fruit large, orange brown russeted fruits.
9.	Shinko	Fruit russet brown, medium to large, resistant to black spot.
10.	Tse-Li	Traditional Chinese pear, large fruit, stored up to 24 weeks.
11.	Kosui	Leading pear in Japan, released in 1959, yellow to golden brown and russeted, sweet flesh.
12.	Hosui	2nd important cultivar of Japan, introduced in 1972, large fruit (300-350 g). Fruits russeted, golden to golden brown.

Source: Mitra *et al.,* 1991

7.1. P448-2: P448-2, an early, productive, large, green/russet pear (*Pyrus communis*) with some tolerance to the fireblight disease incited by *Erwinia amylovora*, originated from a seed derived from a cross of US386 x NJ3 (Star). Developed by the Purdue University Experiment Station (Indiana, USA), P448-2 is characterized by a short storage life but is well suited for home gardens. Its fruits are oblong to ovate, yellow-green (145A) to green-yellow (1B) in colour, pleasant to eat either when firm or when soft upon ripened, and mildly sweet. Its fruits can be kept at room temperature for approximately one week, but can last for ~2 months under refrigerated storage (1 degrees C) (Janick, 2004).

7.2. Blake's Pride: It is a high-yielding pear cultivar that is highly-resistant to fire blight (caused by *Erwinia amylovora*). This cultivar, which produces juicy and aromatic fruits, was derived from US 446 x US 505 in 1965 and released by the Ohio State University, USA, in 1998 for commercial growers and home orchards (Bell *et al*., 2002).

7.3. Turandot: Derived from the cross Dr. J. Guyot x Bella di Giugno made in 1980, and released in 2000 from Forli (Italy). It is very early flowering (late March) and maturing, with harvest in early July (35 days before Williams'). Fruits are medium-small and similar in shape to those of Coscia; they are yellowish-green with a bright red blush over 10% of the fruit surface (Rivalta *et al.,* 2000).

7.4. Carmen: Derived from the cross Dr. J. Guyot x Bella di Giugno made in 1980, and released in 2000 from Forli (Italy). It is mid-late flowering (end of March, beginning of April), with harvest in late July, 15 days before Williams'. Fruits are large and similar in shape to those of Kaiser; they are yellowish-green with a red blush over 20-30% of the fruit surface (Rivalta *et al.,* 2000).

7.5. Norma: Derived from the cross Dr. J. Guyot x Bella di Giugno made in 1980, and released in 2000 from Forli (Italy). Itis late flowering (similar to Conference) and matures in mid-July, 25 days before Williams'. Fruits are large and similar in shape to those of Williams'; they are yellowish-green with a clear red blush over 5-10% of the fruit surface (Rivalta *et al.,* 2000).

7.6. Sabina: It is a cross of Santa Maria Morettini and Doyenne du Comice. Good pollinators are the cultivars Butirra Precoce Morettini, Conference and Decana du Comice. Pears are harvested in early to mid-August, immediately after cv, Williams. They are slightly more red than cv. Doyenne du Comice. It is concluded that cv. Sabina is well suited to Mediterranean conditions (Bellini, 2001).

7.7. Sophia's Gold: This new erect and vigorous pear variety is derived from an open pollinated seedling observed growing in a block of William's Bon Chre<acute>tien pears in 1990. Fruits are large, 96.9 mm long and 90.6 mm wide, convex in shape with very slight/absent russeting. Fruit ground colour at harvest is green (RHS 145A), overcolour absent, margin of eye basin slightly ribbed, eye basin depth and width medium (12.7 and 29.6 mm, respectively). It matures late, around 1 April in Stanhope, Victoria, Australia (Stasey, 2000).

7.8. Delwini: This pear variety, bred from a cross between Williams' Bronzee and Beurre Precoce Morettini and added to the official French catalogue, produces fairly large fruit of a greenish bronze colour and an irregular elongated pyriform shape, with a juicy acid-sweet flesh that sometimes has a slight astringency. It can be stored until January at 0 degrees to -1 degrees C and is suitable for eating between September and January. It is fairly resistant to damage from handling and transport. Delwini begins to bear fruit early and gives good yields (Anon., 1996).

7.9. Erika: Derived from the cross of Beurre Bosc x President Drouard made in 1955, this new *Pyrus communis* cultivar was registered in the Czech Republic in 1995. Fruits are medium large to very large (mean weight 180-250 g), with a short fruit shape similar to that of President Drouard. Ground colour of the fruits is yellow green with an orange blush. Fruit flesh is creamy, fine, very juicy with a good, subacid, sweet taste. Trees are very precocious producing regular, high yields. Harvesting is late (first half of October). In cold storage fruits maintain excellent quality until April. Trees are winter hardy and suitable for all types of training (Paprstein and Kloutvor, 1996)

7.10. Delta: Derived from the cross Beurre Bosc x Comtesse de Paris made in 1955, this new Pyrus communis cultivar was registered in the Czech Republic in 1995. Fruits are medium large (mean weight 179 g) and fruit shape is broadly conical, similar to that of Alexander Lucas. Ground colour of the fruits is yellow green while flesh is cream, fine and very juicy. Taste is good, subacid to sweet with a rich flavour and fruits have good storage quality. Trees are precocious and are regular croppers with high yields. Harvesting is late (first half of October). Trees are winter hardy and tolerant of scab (*Venturia pirina*), lending themselves to all types of training (Kloutvor and Paprstein,1996).

7.11. Jingxiang: Jinxiang is a new, top quality dessert and processing variety which matures in early September in Xingcheng county. Derived from a cross made in 1956 between Nanguoli (*Pyrus ussuriensis*) and Bartlett (*P. communis*), it produces yellow-green fruits with a faint red blush which weigh 130 g. The flesh is soft, creamy and very juicy with a rich, acid sweet flavour and a strong aroma. Canned fruit is of better quality than that of Bartlett. Trees are very hardy, showing no winter damage in this area of China where European pears are not cultivated to any great extent (Pu and Fang, 1990).

7.12. Angelys: The Angelys variety of pear bred by INRA of France shows little susceptibility to fire blight [*Erwinia amylovora*] and is characterized by good yields. However, it appears to have limited compatibility with the quince rootstock E.M.C. Fruits are harvested in late September, are of good size, good flavour and have 50% to 100% bronze surface colouring, the extent depending on the year, in southern France. Fruits of Angelys can be stored until February to April, according to the post-harvest conditions. These characteristics should enable Angelys to gradually capture the market for late pears, replacing the Passe Crassane variety in particular (Masseron *et al.*, 2000).

7.13. Pear variety for processing purposes

Cao *et al.* (2003) selected cultivars suitable for different process products from 61 pear (*Pyrus bretschneideri, P. pyrifolia, P. ussuriensis, P. communis,* and *Pyrus* hybrid) cultivars. To study the juice processing characteristics, the cultivars with crisp fleshed texture could be processed for clear juice. Cloudy juice could be produced from the tender and melting (soft) pear fruit, but some could also be processed for clear juice. The soluble solids content, titratable acidity, and juice percentage averaged 13.1, 0.45, and 54.1%, respectively. Among the processed cultivars, Anli (*P. ussuriensis*) was very suitable for juice processing due to its high soluble solids content, high titratable acidity, high percentage of juice production, high yield, disease resistance, and good storability and shipping quality. Huagau (*P. ussuriensis*), Jinxiang (*Pyrus* hybrid), Dananguo (*P. ussuriensis*), Balixiang (*P. ussuriensis*), Huajin (*Pyrus* hybrid), and Jinfeng (*P. bretschneideri*) were suitable for juice processing with good juice flavour and a potential market.

8. Climate

Pear is found growing in a wide range of climatic conditions. Different Pear cultivars have inherited this wide adaptability from their progenitor, i.e., *P. communis*, *P. ussuriensis* and *P. pyrifolia* which have originated in extremely varied climatic conditions. A large number of pear species require about 1200 hours below 7^0C to complete their chilling requirement. Bartlette requires slightly more chilling hours (about 1500) while as the low chilling cultivars like Patharnakh fruit heavily even at 150 chilling hours and can withstand high temperatures and hot winds during the summers. Spring frosts are detrimental to pear and a temperature of -3.3^{0}C kill the open blossoms.

9. Soil

Pear fruits thrive well in well drained, deep and rich soils with medium texture. They grow in loamy, sandy and clay soils however loamy soils are best for Pear cultivation as they generally hold water without becoming waterlogged, and have more available oxygen and nutrients for the tree's roots to absorb. The crop requires neutral pH of 6.0-7.5. Pear can be grown successfully at 1200 to 1800 meter above sea level. It can tolerate very low temperature below 0°C and can with stand high temperature during summers (47°C). Best temperature for its flowering and fruiting is 2°C in winter and 32°C is summers. Rainfall 100-125 cm is sufficient for its growth. Summers should be less humid.

10. Planting

Successfully establishing a young pear fruit tree starts with planting site and method. It is necessary to prepare the land by carrying out preliminary operations before planting the trees. The layout system depends on the density to be adopted and topography of land. Generally a square system or a rectangular system is followed. In hilly areas contour system is followed on a rolling land. The planting distance depends upon the soil fertility, cultivar, rootstock, training system and climate of the region. In hilly areas the trees on seedling rootstock are planted at a distance of 5m but for clonal rootstocks distance can be reduced to 3m. In plains, the planting distance of 6m and 8m is recommended for pear Baggugosha and Patharnakh respectively. The planting of trees can be done anytime from December to mid-February in plains, however late fall or early spring are the common planting periods in hills (Sharma and Dutt, 2009). At the time of planting a small hole is dug just big enough to accommodate all the roots. Very long roots can be shortened and plant should remain straight in its position when roots are being covered with soil firmly. Immediately after planting the basin should be prepared and irrigation should be given. The level of soil near the trunk should be kept slightly higher than the level of basin to avoid direct contact of water with the trunk. To avoid sunscald, trunk is painted white. Stacking is also useful. No shoots should be allowed to sprout from the ground level to 60 cm of the trunk height.

10.1. High density planting approach in pear

Nowadays, early bearing and smaller tree size in pear orchards can be achieved decreasing spacing among trees and grafting them on quince rootstocks which are particularly suitable for this purpose. To achieve this aim the choice of training system also holds a key role; in fact some training systems prove more suitable for HDP than others. High density planting in pear is expanding due to the widespread use of quince rootstock able to reduce tree size and induce early bearing. The precocity of bearing and the productivity of an orchard is dependent on management designed to meet the physiological requirements of the specific scion/rootstock combination. In order to maximise solar energy conversion into organic biomass per hectare the density of planting is very important. To obtain maximum productivity per hectare it is therefore essential to combine tree density rootstock/scion combination and tree management in the most appropriate way. Like apple, pear is also most suitable fruit species for high-density planting because of the many possibilities to combine scion varieties and rootstocks of different vigour (Suta *et al.*, 1986). Most high-density pear orchards in the world are grown on quince rootstock followed by BA 29. Quince is preferred because it can produce large fruit, bigger than any *Pyrus* rootstock. However OH x F 87 or Pyro 2-33 and OH x F 97 is a good alternative. Tall spindle is the best growing system for high-density pears. At about 1,200 trees per acre, tall spindle's planting density is high whereas super spindle requires more than 2,000 trees per acre. The rootstocks OHF 87 and OHF 97 rootstocks are more vigorous than Quince A (Westwood *et al.,* 1976). Pyrodwarf and Pyro 2-33 rootstocks appear to be intermediate in vigor between Quince A and OHF 87 or 97 (Wertheim, 1998). Pyro 2-33 was classified by Wertheim (1998) as more vigorous than Pyrodwarf. The two OHF rootstocks and the two Pyro stocks had significantly greater yield per acre with 'Bartlett' than does Quince. From a practical perspective, OHF 87 appears to be an excellent choice orchards since it is winter hardy, dwarfing, highly yield efficient and very productive. An interesting alternative is Pyro 2-33 which is a little more dwarfing than OHF 87 (Robinson, 2010). Higher yields were mostly recorded at very high densities (5555 to 7936 trees/ha) which indicates that the planting in hedge-row pattern and trained with open centre systems like V-shape and Y-trellis facilitates maximum sunlight interception and results into maximum yields. Therefore High Density plantation leads to best utilization of land and resources. Increase yield per unit area, produce quality fruits, make interculturing operation, plant protection and harvesting easy thereby increasing chances of export quality harvest.

11. Crop modelling

Crop growth model is a very effective tool for predicting possible impacts of climatic change, Disease intensities on crop growth and yield. Crop growth models are useful for solving various practical problems in agriculture and horticulture. Various such models have been developed in fruit crops & one such in pear, i.e. CSS_Pear: a Model to Simulate Growth, Production and Quality of Pear (*Pyrus communis* L.) during the construction of this model four years of very detailed experiments was conducted in central Portugal. Treatments consisted of training systems and pruning levels. Experimental orchards differed in age (3 to 11 years) and planting density (2700 to 5400 trees ha^{-1}). The data yielded by these experiments were used to construct, calibrate and validate a model to simulate potential and water-limited level of production and fruit grade, for a given range of years so it can be used as a tool for selection of training system and definition of ideal pruning level (Abreu *et al*., 2015).

12. Canopy management in Pear

It is an art of growing fruit growing. Managing a canopy will help to develop a strong tree that will support heavy crop loads, while increasing fruit production and improving fruit quality in the long-term. Canopy management is one of the most important fruit plant management practice. This forms the basis of precociousness and longevity of the fruit trees in an orchard. Pears bear fruit on spurs and from terminal buds of brindles (shoots shorter than 20 cm) or long shoots. Less common is the formation of fruit from axillary buds (buds formed in the axils of leaves in the previous season). In the first years of production, fruit tend to form on terminal buds of shoots (Sansavini 2002). By the time they are in their eighth leaf, many varieties are predominantly spur-bearing but some (e.g. Williams) maintain a high level of production on shoots (Sansavini 2002)

Pear orchard architecture depends mainly on planting density. A density of 400-600 trees/ha demands a wide and tall canopy, 1000-2000 trees/ha is best when using the normal Quince rootstock and 3000-4000 trees/ha is best for the dwarf Quince clones that induce early cropping and reduce vigour in such cultivars as Doyenne du Comice and Blanquilla. Cycocel is used as a growth retardant in some countries, while gibberellins are used to produce parthenocarpic fruits of cv. Conference. An established pear orchard lends itself to mechanical pruning better than other orchard types. There are 5 types of fruiting wood on pear cultivars: (i) brindilla or 1-year-old flowering shoots (e.g. cv. Bartlett); (ii) spurs on 2- and 3-year-old branches as in the cultivars Doyenne du Comice and Abbe Fetel; (iii) spurs on all wood types as in cv. Conference; (iv) spurs on old branches as in cv. Kaiser-Beurre Bosc and (v) the vegetative sink controlled spurs as in cv. Passe Crassane. Long pruning is only suited to the first shoot type; the second requires yearly short pruning and cutting back of the fruiting branches; the third moderate annual pruning to thin shoots and cutting back of branches; the fourth needs a light-to-moderate cutting back to stimulate branch renewal and the fifth must have a severe pruning to enhance fruit set (Sansavini and Musacchi, 1994).

13. Training

Proper training and Pruning is essential for the development of a strong framework, to maintain vigour and growth, spread the fruiting area uniformly, secure fruits of good size and quality, encourage regular bearing and to provide convenience of pruning, spraying and harvesting. The time of training and pruning in pears is December-January, when the trees are still in dormancy. In the first year, no sprout should be allowed to come up on the trunks upto the height of 50-60 cm from ground level. The plant should be head back at 90 cm height. In the second year top most bud sprout shall become the leader. Three to four outgrowing well-spaced branches on the main trunk are selected, by heading back at 50-55 cm, now the total height of the plant shall be in between 1.40 m to 1.55 m. There should be 6-8 scaffolds on the whole tree. During third year secondary branches on the main scaffolds may be headed back to appropriate lengths, so as to promote the sprouting of buds for tertiaries finally in fourth year and fifth year unwanted secondaries/ tertiaries should be removed. Leader if not cut earlier should be headed back to appropriate length. No training is required after developing the desired scaffold system.

13.1. Modified leader system of training

The central leader is allowed to grow to produce 3-4 side branches, then it is headed back at 75 cm height for low headed and at 90 cm for high headed plants. In the next year, the top bud sprouts to take the shape of the central leader, which is again headed back after getting 2-3 scaffolds at the last scaffold giving it an open centre. This can be done after 2-3 years of removal of the central leader that is why the system of training is called modified leader system of training. In all, there can be 5-7 scaffolds on which secondaries are made to develop by removing apical dominance of each scaffold periodically. Thus a tree takes the shape of an umbrella in spread and a cone in height. There is no breakage of limbs due to the load of the fruits.

Tatura trellis and Y frame systems are being adopted in NewZealand for Japannese pear. Palmatte, Espalier and Lincoln canopy systems are also being used in New Zealand, U.K. and USA.

14. Pruning

Pruning intensity varies with the bearing habit of a cultivar and vigour of a tree. Pear bears fruits on 2 year old wood and a spur continues to bear for more than 6 years. After developing the scaffolds no pruning is done. Due to heavy bearing, if supports are not provided to the bearing limbs, the branches can break. Hence, during every December broken limbs/shoots should be removed. Any new shoot emerging in between the main scaffolds should also be removed. While pruning care should be taken regarding pulling of cut branches to avoid spur breakage. Light pruning reduces cork spot and increases yield of desirable fruit size.

Table 10: Effect of different pruning treatments on flowering, fruiting and yield of pear

Treatments	Flowering (%)	Fruit set (%)	Fruit weight (g)	Fruit size (cm 2)	Fruit size (c.c.)	Yield (Kg /plant)
T1=Strapping+Pruning+ Thiourea (5 %)	51.52	09.81	194.66	50.44	193.33	13.13
T2 =Strapping +NonPrunning+ Thiourea (5%)	43.78	08.77	186.50	74.21	190.83	19.14
T3=No Strapping+Prunning+thiourea (5%)	61.07	12.08	236.66	57.19	220.00	20.68
T4=Thiourea(5%)+NonPrunning +Strapping	39.68	08.49	228.68	53.78	219.16	06.97
Control: No treatments and Spraying	11.78	08.81	192.86	72.86	188.12	02.07
CD at 5 %	10.56	1.82	15.64	3.91	8.55	2.48

Source: Jana, 2016

From an experiment (Table 10) maximum fruit weight of 236.66 g was obtained with pruning and thiourea (5%) treatment, the maximum flowering 61.07% and yield (20.64 kg/plant) were also observed from this treatments. Pruning induced the maximum bud burst. It clearly indicated that pruning has much influence on yield of pear.

Gill *et al.* (2011) concluded from a pruning experiment on sub-tropical pear that manual topping of trees at 1.0 m from the ground level during the dormant season showed that the best on the basis of light penetration efficiency, number of fruiting spurs, tree volume and yield efficiency.

15. Rejuvenation

Old, tall overgrowing Pear fruit trees can be rejuvenated to get the fruit from the same trees for a number of years again. To rejuvenate the tree, whole of the tree should be headed back at one time. Keep only 4-5 well placed scaffolds to a length of 15-20 cm. Many sprouts appear on these stubs. Make selection in the same season and keep 7-10 scaffolds on all the stubs not more than two on each stub in any case. Such trees come into bearing again in 3-4 years after rejuvenation. The time of rejuvenation for pear is December-January.

16. Crop Regulation

16.1. Improving fruit set: Dheim and Browning (1987) stated that embryo sac development during the first 45 days after full bloom appeared to depend upon gibberellin synthesized within the ovular region, without which embryo sacs degenerated and rendered the flower incapable of setting. GA_3 treatments on frost damaged flowers induced parthenocarpic fruits and improved fruit setting and fruit quality, whereas GA_3 applications under favourable conditions showed no beneficial effect. GA_3 is the most widely used bioregulator in the chemical induction of parthenocarpic fruit formation. AVG is an ethylene biosynthesis inhibitor that increased fruit set if applied in pears at full bloom or after bloom respectively. Gibberellin increases yield particularly after flower damage by spring frost (Flick and Hermann, 1984).

16.2. Fruit thinning: Thinning is done in pear to obtain fruits of good size and quality and to stimulate floral bud formation for the next year's crop. Thinning out pears when they are about ½ to 1 inch long will allow the remaining young pears to get more sunlight and the branches to get more air. This encourages larger fruit and reduces the chances of pest problems and fungal disease. BA (MaxCel, RiteWay) may be applied when pears are 5-15 mm in diameter to reduce fruit set, increase fruit size, and promote return bloom. NAA (K-Salt Fruit Fix 200, K-Salt Fruit Fix 800, Fruitone) can be used to thinning pears 15-28 days after full bloom. For best results, apply NAA products with a surfactant (wetting agent) when temperatures are between 70 and 75 F (Verma, 2015).

16.3. Fruit drop: NAA is effective in preventing preharvest drop of pears. NAA becomes effective 2 to 3 days following application, and has an effective period of about 2 weeks. Apply 5 to 7 days prior to harvest (Verma, 2015).

Market demand extends throughout the year although any one cultivar in a particular region has only a limited season, excepting some relatively minor tropical production. To meet year-round demand the season of availability is extended by: growing a range of cultivars with different harvesting seasons, from early to late, in the same orchard or locality; growing the cultivars which are in greatest demand in area with different seasons of production and; the use of long-term storage.

17. Irrigation

Water is needed during summer months, which is the active growing period and when young trees are developing. Shortage of water during the months of April, May and June restricts the tree growth and leads to the formation of hard and small fruits of poor quality. During summer months pear should be irrigated at an interval of 5-7 days and at an interval of 15 days during August and September. Although Pear trees can tolerate drought condition for quite long time, they can also withstand waterlogged condition more efficiently than any other temperate fruit crop. But for a good crop, timely irrigation and removal of stagnating water is necessary. Four irrigation systems viz, flood, furrow, sprinkler and drip are generally used in pear orchards. Bearing trees should be irrigated by flooding method however in High Density pear Orchards irrigation is given along with fertilizers called fertigation where in only the rootzone gets wet thereby making efficient utilization of water and fertilizers. A loam soil should be wet 1.5-1.8 m deep. In hills, pear cultivation is mostly done under rain fed conditions but at many places irrigation facility may be available which is an additional advantage. Besides rain fall, 75-100 cm irrigation may be applied annually in some regions. Both excessive and scanty moisture affects colour, composition and keeping quality of fruits. The trees should be irrigated at 20 days interval upto the end of October. Afterwards no irrigation is required upto January except when manures and fertilizers are added in December (Sharma and Dutt, 2009). Semash and Panasenko (1984), observed that in young pear trees, the shoot growth was 40-72% greater under drip irrigation than control but the early cultivar ‘William’s Bon Chretien’ responded best to irrigation at 70% field capacity whereas mid-season cultivar ‘Beurre Bose’ and the late cultivar ‘Beurre’d Hardy’ responded at 20% field capacity.

18. Weed Control

In pear cultivation weeds can be controlled either by hoeing, mechanically or by using herbicides. As pre emergence Diuron 80 WP@ 4 kg /ha should be applied in the first fortnight of March. For post emergence when weeds are 15-20 cm in height, Paraquat 24 WSC@ 3L/ha dissolved in 200L of water is effective to control wide spectrum weeds. Depending on the species, perennial weeds can be controlled mechanically by ploughing repeatedly during the summer, using herbicides, or a combination of the two techniques. The ideal time to control perennial grasses such as dallisgrass, bermudagrass, and johnsongrass is the summer before planting. (Verma, 2015).

Li and Ye (1996) studied effect of annually applying glyphosate on the growth and fruiting of Huanghua pear variety and soil properties in China. Glyphosate was sprayed twice annually (1 kg of 10% glyphosate + 50 kg water, sprayed in mid-April, and 0.5 kg + 50 kg, respectively sprayed in late June to early July) for weed control in pears cv. Huanghua. No negative effects of herbicide treatment were noted; on the contrary, shoot growth, production and fruit quality were a little bit better than the control (although fruit weight was slightly reduced). Soil physical properties were all improved.

A number of herbicides were tested for weed control in an orchard of pear cv. Kaluerka on MA quince rootstock at Kulinovci near Cacak, Yugoslavia. The most effective were Basta (glufosinate) with a split application of 5 and 4 litres/ha, and Herbator (glyphosate) at 8 litres/ha. The split Basta treatment provided a virtually weed-free soil surface for most of the vegetative growth period. None of the herbicides tested induced visible toxic effects on the pear trees (Stanojevic et al., 1997).

Bajwa *et al.* (2003) studied the efficacy of various herbicides at different agroclimatic conditions throughout Punjab state (5 in Amritsar district, 2 each in Hoshiarpur and Jalandhar districts and one each in Patiala and Ludhiana districts), India to control weeds in pear orchards. Diuron at 3.2 and 4.0 kg/ha was applied at pre-emergence (1st week of March) after thorough cultivation of the field. Glyphosate at 1.2 and 1.6 litres/ha and paraquat at 0.75 and 1.00 litre/ha as post-emergence (3rd week of March) were applied when the weeds were growing actively and were 15-20 cm tall. Hand weeding was done at 30-day intervals. Diuron (at 3.2 kg/ha), glyphosate (at 1.2 litres/ha) and paraquat (at 1.0 litre/ha) effectively controlled the dicot and monocot weeds in pear orchards. Paraquat, although significantly better than hand weeding and the control, was comparatively less effective. Fruit yield and quality were highest in herbicide-treated fields. Glyphosate and diuron proved more economical compared to the control and hand weeding treatments. The average fruit yield was highest under glyphosate at 1.6 litres/ha. Fruits recorded the highest total soluble solids upon treatment with 1.6 litres glyphosate/ha and 4.0 kg diuron/ha. Glyphosate at 1.2 litres/ha recorded the highest net return (Rs. 2860/ha).

19. Mulching

Mulch is, usually, used to converse moisture and improve the fertility and health of the soil, reduce weed growth, control soil temperature by acting as an insulating layer for the soil so the temperature of the ground changes more slowly. Mulch also acts as a store house of soil nutrients. *Moniruzzaman et al.,* (2007), while studying the effects of mulching and pruning in pear (cv. BARI Naspati-1) obtained significant effect on yield and some yield components of pear. Mulching in combination with 50% pruning gave the highest number of fruits per plant and the highest fruit yield. Percentage of total soluble solid (TSS %) increased with mulching. The combination of 50% pruning and mulching treatment gave the highest gross return as well as gross margin with maximum benefit-cost ratio 10.60. Hence, confirming the beneficial effects of mulching in pear.

20. Intercropping

Due to long juvenile period in pear, growing of inter crops in necessary for the grower. Only those crops should be grown which adjust to the growth behaviour of pear. For Non-bearing Orchards vegetables can successfully be grown. Avoid growing of potatoes. Pulses like moong, mash, massar grams and peas can also be grown. Do not grow fodders like Jowar, Maize or Berseem. Some plants, e.g. peach, plum can be planted as fillers in pear orchards. For Bearing Orchards growing of rabi season crops should be avoided. Vegetables as radish, carrots and cauliflowers can be grown. Toria is a good crop which adjusted well with pear.

In an experiment conducted in China to determine the effects of pear-wheat intercropping on water and land use efficiency, transpiration in intercropped wheat decreased by 21.6% compared to sole wheat cropping (control). The soil water capacity at 0-200 cm depth and water utilization efficiency by intercropped wheat increased by 18.1 and 8.7% compared to the control. In addition, the yield of intercropped wheat increased by 6.6% compared to the control. The land equivalent ratio was 1.19 (Meng and Zhang, 2004).

21. Mineral Nutrition

Pear needs nutrition for the formation of spurs on the vegetative shoots and for flowering and fruiting later on. Trees produce heavy crops and can remove large quantity of nutrients from the soil (Mitra *et al.*,1991). Singh and Singh

(2004) estimated that 100 kg of pear fruit removed 103.31 g N, 27.89g P and 361.77g K. The dropped foliage during dormancy adds some nutrients back to the soil, while as the portion of the nutrients taken up by the fruit and pruning wood needs to be replenished in the soil.

Jankovic (1994) concluded from results of an 11-year-long fertilizer trial with the 2 cultivars grown on quince MA rootstocks, with interstocks, that for successful pear production, application of 120-150 kg N, 60-80 kg P_2O_5 and 150-200 kg K_2O/ha is needed.

Nitrogen should not be applied in excess as profuse vegetative growth can invite twig die back. Pear usually show- iron and zinc deficiencies during summers, which should be taken care of by spraying ferrous sulphate @ 2g/L or zinc sulphate @ 2g/L respectively. In pear, following doses of manures and fertilizers (Table 11) may be applied for getting high returns from the orchards (Sharma and Dutt, 2009).

Table 11: Fertilizer schedule in pear

Age of Tree (Years)	F.Y.M.(Kg)	Urea (g)	Single Super phosphate (g)	Muriate of Potash (g)
1 – 3	10-20	50 – 150	100 – 300	50 – 150
4-6	25 – 35	200 – 300	400 – 600	200 – 300
7-9	40-50	350 – 450	700 – 900	350 – 450
10 and above	50	500	1000	500

21.1. Method of Application

Proper method and time of nutrient application may enhance the uptake efficiency and reduce the nutrient losses due to leaching. The farm yard manure should be well rotten. Mix farm yard manure, super phosphate and muriate of potash and add by uniformly spreading in the tree basins in December- January.

For the first five years of the age of the plant, the quantity of urea to be applied may be divided into three equal parts. Apply each part in April, June and August. Mix the fertilizers well with soil in the plant basins and apply light irrigation.

For the bearing plants apply farm yard manure, super phosphate and potash during December. But urea may be divided into two halves. Apply first half in February and second in April after the fruit set. The dose of this half can further be reduced if the fruit set is less. Sometimes few trees may not bear at all or bear very little fruit. In such a situation the April dose may not be applied to these trees. These situations do arise in soft pears. To get more advantage from the inorganic fertilizers, green manuring with leguminous crops should be done after every alternate year (Sharma and Dutt, 2009).

21.2. Deficiency of essential elements

Pear orchards irrigated with slug water from cities usually show deficiencies of manganese, magnesium, zinc and iron. The deficiency symptoms resemble that of citrus leaves. Leaf analysis may be done to confirm the deficiency symptoms. However, after identification of the deficient element, the trees may be sprayed with that element. For leaf analysis collect leaf samples at the height of 2.0-2.5 metres. Take 7-8 leaves from mid shoots from all sides of the tree, totaling 50-60 leaves per tree. Wash the leaves immediately place under fan. Put these leaves in respective envelops by fully labeling, transport the samples to the laboratory. The indication of the nutrient concentration in plant tissues whether deficient, low, adequate or high is given in the Table 12 which needs to be consulted before any fertilization programme.

Patharnakh propagated through cuttings can show yellowing of leaves during summers. Such trees of on light soils show severe leaf chlorosis leading to tip burning of shoots and finally bark shredding of the branch. The chlorosis can be checked by spraying micronutrients like manganese sulphate; ferrous sulphate; zinc sulphate @ 2g/L of water during summers. In orchards where Bordeaux mixture is sprayed at dormancy and again in June, the copper deficiency cannot exist (Sharma and Dutt, 2009). DRIS (Diagnosis and Recommendation Integrated System) is a new approach for interpreting leaf or plant analysis which was developed by Beaufils at the University of Natal, South Africa. It is a comprehensive system which identifies all the nutritional factors limiting crop production and in doing so, increases the chances of obtaining high crop yields by improving fertilizer recommendations. The DRIS method expresses results of plant nutritional diagnosis through indices, which represent, in a continuous numeric scale, the effect of each nutrient in the nutritional balance of the plant. These indices are expressed by positive or negative values, which indicate that whether the referred nutrient is in excess or deficiency, respectively. The values closer to zero are the indices for all the nutrients being at the adequate nutritional balance (Beverly, 1987).

Table 12: Leaf nutrient standards for pear

Nutrient % or ppm	Deficient less than	Low range	Normal range	High range	Excess more than
Nitrogen %	<1.8	1.8-2.2	2.3-2.7	2.8-3.5	>3.5
Phosphorous %	<0.10	0.10-0.14	0.15-0.20	0.21-0.30	>3.0
Potassium %	<0.8	0.8-1.0	1.1-1.5	1.6-2.0	>2.0
Calcium %	,<0.7	0.7-1.0	1.1-2.0	2.0-2.5	>2.5
Magnesium %	<0.18	0.18-0.24	0.25-0.35	0.36-0.50	>0.50
Sodium %			<0.02	0.02-0.05	>0.05
Chloride %			<0.4	0.4-1.0	>1.0
Zinc ppm	<10	10-15	16-50	>50	
Manganese ppm	<20	20-24	25-100	101-200	>200
Copper ppm	<4	4-5	6-20	21-100	>100
Boron ppm	<15	15-19	20-60	61-200	>200

Source: Apple and pear nutrition | Prime fact 85, 2005

21.3. Effects of nitrogen application

Yadav and Bist (2003) studied the effects of N fertilizer rate (20, 40, 50, 60, 70, 80, 90 or 100 g per tree) on the performance of 24-year-old trees of pear [*Pyrus pyrifolia* x P. *communis*] cv. Bagugosha grown in Uttar Pradesh, India. N as urea was applied in two splits, i.e. 2/3 prior to balloon bud stage and 1/3 during the active growth stage. The results suggested that 60 g N per tree was optimum for the enhancement of the growth, yield and quality of Bagugosha pear.

21.4. Micro-nutrient application

21.4.1. Effect of boron application on physiological disorders

Under particular CO2 levels in controlled atmosphere (CA) storage 'Conference' pears are susceptible to physiological disorders, such as brown heart and cavities. The effect of pre-harvest boron sprays on several physiological parameters and on the subsequent occurrence of brown heart in pears stored under high CO2 in CA, was tested by (Xuan *et al.,* 2001). Trees of 'Conference' pears were sprayed with boron several times before harvest. After harvest the fruits were stored in CA (5% CO2+ 2% O2) at -0.5 degrees C. During the 4-months storage period, the pears were rated for physiological disorders and the permeability of cell membranes, mineral concentrations, vitamin C content and fruit quality parameters were determined. The results show that brown heart symptoms appeared in the control fruits already after 2 months' storage. On the other hand, fruits treated with boron did not exhibit any brown heart disorder. Boron treatment reduced membrane permeability well in advance of the appearance of browning symptoms in the control fruits, and increased boron and vitamin C content, without affecting macronutrient levels and other fruit quality parameters resulting increased cellular defence system.

21.4.2. Effect of boron application on fruit yield

The findings of Wojcik and Wojcik (2003) indicated that prebloom (at green and white bud stage, and when 1-5% of flowers was at full bloom) and postharvest (approximately 6 weeks before leaf fall) B sprays were successful in increasing pear tree yielding and in improving fruit storability under the conditions of low B availability in the soil. Spray treatments involved application of B at a rate of 0.2 kg ha^{-1} in spring or 0.8 kg ha^{-1} in fall.

21.4.3. Effect of boron and zinc application

The influence of foliar nutrition with zinc and boron on yield quantity was studied by Hudina and Stampar (2004) at different pear (*Pyrus communis* L.) cultivars: 'Williams', 'Conference', 'Abate Fetel' and 'Doyenne de Comice' at the Fruit growing centre, Bilje (Slovenia). The total number of fruits of the cultivars 'Williams', 'Conference', and 'Doyenne de Comice' was higher when the trees were sprayed with zinc and boron. The yield coefficient, yield per tree and yield per hectare were higher when the trees were treated with foliar solution for all cultivars. For the cultivar 'Conference' the yield per hectare was for 14% higher on treated trees, and the percentage of fruits of extra quality increased. For the cultivar 'Abate Fetel' the yield per hectare was 8% higher, for the cultivar 'Doyenne de Comice' 36% higher, and for the cultivar 'Williams' 30% higher on treated trees.

The effect of Ca, Zn and B spray application on preharvest fruit drop, sugars, calcium, zinc, and boron concentrations and as well on some physicochemical fruit properties of five commercial Asian pear cultivars ('KS'7, 'KS'8, 'KS'11, 'KS'12 and 'KS'14) was investigated by Khoshghalb *et al.*(2013) under Tehran Province's environmental conditions. Trees were sprayed one month past fruit set until harvest within two week intervals with Ca, Zn, B and as well with their mixtures at 5% concentrations. Fruit drops were recorded from treatments' start untill harvest. At the time of fruit harvest, fruit sugar and mineral contents were determined using HPLC and ICP, respectively. Results revealed that Ca, Zn and B applications decreased preharvest fruit drop in a way that a combination of Ca, Zn, and B applications rendered a 6.3% fruit drop in comparison with 17.3% occurring in control. Throughout the research, fructose was found out as the dominant sugar followed by glucose, sorbitol and sucrose in the studied cultivars. Spray applications of Ca, Zn and B increased the fruits' fructose, glucose, sorbitol and sucrose contents in all the studied cultivars (1-2.5, 1-2.1, 2-3.7 and 0.7-2.2% fresh weight, respectively). In addition, total soluble solids increased by 1-2 degrees Brix total sugars (1-2% of fresh weight), total acidity

(0.07-0.12 mg 100-1 g) and dry weight (1-1.5 g) of fruits in all cultivars. Spray of Ca increased fruit firmness (1-2 N varied with cultivar). Spray of Ca, Zn and B applications increased Ca, Zn and B mineral contents of both leaves and fruit.

21.5. Fertigation

Namestek *et al.* (2011) studied fertigation on pear crops and concluded that the highest crops (30.8 kg/tree) were recorded at variant with drip irrigation of 3.5 L/h output, with maximum supply of water with fertilizer Kristalon.- Aa special fertilizer (Kristalon violet), supplied by the irrigation once a week in concentration 3 g/tree/L dose.

22. Trunk girdling

Fifty percent of Argentine pear production corresponds to the 'Bartlett' cultivar. One of the principal problems on its production is small fruit size. Excessive vegetative growth of trees can increase this problem because of the competition with fruit growth. The girdling of trunks is a practice that is used to control the vigor of the plants and enhance yields and fruit size. A trial was performed during two seasons, in six locations in the Rio Negro and Neuquen Upper Valley, Argentina. The results show, in some cases, higher yields in girdled trees than in control ones, in both seasons. Although a tendency to an increase in yield has been noticed in girdled trees, no conclusive results were found. Since the response of girdling shows cumulative effects, it is expected that solid results would be more clearly shown by repeating the treatment in successive seasons (Raffo *et al.*, 2011).

23. Limb girdling

The response of 'Patharnakh' pear to girdling was evaluated on 18-year-old vigorous trees growing under uniform cultural practices. The fruits with best quality characteristics in terms of juice content (58.5%), total soluble solids (11.8 degrees Brix), TSS/acid ratio (42.8) and total sugars (8.86%) with low juice acid (0.27%) contents were observed under sub-limb girdling performed on 15 days after flower initiation (DAFI). Girdling treatments also advanced the fruit maturity over control. In conclusion, sub-limb girdling performed on 15 DAFI was the best in enhancing fruit yield (162.0 kg/tree) compared to control (135.3 kg/tree) and physico-chemical composition of 'Patharnakh' pear under sub-tropics of north India (Singh *et al.*, 2014).

24. Effect of plant growth regulators

The use of growth regulators is an important tool in pear production. Fruit trees are complex perennial structures that differ in many respects from annual or herbaceous plants. Growth and reproductive potential of the tree can be manipulated using different compounds. Plant bio-regulators (PBRs) have the peculiarity that in some cases the same active ingredient can induce different responses depending on the application time and rate used. Their effectiveness is influenced by various factors like pear cultivar, tree vigor and yield, weather factors during, before and after the application and application method, e.g., water volume and/or adjuvant usage among others. One of the important aspects that contribute to increasing or maintaining the sustainability of fruit tree agro-ecosystems is the use of substances in a rational way, with high efficiency and low environmental impact to improve fruit quality without toxic residues, preserving human health and productive areas. Many of the growth regulators are hormones or hormone-like substances that promote, inhibit or affect biological or biochemical processes in plants. Endogenous plant hormones have shown the ability to affect different developmental processes in a positive or negative manner but in some processes (e.g., flower induction) there is still a need for a better understanding of these signal interactions, their origin, transport and perception to reliably manipulate them with PBRs.

Chemical and hormonal PGRs have been shown to advance the transition of buds from vegetative to reproductive growth and alter the proportion of bud types to maximize fruit quality and productivity. They can promote lateral shoot development in young trees, slow shoot growth in more vigorous cultivars and improve fruit set to overcome issues with low chill or poor pollination. Conversely other PGRs can act as flower thinners to reduce crop loads and improve fruit size. Most recent PGR research in pear production has focused on breaking dormancy (Table 13), reducing shoot growth and improving fruit set. Alternative dormancy breakers to replace hydrogen cyanamide (HC), commonly known as Dormex®, due to its high toxicity and de-registration in some countries are being investigated. Included here are PGRs that modify ABA metabolism since ABA is known to regulate bud dormancy (Zheng *et al.*, 2015), and mixtures of organic compounds with mineral oils or nitrogen based compounds (such as CaNO3 or KNO_3 + urea) or various micro- and macro-nutrients (Pereira *et al.,* 2011).

Table 13: Bud breaking agents of pear

Dormancy breaking agent	Dose (a.i)	Time	References
Thiourea	2-4%	Dormancy period	Jana and Das, 2014
Promvit, Glutabion	3-5%	Dormancy period	Banno *et al.,* 1985
SADH	-	Active vegetative stage	Peirera *et al.*, 2011

Vegetative growth control using prohexadione-calcium, a shoot growth retardant, may result in different responses depending on the pear cultivar and rate used. Exogenously applied gibberellins (GA) with prohexadione calcium was reported to induce larger fruit than GA alone (Dussi, 2011). Prohexadione-calcium (Pro-Ca), a gibberellin biosynthesis inhibitor, commonly known as Regalis® can be used to control shoot growth (Smit *et al*., 2005). Shoot growth response to the concentration and number of applications of Pro-Ca was highly variable across pear cultivars, supporting the need to test PGR application rates, number of doses and timing on all new cultivars.

6-Benzyladenine (BA), GA $_{4+7}$, aminoethoxyvinylglycine (AVG), Pro-Ca, Thidiazuron (TDZ) and different combinations of them are being investigated to promote parthenocarpic fruit set in a range of pear cultivars (Öztürk and Aºkin 2015, Luzk *et al*., 2014). On 'Conference' pear trees with low to moderate amount of flower buds, GA $_{4+7}$ spray between full bloom and the end of bloom followed by a split application of Pro-Ca, two and three weeks after full bloom was most effective in improving fruit set (Vercammen *et al*., 2015). Whilst on 'Packham's Triumph', GA $_{4+7}$ and BA (Promalin® @1 ml/L) applied at full bloom + AVG (Retain® @2000 mg/L) applied 15 days after full bloom showed the highest yields in trials in Brazil (Rufato *et al.,* 2011). However, in a separate trial in Argentina, 250 mg/L of Retain® applied at full bloom to 'Abate Fetel' and 'Packham's Triumph' reduced yield on both cultivars, and was only effective when applied 2 weeks after full bloom or at a lower concentration (125 mg/L) during bloom (Sánchez *et al.,* 2011).

Messenger is a chemical with plant growth regulating activities which may improve cropping and fruit size in European pears, specifically 'Bartlett' (Stephen *et al*., 2003).

To achieve target yields, crop regulation measures in 'Conference' pears may vary from year to year and may embrace measures to stimulate flower bud development and enhance fruit set, but often also chemical thinning of excess fruit. In 'Conference' the value of the crop at harvest strongly depends on the size distribution of the fruit. The aim is to harvest as many pears as possible with fruit diameters of 65 mm or more, since these sizes provide better economic returns than do smaller fruit. In years with ample flowering and good conditions for pollination, fruit set frequently becomes excessive and thinning is needed to realize the target crop load and the desired fruit size. Hand thinning is labour intensive and thus expensive. Therefore, chemical thinning is being examined as a cheaper alternative. 'Brevis', a new and specially formulated thinning agent containing metamitron, proved to an effective thinning agent. Depending on orchard conditions, desirable levels of thinning were obtained by the use of single or repeated applications of 175 to 350 mg/L metamitron at the 8 to 12 mm fruit diameter stage of development. Higher dosages were needed to thin well-pollinated 'Conference' trees compared to trees in orchards without pollinators. Thinning was also achieved by the use of benzyladenine (BA), tank mixes of BA and naphthalene acetic acid (NAA), and ethephon. Optimum thinning by BA was established at concentrations of 150 mg/L. Thinning efficacy of BA was increased by the addition of 5 to 10 mg/L NAA. The extent of thinning increased with the concentration of NAA in the tank mix. Ethephon applied at 12 to 14 mm fruit diameter also thinned 'Conference' pears, but was somewhat less effective than the combination of BA and NAA. The percentages of fruit having a diameter >65 mm were significantly increased up to 80 to 90% of the yield for those treatments that thinned the trees to the target fruit load of about 110 fruit per tree (Maas *et al.,* 2011).

The effects of gibberellic acid (GA) and NAA at 5 and 10 ppm each, chlormequat and Alar (SADH [daminozide]) at 500 and 100 ppm each, boron at 100 and 200 ppm, cobalt chloride at 10 and 20 ppm, and sucrose at 2.5 and 5.0% on the performance of five-year-old trees of pear cv. Le Conte were studied. The trees were sprayed at the mid-flowering stage. All the growth regulators significantly affected fruit set, yield and most of the quality parameters. The greatest fruit set and yield was obtained with 5 ppm GA (63.3% and 70.8 kg per tree, respectively) in 1st year and with 1000 ppm chlormequat (38.7% and 55.8 kg per tree) in 2nd year. Chlormequat at 500 ppm (37.8%) and SADH (33.1 and 37.7% at 500 and 1000 ppm, respectively) were on a par with 1000 ppm chlormequat in terms of fruit set in 2nd year.

To improve fruit size in low chill pear cultivars, three sprays of CPPU (N-(2-Chloro-4-pyridyl)-N-phenylurea) at 10 ppm at fortnightly interval after fruit set was found most effective while fruit quality in respect of TSS and total sugars and yield was better with the application of 5 ppm CPPU (Prashant *et al.,* 2015).

25. Fruit bagging

The practice of pre-harvest bagging has been extensively used in several fruit crops to improve the commercial value of the fruit. XianMing *et al.* (2000) carried out an experiment over 3 years on 17-year-old trees of Jinshui 2 pear to know the effect of different types of bags viz., LB Single-layer bag, Taishan Double-layer bag, and newspaper bag. Fruits were bagged in early May. Results showed bagging could improve the fruit appearance, decrease the preharvest drop and prolong the storability and shelf-life. No obvious effect on fruit size and quality were noted. Under the high growing temperature and high humidity conditions, the LB Single-layer bags were recommended from a practical point of view.

Hudina and Stampar (2011) bagged the pear fruits cv. Conference with paper bags in mid-June, 65 DAFB, at an early stage of fruit growth. In first growing season they noted that bagging had a negative impact on sucrose and sorbitol content. The bagged fruits were brighter (higher L*) than the control fruit. The reduction of colour (increased L* parameter and hue angle around 95 degrees) by bagging may have a negative effect on consumer appeal. In second growing season bagging reduced fruit weight to such extend, that fruit were under 110 g and therefore not marketable. Bagging of pear fruit increased some phenolic compounds in the skin. Epicatechin and caffeic acid contents in skin were highest in bagged fruit. Removing the bags seven days before harvest decreased arbutin content in the pulp and also total sugars in the whole fruit and therefore fruit may have poorer eating quality. Quercetin 3-O-rhamnoside content in skin decreased in both years among fruit bagged until harvest. From the results, they don't recommend the use of bagging for 'Conference' pear in European growth conditions.

In another experiment with cultivar 'Concorde', Hudina *et al.* (2012) found beneficial effect of use of fruit bagging. Bagging of pear fruit after June drop until harvest significantly reduced fruit length in both observed years. Fruit bagging resulted in increased firmness of the fruit only in 2006. Soluble solids content was significantly lower in bagged fruit in 2006. In 2005, bagging had no influence on firmness and soluble solids content. Bagging improved fruit surface lightness, since L* was higher in the bagged than in the control fruit in both years. Also the a* parameter was higher in bagged fruit. They noted significant influence of bagging on total sugars content. Bagging European pear fruit decreased the content of phenolic compounds in skin (catechin, chlorogenic acid, epicatechin, p-coumaric acid, quercetin 3-O-galactoside, quercetin 3-O-glucoside, quercetin 3-O-rhamnoside). The removal of bags 7 days before harvest significantly increased glucose, shikimic and fumaric acids.

26. Harvesting

The maturity is generally assessed based on flesh firmness, skin colour, sugar content, titratable acidity, sugar/acid ratio, total soluble solids, starch content, appearance of 2-4 decadienic acid, corking of the lenticels, fruit size, occurrence of fruit drop, separation of fruit from spurs, number of days from full bloom and heat unit accumulation during specific periods of growing season (Verma, 2015). Most pear cultivars ripen and develop best dessert quality at 15.6°C-21°C and 80-85% relative humidity. Soft pears take 135 days to mature. At maturity, green colour of the fruit changes to light yellowish green, firmness decreases. TSS of the juice reaches 9-10 percent. Optimum harvesting time for Patharnakh in North India is end July. Similarly for soft pears it is August. Fruit pedicels also should not break from the centre. Either there should be full stalk or there should be no stalk with the fruit. Harvesting should be done with the help of ladders. Trees should not be shaken for fruit harvesting.

27. Physiological disorder and their management

27.1. Core breakdown

It occurs mostly in the cultivars Bartlett, Bosc, and Comice. A brown, soft breakdown of the core and surrounding tissues occurs. In Bosc, symptoms may be preceded by vascular browning as the fruit ripens. In Bartlett, the watery, brown tissue separates easily from the healthy tissue (Blanpied, 1975).Incidence is increased in fruits harvested beyond optimum maturity, promoted by delays in cooling after harvest, warmer storage temperatures than recommended and extended storage. It may develop in storage or soon after transfer to warm temperatures. Avoid storing pears beyond their post harvest life. To extend post harvest life, fruit should be rapidly cooled and stored at the lowest non-freezing temperature (generally -2 to -1°C, 28 to 30°F). Controlled atmosphere storage can extend the post harvest life as much as two-fold. Monitor fruit colour as it approaches its typical storage life. Market any fruit that begins to yellow.

27.2. Flesh Spot Decay

There is no external indication of the disorder. Generally, FSD is more pronounced above the equator of the fruit (towards the stem end), but it can also be observed all the way down to the calyx. It appears along and around the vascular bundles when the symptoms are severe, with partial browning of spots and/or development of cavities in Asian pear flesh. Cavities are usually dry and surrounded by apparently healthy tissue. This disorder can occur in fruit while still on the tree. It is more obvious, however, after 2-6 week cold storage. FSD is more frequent on large sized (300 g) and over mature fruits. Japanese pear cultivars such as 'Shinseiki,' 'Nijisseiki,' 'Kikusui,' and 'Hosui.'are more prone to flesh spot decay, but the cause of FSD is still unknown. However, climatic factors, such as a fluctuating hot and cool summer, or high rainfall right before harvest may enhance the incidence of this disorder (Lallu and Nagin, 1990).

There is no effective way to control FSD since definite causes have not been identified. Avoid whenever possible the following conditions that might induce FSD: low crop load - large fruit; later picking - advanced maturity; extreme temperature changes during the maturation season; sunburn; erratic irrigation or precipitation - frequency, amount and timing; harvesting fruit under warm temperatures and cooling the fruit rapidly.

27. 3. Internal Browning

Internal browning on Asian pears is the main consumer complaint. This is a worldwide problem. Development of brown to dark brown water-soaked areas in the core and/or flesh during storage is occurred. There is no visible external indication of internal browning. The cause of internal browning is still unknown. Ripening predisposes the fruit to the disorder while it is in cold storage. (Beutel, 1990).

The fruit should be picked when most of the pears on the tree are still green, although a few at the top may begin to develop some light-yellow spots. Fruit picked when the skin is completely yellow will develop internal browning within one month after harvest. Prompt cooling of fruit is recommended as delays in cooling increase the incidence of internal browning in pears that are beginning to turn yellow.

27. 4. Senescent Scald

It develops on pears which have been stored beyond their post harvest life. Brown to black discoloration of the skin associated with fruit which have become yellow in storage and lost their capacity to ripen normally. Browning is initially restricted to the skin but progresses rapidly into the flesh, particularly upon transfer of the fruit to warm temperatures. Browned skin becomes weak and is easily sloughed off in later stages of senescence. It often associated with core breakdown. Decay may follow secondarily. Taste and odour of the fruit are very disagreeable even before discoloration begins (Agricultural Research Service, USDA, 1971).

Avoid storing the fruit beyond its post harvest life. To extend the post harvest life maximally, harvest fruit at optimum maturity, cool fruit rapidly and thoroughly after harvest, and store at the lowest, non-freezing temperature (often -1 to -2°C, 28 to 30°F). Controlled atmosphere storage can extend the post harvest life of the fruit up to twice that achievable in air storage. Watch for change in fruit colour from green to yellow. Yellowing is an indication that the fruit's storage life is nearing the end.

27. 5. Watery Breakdown

Watery breakdown can develop in fresh pears, especially the 'Bartlett' variety, during storage and/or ripening following even short periods in cold storage. Watery breakdown can involve any part of the fruit, being most common in the tissue around the blossom end of the fruit. Watery breakdown, as the name implies, is a soft, watery deterioration of affected tissue. When the affected tissue is cut or punctured, juice flows out of the fruit. Typically, the involved tissue is in the outer portions of the flesh, but in severe cases can move into the core tissue. The affected tissue is not discoloured during early stages, but becomes brown with time. It is thought that the random pattern of symptoms development on individual fruits is related to other stresses that occur during handling.The soft, watery breakdown results from softening of the affected tissue. Fruit that are cooled slowly and/or held at too warm a temperature are most severely affected by this disorder. Fruit of any maturity can be affected, but high maturity fruit appear to be most susceptible (Mitchell, 1992).

Cool the fruits as soon as possible following harvest, and hold them at as low a temperature as possible without danger of freezing. To minimize the incidence of watery breakdown, fruit should be cooled to near storage temperature within 1 day of harvest, and held at 0°C (32°F) or below for up to one-month storage, or -1°C (30°F) for longer storage. Check the soluble solids concentration in the core area (lowest soluble solids content) to determine the lowest safe storage temperature for each lot of pears.

27.6. Bitter pit (Anjou pit /Cork spot)

This disorder occurs in Anjou and Packham's Triumph as brown, corky lesions in the flesh towards the calyx end. The skin surface is uneven with dark pit-like depressions and the fruit tend to colour prematurely. Calcium chloride @ 3g/L of water is the most effective control of bitter pit. Calcium chloride should not be applied at temperatures above 25°C nor under slow drying (high humidity) conditions.

27.7. Pink end / Premature Ripening/Pink calyx

The pink colouration of the calyx end of the fruit indicates premature ripening in the field and is not a post harvest related condition. This disorder is caused by abnormal cool growing seasons preceding harvest. Night temperature lower than 7.1^{0}C and day temperature lower than 21^{0}C for few days are sufficient to induce this disorder (Chen and Borgie, 1984). Avoid overdose of stop drop chemical (NAA), a dose of NAA as low as 37 ppm within 6 days before harvest could significantly increase premature ripening. Besides, harvesting should be accelerated in cool seasons and store the fruits quickly and carefully immediately after the harvest.

27.8. Budjump

It is a floral bud disorder. The incidence and severity of bud loss in *P. serotina* [*P. pyrifolia*] as a result of the floral bud disorders collectively called bud jump were investigated from 1986 to 1991 throughout the Waikato region of New Zealand. Large variation was found in levels of bud loss from year to year. Most bud loss was found in the cultivar Hosui, with high levels significantly reducing fruit production in some years. In the cultivars Kosui, Shinsui, Nijisseiki and Shinseiki bud loss was much less severe. In Hosui most bud loss was found on young trees and on young wood on older trees; least bud loss was found at branch tips and on spur buds on older wood; High levels of bud loss were also associated with reduced flower numbers in remaining flower clusters. The type of rootstock and training system used had no effect on the incidence of bud jump (Klinac and Geddes, 1995).

28. Insect pest and their management

28.1. Pear Psylla (*Psylla pyricola*): Nymphs and adults are phloem feeders. Honeydew, produced by nymphs, drips or runs onto fruit, causing dark russet blotches or streaks. This results in downgrading of fresh and sometimes processing fruit.

In large numbers, Pear *psylla* can stunt the growth and defoliate trees and cause fruit drop. A carry-over effect may reduce fruit set in the following year. These symptoms, called *psylla* shock which is caused by toxic saliva injected into the tree during feeding by the nymphs. When *psylla* are controlled to prevent fruit russet, *psylla* shock does not occur.

Pear *psylla* may also transmit a disease called "pear decline" that can slowly kill trees over a number of years. It can be recognized by symptoms like sticky honeydew on leaves and fruit, and sometimes black sooty mould random, scorched appearance to leaves leaf drop and decreased fruit yields.

The best approach to chemical control is to destroy winter-form adults before they lay eggs. It will take 2 or 3 pre-bloom sprays to accomplish this. Apply the initial dormant spray as soon as adults begin to lay eggs. Female winter-form adults can be dissected to monitor egg development in the ovaries. Eggs turn yellow when they are almost mature and about to be laid. Monitor surviving and immigrating adults after spraying. Spray again at delayed-dormant or cluster-bud stages, or both, when density exceeds 5 per 25 trays. The best control is achieved with a dormant oil spray to kill newly laid eggs. If you need to spray during the growing season, use 1% oil spray, insecticidal soap, or spinosad insecticide.

28.2. Codling Moth *(Cydia pomonella)*

Codling moths are the adults of the common "worms" that infest pears. These moths emerge from over wintering sites in the spring and lay their eggs on and near developing fruits. Codling moth has the greatest potential for damage of pear, yet it can be effectively controlled with properly timed treatments. Larvae may enter through the sides, stem end, or calyx (flower) end of the fruit. One or more holes plugged with frass on the fruit's surface is a characteristic sign of codling moth infestation. Calyx entries are difficult to detect without cutting the fruit. Symptoms are characterized by sawdust-like excrement on the outsides of the fruit, small holes in fruit, larvae inside fruit and rot sometimes accompanies entry holes on fruit. (UC IPM, 2016).

An important parasite of codling moth is *Trichogramma platneri,* which parasitizes its eggs. Remove host trees in nearby abandoned orchards (apple, pear, and walnut) to destroy reservoirs of codling moth. Also remove props, picking bins, and fruit piles from the orchard. Proper pruning and orchard sprayer calibration will improve spray coverage. An option for small, organic orchards is hand thinning to remove all infested fruit during each generation, before worms leave fruit, and removal of dropped fruit. Trunk banding to provide places for the larvae to pupate can also be used to help control population levels, provided the bands are placed before pupation and removed before adult emergence. Insecticides include carbaryl, malathion, gamma-cyhalothrin, bifenthrin, acetamiprid, and spinosad were found effective.

28.3. Pear Leaf Blister Mite (*Eriophyes pyri*)

Blister mites attack both foliage and fruit, producing small galls or blisters. They burrow under the lower surface of leaves and live inside small blisters through out summer. The galls can be seen as early as when the leaves start to unfurl. The galls are like green pimples that can change to yellow as the season progresses. Some produce a red colouring that can be seen on the leaf surface. Usually the galls are present along the main vein but can spread as populations increase. Mites can be transported easily on the wind and on clothing. Severely infested leaves can turn brown and drop off. Damage to fruit is less serious but scarring can render the fruit unmarketable.

By their hidden way of living inside the leaf blisters, the mites are protected from plant protection agents. Until now, oil treatments in early spring, when the mites leave their hibernation sites and invade the young leaves, are the only possibility to control these mites. Because of the low temperatures during this period (February/March) the efficacy of pesticides is often not reliable. Daniel *et al.* (2006) effectively controlled the mites with sulphur application after harvest at the time, when mites are leaving the blisters and migrate to the hibernation sites. With an efficacy of 95% (one sulphur application in autumn) or 100% (three sulphur applications in autumn), respectively, even a decontamination of heavily infested orchards is possible.

Other measures included with carbaryl and horticultural oil in early fall, before leaf drop, when mites are migrating from leaves to buds.

28.4. Spider Mites

Mites are very small arthropods that are more closely related to ticks than insects. Spider mites over winter as adults at the base of trees, or in ground cover, and may become a problem during hot, dry conditions in mid and late summer when they reproduce rapidly. As they feed, they remove sap and chlorophyll from leaves. Pears cannot tolerate high mite populations and symptoms are different from those on other trees. Predatory mites that feed on spider mites can provide effective biological control if they are not harmed by pesticides. Washing down trees or plants with a stiff spray of water or applying insecticidal soap or 1% horticultural mineral oil every 5-7 days until mite densities decline can be effective. Avoid applying soaps or oils during the hot part of the day as some leaf may be burnt may (Verma, 2015).

28.5. Oriental Fruit Moth (Grapholitha molesta)

It causes damage of varying importance on pear. The larvae of the first generation are mostly found in buds and shoots. In young trees when terminal twigs are attacked, several lateral shoots will appear below them and grow rapidly. Under severe and continued attack, the tree may become somewhat bushy. Severe attacks result in crooked stems. Agrialoba (phosalone) at 0.2% and Gordona (tetrachlorvinphos) at 0.1% have been found as most effective insecticides against this pest.

28.6. San Jose scale (Quadraspidiotus perniciosus [*Diaspidiotus perniciosus*])

In pears, chlorpyrifos proved most effective against San Jose scale if preceded by an autumn application (Pasqualini and Civolani, 2000).

29. Diseases and their control

29.1. Fire Blight

The main thing to worry about with pear trees is fire blight, and it may be hard to avoid in the healthiest of trees. Fire blight derives its name from the fact that the affected parts look as if someone had gone over them with a blowtorch while you were not looking. Fire blight, caused by the bacterium *Erwinia amylovora,* is the most serious and economically damaging disease of pears. Pears are very susceptible to fire blight. Blossoms, terminal twigs, and sometimes entire limbs or trees are killed. The sign of this disease is appear in spring when new growth turns brown and a sticky liquid appears around affected areas. Environmental conditions favouring infection are warm, rainy days during bloom. Rain, heavy dew, and insects spread oozing bacteria to flowers, where new infections occur. Infected blossoms and shoots turn black; leaves on dead shoots remain attached, the bacterium (*Erwinia amylovora)* over winters in sunken cankers in the wood, which begin to ooze in spring when temperatures warm. Infected twigs wilt over at the tip, in the shape of a shepherd's crook.

All infected shoots, twigs, and limbs should be pruned out of the tree. Cut 12 inches below the canker into healthy wood to be certain that the bacteria are removed. Keep all infected trimmings away from the tree trunk and burn them in a hot fire away from the orchard. The worst blight season is from bloom till fruit set and is the time to keep a sharp lookout for the culprit. Do not prune during moist conditions, as this can contribute to disease spread. Protect flowers from infection by applying an antibiotic spray just before, or 24 hours after, a potential wetting event. Use a mix of streptomycin and oxytetracycline to prevent resistance (Verma, 2015). Once fire blight infection is occurred, there is no spray or other treatment, beyond quickly cutting out newly infected limbs that will minimize damage.

29.2. Crown Rot and Root Rot

Crown rot and root rot are caused by a soil-borne, fungus-like organism *(Phytophthora*). This pathogen is present in most soils, but only causes infection when soils are saturated with water and a host tree is present. The pathogen quickly grows in the cambium, girdling the tree. Once trees are infected, there is no cure. Select sites with good water drainage for planting. Prevent standing water and do not over-irrigate. Do not replant in areas where root and/or crown rot occurred previously. Protect trees adjacent to infected trees by spraying foliage with phosphorus acid (Agri-fos, Fosphite) because phytophthora can spread by root-to-root contact. (Verma, 2015).

29.3. Pseudomonas Blight (Blossom Blast)

Another bacterial disease, blossom blast (causal organism: *Pseudomonas syringae*, not to be confused with *Pseudomonas fluorescens* discussed above), may afflict pears, usually as a blossom blight resulting in reduced fruit set. It can also cause twig dieback and bark cankers and may lead to severe wood damage of Asian-pear cultivars in particular (see the Boullioun "Grower Profile" for more information on pseudomonas and Asian pears). Because the presence of blossom-blast bacteria allows ice crystals to form at higher-than-normal temperatures, the disease increases the incidence of freeze damage during cold, wet weather. Asian pears are especially affected because their early bloom makes them more susceptible to frost injury (Elkins, 2001). Among Asian-pear cultivars, Shinko and Ya Li are moderately resistant to *P. syringae.*

Controlling this disease is difficult because its occurrence is widespread on many plant species and not easily predicted; once symptoms appear, control efforts are too late. Protecting orchards from frost damage can limit injury. An early application of Blight Ban A506 can help reduce frost-damage potential by excluding the ice-nucleating bacteria.

29.4. Pear Scab

Pear scab resembles apple scab in nearly all respects and is caused by the closely related fungus *Venturia pirina*, it can cause economic damage by marring the appearance of the fruit. Pear scab causes lesions on leaves, shoots, and fruit, and—unlike apple scab—infects twigs, where it can over winter. Pear cultivars differ in their susceptibility to scab. Although it is not particularly common, pear scab is very destructive when it does occur. Its symptoms and disease cycle are similar to those of apple scab. A major difference is the frequent appearance of pear scab on twigs, where it can over winter and start new infections in spring. Infection of pear leaves is not as common as apple scab on apple leaves.

Fungicides such as dodine (Syllit) and the ethylenebisdithiocarbamates (mancozeb) are good protectant but have limited eradicate properties. Thiophenate methyl (topsin-M) give protection for longer duration (Pest Management Guide for Tree Fruits in the Mid-Columbia Area, 2010).

29.5. Pear decline

Garcia *et al.* (2003) reported pear decline disease and its relative importance in northeastern Spain. The presence of pear decline (PD) phytoplasma in surveyed plots was confirmed by polymerase chain reaction (PCR) amplification of phytoplasma DNA with universal or group-specific primers.

30. Fruit maturity

Flesh firmness is considered as one of the indicators of fruit maturity and flesh firmness of fruit of a variety is governed genetically. Early harvested fruits ripened with melting juicy texture but lacked flavour. Fruits harvested later were capable of developing high dessert qualities including melting juicy texture, distinct mixed Bartlett and Comice pear aroma and tart sweet taste upon ripening.

Fruits with FF (Flesh firmness) between 63.0 and 55.1 N were considered to be at optimum maturity for commercial harvest (Ma *et al.,* 2000).

31. Yield

The yield of fruits varies greatly with the cultivar, age of the plant, orchard management and the climatic conditions. The common pear yields 100 to 120 kg per tree per year, New Pear and Kieffer yield 70 to 80 kg per tree per year and Jargonelle and William pear yields 30 to 40 kg per tree per year. Yields of Asian pears are lower per acre than for Bartlett or Bosc because heavy fruit thinning is necessary to get sizes the market demands and pays for with premium prices. Orchards may produce a few pears in the third year, better in the fourth season but yields in the fifth to seventh years range from 200 to 400 packed boxes per acre. When trees are full-sized (12 feet wide and high) at age 10-14 years old, yields may reach 800 to 1,000 packed boxes per acre of 3 to 4 inch diameter pears. A well managed orchard of Bartlette pear gives about 30-40 tonnes of fruit per year.

32. Post harvest management

To secure better price and goodwill, proper grading of fruits is desirable. Bartlett pears are graded as extra large, large, medium and small with diameters of 8, 7, 6.5 and 5 cm respectively. Sometimes these grades are also known as Extra Class, Class I, Class II and Class III. All damaged and misshapen fruits should be excluded from grading. The fruits after wrapping need to be arranged in layers in the containers. Both top and bottom portions of containers should be cushioned with dried grass or paper along with labeling the container.

Studies were conducted on the effect of postharvest handling on quality and ripening of Bartlett pears. It was found that packed fruit from commercial pack houses often ripened without ethylene treatment or cold storage while fruit carefully picked and not packed required at least 24 h of ethylene conditioning or cold storage. Fruit softened and changes in skin colour from green to yellow occurred faster in commercially picked samples put through a packing line compared to fruit picked commercially but not put through a packing line. Increased postharvest handling produced increased rates of ethylene production and respiration in the fruit. Suggestions to reduce postharvest damage and consequently improve fruit quality were made (Agar and Mitcham, 2000).

33. Storage

Pears should be stored for 20-30 days to tide over the glut period. Pears have a definite period of storage life. If kept in storage for a longer period they will not ripe normally even on removal of high temperature. Pears are stored in cold store or controlled atmospheric storage to prevent the decay before consumption (Table 14). Hydro-cooling reduces the incidence of shrivelling and brown core. Storage life of the fruits of different cultivars ranges from 120-245 days at – 1^{0}C.

Table 14: Controlled atmospheric storage of important pear cultivars

Cultivar	Temperature °C	Oxygen (%)	Carbon dioxide (%)	Storage period (months)
Anjou	11	1.5	1.0	8
Bartlett	-0.5	1.5	1.5	2
Blanquilla	0.5	3.0	5.0	8
Comice	0	5.0	5.0	6
Conference	-0.5-0	2-3	0.5	8
Kaiser	0-1	4-5	2-3	5
Krystalli	0	3.0	5.0	6
Packham's Triumph	0	5.0	3.0	7
Passe Crassane	4.5-5.0	3.0	4.0	5

Source: Mitra *et al.* (1991)

Shelf life of pears is recorded by means of fruit firmness and incidence of internal breakdown in European pears (*Pyrus communis*). Maturity at harvest characterized by fruit firmness was highly variable, and influenced the shelf life to a high extent in such a way that ripe fruits had a short shelf life. Fruit firmness was not directly related to the incidence of internal breakdown, and therefore it is suggested that records of internal breakdown following a specific procedure of ripening could be adopted as selection criteria for shelf life (Hjeltnes, 2000)

To increase shelf life of pear fruits and to maintain / or improve fruit quality in storage several suggestions have been given by various workers worldwide. Ahmad and Khajawall (1999) from Jammu and Kashmir (India) observed that harvesting of fruits at 105 or 107 DAFB (days after full bloom) and dipping them in 8-12% CaCl2 proved the best for enhancing the keeping quality of the fruits at ambient temperature (29+or-2 degrees C) for 24 days.

Wang *et al.* (2014) showed that, 1-MCP treatment could inhibit the superficial scald of pear cv. Dangshansu completely during the shelf-life. It was also conductive to keeping desired colors and inner qualities, as well as reducing the respiration intensity, ethylene release, total phenol content and PPO activity of the organizations of the fruits apparently. 1-MCP treatment could be used as a main technology for the control of browning (especially superficial scald) and quality maintenance, the suggested concentration of 1-MCP was 0.5-1.0 micro L/L.

The plant hormone ethylene is known to be important during fruit ripening, especially in typical climacteric fruits, including pear (*Pyrus communis*). It is currently thought that the inhibition of ethylene biosynthesis via genetic engineering should improve the quality and shelf life of fruits. 1-Aminocyclopropane-1-carboxylate (ACC) synthase and ACC oxidase [1-aminocyclopropane-1-carboxylate oxidoreductase] are two key enzymes in the ethylene biosynthetic pathway. Therefore, inhibiting the expression of genes encoding these two enzymes should reduce ethylene synthesis (Gao *et al.,* 2003) resulting increase storage life of fruits.

The pears developed flesh browning, core browning and cavities under all CA conditions (controlled atmosphere). Severity of these disorders depended on the CO2 and O2 concentrations and the length of storage. CA-stored pears showed lower ATP concentrations than in air, especially in ultralow oxygen combined with high CO2. At these conditions, a low energy charge and an accumulation of ethanol and acetaldehyde as well as a strong decrease in total and free fatty acids was observed. In particular, the contents of C-16 and C-18 fatty acids were strongly reduced under CA conditions. Most fatty acids of the polar lipids showed a slight increase followed by a decrease towards the end of storage. Neutral lipids declined initially and remained at low levels thereafter. These results suggest the involvement of ATP in the development of pear browning possibly by limiting fatty acid biosynthesis. This in turn may alter membrane permeability and function resulting in breakdown and browning of tissues. The increased ethanol and acetaldehyde may also contribute to this process (Saquet *et al.,* 2003).

It is supposed that the development of controlled atmosphere (CA)-induced browning disorders in pears cv. Conference is related to the "energy charge" of tissues which decreased during storage period. It is known that the delay in the establishment of CA conditions can reduce the incidence of flesh browning in pears. Saquet *et al.* (2001) carried out an experiment to test the effectiveness of a delayed CA storage treatment (21 days at 0 degrees C and ambient atmosphere before CA storage) in order to adapt fruits to CA. Thereafter, fruits were stored for up to six months in 3% CO2 + 1% O2 at 0 degrees C. Fruits without delay were immediately stored under CA and developed 80, 90 and 72% higher incidence of flesh browning, core browning and cavities, respectively, compared to delay-treated fruits. Fruits subjected to delay CA showed higher ethylene production and higher respiration activity during the first two months of storage, and was associated with higher concentrations of ATP in the fruit tissue. Higher ATP levels remained until the fourth month of storage. A higher ATP: ADP ratio of delay-treated fruits was also found, indicating a higher energy charge of tissue during this period. Results suggest that 'Conference' pears can become adapted to CA during the early stages of the storage period, making them more resistant against brown heart.

Zerbini *et al.* (2002) noted that brown heart (BH) in Conference pears was appeared in different treatments when AA (ascorbic acid) decreased below a threshold, which was about 2 mg kg^{-1} fresh weight, corresponding to 5% of the AA content at harvest. The frequency and severity of symptoms depended linearly on the time spent by fruits in low-ascorbate conditions. Advanced maturity at harvest and storage in high CO2 determine the appearance of BH and increase the rate of AA loss, probably through a reduction in the turnover of antioxidants.

One of the major disorders that develop during storage in Asian pears is fruit internal browning (IB) of the core and/or flesh. The relationship between IB, ascorbic acid (AA) concentrations, total phenolic content (TPC) or polyphenoloxidase (PPO) activity during five months storage was investigated in Asian pear fruit by Arzani *et al.* (2009). Mature Asian pear (*Pyrus serotina* Rehd.) trees cultivar 'KS'13 grown on European pear seedling rootstocks were used for the study. Fruits were harvested on August 1 (H1) and August 15 (H2), representing early and late harvest for this cultivar. Fruit were picked according to the maturity standard based on soluble solid content (SSC), titrable acidity (TA), firmness and background colour and stored at 1 degrees C and 80-85% relative humidity (RH). Results indicated that IB incidence was higher in fruit from H2 than H1. The activity of PPO in disordered fruit varied with the intensity of the disorder. Ascorbic acid decreased during long-term storage and was lower in H2 fruit compared with H1 fruit. There was a trend for IB to develop when the AA concentration dropped below a threshold. This threshold value depended on picking date and time elapsed from storage and varied between about 3.3 to 9.4 mg 100 g-1 FW. The amount of TPC declined during storage in both H1 and H2 fruit as PPO activity and intensity of IB damage increased. It is suggested that the lower incidence of IB in earlier harvested fruit was a result of the presence of a higher level of AA and a lower level of PPO activity than in later harvested fruit.

34. Ripening

Pear fruits require exposure to any chilling temperature for proper ripening regardless of harvest maturity. After one month of air storage at -1 degrees C or 2 degrees C, all fruits were capable of ripening normally (Ma *et al.,* 2000). Fruits stored at -1 degrees C softened at a much slower rate than fruits stored at 2 degrees C during 5 months of air storage, regardless of harvest maturity. Fruits stored at -1 degrees C developed little internal browning (IB) or senescent scald (SS) until after 5 months of air storage, while fruits stored at 2 degrees C developed IB and SS after 3 months of air storage regardless of harvest maturity.

It was concluded that the proper temperature for long-term storage of Cascade pears is -1 degrees C (Ma *et al.,* 2000).

Pear fruits exhibit a typical climacteric pattern of respiration. Among the various cultivars Bartlett has the highest respiration and ethylene evolution. Ethephon application (100-300 ppm) hastens ripening by about two weeks, increase colouring but reduce firmness (Mann and Singh, 1988). Ripening process has been delayed by the treatment of Bartlett pear with aminoethoxyvinylglycine (AVG) and mannose. Watkins and Frankel (1987), inhibited pear fruit ripening like softening of flesh, ethylene evolution, respiration by vacuum infiltration of mannose into intact fruit of cultivar Bartlett.

35. Quality improvement

Triazole chemical (Bayleton) which has both retardant and fungicidal properties is of special interest to pear growers as it increases frost hardiness of flowers and buds. Gibberellin(GA) has been reported to have good effect on frost suffered bloom of Conference pear. The concentration of its use range from 10-15 ppm but for older trees 10 or 12.5 ppm is adequate for increasing fruit set. The spray should be done within 40 hrs of frost occurrence. A single spray of CCC at 2000 ppm when 8-10 leaves per shoot had unfolded increase fruit bud differentiation for next year and also eradicates the harmful effect of Gibberellin application if any. Inhibition of shoot extension and stimulation of flowering in the following year has been observed with CEPA (Etheral) of non bearing trees at 250 ppm for 2-4 weeks after full bloom. Use of water sprinkling for frost protection has been commercially adopted in many frost prone areas (Gautam, 2004).

Application of foliar fertilizers which contained 15% of P2O5, 20% of K2O, 0.1% of Mn, 0.1% of B and 0.1% of Mo, sprayed for five times (from May 22 to July 7) on pear tree (*Pyrus communis*) cv Williams resulted in higher quantities of sugars (glucose, sorbitol, soluble solids) and organic acids (malic, citric). The foliar nutrition treatment resulted in a gradual decrease in the contents of fructose, ucrose, boron, and zinc (Hudina, 2004).

Increasing market competition has forced fruit growers to produce high quality fruits. It has been reported that seaweed extract (SE) contains nutrients, aminoacids, polysaccharides, vitamins, betaines and phytohormones that affect cellular metabolism and that it has a positive impact on plant health, growth and yield. Colavita *et al.*, (2011) suggested that SE applied at appropriate phenological timings increases quality and productivity of pears. Applications of SE at different phenological phases were made on *Pyrus communis* 'William's' planted (1250 tree/ha) in Alto Valle de Rio Negro (38 degrees S; 67 degrees W), Argentina.

Untreated controls were compared with treatments consisting of 3 foliar sprays of 3 L/ha SE (BMReg., Goemar, basically Ascophyllum nodosum) at different stages of pear phenology: BM1 (white bud, full bloom, petals fall) and BM2 (full bloom, petals fall, fruit size 3-4 mm). A positive results were noted on fruit yield and quality due to application of SE.

36. Processing and value addition

Pear has a milk flavour rich in phenolics. Its low acid content and low sugar content contributes to its suitability for consumption by person suffering from acidity and diabetics. Various processed products from pear such as dehydrated slices, juice and powder as value added products and for extended storage life can be prepared.

Dehydrated pear slices, is an easy way to preserve pear harvest for up to 30 days or longer if frozen, also the pear juice is one of the best to beat sunstroke and dehydration and reduce stomach problems like acidity, ulcers and constipation. Pear fruit powder is rich in vitamin C, phytochemicals and antioxidants for boosting the body's immune system, improves the blood circulation and keep heart health, nourish the eyes and protects it from eye diseases, promote colon health, cardiovascular health, help to protect against mascular degeneration. Various pear syrups like Cinnamon pears, Pineapple pears, Mint pears above all pear apple jams and pear apple jellies are common processed products of pear (Hui *et al.*, 2006).

37. Organic cultivation

The organic market has grown exponentially in Europe during the last ten years. However, the organic fruit industry has shown the lowest growth rates (1-5% market share) compared to other commodities. One major reason is the high production risk due to high disease and pest pressure in humid climates. Key pests and diseases in apple and pear include scab (*Venturia inaequalis* and *V. pyrina,* respectively), sooty blotch (complex of microorganisms), and fire blight (*Erwinia amylovora*) as well as rosy apple aphid (*Dysaphis plantaginea*), pear sucker (*Psylla pyri* [*Cacopsylla pyri*], *P. pyricola* [*Cacopsylla pyricola*], *P. pyrisuga* [*Cacopsylla pyrisuga*]), and codling moth (*Cydia pomonella*) make organic production almost unfeasible. In an ideal organic system, all possible measures that lead to improved stability of the system must be implemented. In fruit production, such measures include (i) use of resistant varieties (scab), (ii) active promotion of predators (rosy apple aphid), (iii) microclimate management (planting density, covering trees), and (iv) strengthening the plant self-defence system (e.g. building up of soil fertility, bio-dynamic preparations). At present, most organic orchards have been designed for conventional production and, therefore, lack most stabilizing components. In consequence, organic fruit production still depends largely on direct pest and disease control methods. These include traditional pesticides such as copper, sulfur, and lime sulfur [calcium polysulfide], pyrethrin, oil and soap preparations. More recently, new pest control compounds and techniques have been developed including neem products, granulosis virus, and mating disruption. New tools for disease control are less spectacular and focus on the replacement of copper (e.g. with clay powders and resistance inducers). The introduction of new equipment, materials, and Decision Support Systems will further improve yield stability. However, other limiting factors such as weed competition, crop load regulation, and conventional market demands need innovative solutions (Tamm *et al.,* 2004).

Growing organic pears (*Pyrus* spp.) differs from regular pear production in three critical areas: fertilizing, controlling pear pests — most significantly the codling moth — and controlling common pear diseases like fire blight. With the onset of technology, the synthetic chemicals are easily available to the farmers, which on one hand has made the farming very easy but on the other hand has deteriorated the fruit quality and diminish the condition of the planet earth.

37.1. How to grow a pear tree organically

1. Choose and purchase a pear cultivar; organic practices will work for all cultivars. Plant the tree in a sunny location in your garden. Dig the planting hole in spring, place the seedling's roots or root ball into the hole with its graft line, if any, several inches above soil level and refill the hole with extracted earth. Place the tree near another pear for pollination or plant a second tree within 40 feet of the first.

2. Spread a 3-inch layer of organic compost over the root area of the tree. Keep the mulch 4 to 6 inches from the tree trunk. When the compost diminishes, renew it with another few inches and repeat all summer long. In winter, add a layer of shredded wood to protect against colder weather.

3. Prune the pear tree, starting by taking off the top 24 inches from the ground at the time of planting. Trim to the multiple leader system, keeping the center clear to favour the growth of three or four selected upright shoots that are evenly spaced around the tree; each of these becomes one of the tree's foundation branches. Leave several large lateral branches on each of these new leaders and remove all competing branches.

4. Spray the pear tree with dormant spray to control mites as well as the destructive pear psylla. Prepare the organic spray as: Mix 1 pound of organic oil-based soap, 1 gallon of organic horticultural oil and one-half

gallon of water in a saucepan on the stove and bring to a boil. Dilute with water, one part solution to 19 parts water, and spray it on the tree in January, June, August and October. Cover the trunk and each side of every leaf and branch.

5. Control fire blight by pruning out the infected branches in the spring. Look for watery ooze seeping from areas of dead bark in new shoots that darkens and streaks the branches or trunks. Always sterilize the pruners with denatured alcohol between clips. Cut more rather than less. Find the lower edge of the visible infection in the branch, trace the branch back to where it attaches to the trunk, and prune at the next branch juncture down, taking out both the infected branch and the branch to which it is attached. Burn the pruned pieces (Ames and HollyBorn, 2012).

6. Control codling moth, a pear tree's primary pest, in April. Hang pheromones to disrupt the moths' mating behaviour and traps to catch the moths before they can lay their eggs on the tree.

7. Pear scab, caused by *Venturia pirina*, can be a limiting factor for successful organic production of pears in climates conducive to scab development. Sugar and Hilton (2011) noted that potassium bicarbonate 1% and copper hydroxide (Kocide 3000 1.0 g L-1) treatments reduced fruit scab infections by 94.3 and 87.5%, respectively. Both potassium bicarbonate and copper hydroxide treatments caused unacceptable levels of fruit russet when applied to 'Comice' pears. Hydrated lime treatments did not cause russet in 'Comice' pears. While there were no apparent negative effects from potassium bicarbonate or copper hydroxide in the naturally russet 'Golden Russet Bosc' trees, their use may not be appropriate in some green-skinned cultivars where a russet-free finish is desired. The effects of repeated applications of hydrated lime on insect and mite populations were assessed. Natural enemies of pear pests were slightly reduced in the hydrated lime treatment versus an untreated control but the application of kaolin (Surround), either alone or in combination with hydrated lime, reduced the presence of natural enemies. However, pear psylla (*Cacopsylla pyricola*) populations rose to well above the economic injury level in all treatments.

38. Trade and marketing

In general, pear prices tend to fluctuate based on crop size—which can be highly variable on account of weather and fruit set rates—and on the number of growers producing pears in a given year. It is found grown in many places of India like Karnataka, Tamil Nadu, Kerala, Uttar Pradesh, Punjab, Jammu and

Kashmir and Himachal Pradesh. The marketing season for pears differs among the states and their varieties. Many consumers are concerned about where the produce comes from; therefore U-pick operations have the ability to make consumers feel more secure and connected to their food. The market for pear is weaker than it could be, largely because of consumers' unfamiliarity with the different types of pears and how to use them. China has become the world export leader for pears, producing 15 million tons in 2010 and depressing wholesale market prices for domestic producers (Foreign Agricultural Service, 2011). Asian pears are breaking out of their niche to join the mainstream market and may represent an opportunity for producers as market awareness and demand continue to grow.

In India the fruit growers normally do not conduct direct sale but generally auction the crop to the contractors when the fruit is on the tree. The contractors pick fruit according their convenience and sell it in the market. Only few orchardists market their fruit directly in the main market. In order to ensure due enumeration to the orchardists, sales cooperatives need to be organised and such organisation should be helped in grading, packing and transportation of the fruit. The facility of storage chambers can also play an important role in improving the marketing of pear fruit.

39. Export

India is a major exporter of fruits to the world. The country has exported 3,08,261.23 MT of fresh fruits other than Grapes and Mango to the world for the worth of Rs. 1,538.16 crores during the year 2015-16.

39.1. Major Export Destinations

United Arab Emirates, Nepal, Saudi Arabia, Bangladesh and Iran (APEDA, 2015-2016).

In India, pear export is done by two types, viz., direct exporting and indirect exporting

39.1.1. Direct Exporting

Pear is directly exported to foreign buyers. It is usually under taken by large sized manufacturing companies This type of exporting has high profit margin, benefits from Government incentives, no dependence on middle man, optimum use of production capacity.

39.1.2. Indirect Exporting

The produce is not directly exported to foreign buyers, organisation generally uses indirect exporting if it is small scale business, availability of limited resources. But there is limited business risk, less overheads and limited investment.

In India, Punjab & Himachal Pradesh is the prime pear-growing areas, but in order to fulfil the surging demand for pears, imports appear to be the only option. Fresh imports from US, South Africa and China have soared dramatically. As per stats available with the Agriculture and Processed Food Products Export Development Authority (APEDA), India imported 14,571.90 tonnes of fresh pears in 2013-14, valued at around Rs 78.93 crore. Home grown varieties of pears are usually grown in states like Jammu and Kashmir, Himachal Pradesh, and Uttar Pradesh. Despite the imports and increased world pear production, the global trade of pear is forecasted down at 7 percent mainly due to the Russian ban from the European Union. For India, exports are expected to jump slightly to 320,000 tonnes, prompted by bump in demand from the rich Indonesia market.

40. Economics of cultivation

Pear plantations have a gestation period of about seven to eight years. The initial investment is very high in pear plantations. This is due to huge expenses incurred in digging pits, applying manures and fertilizers, cost of pear plant, transplanting and so on. Two hundred trees, on an average, are planted in one hectare of cultivable land. Farmers have to incur costs on maintenance for about seven years. The main items of maintenance (Table 15) and cost benefit ratio is (Table 16) are presented which states that maximum amount is utilized on the costs of manures, fertilizers, pesticides, pruning and preparation of basins etc.

Table 15: Per hectare initial year cost of pear-orchard of small farmers

Cost Component	Unit	Cost/ Per unit	Quantity	Cost (Rs.)	Percentage
A. Labour Cost					
1. Land clearing and development	Mandays	250	16	4000.00	33.90
2. Digging of pits	Pit	15.00	200	3000.00	25.42
3. Filling of pits	Pit	6.00	200	1200.00	10.17
4. Manures and fertilizer application	Mandays	250	6	1500.00	12.71
5. Plant protection	Mandays	250	4	1000.00	8.47
6. Planting and plant support	Pit	5.50	200	1100	9.32
Total labour cost				11800 (52.92)	100.00
B. Material Cost					
1. Plant material (including transportation)	Tree	10.50	200	2100	25.89
2. Manures	Tree	6.25	200	1250	15.41
3. Fertilizers and other materials	Tree	12.60	200	2520	31.07
4. Plant protection	Tree	8.70	200	1740	21.45
5. Miscellaneous	Hectare	1	500	500	6.17
Total Materials Cost				8110 (36.37)	100.00
C. Interest on working capital		12%	19910	2389.20 (10.71)	
Total cost (A+B+C)				22299.20 (100.00)	

Source: Kumar *et al.* (2007)

Table 16: Cost return structure of small farms

S.No.	Stages of the Orchard	Average Cost of Production (Rs.)	Average Gross Income Per Hectare (Rs.)	Benefit Cost Ratio
1.	8th to 10th year	25,322	80,407	3.18
2.	11th to 15th year	26,584	1,08,608	4.08
3.	16th to 20th year	29,087	1,11,536	3.83
4.	21st to 25th year	31,960	1,03,523	3.24
5.	Beyond 25th year	32,949	97,264	2.95

Source: Kumar *et al.* (2007)

41. Stress factors affecting the pear fruit production included

41.1. Biotic factors: It included bacteria, fungi, nematodes, viruses and insect-pests that are responsible for hampering the production and reducing post harvest shelf life of fruits.

41.2. Abiotic factors: Abiotic stress is defined as the negative impact of non-living factors on the living organisms in a specific environment. These factors included salt, drought, flooding, low and high temperature, winds, etc.

41.2.1. Impacts of abiotic stresses on pear

1. Germplasm extinction
2. Poor vernalization
3. Unsatisfactory chilling hours
4. Poor activity of pollinators
5. Advanced bud burst – frost injury
6. Fruit cracking
7. Appearance of New spectrum of pests and diseases
8. Changes in metabolites viz. amino acids, amines, sugars etc.
9. Increased / decreased vegetative growth
10. Shifting fruit cultivation from lower areas to high altitudes vis –a- vis

41.3. How to combat different stress situations

1. Use of Drought tolerant cultivars example- Starkrimson, Victoria, Beurre Hardy, Clapp's Favourite, William Rouge etc
2. Overhead sprinkler irrigation to reduce frost damage

3. Use of reflective fabric and bagging of fruit to minimise high temperature stress
4. Use of Shelter belts/ wind breaks to reduce wind velocity.

42. Special problem

Low fruit set and cropping are common characteristics of [Doyenne du Comice] pear (*Pyrus communis*) trees in spite of a high flower density. Khemira *et al.* (1998) conducted an experiment to study the effect of late season N application on ovule longevity of this cultivar. Mature trees were given a single foliar application with 5 or 10% (w/v) aqueous urea immediately after harvest. Ovule longevity was extended and fruit set was increased by autumn-applied urea. However, flower density and fruit fresh weight were not affected

References

Abreu. J. P., Sousa, M. L. and Lopes, J. S. 2015. CSS_Pear: a Model to Simulate Growth, Production and Quality of Pear (*Pyrus communis* L.). *Acta Horticulturae,* **218**: 175-179.

Agar, I. T. and Mitcham, E.J. 2000. Commercial handling influences quality and ripening of Bartlett pears. *California Agriculture*, **54**(3): 34-37.

Agricultural and Processed Export Development Authority, 2015-16.

Ahmad Farooq and Khajawall, M. H. 1999. Effects of harvest dates and post-harvest calcium chloride treatments on physico-chemical characteristics of pear cv. Bartlett. *Applied Biological Research*, **1**(1): 43-46.

Ames, G. K. and Holly, B. 2012. In : *Pears : Organic Production*. National Center for Appropriate Technology (NCAT) (pub.)

Anonymous. 1969. The Wealth of India. VIII. Publication & Information Directorate, CSIR, New Delhi. pp: 327-334.

Anonymous. 1996. Varietal description. Delwini. *Arboriculture Fruitiere*, **497**: 53.

Arzani, K; Khoshghalb, H; Malakouti, M. J. and Barzegar, M. 2009. Polyphenoloxidase activity, polyphenol and ascorbic acid concentrations and internal browning in Asian pear (*Pyrus serotina* Rehd.) fruit during storage in relation to time of harvest. *European Journal of Horticultural Science*, **74**(2): 61-65.

Bajwa, G.S; Bal, J. S. and Minhas, P. P. S. 2003. Weed management studies in pear orchards at different locations in Punjab. *Haryana Journal of Horticultural Sciences.* **32**(3/4): 180-182.

Banno. K., Hayasim. S. and Tanabe. K. 1985. *J. Jap. Soc. Hort. Science*., **54**:15-25.

Bell, R. L; Zwet, T van der and Blake, R. C. 2002. 'Blake's Pride' pear. *Hort. Science*, **37**(4): 711-713.

Bellini, E. Sabina, a new summer pear cultivar. 2001. Sabina, a new summer pear cultivar. *Informatore Agrario,* **57**(17): 61-63.

Beutel, J. 1990. *Asian pears*, p. 304-308. In J. Janick and J.E. Simon (eds.). Advances in new crops. Timber Press, Portland

Beverly, R.B. 1987. Modified DRIS method for simplified nutrient diagnosis of 'Valencia' oranges. *Journal of Plant Nutrition*, **10**: 1401-1408.

Blanpied, D. G. 1975. *Core breakdown of New York 'Bartlett' pears. J. Amer. Soc. Hort. Sci.,* 100:**198-200.**

Campbell, J. 2003. Pear rootstocks Agfact H4.1.15, 1st Edition, Agricultural Intitute, Orange. pp 1-11

Cao YuFen; Sun XiSheng; Jiang ShuLing; Zhang Wei; Wang WenHui; Tan XingWei and Ma Li. 2003. Study on the characters of pear varieties for juice-processing. *Journal of Fruit Science,* **20**(6): 450-454.

Carrera, M. and Ortiz, L. 1984. Performance of three quince rootstocks for pear. *Acta Horticulturae,* **161**: 231-4.

Chen. P. M. And Borgic. D. M. 1984. *Good Fruit Grower*, 35: **14-15.**

Daliwal. G. S., Aulakh. P. S. and Jawanda. J. S. 1982. *Punjab Hort. J.*, **22**: 164-168.

Daniel, C; Linder, C. and Wyss, E. 2006. Autumn applications of sulphur - a new approach to control the pear leaf blister mite *Eriophyes pyri. Ecofruit 12th International- Conference on Cultivation Technique and Phytopathological Problems in organic Fruit Growing. Proceedings of the conference, Weinsburg, Germany, 31st January-2nd February-200 pp.* 47-54.

Delaplane, K.S. and Mayer, D.F. 2000. Crop pollination by bees CABI Publishing, New York.

Dheim.M. A. and Browning. G. 1987. The mode of action of (2RS, 3RS)-paclobutrazol on the fruit set of Doyenne du Comice pear. *Journal of Hort. Sci.*, **62**: 313-327.

Dheim.M.A. and Browning. G. 1987. *Journal of Hort. Sci.*, **62**: 313-327.

Dussi, M. C. 2011. Sustainable use of plant bio-regulators in pear production. *Acta Horticulturae,* **909:** 353-367.

Dzhangaliev, A.; Salova, T. and Turekhanova, P. 2003. Pome fruits. In. The Wild Fruit and Nut Plantas of Kazakhstan, by Janick, J. (ed.). *Horticultural Reviews*. 311-324.

Elkins, R.B., R. A. Van Steenwyk, L. G. Varela, and C. Pickel. 2001. UC Pest Management Guidelines: Pear Psylla. www.ipm.ucdavis.edu/PMG/r603301111.html

El-Sharkawy, I; Jones, B; Gentzbittel, L; Lelievre, J. M; Pech, J. C. and Latche, A. 2004. Differential regulation of ACC synthase genes in cold-dependent and independent ripening in pear fruit. *Plant, Cell and Environment,* **27**(10): 1197-1210.

El-Shazly, S. M. and El-Sabrout, M. B. 1994. Root formation on hardwood cuttings of Le Conte pear as influenced by auxin treatments and time of application. *Alexandria Journal of Agricultural Research,* **39**(3): 545-558.

Erhardt, W.; Gotz, E.; Bodeker, N. and Seybold, S. 2002. Zander. Handworterbuch der Pflanzennamen, 1. Eugen Ulmer Verlag, Stuttgart, Germany.

Erig, A. C; Schuch, M. W. 2004. In vitro rooting of quince cv. MC as rootstock for pear and acclimatization of the rooted microcuttings. *Ciencia Rural*, **34**(5): 1443-1449.

Fideghelli, C. 2007. Origine ed evoluzione. In. *Il Pero*, Vol. 1, by Angelini, R. (ed.). Bayer/ Crop Science, Milano, Italy.

Flick, J.D. and Hermann, L. 1978. Effects of Gibberellic acid on fruit set of Passe Crassane pear. *Acta Hort.*, **80**:143-148.

Foreign Agricultural Service. 2011. Fresh Deciduous Fruit: World Markets and Trade. FAS Online. http://www.fas. usda.gov/htp/2011_Fruit.pdf

Gao, M; Matsuta, N; Nishimura, K; Tao, R; Murayama, H; Toyomasu, T; Mitsuhashi, W. and Dandekar, A. M. 2003. Transformation of pear (*Pyrus communis* cv. 'La France') with genes involved in ethylene biosynthesis. *Acta Horticulturae*, **625**: 387-393.

Garcia Chapa, M; Lavina, A; Sanchez, I; Medina, V. and Batlle, A. 2003. Occurrence, symptom expression and characterization of phytoplasma associated with pear decline disease in Catalonia (Spain). *Journal of Phytopathology,* **151**(11/12): 584-590.

Gautam, D. R. 1996. Chip-budding: new promising technique for summer and autumn propagation of temperate fruit plants. *Journal of Hill Research*, **9**(1): 18-22.

Gautam, R. D. 2004.Pear.In: *Recent Trends in Horticulture in the Himalayas*. K.K. Jindal and R. C. Sharma (Ed). Indus Publishing Co, New Delhi.

Gill, P. P. S; Dhillon, W. S; Singh, N. P. and Singh, S. 2011. The influence of canopy architecture on light penetration, soil temperature and fruiting in sub-tropical pear. Journal of Agrometeorology, **13**(2): 157-159.

Gulen, H; Arora, R; Kuden, A; Krebs, S. L. and Postman, J. 2002. Peroxidase isozyme profiles in compatible and incompatible pear-quince graft combinations. *Journal of the American Society for Horticultural Science*, **127**(2): 152-157.

Hancock, J. 2008. *Temperate fruit crop breeding, Germplasm to Genomics*, Michigan State University, Dept. Horticulture. Springer Science.

Hayes. W. B. 1957. Fruit Growing in India. Kitabistan. Allahabad.

Hedrick.W. B., Howe. G. H., Taylor. O. M., Francis. H.E. and Turkey. H.B. 1921. The Pears of New York. N. Y. Dept. Agr. 29th Ann. Rep.2, part 2.

Hjeltnes, S. H. 2000. Selection for shelf life in pears. *Acta Horticulturae,* **538** (1): 339-344.

Horticultural Statistics at a Glance 2015. Ministry of Agriculture & Farmers Welfare Government of India. Oxford University Press.

Hudina, M. and Stampar, F. 2004. Does zinc and boron have an influence on higher yield of pears (Pyrus communis L.)? *Zbornik referatov 1 Slovenskega sadjarskega kongresa-z-mednarodno udelezbo, Krsko, Slovenia, 24-26-marec-2004-Del-1,* pp.169-174.

Hudina, M. and Stampar, F. 2011. Effect of fruit bagging on quality of 'Conference' pear (*Pyrus communis* L.). *European Journal of Horticultural Science,* **76**(5/6): 176-181.

Hudina, M; Stampar, F; Orazem, P; Petkovsek, M. M. and Veberic, R. 2012. Phenolic compounds profile, carbohydrates and external fruit quality of the 'Concorde' pear (*Pyrus communis* L.) after bagging. *Canadian Journal of Plant Science*, **92**(1): 67-75.

Hudina,-M. 2004. How to increase the pear fruit quality? *Zbornik referatov 1 Slovenskega sadjarskega kongresa z mednarodno udelezbo, Krsko, Slovenia*, 24-26-marec-2004, Del-1. 2004; 163-168.

Hui, Y., Barta, J., Cano, M., Gusek, T., Sidhu, J. and Sinha, N. (eds). 2006. *In :Handbook of Fruits and Fruit Processing.* Blackwell Publishing.

Jacob, H. 1996. Pyrodwarf, a new clonal rootstock for high density pear orchards. *Erwerbsobstbau,* **38**(6): 166-169.

Jana, B. R. 2016. Effect of pruning, strapping, dormancy breaking chemicals and irrigation on Asian Pear (Pyrus pyrifolia L.) growth under eastern plateau of India. *International Journal of Current Microbiology and Applied Science,* **5**(10):783-791.

Jana, B.R. and Das, Bikash. 2014. Effect of dormancy breaking agents on flowering, fruit set and quality of Asian Pear (*Pyrus pyrifolia* L.). *African Journal of Agricultural Research,* **9**(1): 56-60.

Jana. B. R. 2016. Effect of Pruning, Strapping, Dormancy Breaking Chemical and Irrigation on Asian Pear (Pyrus pyrifolia L.) Grown Under Eastern Plateau of India. *International Journal of Current Microbiology and Applied Sciences,* 5**(10): 783-791.**

Janick, J. 2004. P448-2 (Green JadeTM) pear. *Hort. Science*, **39**(2): 454-455.

Jankovic, R. 1994. The effect of different rates of NPK nutrients on the productivity of pear cultivars William's and Beurre Bosc. *Jugoslovensko Vocarstvo,* **28**(3/4): 23-30.

Jawanda. J. S. 1981. Pear has unique success in Punjab. *Punjab Horticulture Journal*, 21: **131-135.**

Jawanda. J. S. 1981. *Punjab Horticulture Journal*, **21**: 131-135.

Jill Campbell, 2003. Pear rootstocks Agfact H4.1.15, Agricultural Intitute, Orange.

Kajiura, I. 1994. Nashi (Japanese pear). In. *Horticulture in Japan*, by Konishi, K.; Iwahori, S.; Kitagawa, H. and Yakuwa, T. (eds.). Asakura Publishing Co. Ltd. Tokyo, pp. 40-47.

Khemira, H; Azarenko, A.N; Sugar,D; Righetti,T.L. 1998. Postharvest nitrogen application effect on ovule longevity of 'Comice' pear trees. *Journal of Plant Nutrition,* **21**(3): 405-411.

Khoshghalb, H; Arzani, K; Malakouti, M. J. and Bazegar, M. 2013. Effect of Ca, Zn and B spray application on preharvest fruit drop, sugar and nutrient contents and on some quantitative and qualitative fruit characterestics in some Asian pear (*Pyrus serotina* Rehd.) cultivars. *Iranian Journal of Horticultural Science,* **44**(2): 149-159.

Klinac, D. J. and Geddes, B. 1995. Incidence and severity of the floral bud disorder "budjump", on nashi (*Pyrus serotina*) grown in the Waikato region of New Zealand. *New Zealand Journal of Crop and Horticultural Science,* **23**(2): 185-190.

Kloutvor, J. and Paprstein, F. 1996. New pear cultivar 'Delta'. *Vedecke Prace Ovocnarske,* **15** : 153-155.

Kondo, S. and Seto, H. 2004. Changes of jasmonic acid during ripening in pear fruit and interactions between jasmonic acid and abscisic acid. *Acta Horticulturae,* **636:** 537-543.

Kumar, S.; Karol, A.; Singh, R.; and Vaidya, C.S. 2007: Cost and Return from Apple Cultivation: A Study in Himachal Pradesh, Agriculture Situation in India. pp.307-313.

Lallu and Nagin. 1990. *Fruit growth, handling and storage*, p. 53-74. In A.G. White (ed.). Nashi, Asian pear in New Zealand, DSTR Publishing, Wellington.

Layne, R.E.C. and Quamme, H.A. 1975. Pears. In. *Advances in Fruit Breeding*, by Janick, J. and Moore, J.N. (eds.). Purde University Press, West Lafayette, Ind, USA, pp. 38-70.

Lelievre, J.M; Tichit, L; Dao, P; Fillion, L; Nam YoungWoo; Pech, J.C. and Latche, A. 1997. Effects of chilling on the expression of ethylene biosynthetic genes in Passe-Crassane pear (*Pyrus communis* L.) fruits. Plant Molecular Biology, **33**(5): 847-855.

Li-XianMing; Qin-ZhongQi and Liu-XianQin 2000. Experiment of bagging culture for Jinshui2 pear variety. *South China Fruits*, **29**(2): 45.

Luzk A.R., Muniz J., de Souza D.S., Petineli R., de A. Meyer G., Rufato L., Kretzschmar A.A. 2014. Plant Growth Regulators Increase Yield of Pyrus communis L.'Williams' Pear in Southern Brazil. *Acta Hort.*, **1042**: 325 – 330.

Ma ShuShang; Chen, P. M. and Mielke, E. A. 2000. Storage life and ripening behavior of 'Cascade' pears as influenced by harvest maturity and storage temperature. *Journal of American Pomological Society*, **54**(3): 138-147.

Maas, F. M; Steeg, P. A. H van der. 2011. Crop load regulation in 'Conference' pears. *Acta Horticulturae*, **909**: 369-379.

Mahammad M.U., Kamba A. S., Abubakar L. and Bagna, E.A. 2010. Nutritional composition of pear fruits (*Pyrus communis*). *African Journal of Food Science and Technology.* Vol 1(3): 76-81.

Mann, S. S. and Singh, B. 1988. *Acta Horticulturae*, **218**: 175-179.

Mann, S. S. and Singh, B. 1990. Effect of ethereal on ripening of fruits of 'Pathernakh' pear, harvested on different dates. *Acta Horticulturae,* **279:** 529-32.

Masseron, A. 1992. El Nashi, Vol. 2. Mundi Prensa, Madrid, Spain.

Masseron, A; Roche, L; Hilaire, C; Vaysse, P; Reynier, P; Westercamp, P; Seigneur, K; Coureau, C. 2000. Angelys pear: early results of Ctifl research. *Infos Paris,* **163**: 38-41.

Mehri Kamoun, R; Mehri, H; Faidi, A. and Polts, V. 2004. Micropropagation of six OH x F (Old Home x Farmingdale) pear rootstocks. *Advances in Horticultural Science,* **18**(2): 53-59.

Meng Ping and Zhang JinSong. 2004. Effects of pear-wheat inter-cropping on water and land utilization efficiency. Forest Research, **17**(2): 167-171.

Mitchell, F. G. 1992. *Postharvest handling systems: Temperate Zone Tree Fruits (Pome Fruits and Stone Fruits)*. Chap. 24, pp. 215-222. In: Kader, A. A., ed., Postharvest Technology of Horticultural Crops. Univ. Calif. Div. Agr. & Nat. Resources. Pub. 3311.

Mitra, S.K., Bose, T.K. and Rathore, D.S. 1991. Pears. *In*: *Temperate Fruits*. Horticulture and Allied Publishers, Calcutta, India.

Moniruzzaman, M., Mozumder, S.and Islam. M. 2007. Effect of mulching and pruning on yield and quality of pear. *Bangladesh Journal of Agricultural Research*, 32 **(2):225-233**

Musacchi, S; Pagliuca, G; Kindt, M; Piretti, M V. and Sansavini, S. 2000. Flavonoids as markers for pear-quince graft incompatibility. *Journal of Applied Botany*, **74**(5/6): 206-211.

Nadosy, F. 1997. Micropropagation of pear rootstocks. *Horticultural Science*, **29**(1/2): 17-21.

Namestek, J; Prazak, M; Litschmann, T; Knezacek, L. and Prosa, S. 2011. Effect of drip irrigation with fertigation on pear crops. *Vedecke Prace Ovocnarske*, **22:** 79-88.

Öztürk G., A°kin M.A. 2015. Various gibberellin treatments on parthenocarpic fruit development in 'Deveci' and 'Ankara' pear cultivars. *Acta Hort.,* **1094:** 373 – 382.

Paprstein, F. and Kloutvor, J. 1996. A new cultivar of pear, 'Erika'. *Vedecke Prace Ovocnarske,* **15**: 157-159.

Pasqualini, E. and Civolani, S. 2000. Protection against the San Jose scale. *Informatore Agrario,* **56**(6): 93-95.

Pereira N., Oliveira C.M., Mota M., Sousa R.M. 2011. Evaluation of five dormancy breaking agents to induce synchronized flowering in 'Rocha' pear. *Acta Hort.*, **909**: 423 – 428.

Pio, R; Ramos, J. D; Chagas, E. A; Gontijo, T. C. A; Carrijo, E. P; Mendonca, V; Alvarenga, A. A; Abrahao, E. 2006. Rooting of cutting of cv. Portugal quince submitted to application of indolebutyric and acid naftaleno acetic acid. Revista Cientifica Rural, **11**(2): 115-119.

Prashant K. Nimbolkar, Rai, P.N. Mishra, D.S. and Singh, Saurabh K. 2015. Influence of plant bio-regulators on yield, physico-chemical attributes and post harvest quality of pear cv. Gola under Terai region of Northern India. *Indian J. Hort.*, **72**(4): 485-488.

Pu, F. S. and Fang, C. Q. New dual-purpose pear variety Jingxiang, *China Fruits*, **4**: 16-17.

Raffo, M. D; Calvo, P; Angelis, V de; Manueco, L; Ziaurriz, S. and Menni, F. 2011. Effects of trunk girdling on fruit production, fruit size and tree vigor on 'Bartlett' pears in Rio Negro and Neuquen Valley, Argentina. *Acta Horticulturae*, **909** : 645-650.

Rivalta, L; Dradi, M. and Rosati, C. 2000. Three new, early maturing cultivars of pear. *Informatore Agrario,* **56**(50): 63-65.

Robinson, T.2010. High Density Pear Production: An Opportunity for NY Growers. New York Fruit Quarterly 18(4).

Rufato L., Kretzschmar A.A., Brighenti, A.F., Machado. B.D. 2011. Plant growth regulators increase productivity of 'Packham's Triumph' pear in Southern Brazil. *Acta Hort.*, **909**: 429 – 434.

Sánchez E., Curetti M., Retamales J. 2011. Effect of AVG application on fruit set, yield and fruit size in 'Abate Fetel' and 'Packam's Triumph' pears in a semi-commercial statistical trial. *Acta Hort.*, **909**: 435 – 440.

Sansavini S. 2002 Pear fruiting-branch models related to yield control and pruning. *Acta Hort.*, **596**: 627–633.

Sansavini, S. and Musacchi, S. 1994. Canopy architecture, training and pruning in the modern European pear orchards: an overview. *Acta Horticulturae*, **367**: 152-172.

Saquet, A. A; Streif, J. and Bangerth, F. 2001. On the involvement of adenine nucleotides in the development of brown heart in 'Conference' pears during delayed controlled atmosphere storage. *Gartenbauwissenschaft*, **66**(3): 140-144.

Saquet, A; Streif, J. and Bangerth, F. 2003. Brown heart incidence in 'Conference' pears as affected by ATP and ADP levels, and membrane lipid alterations during controlled atmosphere storage. *Acta Horticulturae*, **600**: 839-842.

Semash. D.P. and Panasenko, I. N. 1984. Sadovodstvo Vinograderstvo I Vinodalie Moldavii, No.6, pp.12-14.

Sharma, R.P. and Dutt, S. 2009. Handbook of Horticulture. Published by Dr. T.P.Trevedi. Lasertypeset Xpedite Computer system . Ranjit Nagar Commercial Complex, New Delhi.

Singh Damandeep; Dhillon, W S; Singh, N.P; Gill, P.P.S. 2014. Impact of girdling on quality and maturity of Patharnakh pear. *Indian Journal of Horticulture*, **71**(3): 335-339.

Singh, A.K; Raghbir Singh and Mann, S. S. 2003. Effects of plant bioregulators and nutrients on fruit set, yield and quality of pear cv. Le Conte. *Indian Journal of Horticulture,* **60** (1): 34-39.

Singh, G.R.P. and Singh, Raghbir. 2004. Studies on nutrient removal by leaves and fruits of pear cv. Patharnakh. *Indian J. Hort.,* **61**(4): 357-58.

Smit M., Meintjes J.J, Jacobs G., Stassen P.J.C., Theron K.I. 2005. Shoot growth control of pear trees (*Pyrus communis* L.) with prohexadione-calcium. *Scientia Horticulturae* **106:** 515 – 529.

Stanojevic, V.; Jankovic, R. and Rakicevic, M. 1997. Results of an investigation into the efficacy of herbicides on weeds in a pear mother plantation on quince. *Jugoslovensko Vocarstvo,* **31**(1/2): 167-175.

Stasey, V. J. 2000. Variety: 'Sophia's Gold'. Application no. 1995/161. *Plant Varieties Journal,* **13**(3): 82-83.

Stephen, S., Kitren, G., Randy, H. and Khan, Z. 2003. Effects of Messenger on 'Bartlett' Pear Fruit Size and Cropping. US bulletin 2003.

Sugar, D. and Hilton, R. J. 2011. Potential organic methods for management of pear scab. *Acta Horticulturae,* **909** : 527-530.

Suta, A., Cotorobai, M. and Stan. S. 1986. The yield potential of some apple high-density planting systems. *Acta Horticulturae,* **160**: 305-314.

Tamm, L; Haseli, A; Fuchs, J G; Weibel, F P. and Wyss, E. 2004. Organic fruit production in humid climates of Europe: bottlenecks and new approaches in disease and pest control. *Acta Horticulturae,* **638** : 333-339.

UC IPM Pest Management Guidelines: Pear, 2016. Agriculture and Natural Resources, University of California.

Vadivel, V. and Janardhanam, K. 2005. The wonder of pears" free diets. http//:www.freediets.com/fruits-vegetables/thewonderof pears. *Plant foods for Human nutrition* Pp 60: 69-75. Access 12 August , 2010.

Vavilov. N. I. 1951. The Origin, Variation, Immunity and Breeding of Cultivated Plants. Ronald Press. New York.

Vercammen J., Gomand A., Bylemans D. 2015. Improving the fruit set of 'Conference' with gibberellins or Regalis. *Acta Hort,* 1094: 257 – 264.

Verma, M. K. 2015. Pear Production Technology. *In. Training manual on teaching of post-graduate courses in horticulture.* Editors: SK Singh, AD Munshi, KV Prasad, AK Sureja. Department of Fruit Science, Indian Agricultural Research Institute, New Delhi pp. 249-255.

Verma, M. K., Ahmed, N., Singh, A.K. and Awasthi, O.P. 2010. *Temperate tree fruits and nuts in India. Chronica Horticulture,* **50**(4):43-48.

Wang BaoLiang; Wang WenHui; Jiang YunBin; Jiang XiuCheng and Wang ZhiHua. 2014. Effect of 1-MCP treatment on inhibition of browning and preservation of Dangshansu pears growing in different regions during shelf-life after cold storage. *Storage and Process,* **14**(5): 24-30.

Wang SuKe; Yang Jian; Wang Long; Zhou HouCheng; Li XiuGen. 2010. Studies on tissue culture and rapid propagation of dwarf rootstock BA-29 of European pear. *Journal of Fruit Science,* **27**(6): 1002-1005.

Watkin, C. B. and Frankel. C.1987. Inhibition of pear fruit ripening by mannose. *Journal of Plant physiology,* **85**:56-61.

Weeden, N. and Lamb, R.C. 1987. Genetics and linkage analysis of 19 isozyme loci in apple. *Journal of the American Society of Horticultural Science,* **112**: 865-872.

Wertheim, S. J. 1998. Rootstock guide: apple, pear, cherry, European plum. Fruit Research Station, Wilhelminadorp, Netherlands.

Westwood, M. N., Lombard, P.B. and Bjorstand, H.O. 1976. Performance of 'Bartlett' pear on standard and Old Home X Farmingdale clonal rootstocks. *J. Amer. Soc. Hort. Sci.*, **101**:161-164.

Westwood, M.N. 1982. Pear germplasm of the new national clonal repository: its evaluation and uses. *Acta Horticulturae*, **124**: 57-65.

Wojcik, P. and Wojcik, M. 2003. Effects of boron fertilization on 'Conference' pear tree vigor, nutrition, and fruit yield and storability. *Plant and Soil,* **256**(2): 413-421.

Xuan, H; Streif, J. and Bangerth, F. 2001. Effect of boron applications on physiological disorders in 'Conference' pears during CA storage. *Acta Horticulturae,* **553** (Vol. 1): 249-253.

Yadav, Ashish and Bist, L.D. 2003. Effects of nitrogen on shoot growth, flowering, fruiting and fruit quality in pear cv. Bagugosha. *Indian Journal of Horticulture,* **60**(1): 40-44.

Yuda. E., Nakagawa. S., Matsui, H., Wada, K. and Imai, T. 1981. The effect of 15-â-OH gibberellins on the fruit set and development of pears. *Hort. Science*, **16:** 436.

Zerbini, P. E; Rizzolo, A; Brambilla, A; Cambiaghi, P. and Grassi, M. 2002. Loss of ascorbic acid during storage of Conference pears in relation to the appearance of brown heart. *Journal of the Science of Food and Agriculture,* **82**(9): 1007-1013.

Zeven. A. C. And Zhukovsky, P. M. 1975. Dictionary of cultivated plants and their Centres of Diversity. Centre for Agricultural Publishing and Documentation. Weginengin.

Zheng, C., Halaly, T., Acheampong, A. K., Takebayashi, Y., Jikumaru, Y., Kamiya, Y., Or E. 2015. Abscisic acid (ABA) regulates grape bud dormancy, and dormancy release stimuli may act through modification of ABA metabolism. *Journal of Experimental Botany,* **66**(5), 1527 – 1542.

10

Pecan nut [*Carya illinoinensis* (Wangenh.) K. Koch]

Amit Kumar, M.U. Rehman, G.H. Rather, T. Angmo and Sayan Sau

1. Introduction

Pecans (*Carya illinoensis*) are one of the most desirable trees to plant in home yards. Their size and natural beauty as a shade tree and its capacity to produce a nutritious food make them valuable. Pecan is the most widely cultivated and the most economically important member of the North American hickory group. Beyond its native range, pecan is commercially grown from the south-eastern USA to the south-west (California) (Hanna, 1987, Peterson, 1990, Harlow *et al.*, 1991). Pecan is from an Algonquian word, meaning 'a nut requiring a stone to crack'. In Mexico, pecans and walnuts share the same Spanish name, neuz, which is a cognate of the English word nut. In the USA, pecan is considered the queen of nuts because of its value both as a wild and as a cultivated nut and acceptable quality of meats (Woodroof, 1979). Pecan is also considered as 'queens of nuts' because of its value both a wild and as a cultivated nut and acceptable quality of meats. As compared with other nut fruits which have been grown since time immemorial, pecan has been under cultivation only for about one hundred years but has gained enormous popularity.

1.1. Nutritive value

Pecan is a rich source of energy carries 690 calories, fat, protein, carbohydrate and minerals (Table 1). The proximate composition of 100 g kernel has been determined as fat 71.9 g, protein 9.17 g, carbohydrate, 13.86 g, dietary fiber 9.6 g, calcium 70 mg, phosphorus 277 mg, potassium 410 mg, magnesium 121 mg, iron 2.53 mg, Vitamin A 56 IU, thiamine 0.660 mg, riboflavin 0.130 mg, niacin 1.167 mg, ascorbic acid 1.1 mg (USDA, 2000). The oil composition of the cultivar Stuart is oleic acid, 68.8 %, linoleic acid 21 %, palmitic acid 6.6 %,

Table 1: Nutritive value of per 100 of nuts

	Nutritive value	% of RDA
Constituents		
Energy	690 Kcal	34.5 per cent
Carbohydrates	13.86 g	11 per cent
Protein	9.17 g	17 per cent
Total fat	71.9 g	360 per cent
Cholesterol	0 mg	0 per cent
Dietary fiber	9.6 g	25 per cent
Vitamins		
Folates	22 mcg	5.5 per cent
Niacin	1.167 mg	7 per cent
Pantothenic acid	0.863 mg	17 per cent
Pyridoxine	0.210 mg	16 per cent
Riboflavin	0.130 mg	10 per cent
Thiamine	0.660 mg	55 per cent
Vitamin A	56 IU	2 per cent
Vitamin C	1.1 mg	2 per cent
Vitamin E	24.44 mg	163 per cent
Minerals		
Sodium	0 mg	0 per cent
Potassium	410 mg	9 per cent
Calcium	70 mg	7 per cent
Copper	1.2 mg	133 per cent
Iron	2.53 mg	32 per cent
Magnesium	121 mg	30 per cent
Manganese	4.5 mg	196 per cent
Phosphorous	277 mg	40 per cent
Selenium	3.8 mcg	7 per cent
Zinc	4.53 mg	41 per cent
Phyto-nutrients		
Carotene β	29 mcg	
Crypto-xanthin β	9 mcg	
Lutein-zeaxanthin	17 mc	

Source: USDA, 2000

stearic acid 2.2 % and arachidonic acid, 0.2 % (French, 1962). The predominant amino acids recorded in the matured pecan nut are arginine, phenylatanine, valine, glutamic acid, histidine, cystine andhysine in decreasing order (Wood and Reilly, 1984). It has acceptable levels of sulphur amino acids except the lysine content which is low (Elmore and Polles, 1980).

1.2. Uses

Raw pecans, like in walnuts, can be eaten either alone, salted or sweetened. Pecan nuts are buttery in consistency yet pleasantly sweet in taste. High fat content in them makes them wonderful additions in the preparations of delicious recipes like pecan caramel puddles, pie, fudge, baklava, and muffins. The nuts are often sprinkled over desserts, particularly sundaes, and bourbon ice creams. They are widely used in the confectionery, as an addition to biscuits, sweets and cakes. The nuts are also used to make pecan nut-butter, which is popular spread over bread, toast, etc. The shell flour of various sizes is also used as soft grit in non-skid paints, adhesives, dynamite and polishing metals. Pecan tree is an exceedingly beautiful plant which can be used for shade and ornamental purposes. The tree has a surpassing value on timber market due to its strength and hardness ranking third only after bland walnut and wild cherry, respectively (Woodroof, 1979). Pecan wood is used for furniture, cabinetry, paneling, pallets and veneer. The wood has good machining properties, resembling those of true hickories (Adams and Thielges, 1977).

Popularly, the leaves of *Carya illinoensis* have been used in the treatment of smoking as a hypoglycemic, cleansing, astringent, keratolytic, antioxidant, and antimicrobial agent. The phytochemical compounds such as phenolic acids (gallic acid and ellagic acid), flavonoids (rutin), and tannins (catechins and epicatechins) were identified in the leaves which were responsible for the antimicrobial activity against Gram-positive, Gram-negative, and yeast (Bottari, *et al.*, 2017).

2. Area, Production and Trade

Reliable data of the world and even of the states of India are not available, as pecan is a very minor crop. Pecans are lumped together with other nut crops, including macadamia nut, in FAO statistics. Pecan trees are cultivated over a wide geographic area spanning from California to Virginia, and contributes to the economy of 24 states (Wood *et al.*, 1990). Pecan production can be separated into four broad regions: the southeastern spanning from Virginia to Louisiana and Arkansas, the south central consisting of east and central Texas and southern Oklahoma, the northern containing northern Oklahoma and the Midwest, and the west which includes far west Texas and southern areas of New Mexico, Arizona, and California.

According to NASS (2015) during the year 2014 commercial pecan production was reported in fourteen states of United States which includes 3/4th production in Georgia, New Mexico and Texas. The entire 2014 pecan nut production in United States was 132, 075 tons which contribute $ 517 million with an increase of 12 per cent from 2013. Georgia produces 76 million pounds, followed by New Mexico at 67 million pounds, Texas at 61 million pounds and Arizona at 21 million pounds.

Pecan is the most widely cultivated and the most economically important member of the North American hickory group. Beyond its native range, pecan is commercially grown from the south-eastern USA to the south-west California (Harlow *et al.,* 1991). Commercial production also occurs in Australia, Brazil, Canada, Israel, Mexico and South Africa, although the US is the leading producer. During the year 2016 the U.S.'s domestic production was estimated as 121912 MT (Anon., 2017).

With the production of 80,000 MT of Pecan nuts per year on an approximate area of 64,376 ha., Mexico is the second largest producer behind the United States (Camarena *et al*., 2012). As 45% of production is exported, there is an ample supply and a wide variety of product formats in the domestic market. In 2015 Maxico became the main exporter of pecan nuts, as it exported 22,738 MT of nuts internationally (Anon., 2016a).

Both imports and exports have become important for the US pecan industry. The export market holds particular importance, especially in light of flat domestic consumption over many years, hovering around 227g per capita. During the year 2014-15 the domestic utilization of pecan was recorded as 70734 MT. The pecan imported and exported was recorded as 47157 MT and 45334 MT respectively (Blayney and Gutierrez, 2017). U.S. pecan exports were valued at $446.3 million in 2014, up 28% from 2013.

Hong Kong ($118 million) is the primary destination for US in-shell pecan exports followed by Vietnam ($87.6 million), whereas Canada ($45.9 million) remained largest market for shelled pecans followed by Netherlands ($30.7 million) (FAS, 2015).

In India, pecan is mainly grown in Jammu and Kashmir, and Himachal Pradesh. Area under Pecan in Himachal Pradesh is about 700 ha. (Mehta and Thakur, 2004). While in Jammu and Kashmir it has been 415 ha. with the production of 81 MT during 2015-16 (Anon, 2016b). The total area under pecan nut production is increasing due to high economic returns and adaptation to intermediate zone of Jammu (Ravindran *et al*., 2008).

3. Origin and distribution

Pecans and hickories members of the walnut family, Juglandaceae constitute the genus *Carya*, which is the only genus in the tribe Hicoreae (Manning, 1978). Based on fossil remains of pecan and other Juglandaceae, Stuckey and Kyle (1925) proposed that the pecan first appeared on the North American continent during the Cretaceous period so pecan is a native of North America but the opinions differ whether it originated further north and was water borne to the south or appeared simultaneously over a wide region. Sargent (1965) and Romberg (1968) felt that the native Indians played a dominant role in disseminating the pecan northward. On receding of the last caps, the North American crow and Indians played a significant role in its spread over a wide region. The early Spanish explorers on their arrival found the pecan growing along the Mississippi river and its tributaries. The native habitat of pecan ranges from Texas in the southwest to Alabama in the southeast extending up to southern Illinois (Hanna, 1987). Its cultural range, however, overlapped the native range, extending beyond into New Mexico, Alabama, Georgia, Florida, South Carolina and to a lesser degree into other fringe states (Madden *et al*., 1969).

The Spaniards brought the pecan into Europe, Asia and Africa in the beginning of the 16th century. Most of our native pecan species are named for Indian Tribes including Cheyenne, Mohak, Choctav and Shawnee. Besides the USA, the worldwide distribution of pecan is confined to only Australia, Brazil, Canada, Egypt, India, lsrael, Mexico, Morocco, Peru, Turkey and South Africa probably because the crop is rather new and have a longer gestation period than other fruits. It can be successfully grown in the Himalayan regions in the areas which are free from spring frost and excessive heat in summer. In India, it is grown in Jammu and Kashmir (Baramulla, Srinagar, Poonch, Bhadarwah, Anantnag), Himachal Pradesh (nearby areas around Palampur, Solan) and hills of Uttarakhand and climatically similar areas of North-Eastern states and Niligiri hills of Tamil Nadu.

4. Taxonomy

The *Carya* genus is a member of the walnut family, Juglandaceae, and comprises 20 species (Grauke and Thompson, 1996). The hickories are large, deciduous trees which tend to form upright, cylindrical crowns when grown in the open. All species have pronounced taproots which securely anchor the trees, if soil conditions allow deep root development.

The genus *Carya* is divided into three sections: *Rhamphocarya, Apocarya* and *Carya*. A new section, *Sinocarya*, has also been proposed. The section is characterized by species which lack terminal bud scales and includes only the four Chinese members of *Apocarya* (excluding *C. poillanei*). More thorough evaluation of the characteristics of all species is needed prior to the revision of the organization of sections.

Section *Rhamphocarya* contains only one Asian species, *C. sinensis* Dode. This species in the monotypic genus *Annamocarya*. The section is distinguished from others in the genus by its entire leaf margins, more numerous catkins per male inflorescence (5-8 vs 3), more numerous stamens per male flower (5-15 vs 3-8), and more numerous bracteoles of the pistillate flower (3-5 vs 3).

Section *Apocarya* contains the "pecan hickories", while section *Carya* contains the "true hickories". The best method for distinguishing between the two sections found in the United States of America is by comparing terminal buds; pecan hickories have 4-6 valvate bud scales that touch at the edges, but do not overlap; true hickories have 6-12 overlapping or imbricate terminal bud scales. The increasing development of bud scales is an adaptation to cold climates. The Asian members of *Apocarya* may have naked buds, completely lacking scales.

Other features, such as the number of leaflets per leaf and the presence of wings on the sutures of the husks, are helpful, but not as reliable. Pecan hickories tend to have greater numbers of leaflets than true hickories (5-13 in *Apocarya* vs 5-9 in *Carya*). Pecan hickories tend to have more pronounced wings on the sutures of the husk than do true hickories. If nuts are cut in cross-section, those in section *Carya* do not have lacunae in the walls or internal ridges of the nut, and all have secondary septa, making the nuts 4 called at the base. Species in section *Apocarya* have prominent lacunae (except *C. myristiciformis* and the Asian hickories) and may have secondary septa (as in *C. cordiformis*) or lack them (as in pecan). All species in *Apocarya*are diploids (n=16) while section *Carya* includes both diploids and tetraploids (n=32).

5. Cytogenetics

Pecan is a diploid (n = 16), monoecious, long-lived tree species. Owing to its hetero-dichogamy, pecan is primarily cross-pollinated, resulting in high heterozygosity with severe inbreeding depression when selfed. Establishment of commercial pecan orchards during the nineteenth century was mainly by planting open-pollinated nuts from mother trees possessing desirable characteristics. These orchards consist of trees with widely varying production and quality attributes due to the heterozygosity of pecan. The *Carya* genus is a member of the walnut family, Juglandaceae and comprises 20 species. Thirteen *Carya* species, including pecan, are native to the USA. Of all *Carya* species, seven are reportedly cultivated for their nuts, but pecan is the only economically important crop.

6. Botanical description, flowering, fruiting

Caryas are commonly known as hickory or pecan trees. Trees are monoecious (male and female flowers on the same tree), deciduous and are large, long-lived with upright, vase-shaped crowns when open-grown. The pecan may grow upto 150 feet high in the south. It makes a wonderful shade tree in the north east even though the nuts don't mature there. The leaves are large 12 to 18 inch long and they turn golden yellow in the autumn. Leaves are pinnately compound, from 5 to 11 leaflets are alternately arranged along the rachis. The leaves are fragrant when crushed. Leaflets are lanceolate to obovate, with the proximal half of the blade generally smaller than the distal half (asymmetrical); margins serrate. Male flowers are located on 4-5 inch long catkins. Female flowers are small, yellowish-green, and grow on spikes at the tips of shoots. Male and female flowers are dichogamous and mature at different times. Some cultivars are protandrous with male flowers that mature before females, and some are protogynous with female flowers that mature before males. Because pecan is wind pollinated, at least two different cultivars should be planted in close proximity to ensure adequate fruit set (Wells and Conner, 2012). The flowers are borne in the spring along with the foliage. The male flowers are produced in pendant, 3-branched catkins near the base of the current season's growth. The female flowers are unnoticeable and grow in spiked clusters towards the end of the growth. The bony shelled nuts are encased in a 4-valves husk. In some kinds, the husk is thin and splits partially open and in others, the husk is thick and splits at the base at maturity. The nuts ripen in October and fall from the trees. Several varieties of *Carya* have edible nuts though the pecan nut is by far the most important kind. In fruit producing regions, the tree produces their first crop of nuts four to seven after planting. Eventually, they will bear large crops of 1.5 to 2.5 inch nuts. The pecan will fall on the ground when they are ripe in the fall.

6.1. Floral Biology

6.1.1. Floral bud development

The pecan has a compound bud. The apical bud is mixed with catkin and vegetative buds enclosed by separate bud scales, within a common outer scale (Isbell, 1928). Following bud break, the mixed bud produces a shoot which, if continues to elongate, terminates in either a vegetative bud or a single pistillate inflorescence. While pistillate flowers are completing differentiation, leaves, nodes and internodes develop rapidly, and rudimentary buds form in the axils of the leaves. The remaining buds produce catkins if developed. In spring, when the conditions are favourable for growth, both the vegetative centre of the bud and the rudimentary catkins continue their development but the vegetative centre develops more rapidly. Frequently, the vegetative shoot will abort completely, leaving a short green stub; or if it grows, its apical bud will abort 4 to 6 weeks after bud break (Sparks, 1967). Some of the latter shoots can develop a new or false apical bud in late summer. Buds from long one year shoots produce more catkins and clusters of pistilllate flowers and set more nuts than buds on shorts shoots of some type and in fact, there is a direct relation between flower formation and fruit set up to 25 cm shoot length (Spark and Brack, 1972). In Money Maker pecan cultivar, trees with longer shoots consistently produced more pistillate flowers and retained more nuts. It was also observed that the shoots growing from one year old wood, which was previously vegetative, produce 26 per cent more flowers and 32 per cent more nuts then those growing on former fruiting shoots (Malstrom and McMeans, 1982). Pistillate flower production increased and subsequent abortion decreased with the increasing shoot vigour. Treatment with 2000 ppm Daminozide in late May or early June increased the number of pistillate inflorescences but its application in July gave opposite result (Maximos *et al.*, 1984a).

There are 7 stages of pistillate bud development that take place between early February to late April (Maximos *et al.*, 1984b). The morphological evidence of pistillate flower bud is visible in early spring after growth has initiated and proceeds rapidly after the bud scales abscise and growth starts (Wetzstein and Sparks, 1983). The female inflorescence grew terminally on shoots produced in the current year (Torikata *et al.*, 1988). The time of initiation as well as the rate of development of terminal and sub-terminal inflorescences varies in different cultivars (Mullenax and Young, 1972). Normally, one to nine pistillate flowers are born sessile on spikes that terminate the current year growth. An individual flower measures from 5.5 to 8.0 mm in length, about half of which is enclosed in a light green pubescent calyx and the rest is stigma. The calyx is marked by 4 ridges that divide it into four sections and each section is terminated by a tapering bract from 3 to 5 mm in length, varying with cultivar. These may be erect, extended laterally or reflexed downward along the sides of the flower. The receptive stigmas are sessile, variously shaped and colored, and a viscid fluid covers the uneven surfaces. Various cultivars differ in size, colour and shape of the stigmas.

The vegetative regions of the flower resemble those young growing nut with vascular systems fairly complete. About two third of the ovule is enclosed by the integument when flower become receptive and the vascular system of the septum develop independently from vascular traces and later becomes connected with the series of bundles next to the shell region. Each of the two parallel bundles of septum becomes divided near the apex into a semicircle distinguishable in cross section of the integument.

6.1.1.1. Physiology of flowering at molecular level

Randall *et al.* (2015) described that flower development initially requires reprogramming of the shoot meristem resulting in a transition from vegetative to floral tissues. Specific genetic switches within the plant that respond to environmental cues such as photoperiod, light quality, temperature, and hormones regulate this reprogramming. Several of these genetic switches were elucidated in the model plant an intricate signaling pathway is initiated when the CONSTANS (CO) protein responds to lengthening days, activating expression of the FLOWERING LOCUS T (FT gene). The FT protein then is transported via the phloem to shoot meristems where it activates expression of genes responsible for floral induction (AP1 and LFY).

6.1.2. Pistillate inflorescence

Pistillate flowers are borne in a spike (cluster) at the end of the currents season shoots. The number of flowers per cluster varies from 1 to 10 or more and is cultivar dependent. Pistillate flowers are differentiated in the early spring, when outer bud scales are shed and buds swell, but prior to inner scale pit (Wetzstein and Sparks, 1983). Low temperature exposure was required for pistillate flower initiation. Cultivars have not been distinguished in terms of the amount of chilling required for flower development, but cultivars are known to differ in chilling required for initiation of bud growth.

Pistillate flowers in pecan are first visible as a flattened area at the shoot terminal. As growth continues, individual flowers become visible. Flower development proceeds from the base of peduncle toward the apex (acropetally). Amling and Amling (1983) noted that exposure to low temperature is essential for pistillate flower induction, which take place in late July or early August. The pistillate flower development occurs few weeks before anthesis on the resumption of current season growth in spring. Wetzstein and Sparks (1983) reported that the sequence of development has been categorized into 9 stages. The number of female flowers on a spike varies from 2 to 10 determined by the cultivar and vigour of the shoot.

Each flower consists of a 4 to 5 segmented floral envelope (the involucre, formed by an abaxial bract and bracteoles) surrounding the ovary and surmounted by a stigma. Flowers from different cultivars are extremely variable in size, stigma color and stigma shape at receptivity. Stigma color ranges from green (e.g, Desirable) to deep red (e.g, Pawnee) and is often somewhat variable between flowers of a single tree stigma color is not a reliable index of receptivity. Stigma are typically bilobed, but lobes may be inconspicuous or pronounced and either upright or reflexed. Receptivity has been determined by the presence of a viscous fluid on the stigmatic surface and by adherence of applied pollen to the stigmatic surface. It can also be judged on the basis of stigma drying. The reliability of the data depends both on the method used and the experience of the observer, and no method is fool proof. Adherence of applied pollen is the method recommended for evaluating receptivity in breeding orchards. The data on which pollen shed and pistil receptivity occur for a particular cultivars may vary as a function of season and location, thus affecting the extent of separation or overlap of pollen shed and pistil receptivity. Cultivars generally maintain the same dichogamy pattern across seasons and locations. Reported changes in dichogamy pattern between locations may be due to different methods of evaluating bloom or to problems in cultivar identification.

6.1.3. Staminate inflorescence

The staminate inflorescence (catkin group) is initiated during the spring prior to pollen shed. Two catkin groups are differentiated at the same time, on opposite sides of the shoot bud. Sometimes another pair of catkin groups is formed, opposite to each other and at right angles to the first two groups. Each catkin group is enclosed by bud scales and the entire bud, including all catkin groups and the shoot bud, is enclosed by the outer bud scale.

The male catkins appear laterally on one year old shoot. The staminate primordia are differentiated the year prior to their appearance. There are five stages of staminate bud development taking place between early June and early August. Differentiation in buds of the expanding leaves within 2 to 3 weeks after initial expansion. The number of staminate inflorescences per compound bud range from one to four but the usual number is two. The number of flowers per catkin varies (60 to 160) in the same as well as in different cultivars.

Catkin development progresses to different stages in protandrous and protogynous cultivars: Protandrous cultivars initiate anthers on catkins in the buds of the staminate inflorescence during the summer prior to pollen shed; protogynous cultivars initiate anthers on catkins during the spring that pollen is shed (Wetzstein and Sparks, 1985~ Haulik and Holtzhausen, 1988). The staminate inflorescence or catkins groups are composed of three aments or catkins, joined to a common stalk, or peduncle. Protogynous cultivars typically have long thin catkins while protandrous cultivars typically have catkins that are shorter and of greater diameter. Regardless of dichogamy class, the central catkin of a catkin group is usually the longest. Each catkin is composed of many individual staminate flowers; 72/catkin in protandrous cultivars; 122.8/catkin in protogynous cultivars. Each individual staminate flower is composed of a central bract and two lateral bracteoles and bears 3 to 7 stamens. Protogynous cultivars typically have flowers with short broad bracts.

6.1.4. Pollen biology and production

During development each anther has four pollen sacs, or thecae. The filament is so short that it is inconspicuous. Each pollen sac contained an average of 365 pollen grains. Pecan pollen is produced in large quantities, as is typical of anemophilous species. Microspores are free within the pollen sac 15 to 20 days before pollen is shed after the tapetum disintegrates. About the time anthers are externally visible, the four pollen sacs fuse to form two chambers by the dissolution of the separating wall. This coincides with the development of binucleate pollen grains. Lipids and polyphenols are first recovered from binucleate pollen grains at that period. For the greatest chance of success in haploid anther culture, catkins should be collected when bracts and bracteoles are visible, but prior to spreading of the bracts. This coincides with the earliest period of free microspores and precedes starch accumulation and nuclear enlargement.

Carya pollen is triporate paraisopolar (the pores are never exactly equatorial in position, but are drawn toward the distil pole) and sub-oblate with a textate surface. Pollen diameter averages 46 μm and ranges in diameter from 38 to 55 μm. Pollen from the diploid species of *Carya* is generally smaller than that of the tetraploid species. The anther wall is two cell layers thick, with an outer epidermis and an inner endothecium. Pollen is shed when the pollen sac splits open along a longitudinal slit. The opening of the anther is caused by drying and contraction of the outer layer in relation to the inner layer. When moistened, the anther has the ability to reclose. Anther dehiscence is hastened by increasing temperature under dry conditions, but does not occur under moist conditions regardless of temperature. Pecan pollen continued maturation but did not dehisce if relative humidity exceeded 85 per cent with subsequent dry conditions resulting in periods of very heavy shed. Sustained high winds coupled with low humidity tend to shorten the period of effective pollination, both by speeding pollen dehiscence and by reducing the period of pistillate receptivity. Conversely, high humidity delays pollen dehiscence and extends the period of pistil receptivity.

Catkins must be collected at the proper time for maximum pollen recovery and viability: catkins collected too early will wilt without dehiscing pollen; those collected too late will have already shed pollen, reducing the amount recovered. Catkins at the ideal stage of development for collection will be 7.6 to 15.2 cm long, with plump, yellow green anthers some of which are splitting. The best way to monitor catkin development is to visit trees several days prior to collection and observe catkin length and anther color. Since pollen shed is cultivar dependent, the optimum date for collection will vary among cultivars. On an individual tree, catkins on low limbs near the interior of the tree usually shed first. Catkins at the top of the tree are the last to shed. Since seasonal growth of catkins begins with bud growth, a tree that has uniform bud break throughout its canopy will have relatively uniform pollen shed also. Alternatively trees that are uneven in bud break will also have uneven pollen shed. However, rates of development are increased by warm temperatures. Once catkins are mature, pollen dehiscence may occur within a short period, especially if conditions are dry and windy. Pollen shed is delayed by rainy conditions or high relative humidity. Routine monitoring of target trees is therefore very important for successful collection. Given typical diurnal fluctuations in temperature and relative humidity, the best time to monitor catkin development and collect catkins is in early morning before pollen shed begins. Pollen shed often begins in midmorning after relative humidity decreases.

Zhang *et al.* (2014) studied the flower morphological characteristics, pollen viability and stigma receptivity of nine excellent pecan (*Carya illinoinensis*) cultivars. The results showed that optimal period for pollens collection was "stage V" when the angle between bracts and catkin rachises were about 90 degrees and anthers turned yellow. Among the tested cultivars, 'Pawnee' had the highest pollen viability. For the pollen collection, some cultivars like 'Mahan' was collected more easily when the open angles of stigmas from 90 degrees turned to more than 90 degrees, other cultivars like "Pawnee" had receptivity from the beginning when stigmas started open and got better when open angles got bigger from 15 degrees to 90 degrees, which were relatively smaller than the former. The optimal receptivity time for all cultivars was only lasted 2-3 days. Receptivity would lose when black spots appeared on the stigmas. Pollen spreading time of 'Pawnee' overlaid with best stigma receptivity time of 'Mahan', and the pollination time of 'Pawnee' and 'Mahan' was longer than others, so they were suggested to be optimal pollination cultivars.

6.1.4.1. Pollen storage

Often pollen is shed from protogynous cultivars after other protogynous pistllate parents have passed receptivity. Such pollen can be stored for one year in desicator at -10°C and 25 % relative humidity (Romberg and Smith, 1968). Pollen maintained viability when stored for almost two years at -80 and -196°C but pollen stored at -10°C did not maintain viability (Yates and Sparks, 1989). Pollen dried in a mechanical convection oven at 35°C, placed in air tight containers and stored in a freezer at -12°C for two years was comparable to fresh pollen in germination efficiency (Yates *et al.*, 1991). Pollen stored in liquid nitrogen for 'Desirable' pecan cultivar not diminished for years in comparison to fresh pollen (Sparks and Yates, 2002).

6.1.4. 2. Pollen germination

Germination of pecan pollen is very sensitive to changes in relative humidity with maximum germination occurring at high relative humidity. Stored pollen benefits by rehydration prior to incubation in germination media for *in vitro* determinations of pollen viability. Pollen can be rehydrated at 97.2 per cent relative humidity and 25° C for 4 hours. Constant 97.2 per cent relative humidity is maintained with $CuSO_4$ in desiccation vessels. If placed under appropriate conditions pecan pollen rehydrates rapidly, changing from a concave to a spherical shape and germinating in as little as 1 hour. Per cent germination of fresh pollen was variable. Tube lengths were about equal to pollen grain diameter after 4 hours. It was also estimated *in vivo* pollen growth to be 150 µm/h from 3 to 8 hours after pollination and was not significantly influenced by self versus cross pollination. Most germination tests were conducted at 20° C or at 25° C. It was also reported that tube growth was strongly influenced by temperature, with 27° C being optimal. Pollen germination was reduced at the temperature of 30° C.

The development of male flowers and pollen germination of pecan, 'Mahan' was systematically observed by Zhang *et al.* (2016). The results showed that the male flower primordium took shape on the top of stamen in late March and began to differentiate in early April. In late April, the pollen sac formed in anther and the pollen mother cell began meiosis to tetrad stage. Then the tetrad produced pollen grains of single cell nucleus and the tapetum disappeared gradually in the developmental process. The pollen grain was fully developed when it had two nuclei. A stigma in the receptive period after pollination showed wave crests and papillary cells at the top of the carpel. Pollen germinated on a stigma four hours after pollination. More pollen germinated on stigma and pollen tubes grown toward ovary 12 hours after pollination. Pollen began collapse 24 hours after pollination and pollen tubes began collapse, and papillary cells at the top of the carpel began depression and collapse 3 days after pollination.

6.1.5. Pollination

The period of pollen shedding is five to twenty five days and that of stigma receptivity is 6 to 14 days, depending upon cultivar, temperature and humidity (Madden and Brown, 1973). Pollens shed only under dry condition and none when relative humidity is above 85 percent, (Yates and Sparks, 1993) rains followed by slow drying condition can prevent shedding upto 24 hours. But prolonged high humidity or rainy periods are usually followed by exceptionally heavy pollen shedding. Heavy dew occurring in most pecan growing areas cause very little shedding during night and most shedding occurs during day light with a peak in afternoon. Catkins normally abscise a few days after pollen shedding. The degree of defective pollen production varies among cultivars and years and is probably due to factors that cause a failure of development of staminate flowers but percentage of defective pollen is reported low. The pollen is light, dry and pale yellow in colour. Pollen is produced in abundance, with about 2.5 million pollen grains by single catkin (Madden *et al.*, 1969). The pollen produced by a single 13 years old Muntura pecan trees is capable of producing sufficient pollen for the estimated production of 29 million nuts weighing 2,99,000 tones. As the pollen is distributed by gravity and wind or rain, this extravagance is necessary to ensure effective pollination.

The pollinizer rows should not exceed 150 m from trees to be pollinated, but the pollen dispersal was up to 300 m however, the pecan pollen could be carried to 900 m by wind. On the basis of these statistics, the recommended planting of 5 percent of the orchard to a pollinizer is sufficient to provide adequate pollen (Romberg, 1931~ Woodroof, 1930). The multiple cultivars are often planted in long, narrow blocks. In newer plantings, the pollinizer cultivar often comprises 10 percent of planting. The 9th and 10th rows are often planted with a pollinizer. Except in case of an isolated tree or cultivar, lack of pollen is seldom a problem in new orchard. This is because of (1) the practice of planting the proper protogynous and protandrous cultivated in a given orchard, (2) frequent presence of pollen from nearby orchard and /or seedling in the wild, (3) the excess of pollen produced, and (4) because of lack of staminate inflorescences and pollen production as such is rarely a problem in any season.

Pecan is anemophilous and excessive rainfall during the flowering period may prevent pollination. Pollination is not a problem in pecan under normal condition. The duration of both pollen dehiscence and stigma receptivity ranges from about 8 to 14 days depending on the seasons and cultivars (Sparks, 1986b). Pollen dehiscence is hampered by wind, high relative humidity (above 85 %) and very low temperature. The pecan stigma is of dry type. The pollen tube emergence is visible within three hours of pollination (Wetzstein, 1989).

Most of the pecan cultivars require cross pollination for good fruit set and there is a need to plant 3 to 4 pollinizer cultivars in an orchard so that the stigma, when receptive, be pollinated. Early pollination of stigmas has been reported for higher fertilization but arrival of pollen 24 hours after first pollen was less than 3 per cent (Marquard, 1992a). Pollen germination decreased markedly after anther dehiscence and less than 1 percent of the pollen germinated 5 days after collection (Wetzstein and Sparks, 1985). So as to provide timely adequate pollination of stigma, some workers have conducted studies on the storage of pollen and its later use in pollination.

6.2. Flowering and fruiting habits

The flowering and fruiting habits of 'Mahan' pecan were studied by Xu *et al.* (2015). The results showed that the pecan buds on the fruiting spur are usually mixed buds in near-round shape and tightly packed with scales. After germination, fruit bearing-branches elongated and compound leaves appeared. The male inflorescence was developed from the lateral buds on the low portion of the branch and the female inflorescence was developed from the proximal top of the branch. From mid to late March, buds swelled and their outer scales shed. The inner scales split and the male inflorescence appeared and bloomed on an extended fruiting branch from the apical bud in late March. In early April, leaves expanded and reached the maximum size after 20-d growth. The growth patterns of the fruit-bearing branch, compound leaf, leaf length and leaf width all showed an "S" curve, which highly fit in the logistic growth model. A correlation analysis indicated a significantly positively correlation between the growth and the effective accumulated temperature. The first time of the fruit dropping occurred at early to late June with a fruit dropping rate of 16.27%. The second time of the fruit dropping occurred from early to mid July with a fruit dropping rate of 68.75%. The third fruit dropping time was in August with a fruit dropping rate of 2.2%.

6.3. Phenology of fruit development

Reid *et al.* (2015) studied twenty-two pecan [*Carya illinoinensis* (Wangenh.) K. Koch] cultivars fruit development to establish how differences in fruit phenology ultimately define time of ripening. The earliest fruit ripening cultivars entered into the period of rapid fruit expansion over two weeks before cultivars that ripen late in the season. Once rapid fruit expansion began, the pace of fruit development for early-ripening cultivars was faster than later maturing pecans. Early ripening cultivars took the least amount of time to fully expand their fruits and achieve maximum ovule expansion (the water stage). Early maturing cultivars were also quicker to split shucks (ripen fruit) once kernel deposition was completed. The time required to deposit solid kernel tissue inside a fully expanded ovule was roughly the same for all early ripening cultivars (two weeks). Only extremely late maturing cultivars, or those cultivars generally not adapted to the 200 day growing season at this study's location, took longer to fill their kernels (18-21 days). Charting the key phenological events that define time of fruit ripening for all 22 cultivars studied illustrates that pecan fruit phenology is a continuously variable genetic trait.

Jia *et al.*(2015) elucidated the dynamic change and quality mechanism of pecan nut development. The results indicated that fruit weight, kernel weight, kernel percentage, fruit length, nut length, fruit width, nut width, nut shape index showed a first increase and then decrease pattern; Peel thickness increased and fruit shape index decreased gradually, shell thickness first decreased and then increased during nut development. From the investigation it was concluded that during 145-155 d after full blossom, most traits especially key traits which determine the nut quality and marketability reached the optimum value.

7. Climate

Pecan needs warm temperate climate meaning thereby that the extremes of both temperate and subtropical climates are undesirable. Wolstenholme (1979) suggested the following requirements for an ideal pecan environment: (i) growing season of 240-280 days (ii) mean temperature above 26.7°C, (iii) heat unit accumulation (base 10°C) of 5000 degree days for the 7 months nut growing period, (iv) mean temperature rapture for the 3 coldest month between 7.2° and 12.8°C with at least 400 h of chilling temperature at or below 7.2°C (v) medium to low rainfall, (vi) high levels of solar radiation during the growing season and (vii) deep fertile soil with pH between 6.7 and 7.0. It can be successfully grown at an elevation 914 to 1829 m above mean sea level in Himalayan regions are as which are free from spring frost and excessive heat in summer (Singh *et al*., 1967).

Pecans require 200 or more frost free days to produce a crop on the earliest maturing varieties. Pecan trees break dormancy late in the spring and it is unusual, although not unthinkable for the flower crop to be damaged by late spring freezes. Cold pockets that place plants at higher risk for freeze damage should be avoided for young trees and reduced spring frost damage. Rainfall decreases irrigation water pumping requirements, but also has higher incidence of diseases (Pecan Scab, Downy Spot) and greater difficulty harvesting in rainy years. Pecan requires the minimum average rainfall 760 mm. As compare to east regions there is less concerns of diseases in the pecan growing in the West region. An annual of 55 inches of water per acre is requires for good growth and production. Like other temperate fruits, pecan also needs adequate chilling hours at a specific threshold temperature to foliate and fruit normally. Insufficient winter chilling results in delayed foliation and uneven development of fruit buds. McEachern *et al*. (1978) determined a chilling requirement of 500 h below 7.2°C for the cultivars Desirable and Mahan, and 600 h for Stuart. North American cultivars require more chilling than southern cultivars.

High humidity affects adversely the growth, fruiting and regularity of bearing in pecan by preventing pollination, increasing the incidence of diseases on leaves and nuts, premature sprouting of nuts in husks and delaying husk splitting (Madden *et al*., 1969). Hailstorms, floods and winds of high velocity are harmful to the trees.

Kuden *et al.* (2013) determined the chilling requirements of the male and female flower buds of 23 pecan nut cultivars. Calculating the chilling hours below 7 degrees C and chill unit (CU) values during the experimental winter periods, 555 to 660 h and 407 to 416 (CU) of chilling duration were obtained under subtropical conditions of Adana (Turkey). There are some differences on dormancy release dates and chilling requirements of the female and male flower buds as well as some little differences between the chilling methods. In 2003 to 2004 winter period, the chilling requirements were changed between 208 to 338 CU and 250 to 436 h in male flowers, whereas, 203 to 389 CU and 250 to 484 h in female flowers. In this year, similar to the previous year, Moneymaker broke endodormancy of female buds primarily on 22nd January and showed the lowest chilling requirements (203 CU and 250 h), while male flower bud dormancy release occurred on 12th February with 300 CU and 346 h of chilling. The male flower buds of Kiowa, Schley, and Texhan were found to be the lowest chilling requiring cultivars suitable to warm climatic conditions, whereas, the others require more. For the female flower buds, Harris Super, Mahan, Moneymaker and Wichita gave the lowest chilling requirements, while Desirable, Schley and Woodard gave the highest. Considering the tendency of the male and female flower buds to protogeny and protandry, Kiowa, Schley, Texhan, Pawnee and Woodard showed protandry, while Choctow, Moneymaker, Harris Super and Western showed protogeny in both years.

8. Soil

Good soil is an essential factor for the successful cultivation of pecan (Madden *et al.*, 1969). Pecan orchards should be established on deep and well drained soils that contain proper amounts of air, water and nutrients. Since pecan has a long taproot, a good soil depth of 6 m is desirable (Hanna, 1987). The soil pH of 6.4 is the best (Woodroof, 1979). Poor soil drainage is one of the most serious limiting factors in pecan orchards. Pecan trees can grow on very shallow soil, but for commercial production of more than 1,000 pounds per acre, the soil should be 32 inches deep or more. Soil aeration is another important factor in pecan growing (Smith *et al.*, 1989). They reported that restricted soil aeration induced partial stomatal closure and decreased leaf number, leaf area, and leaf, trunk and root dry weight. Restricted aeration increased N and F concentrations in the roots by decreasing their translocation to the leaves. Leaf concentrations of Ca, Mg and Mn were lower whereas that Zn and Fe in roots higher due to lower aeration.

9. Varieties

Description of important cultivars

9.1. Aggie: Trees are vigorous and prolific bearer which bears upto eight nuts per cluster. Nut are elliptic with obtuse apex and base, laterally compressed (taller than wide) to round in cross section; 69 nuts/lb, 63 per cent kernel; kernels are golden in color with smooth texture, wide dorsal grooves, and very tight basal cleft; shells easily, quality good.

9.2. Caddo: Nuts are oblong elliptic to oblong, with acuminate apex and base; round in cross section; 60 nuts/lb, 56 per cent kernel; kernels are golden in color with shallow dorsal grooves, wide and extended dorsal ridge; excellent shelling characteristics. This variety is precocious and prolific bearer, protandrous in nature with early pollen shed and mid-season pistillate receptivity. It ripens in mid season and moderate in scab resistance.

9.3. Apache: Trees are protogynous in nature with mid to late season pollen shed and mid season receptivity, precocious and prolific bearer. Nuts are oval elliptic to elliptic, with acute apex and obtuse base, round in cross section, 45 nuts/lb, 59 per cent kernel, kernels are golden in color, with prominent basal cleft. Nut ripens late in the mid-season. It is susceptible to scab. It is also used as a seed stock.

9.4. Barton: Late to break buds in spring, with Stuart. Protandrous in nature with mid-season pollen shed and mid to late-season receptivity. It is precocious and prolific bearer, with a tendency to overbear with maturity which ripens early mid-season. Nuts are elliptic with obtuse apex and acute base; round in cross section; shell suture usually dark at base; 48 nuts/lb, 57 per cent kernel; kernels are golden in color, with deep secondary dorsal and ventral grooves. It is resistant to scab.

9.5. Brooks: Nuts are oblong with acuminate apex and base; slightly compressed in cross section; markings on shell dark and rough; 87 nuts/lb, 52 per cent kernel; kernels are cream to golden in color, with wide dorsal grooves. It has protogynous bloom pattern, with mid-season pollen shed and mid-season receptivity. Used as a female parent of 'Caddo'. It has large leaflets and is scab resistant.

9.6. Burkett: Nuts are orbicular, with obtuse apex and rounded base; round in cross section; 43 nuts/lb, 55 per cent kernel; kernel is golden to light brown in color with prominent dark brown speckles. It is protogynous in nature with mid-season pollen shed and receptivity. It ripens late mid-season and is very susceptible to scab and downy spot (*Mycosphaerella caryigena*). It is prone to pre-harvest germination and is widely used as a nursery seed-stock for the western pecan growing region. It is used as a female parent of 'Apache' and 'Comanche'.

9.7. Candy: Protogynous in nature with mid-season pollen shed and receptivity. Initiates growth early and makes dark green, attractive leaves. Nuts are oval, with obtuse apex and base; round in cross section; 66 nuts/lb, 48 per cent kernel; kernels are golden in color, smooth textured, with wide dorsal grooves; shells to produce kernel halves despite thick shell. It bears at 5-6 years of age, prolific in bearer and ripens early. Candy often produces clusters with up to 8 nuts. Shucks have a distinctive rusty stripe in the middle of each segment. It is resistant to scab in Texas and most of North Louisiana but moderately susceptible to scab in South Louisiana. 'Candy' developed severe scab when unsprayed in an orchard near Monroe, Louisiana, in which 'Wichita', a very scab susceptible cultivar was being grown and sprayed for scab.

9.8. Cape Fear: Very precocious and prolific in bearing with a tendency to overbear with maturity. Protandrous in nature with early pollen shed and mid to late season pistillate receptivity. Nut are elliptic with obtuse apex and base; shell dark striped; round in cross section; 45 nuts/lb, kernel 54 per cent; kernels cream to golden in color, with wide dorsal grooves, deep secondary dorsal grooves and ripens in late mid-season, about three days after 'Stuart'. It is susceptible to fungal leaf scorch complex (*Alternaria, Pestolotia* and other fungi) which can defoliate trees. It is variable in scab susceptibility by location, being moderately resistant in some orchards, very susceptible in others.

9.9. Cherryle: Nuts are oblong elliptic to oblong, round in cross section, with obtuse apex and base; 43 nuts per pound, 55 per cent kernel; kernels with medium dorsal grooves that do not trap packing material, slightly tapered dorsal ridge, shallow basal cleft. Moderately susceptible to pecan scab (*Fusicladosporium effusum*).

9.10. Curtis: Protogynous in nature, with late-season pollen and mid-season pistillate receptivity. Nuts are oblong elliptic with acute apex and base; slightly compressed in cross section; shell with few markings; 89 nuts/lb, 57 per cent kernel having cream to golden in color, but often with darker speckles that detract from appearance; tight dorsal grooves. It is resistant to scab, but susceptible to powdery mildew (*Microsphaera alni*). It is used as a nursery seed stock in the southeastern United States.

9.11. Cheyenne: Very precocious and prolific in bearing which ripens in late mid-season, shortly after 'Stuart'. It is a slow growing and dwarf type tree. Protandrous in nature with mid-season pollen shed and mid to late season pistillate receptivity. Nuts are oval elliptic to elliptic, with acute apex and obtuse base; round in cross section; 48 nuts/lb, 58 per cent kernel; kernels are light cream to cream in color; very wrinkled texture; wide, shallow dorsal grooves. It is very susceptible to yellow aphids and less susceptible to scab in south LA and somewhat resistant in Texas.

9.12. Chickasaw: Precocious and prolific bearer with a tendency to alternate bear with maturity which ripens early mid-season, about 18 days before 'Stuart'. Protogynous in nature with mid-season pollen shed and early pistillate receptivity. Nuts are elliptic with obtuse apex and base; flattened in cross section; 55 nuts/lb, 54 per cent kernel; kernels are flattened in cross section. Nutrient deficiencies in nitrogen, phosphorus and potassium associated with heavy production. It is resistant to scab.

9.13. Choctaw: Prolific bearer which is protogynous in nature with mid to late season pollen shed and early to mid season pistillate receptivity. Nuts are oval elliptic to elliptic, with obtuse apex and acute base; round in cross section; weak suture on shell; 37 nuts/lb, 58 per cent kernel having cream to golden in color with darker veins; wide, shallow dorsal grooves. It is resistant to scab in most locations, although very susceptible at Hanna, Louisiana. This variety requires good site and management to produce large crops of well filled nuts.

9.14. Comanche: Protogynous in nature with mid to late season pollen shed and early to mid-season pistillate receptivity. Nuts are round with obtuse apex and base; round or slightly flattened in cross section; very similar in shape to 'Burkett', but with darker shell markings; 37 nuts/lb, 56 per cent kernel which are golden in color, with very wide dorsal grooves; lacking the dark flecks that mark 'Burkett' kernels. This variety is very susceptible to scab.

9.15. Delmas: Prominent cultivar of Israel and is also known as 'Nellie' in Australia. This selection is protogynous with mid to late season pollen shed and mid season pistillate receptivity. Nuts are elliptic, with obtuse apex and base; round in cross section; prominent ridges on shell, which is smooth, with few stripes; 50 nuts/lb, 48 per cent kernel having golden colour with wide dorsal grooves; wrinkled on ventral surface. It is susceptible to scab.

9.16. Desirable: Bears in about 6 years and makes consistent, moderate production of high quality nuts, due in part to self-thinning fruit drop that reduces the number of nuts per cluster. It is protandrous in nature with abundant early pollen shed and mid to late season pistillate receptivity. Nuts are elliptic with obtuse apex and obtuse to rounded base; round in cross section; shell rough, with elevated suture; 38 nuts/lb, 54 per cent kernel which is golden in color, with wide dorsal grooves. This variety ripens in late mid season, shortly after 'Stuart'. It is susceptible to scab (Thompson and Young, 1985).

It is an excellent cultivar, producing higher than average yields on a consistent basis, with good nut size, dependable kernel quality, and nuts suitable for mechanical shelling. However, it is very sensitive to environmental conditions, and may suffer as a result of cold/freezing temperatures, nutrient imbalances, low light, high winds, drought and waterlogging, thus it requires careful crop management techniques (Sparks, 1997).

9.17. Edisto: Nuts are ovate with obtuse apex and rounded base; round in cross section; shell smooth, with few markings; 91 nuts/lb, 49 per cent kernel having golden colour, with broad, triangular dorsal ridge, relatively open dorsal grooves, and a prominent basal cleft and ventral groove.

9.18. El Mart: Trees are precocious and like all 'Mahan' progeny, it is also protogynous in nature. Nut are oblong with obtuse apex and acute base; round in cross section; 41 nuts/lb, 60 per cent kernel; kernels golden, with broad, triangular dorsal ridge, relatively open dorsal grooves, and a prominent basal cleft and ventral groove. Its nuts mature in October.

9.19. Fisher: Tree reputed to be a heavy bearer which ripens a few days after 'Starking Hardy Giant'. Nuts are oblong elliptic with an obtuse apex and acute base, round in cross section. There are about 74 nuts per pound, with 48 percent kernel having cream color, with medium dorsal grooves and a narrow dorsal ridge.

9.20. Forkert: Protogynous in nature, with mid- to late-season pollen and early pistillate receptivity. Nut are oblong elliptic with acuminate apex and obtuse base, surface very rough, with prominent dark stripes; round in cross section; 49 nuts/lb, 62 per cent kernel which is cream to golden in color, with deep, relatively narrow dorsal grooves. This variety is susceptible to scab in Los Angeles.

9.21. Giles: Protandrous in nature, with mid-season pollen shed and pistillate receptivity. Nuts are oblong with acute, asymmetric apex and rounded base; laterally compressed in cross section; shell darkly marked with stripes, raised at suture; 74 nuts/lb, kernel is 53 per cent having golden colour, with wide dorsal grooves, narrow dorsal ridge, deep, but wide basal cleft. It is used as a seed-stock in parts of Oklahoma and Kansas.

9.22. Gloria Grande: Protogynous in nature with mid- to late-season pollen shed and early to mid-season pistillate receptivity. Nuts are elliptic with obtuse often asymmetric apex and obtuse to rounded base; round in cross section; shell thick, smooth, darkly striped; 44 nuts/lb, 48 per cent kernel with golden colour, with shallow, wide dorsal grooves and a deep basal cleft. It is resistant to scab, fungal leaf scorch and downy spot, however, susceptible to black pecan aphids.

9.23. Greenriver: Nature protogynous, with mid season pollen shed and early pistillate receptivity. Nut are oval elliptic with obtuse apex and base; laterally compressed in cross section; ridges on shell; 70 nuts/lb, 49 per cent kernel with golden color, with wide dorsal grooves and deep, wide basal cleft. Not appropriate for the far north, due to late nut maturity. This was scab resistant in north. It has a unique isozyme genotype: cc for phosphoglucose isomerase.

9.24. Hastings: Protogynous in nature. Nuts are oblong elliptic, with acute apex and obtuse base; 68 nuts/lb, 48 per cent kernel with golden color, with wide dorsal grooves, deep and wide basal cleft, wrinkled texture.

9.25. Houma: Nature protandrous, with early mid-season pollen shed and mid-season receptivity, similar to 'Desirable'. Nuts are elliptic, with obtuse apex and base, often asymmetric; round in cross section; 55 nuts/lb, kernel 51 per cent with golden color, with wide, shallow dorsal grooves and a deep, wide basal cleft.

9.26. Iago: Trees are protogynous with overlap of pollen shed with pistil receptivity in mid bloom season. Nuts are oblong elliptic with obtuse apex and rounded base; round in cross section; shell smooth, prominently marked by black stripes at apex; 41 nuts/lb, 60 per cent kernel having golden colour, often with a prominent vein in the integument on the broad dorsal ridge, shallow and wide dorsal grooves, prominent basal cleft and ventral groove. Nuts shell well. It is somewhat resistant to scab (*Fusicladosporium effusum*)

9.27. Kanza: Protogynous in nature, with late pollen shed and early pistillate receptivity. Nuts are ovate, with an acute apex and obtuse to rounded base; round in cross section; 77 nuts/lb; 54 per cent kernel with golden color, with dorsal grooves that do not trap packing material. Released for its ability to produce good yields of high quality nuts in the northern pecan producing regions. Kanza has excellent scab resistance.

9.28. Kiowa: Nature of variety is protogynous, with mid-season pollen shed and early to mid-season receptivity. It cross pollinates well with 'Desirable'. It is more precocious than 'Desirable'. Nuts are oblong elliptic, with obtuse apex and base; round in cross section; 38 nuts/lb, 58 per cent kernel having golden color, with wide dorsal grooves, narrow dorsal ridge, deep basal cleft and ventral groove. It ripens in late mid-season, with 'Desirable' and is susceptible to scab.

9.29. Imperial: Nuts are elliptic or ovate elliptic with obtuse apex and base, base often with short tip; round in cross section; shell rough, with very weak suture; 32 nuts/lb, 56 per cent kernel; kernels are golden, with wide, shallow dorsal grooves beside wide, triangular dorsal ridge; deep basal cleft and ventral groove.

9.30. Jackson: Nature protandrous, with incomplete dichogamy; early to mid season pollen shed and mid to late season receptivity. Nuts are elliptic with obtuse (almost truncate) apex and rounded base, round in cross section with prominent dark markings at apex. There are about 33 to 38 nuts per pound, with 52 to 56 percent kernel. Kernels are typically quite rectangular in shape, golden to light brown in color, with wide dorsal grooves and a wide, triangular dorsal ridge. Kernels often have secondary dorsal grooves and a wide ventral groove but seldom have a basal cleft. Trees have some resistance to bunch disease and scab but are susceptible to vein spot (*Gnomonia nerviseda* Cole).

9.31. Lakota: Vigorous, upright growth habit and develop strong limb angles and a wind-resistant tree structure. Time of spring bud break is similar to 'Kanza' and 'Pawnee'. It is protogynous, with early to mid-season receptivity and mid to late season pollen shed (similar to Kanza). It should be a good pollinizer for, and be well pollinized by 'Pawnee', 'Osage', and 'Giles'. The nut is oblong elliptic with an acute apex, and rounded base and is round in cross section. There are 59 nuts/lb, with 62 per cent kernel with cream to golden in color, with medium, non-trapping dorsal grooves and rounded dorsal ridge. Nuts shell out easily into full halves and are very attractive. Time of nut maturity is early, similar to 'Giles', and about two weeks after 'Pawnee'. It has performed well in tests in the northern pecan production areas of Kansas, Missouri, Illinois, Oklahoma and Texas. It is very resistant to scab disease, with medium susceptibility to yellow and black aphids.

9.32. Lipan: This variety was released because of its high nut quality, high yield potential, medium early nut maturity, and excellent scab disease resistance. Pecans from this cultivar can be sold in-shell or shelled to produce a large proportion of halves and large pieces. It appears to be regular in its yield, increasing yields with increasing age with little alternate bearing. Time of spring bud break is comparable to 'Pawnee' and 'Desirable'. It is protandrous, with early to mid-season pollen shed and mid to late season pistil receptivity (similar to Pawnee). It should be a good pollinizer for, and be well pollinized by Kanza, Wichita and Lakota. Nuts weigh about 10.4 g each (44 nuts/lb) and shell out about 55 per cent kernel. Kernels are cream to golden in color, with open, non-trapping dorsal grooves and a rounded dorsal ridge. The nut is oblong elliptic, with a slightly pointed apex and rounded base. The shell suture is very strong and should be very resistant to splitting if harvest is delayed. Time of nut maturity is medium early or about two weeks after 'Pawnee' and two weeks before 'Desirable'. It is resistant to scab disease with medium susceptibility to yellow and black aphids.

The pecan cultivar released by the U.S. Department of Agriculture, Agricultural Research Service. The Lipan are a Native American Apache tribe. It is a progeny from a cross between cultivars 'Cheyenne' and 'Pawnee' made by T. E. Thompson in Brownwood, Texas, in 1986. 'Lipan' was released because of its high nut quality, high yield potential, medium-early nut maturity, and scab disease (*Fusicladium effusum*) resistance. 'Lipan' should be adapted to all pecan-growing areas of the world except the extreme northern production area of the USA. Pecans from this cultivars can be sold in-shell or shelled to produce a large proportion of halves and large pieces (Thompson and Grauke, 2012).

9.33. Mahan: Very precocious and prolific bearer with a strong tendency to overbear as a mature tree. Homozygous for protogyny (PP) which is a dominant allele, making all 'Mahan' progeny protogynous. Mid to late season pollen shed, with early to mid-season receptivity. Nuts are oblong, with acute apex and base; nut often asymmetric, appearing pinched in middle due to flattening of abaxial and adaxial surfaces; flattened in cross section; 32 nuts/lb, 58 per cent kernel; kernels with deep secondary dorsal grooves and basal cleft, often poorly filled to base, woody in texture. It ripens late, about 12 days after 'Stuart' and is very susceptible to scab. This is the female parent of 'Tejas', 'Kiowa', 'Harper', 'Mahan-Stuart' and 'Maramec' and the pollen parent of 'Choctaw', 'Wichita', and 'Mohawk'.

9.34. Mahan-Stuart: Precocious bearer and protogynous in nature. Nuts are oblong with obtuse apex and base; flattened in cross section; 30 nuts/lb, 52 per cent kernel; kernels with extended dorsal ridge, tight dorsal grooves, deep secondary dorsal grooves, and wrinkled texture.

9.35. Mobile: Nature protandrous with mid season pollen shed and pistillate receptivity. Nuts are oblong with right angled apex and obtuse base; round in cross section; shell surface with ridges; 53 nuts/lb, 43 per cent kernel; wrinkled, with very broad, shallow dorsal grooves, and a prominent basal cleft. It is resistant to scab.

9.36. Mohawk: This variety is very precocious and prolific with a tendency to alternate bear on mature trees. Protogynous in nature with mid-season pollen shed and early to mid-season pistillate receptivity. Nuts are oblong with obtuse apex and base; slightly flattened in cross section; shell rough, dark striped; 32 nuts/lb, 59 per cent kernel with golden to light brown in color with narrow dorsal grooves when well-filled. Freeze susceptible, especially after large crop. It ripens mid-season, about 9 days before 'Stuart'. It is susceptible to scab.

9.37. Nacono: Trees are vigorous, with branched central leader form with strong limb angles. Bud growth commences shortly after 'Desirable' in the spring. Leaves are characterized by large leaflets, a feature common to other members of this cross family. Protogynous in nature with late season pollen shed and early to mid season receptivity. Nuts are oblong with acute apex, and an acuminate (pointed) base; round in cross section; 44 nuts/lb, 56 per cent kernel having cream to golden color, with shallow dorsal grooves and rounded dorsal ridge. Nuts shell out easily into full halves and have excellent eye appeal. It is resistant to scab while moderately susceptible to yellow and black aphids. Nuts average 3 per cluster. Shucks are glossy yellow green with medium suture wings near the apex. Nuts mature mid-late October, about one week prior to 'Desirable'.

The pecan cv. Nacono has superior nut quality compared to other protogynous cultivars, moderate resistance to scab (caused by *Cladosporium caryigenum*) and is suitable for planting in most U.S. pecan production areas. Nacono pecans are large enough to be sold in-shell, or they can be shelled to produce a high proportion of intact halves and large pieces of very high quality. The pedigree and characteristics are further described (Thompson and Grauke, 2001)

9.38. Navaho: Trees are vigorous, precocious and prolific bearer with strong structure, and begin growth early in the spring. Protandrous in nature with early pollen shed and mid-season pistillate receptivity. Nuts are oblong elliptic with an acuminate apex and obtuse base; round in cross section; 62 nuts/lb, 61 per cent kernel with golden to light brown in color, with deep, relatively narrow dorsal grooves and a prominent basal cleft. It is susceptible to scab and vein spot. Nut matures in mid October.

9.39. Owens: Nuts are oval elliptic with obtuse apex and base; round in cross section; shell rough, ridged, with silver sheen; 60 nuts per pound; 48 per cent kernel with wide dorsal grooves, narrow dorsal ridge, deep basal cleft; cracks out well. Protandrous in nature with early to mid season pollen shed and mid to late season receptivity. Strong tree structure which bear annually a moderate crop. It is resistant to scab and black pecan aphid.

9.40. Pawnee: Trees are precocious and production medium with some tendency to biennial bearing. Nature protandrous with early to mid season pollen shed and mid to late season receptivity. Nuts are elliptical with obtuse apex and rounded base; laterally compressed in cross section; 44 nuts/lb, 58 per cent kernel with golden color, with wide dorsal grooves and deep basal cleft. Nut matures early and is susceptible to scab, fair resistance to downy spot, outstanding resistance to yellow aphids (Thompson and Hunter, 1985).

9.41. Peruque: Nuts are ovate, with obtuse apex and rounded base; laterally compressed in cross section; dark suture line at base; apex slightly furrowed on abaxial and adaxial surfaces; 81 nuts/lb, 59 per cent golden colour kernel with tight dorsal grooves and a deep basal cleft. Nature protandrous, with early to mid-season pollen shed, and mid-season receptivity. It ripens very early and good scab resistance.

9.42. Schley: Nuts are oblong with acute apex and base; asymmetric; 54 nuts/lb, 62 per cent kernel having golden with narrow dorsal grooves. It is protogynous in nature with mid-season pollen shed and mid-season pistillate receptivity. Long considered the standard of nut quality for the pecan industry. It is used extensively in breeding. Female parent of 'Sioux', 'Shawnee', 'Cherokee', 'Oconee', 'Woodroof' and 'Cape Fear'. Male parent of 'Apache'. It is susceptible to scab, black pecan aphid, and a trunk disorder described as phomopsis canker.

9.43. Wichita: Tree moderately upright, vigorous, often with a late flush of growth. Trees are precocious, prolific bearer and is protogynous in nature with mid-season pollen shed and early to mid-season receptivity. Nuts are oblong, with acute to acuminate, asymmetric apex and rounded apiculate base; round in cross section; 43 nuts/lb, 62 per cent kernel with golden to light brown in color with narrow dorsal grooves and a wide, shallow basal cleft. It ripens in mid-season with 'Western', 4-20 days before 'Stuart'. It is very susceptible to scab.

9.44. Shawnee: Trees are with medium precocity, productive and protogynous in nature with mid-season pollen shed and early to mid-season receptivity. Nuts are oblong, with obtuse apex and base; round in cross section; shell light brown with few dark stripes; 48 nuts/lb, 58 per cent kernel with golden color, with very narrow dorsal ridge. It ripens early mid-season, one week before 'Stuart'. It is susceptible to scab, but manageable with fungicides.

9.45. Sioux: Trees are medium in precocity, good productivity and protogynous in nature with mid season pollen shed and early to mid-season receptivity Nuts are oblong, with acute apex and right angled base; round in cross section; 60 nuts/lb, 60 per cent kernel having cream to golden in color, with narrow dorsal grooves, smooth texture, high oil. It is susceptible to scab.

9.46. Stuart: The trees are late to begin growth in spring, making the cultivar hardy and slow to bear. Nuts are oblong elliptic with obtuse apex and rounded base; round in cross section; dark stripes on shell; 51 nuts/lb, 49 per cent kernel with golden to light brown in color with wide, shallow dorsal grooves, deep secondary dorsal grooves and a pronounced basal cleft. Nature of the tree is protogynous with late pollen shed and mid-season pistillate receptivity. It is resistant to scab and susceptible to downy spot, black pecan aphids, and yellow aphids (Thompson and Young, 1985).

9.47. Success: The trees are protandrous in nature with mid season pollen shed and mid to late season receptivity. Nuts are oval elliptic with obtuse asymmetric apex and obtuse to rounded base; round in cross section; shell dark striped at apex; 48 nuts/lb, 50 per cent kernel with golden to light brown in color, with wide, shallow dorsal grooves, prominent ventral groove and wide, secondary ventral grooves. This is a female parent of 'Choctaw', 'Mohawk' and male parent of 'Barton', 'Comanche' and 'Gratex'. It is very susceptible to scab, shuck dieback and stem end blight.

9.48. Maoshan 1: It was obtained by seedling selection. The nut is short rounded with perfectly round base. The average fruit weight is 8.6 to 8.9 g with 47.56% to 48.67% kernel. The crude fat, protein, oleic acid and linoleic acid are 65.20% to 66.10%, 10.32% to 10.50%, 58.91% to 59.59% and 30.69% to

31.14%, respectively. The nut has excellent quality and sweet taste. Male and female flowers almost bloom at the same period. The yield is high and the disease resistance is strong. The fruit ripens in late October and its development period is about 160 days (Liu *et al*., 2011).

10. Propagation and rootstock

10.1. Rootstock

Due to open pollinated nature of the crop and in the absence of any clonal rootstock, the most commonly used rootstocks is of seedling origin of the commercial grown varieties (Thompson, 1984). The rootstock effect on scion is very well known in fruit trees and the marked variation found in pecan cultivars may be ascribed to it because all seedlings rootstocks that have great heterozygosity. Most commonly used seeds are of the Elliott and Curtis cultivars as Elliott has good germination and develops good caliper relatively quickly, while Curtis is colder tolerant (Hanna, 1987).

The rootstocks are propagated by sowing stratified seeds (12 to 16 weeks) in nursery beds. For stratification, the seeds were placed in boxes with moist sand, in a cold chamber at 4°C, for 90 days. Seed stratification significantly improves the germination potential and morphological traits (Poletto *et al.,* 2015). After attaining suitable size by seedlings the cultivar is grafted/budded onto it. The germination and seedling growth of pecan nuts could be improved by dipping the nuts in 10% solution of sulphate of potash for 10 hours (Muhammad Sajid *et al*., 2017).

10.2. Propagation

Pecan is commercially propagated by budding or grafting onto seedling rootstocks but the vegetative propagation has been less efficient in pecan when compared with other common fruits due to high mortality during transplanting. Most of the plantations in the hilly tracts of India are of seedling origin.

10.2.1. Seed propagation

For raising or propagating the plants from seeds, the nuts are collected from the ground after natural seed fall or after shaking the trees. Outer hull of the nuts are removed by trampling or with the help of a walnut or corn huller. Larger sized nuts of a cultivar always produce taller and thicker seedlings than smaller nuts (Adams and Thielges, 1977; Herrera and Martinez, 1983) so sizing of nuts may be beneficial. After extraction, the nuts should be dried to low moisture contents and refrigerated. In late October or first fortnight of November the nuts are put in pits for stratification. The temperature during the stratification must be between 1 to 4°C for 30 to 150 days. Stratification of imbibed nuts in plastic bags without medium is suitable for most species (Bonner and Maisenhelder, 1974) and good results have been reported for pecans from southern sources by soaking the nuts at 20°C for 64 hours (Goff *et al.*, 1992). There are indications that stratification should be shortened for stored nuts (Bonner, 1976b). However, different opinions were their regarding stratification need for pecan nuts. van Staden *et al.*, (1976) reported that the shell act as a mechanical barrier to germination which can be overcome by germinating the nuts between 30° and 35° C. Caminada (1979) got 9.6 per cent germination on incubating the nuts at 30° C whereas 86 per cent germination resulted after stratification at 4°C for 60 to 90 days. Sometimes the nuts germinate while still on the tree. The germination is definitely higher and quicker at higher temperatures irrespective of stratification. It is believed that stratification is required for the nuts of northern cultivars and not for the nuts of southern cultivars. Cold stratification of the nuts for 45 days followed by soaking in GA_3 (500-1000 ppm) for 48 hr has given the best germination and subsequent seedling growth (El-Nabawy *et al.*, 1980; Knox and Smith, 1981). During the stratification, inhibitors gradually decreased and GA-like substance increased, reaching a maximum at the end of the 6-week stratification (El-Nabawy *et al.*, 1980). Stratification for 60 days is generally recommended. Adequate germination can also be made on stratified nuts in flats of sand, peat or soil at the same temperature regime.

Stratified seeds should be kept moist at a temperature of 35°F. Stratified seeds must be put or transplanted in deep, fertile sandy soils for good seedling growth. The soil should be well prepared by ploughing. Stratification does not hasten germination once the nuts are planted, but helps the seedlings to emerge more uniformly, facilitating the ease of weed control in the nursery. Sow the stratified seeds 3 to 4 inches deep in soil at 4 to 6 inches apart in rows. Planting depth should be more shallow in clay soils than in sandy soils. Apart from stratified seeds, nuts of good size and shape collected from the healthy and free from insects-pests and diseases may also be directly sown in well prepared nursery beds at proper spacing. Prior to directly sowing the nuts are soaked for 2 to 3 days in a container of water in order to activate enzymes that stimulate growth.

However, the plants raised (either through stratified or directly sowing in beds) and taken as for establishing the orchards showed lot of variation as each seed from an individual is not identical to the parent which accounts for much of the variation in tree performance when it comes into full production and it takes many years for trees planted from seeds to begin producing nuts. However, such seedlings or raised plants can be used as rootstock for grafting and budding purpose. The plants remain in the nursery for one or two years in order to attain proper size for budding and grafting and subsequent growth but they are transplanted after one year by shortening the taproot to 20-25 cm length which enhances vigorous fibrous root system and high survival rate (Pichkhaya, 1987).

10.2.2. Vegetative method

The most efficient, feasible and commercial way to propagate pecan is by grafting or budding. Grafting and budding are the easiest and most widely practiced propagation methods, especially in commercial production of pecan nuts. However, there are a number of other viable propagation methods available for the pecan tree. Scion wood should be collected from the previous season' growth of a productive tree during the dormant period (between January and March). Scions should be straight and smooth, measuring ¼ to 5/8 inches in diameter. They should be 6-8 inches long having three to four healthy buds and should be insect and disease free. Scions from older wood may also be used, but are generally less successful. Scions cut from end of the shoot should be avoided because the buds are often immature and damaged by cold, have pithy wood. If the scion is stored, it should be packed in moist wood shavings in a plastic bag to prevent drying out, and placed between 30° and 38°F (Wells, 2010).

10.2.2.1. Grafting

The most common and commercial method of grafting pecan nut plants is whip grafting. Two year old dormant seedlings are typically whip grafted in February and March. Make a slanted cut of the same length on both the rootstock and the scion; then place them together so the cambium layers align and wrap with parafilm, grafting tape, or a rubber band to hold the graft in place (Wells, 2010). Mostly scion used for grafting should be of the same diameter of the rootstock, nut in case of whip grafting thinner rootstocks are also used. Whip grafting gives about 90 per cent success in late dormant season. Successful grafts will produce new shoots in 4 to 6 weeks (Peterson, 1990). The scion is grafted either on underground of the stock in the field or on a root piece as bench grafting (Gustafson *et al.*, 1982; Sampaio and Barbin, 1980, 1981).

The four flap or banana graft is another easiest graft to perform with a high success rate because it allows more cambium contact than other grafts. The scion and rootstock should be similar in diameter, with the scion slightly larger than the rootstock. Some commercial growers prefer the four-flap or banana graft because it is one of the easiest grafts, even for beginners.

Various grafting methods and grafting schedules for pecan propagation were evaluated by Joolka *et al.* (2001). Two-year-old seedlings of pecan cv. Burkett were grafted on cv. Mahan by tongue, left, or crown method. Grafting started from 15 February to 18 March at 15-day interval. Tongue grafting conducted on 3 March gave the highest bud sprouting (39.55%), followed by grafting using the same method on 15 February (29.75%). Among the grafting methods, crown grafting gave the earliest bud sprouting (average of 25.33 days). However, cleft grafting on 18 March required only 23.33 days for bud sprouting. The highest survival rate was observed in tongue-grafted plants (7.48% mean survival and 13.16% for grafting on 3 March). Tongue-grafted plants had the highest proportion of saleable plants (71.33%).

Ajamgard *et al.* (2016) reported new improved technique of grafting in pecan. Scion wood of 'GraTex', 'Wichita', 'Choctaw', '10J' and 'GraKing' pecan cultivars were grafted onto two-year old seedling rootstocks 1-1.5 cm in diameter. The best graft time was from late January till late February using the cleft graft with hot cable system graft method and from mid-February till late March using the cleft graft with polyethylene bag grafting method. Also, the highest percentage of graft success was achieved using 'Wichita' scion (92%), while '10J' had the lowest percentage.

10.2.2.1. 1. Four flap or banana grafting

The four flap or banana graft is one of the easiest grafts to use and is ideal for small caliper trees up to 1 inch in diameter. This graft is most successful when the scion and rootstock are of similar size. The high degree of success with this method can be attributed to the large amount of contact between the cambium of the rootstock and scion. Whereas most grafting techniques only allow for the cambiums to connect at two locations, the four flap allows eight opportunities for cambium contact with each graft. Therefore, a precise cut and fit is less critical with the four flap graft.

One year old healthy scion wood with well-developed buds of desired cultivar should be collected in dormant season having a size of 3/8 inch to 1 inch at maximum. The four flap graft can also be used on older trees. Use limbs less than one inch in diameter. Large limbs should be de-horned, allowing new growth to develop for one year. The four flap graft should be made in early spring after the bark begins to slip. Remove all lateral growth on the stock to about 6 inches. Cut the stock straight across with sharp pruning shears at a comfortable working height with a trunk or limb diameter of ½ to one inch. Place a small rubber band around the stock and place it 3 to 4 inches below the cut. Place the knife blade straight across the cut surface of the stock and slice down through the center about ¾ inch. After this cut is made, make a second cut perpendicular to the first.

Choose a smooth, straight piece of scion wood of 6 inches long, having same diameter as the stock. Each scion stick should have two to three plump, healthy buds. With a sharp knife, make a flat cut through the bark on three to four sides of the scion, beginning 1 to 2 inches from the bottom end. Gradually increase pressure on the knife so that the cuts taper to join in a square at the base. These cuts should expose a long line of cambium around the edges of the bark. Pull down four flaps of bark on the stock to expose 2-2½ inches of bare wood. Carefully cut and remove the plug of exposed stock so as not to damage the four flaps. Insert the scion upright on the stock and pull up on the four flaps to cover the cut surface of the scion. Push the rubber band up near the ends of the flaps holding them in place. Wrap the cut areas with a ¾ inch rubber band or grafting tape.

10.2.2.1.2. Epicotyl grafting

Scion wood of 'Caddo' and 'Desirable' pecan (*Carya illinoinensis*) was grafted onto the epicotyl of 1-month-old, open-pollinated 'Shaoxing' pecan seedlings for evaluation as a grafting technique to reduce the time to produce grafted trees. The results showed that seedlings grafted with "base scions" had higher survival than those grafted with "terminal scions" for both 'Caddo and 'Desirable'. Also, grafting with paraffinic tape could achieve greater success rate than that with medical tape. The most ideal time to perform this grafting was late April in Nanjing, China, when pecan seedlings were about 35 days old. This study demonstrated that the technique yielded successful epicotyl grafting of >70%, and it could thus be applied in practice (Zhang *et al.*, 2015).

10.2.2.1.3. Hot-water treatment of pecan scions

Pecan (*Carya illinoinensis*) bacterial leaf scorch is a chronic disease of pecan that reduces yield, causes defoliation, and retards growth. The causal agent of this disease, *Xylella fastidiosa* is a xylem-inhabiting bacterium that can be transmitted in pecan by grafting with infected plant tissue or by insects in the families Cercopidae and Cicadellidae that feed exclusively in the xylem. No economically effective treatment for this disease exists once a tree becomes infected, and no immune pecan cultivars have been identified. Infection is persistent and disease and yield loss will occur annually. A simple, hot-water treatment in which pecan scions are submerged in water at 46°C for 30 min prior to grafting provides a means for reducing the potential introduction of *fastidiosa* into orchards through grafting without harming the viability of the scions (Melanson *et al.*, 2015).

10.2.2.2. Budding

Patch budding is most successful method of budding. The rootstocks of more than 2 cm diameter are usually patch budded. Patch or ring budding may be used to propagate nursery stock or top work larger pecan trees. Scion wood should be selected before the buds begin to swell. Budding may be done as late as June or July however, earlier buds have more time to produce a stronger union with the rootstock but will produce more vegetative growth during the first year. Maximos *et al.* (1979) investigated patch budding and reported 90-100 per cent bud take in August, whereas Bhardwaj (1983) got 75 per cent bud take in May. The operation should be accomplished by a double blade budding knife when bark peels easily in spring or summer season. The buds of preceding season are used for spring budding while well-developed mature buds of current season are patch-budded in late autumn season. The bud is inserted more than 20 cm high on the stock from the ground and secured tightly with polyethylene or waxed cloth strips.

Joolka and Rindhe (2000) standardized the method and time of budding in pecan. Two-year-old seedlings of pecan cv. Burkelt were used for budding. Budsticks were taken from trees of cultivar Mahan. Treatments comprised 3 budding methods, namely annular, forkert and chip, and 7 timings of budding, starting from 20 May to 20 September at 20 days intervals. Annular budding gave the highest sprouting compared to forkert and chip budding. The highest bud sprouting was recorded in annular budding performed on 30 June, followed by the same method performed on 20 July. Annular buding on 30 June was the earliest to sprout. It had the highest survival percentage followed by forkert budding performed on the same date. The proportion of saleable plants was highest (98.56%) in chip budding, followed by forkert budding (98.32%). All the plants budded up to 20 July were fit for sale under Solan (Himachal Pradesh) condition.

Srivastava *et al.* (2007) found that modified ring budding and patch budding gave maximum success in respect of bud take, bud sprout and proportion of saleable plants. Irrespective of budding methods bud take bud and sprout was highest in mid July and mid August budding followed by mid September budding. In the Kashmir valley the modified ring budding and patch budding performed in mid August and mid July on the basis of bud take (respectively), bud sprout and proportion of maximum of saleable plants and scion growth potential.

10.2.2.3. Cuttings

Attempts to propagate pecan by hardwood cuttings have not been encouraging because the cutting fail to establish and survive even after rooting (Hanna, 1987). In fact, roots emerged from cuttings are thick and long instead of being fibrous. Allan *et al.* (1980) obtained 90 per cent rooting and 70-80 per cent survival when the shoots of top-worked stumps were treated with IBA (1%) and planted in a mist bed with bottom heat. The cutting made from etiolated shoots gave 96 per cent rooting in mist after 5000 ppm IBA treatment (Rathore, 1985). Assaf (1977) reported 100 per cent rooting under mist in dormant cuttings treated with 3000 ppm IBA. The cuttings were inserted between the cotyledons of germinating nuts from which the radical and epicotyls had been removed. Root cuttings (20 cm long and 1.0-1.5 cm in diameter) of 2 year Barton seedling when taken in dormant season and planted under 50 per cent shade gave 100 per cent rooting and 250 mm long shoot growth after one year (Bolt, 1983).

Zhang *et al.* (2014) showed that the rooting rate of annual cuttings of *Carya illinoensis* was significantly higher than that of biennial cuttings either soaking or speed dipping treatments. The survival rate of cutting growth in greenhouse was higher than that in the field. The survival rate of cutting with speed dipping treatment was relatively higher than that with soaking treatment, and the rooting rate was highest, which was 80%, with speed dipping treatment combined with IBA 600 mg kg^{-1} and NAA 900 mg kg^{-1}.

10.2.2.3.1. Physiological basis of rooting in hard wood cuttings

The results showed that the exogenous hormones treatment changed four endogenous hormone (IAA, ZR, GA and ABA) content of pecan hardwood cuttings in the process of adventitious root formation. (1) The endogenous IAA content was relatively low and it was beneficial to the growth of adventitious roots. (2) The low concentration of endogenous GA content could delay the bud break of cuttings. (3) The low concentration of endogenous ABA content was favorable for the formation of root primordia; High level of endogenous ABA content improved the resistance of cuttings and reduced decay phenomenon. (4) The formation and development of callus or the adventitious roots of pecan hard wood cuttings were mainly regulated by the IAA/ZR ratio (Wu *et al.,* 2016).

10.2.2.4. Stooling

Clonal propagation of pecans by stooling has been quite successful (Sharma and Rosanghura, 1987; Wood, 1989). Girdling and IBA application (3000 to 6000 ppm) are necessary to induce good rooting in stools. The mother stools should, however, should be established from the cuttings or air layers of the commercial cultivars before resorting to propagation by this method.

10.3. Top working

Pecan seedlings and inferior cultivars are commonly converted to improved cultivars by grafting in situ and long life of the tree makes the venture more fruitful than planting young plants. The plants are beheaded in dormant season by cutting the limbs which are less than 10 cm in diameter but not more than two-third of the top should be removed at a time.

The covering of the inlay bark graft with aluminum foil and plastic bags has been found very satisfactory for top working (Madden *et al.*, 1969). The early growing season is the proper grafting time. Precut dormant scions can be used as it has many advantages. If grafts fail, the new sprouts on the stubs can be patch budded. Likewise, native sprouts emerged on the thicker branches (more than 10 cm diameter) after cutting back should be budded in the summer or spring of the following year. A good care of the grafts is required after they start growing. All the sprouts on the stock before the scion should be removed.

10.4. Micropropagation

Hansman and Novoa (1986) reviewed the literature on micropropagation of temperate nut trees. Smith (1977) initiated the work on micropropagation of pecan but he could only obtain callus growth. Significant progress in this regard was however, made by Knox and Smith (1981) when they produced shoots and roots from the spring flush of the cultivar Riverside.

Phillips and Ramirez (1983); Ramirez-Martinez (1983) and Cortes-Olivares *et al.* (1990) obtained successful shoot elongation using apical and axillary buds from mature trees and rooting of shoots was 40 per cent while establishment of plantlets were unsatisfactory. Shoot multiplication was obtained with nodal cuttings placed on a WP medium enriched with 2 per cent glucose and 3 mg IBA/1. The excised shoots were placed for rooting in test tubes containing WP medium with 2 per cent glucose and 3mg IBA/1. The rooted shoots were then transferred to peat pellets covered with plastic caps. Later, these plantlets grew vigorously and got acclimatized to greenhouse conditions. Thus, the commercial micro propagation of pecan appears feasible in near future (Preece and Read, 2003).

Obeidy and Smith (1990, 1993) addressed the problem associated with fungal contamination by manipulating the water availability of the medium and established plantlets in soil. Micropropagation has limited utility due to poor rooting, high rates of contamination and low regeneration. Hassanen and Gabr (2013) obtained best rooting percentage (75%) on half strength WPM medium (Mc Cow's Woody Plant Medium) supplemented with 3.0 mg/l IBA + 1.5 mg/ l $AgNO_3$ and with this eighty per cent plantlets were successfully acclimatized in greenhouse.

Embryogenic and organogenic responses in pecan (*Carya illinoinensis* [Wangenh] K. Koch) were observed as a result of the in vitro cultivation of segments of leaves, axillary buds and zygotic embryos. Necrosis was controlled through the use of activated carbon (AC: 1%), polyvinylpyrrolidone (0.1%), silver nitrate (AgNO3: 1%), citric acid (150 mg L^{-1}) and ascorbic acid (100 mg L-1), in both light and darkness. Murashige and Skoog base medium (MS) was used, supplemented with 0.40 mg.L^{-1} of thiamine, 100 mg L^{-1} of myo-inositol, 3% saccharose, incorporating 2,4-D for leaves, thidiazuron (TDZ) for embryos, and combinations of benzyladenine (BA), kinetin (KIN) naphthalenacetic acid (ANA) and indolebutyric acid (AIB) for axillary buds. Tissue necrosis was reduced by 75% and 83% adding CA and AgNO3, respectively. 33% and 66% of embryogenic callus originated from leaves, using 1 and 3 mg L^{-1} of 2,4-D. The highest callus production (58%) from embryos was obtained from the concentration of 3 mg.L-1 of TDZ. In axillary buds, the combination of KIN (3.0 micro M), BA (1.0 micro M) and AIB (0.3 micro M) increased the number of leaves and seedlings, as well as shoot length (Avila Trevino *et al.*, 2013).

11. Cultivation

11.1. Land preparation and pit digging

Preparing the orchard site for planting involves clearing, leveling, improving drainage, fertilizing, liming, and sub-soiling. Clear the site of all brush and trees. As weed competition greatly reduce growth and survival of young trees, weeds should be eradicated in the tree row prior to planting. Efficient harvesting and equipment movement requires a smooth surface, so land should be leveled reasonably well with high spots cut down and low areas filled. Soil samples should be taken for nutrient availability before planting. This will give an indication of the amendments required to make the soil as suitable as possible for pecan growth.

Dig a pit of size 1 x 1 x 1 m at least two months before the actual planting of the plants. Before planting, examine the roots and remove all broken or injured roots. Occasionally the taproot and/or some of the lateral roots will need to be pruned to fit in the pit. Never twist lateral roots in the hole, as this could eventually cause death to that part of the root system. Plant the tree at the same depth that it grew in the nursery. Fill the pit with the mixture of the top soil and the recommended dose of the well decomposed FYM. Light irrigation must be given soon after the planting in the pits.

11.2. Planting and Spacing

Location is important because of the ultimate size of the tree. Plant trees well away from the residence and other buildings. Space and ample sunlight are essential for maintaining productivity and profitability. Crowding in pecan orchards causes nut production to decline (Wells *et al.*, 2011), nut size to decrease and alternate bearing to be more severe. Crowding should be prevented rather than corrected, by tree removal or canopy hedging. The initial tree spacing determines how soon crowding will occur in a pecan orchard. Pecan orchards are planted at a spacing of 30 to 35 feet between trees (30 x 30 ft., 35 x 35 ft.).

Trees at such spacing crowd in 12 to 18 years after planting, depending on site, water and management inputs. Closer spacing is uncommon due to the fast rate of growth (Worley, 1991). Daniell and Moss (1985) reported that trees spaced 9.1 x 9.1m were crowded at 12 years. A model that has been created using existing data as a base shows maximum yield for 'Desirable' at 10.7 x 10.7 m and 'Stuart' at 13.7 x 13.7 m (Witt *et al.*, 1989).

The trees will generally be bare-root if directly taken from the nursery or may be packaged if bought from a garden center. Pecan trees should be planted as soon as they are received. Bare-root or packaged pecan trees can be transplanted anytime during the dormant season, but late fall and early winter is the best time. In recent years, container-grown pecan trees have become available for the home gardener. Container-grown trees can be planted anytime of the year, but early in the dormant season is best. Care must be given to provide sufficient moisture, especially when planting during the summer months.

Recent study indicated that severe root pruning when three to five leaves had emerged resulted in root systems with more taproot branches and the greatest root dry weight after one growth season, which may increase survival and reduce transplanting shock (Zhang *et al.* (2015).

11.3. Training

Proper training of a young tree will promote a strong, attractive tree. The main branches, known as scaffold limbs, will grow and define the tree's shape. As the pecan trees advance in age, they become larger and get crowded making pruning, spraying and harvesting operations difficult. The first year after planting, pecan trees grow slowly, and only the top needs to be pruned to one limb (trunk). Pecan trees grow in stages and the juvenile or seedling stage lasts the first five to six years or more after emergence of the seedling. During this time, most pecan trees tend toward a central leader growth habit. There is strong apical dominance and no fruiting wood is produced. After juvenility, the wood becomes mature. From the second year the plant is trained to central leader system and branches are allowed to emerge as high as 1.5 to 2 m from the ground on the trunk. The tendency then is now to produce more branching and production of fruiting wood. Grafted trees show this tendency from the first year after grafting. They require training from the outset to build a sturdy framework.

Leaving branches on the lower portions of the trunk will produce a "trashy trunk." A trashy trunk tree will produce a larger tree faster than a "limbed up" tree. Begin removing these limbs the winter after the third-year after planting. Remove all limbs that are dead, diseased, broken, crossing or rubbing at any time. In subsequent years, continue the process of selecting scaffold branches.

Training procedures may be repeated each year until the desired tree height is obtained or the tree height becomes too tall to continue training.

Smith and Hinriches (1980) described the training of high density plantings of the cultivar Western. Initially, the trees were planted at 5.83 x 8.3 m spacing and trained to central leader system and after 8 years were pruned to a hedgerow at spacing of 2 m and topped at different heights ranging from 3.23 to 6.66 m. Nut yield increased with tree height to a maximum of 6 m and declined thereafter at 6.66 m.

11.4. Pruning

Likewise other fruit crops, pecans also required annual dormant pruning which is desirable but is often neglected. Winter or dormant pruning should be done from time to time. This allows light into the center of the tree and to sustain nut quality. During the life of the orchard, occasional pruning will be required to maintain tree health and vigor. Limbs that cross over other limbs or compete for the same space should be removed. Broken limbs resulting from wind or ice storms will require removal prior to harvest. Limbs that die as a result of shading or pest infestation are unproductive and require removal.

Thinning cuts are used to remove an entire branch at its point of origin. In order to prevent damaging the trunk when large branches are removed, a three-step cut (also known as a jump cut) is used. The first cut is made at least 12 inches from the trunk on the bottom of the limb to be removed, about one-third through, stopping before the saw pinches. The second cut is made on the top of the limb and 6 inches outside (towards the branch tip) the first cut. This allows the limb to split off at the first cut. The third cut is made by jumping over the first cut and removing the remaining stub at the point of attachment just outside the collar (Upson *et al.*, 2012). Old large trees may reach a point where they are not productive or produce nuts that are not filled. Many times, severe pruning will help these trees to become productive. In these cases remove one-third of the large limbs (three to six inches in diameter) while keeping the tree symmetrical.

To obtain a well-balanced tree, all scaffolds must develop evenly around the main stem, about 300 to 360 mm apart. Scaffolds must be pruned back during the second, third and fourth seasons. The primary bud is not removed so that the scaffold can continue growing upwards and outwards. Primary buds are only removed on the leader shoot to form scaffolds. The experiments have proved that pecan pruning does not adversely affect the production of nut quality (Malstrom *et al.*, 1982; Witt *et al.*, 1989). On the contrary, judicious annual pruning increases production besides controlling the tree size. Since pecan is a terminal bearer, it requires only light to moderate pruning. Heavy pruning results in vigorous growth with no commercial production in the following years (Hanna, 1987).

Dried and broken branches should be removed and overcrowding branches be thinned.

Upadhyay *et al.* (2011) studied the effect of different pruning intensity on pecan. Pecan trees (25-year-old) were pruned to pruning intensities where height to spread ratio was kept at four different levels viz., 1:1.50, 1:1.25, 1:1.00 and 1:0.75. Pruning was done in the month of February during dormancy. The data revealed that 1:1.50 height to spread ratio proved to be the optimum pruning intensity as it did not reduce the yield significantly and also improved growth, nut quality and leaf nutrient status compared to unprunned trees.

Zhang *et al.* (2014) conducted a pruning experiment with another angle. Taking 5-year-old grafted seedling of *Carya illinoensis* 'Mahan' as experimental materials, effects of short-cutting (1/4, 1/3 and 1/2 short-cuttings), girdling and ringing of branch and trunk on branch growth and carbon-nitrogen metabolite accumulation in branch and leaf were researched. It is suggested that different short-cutting measures all can enhance sprouting rate, promote new branch elongation and its diameter increasing, while girdling and ringing treatments of branch and trunk have no obvious promotion effect on branch sprouting rate but are benefit to its diameter increasing. Different short-cutting measures are generally benefited to synthesis and accumulation of carbohydrate in branch and leaf. In general, 1/3 short-cutting and proper girdling of branch and trunk can promote fruiting branch formation of cultivar 'Mahan'.

11.4.1. Summer pruning

Scaffolds are pruned in summer (Nov-Dec) by removing 100 to 200 mm of the growth. This forces a branch to develop numerous lateral branches which can bear fruit during the same year in which they were pruned. Summer pruning dwarfs a tree and will increase production considerably during the first 10 years. After 10 to 15 years the producer can start pruning adult trees, especially if they begin crowding one another.

11.4.2. Rejuvenation pruning

Many old trees with declining production and nut quality can be stimulated to more active growth and increased nut production by pruning. By completely pruning back a big tree, the production of 1 or 2 years is lost, but there is new growth later and increase in the production of nuts with improved quality and compensated the loss.

11.5. Irrigation

Irrigation is essential for survival and rapid growth of young trees and for consistent production of quality nuts on bearing trees. Pecan trees need an ample supply of soil water for good growth and nut production. Almost all growth and productivity parameters in pecan have benefited from irrigation. Supplemental water treatments significantly increased kernel yield, nut weight and diameter kernel specific gravity, oil content and appearance (Heation *et al.*, 1982). On bearing trees, irrigation will help to produce larger nuts and improve nut fill. Generally, if water is scarce before nut shells form, nut size will be reduced. Lack of water after shell hardening inhibits kernel filling. Pecan trees do not like wet or saturated soils, so do not over water; however irrigation should be sufficient to thoroughly soak the soil.

Newly planted trees must be irrigated immediately. Thereafter, irrigation should be applied carefully, because too much water given before the tree starts growing, may cause the roots to rot. Young pecan trees will need 10 to 15 gallons of water weekly by irrigation or rain. The pecan-nut tree has a deep-tap root system, but for optimum irrigation purposes it will be adequate to supply the top 1 m with water. Water regularly from mid-March to September. Ideally, water the plants at least weekly, especially in mid-summer. Water the entire roots soil area to stimulate growth and production. Stein *et al.* (1989) reported increased yield, nut size, trunk diameter and kernel percentage due to summer and fall irrigation. Interestingly, they reported the need of water even before shuck opening as late season irrigation reduced stick tights and viviparous nuts.

Irrigation scheduling (applying the proper amount of water at the proper time) and method saves water, reduces nutrient leaching, improves crop (nut) quality and saves money. Irrigation scheduling requires an understanding of the crop rooting depth, the soil water-holding capacity, the crop water use rate and the irrigation water delivery system. Frequent application of relatively small amounts of water can reduce the soil volume explored by the root system, leading to increase irrigation dependency. The soil water-holding capacity varies with soil type. Flood or furrow irrigation is the preferred methods of water application. When water is available in plenty and the land is fairly leveled the flood and furrow irrigation systems are used which are quite popular all over the world. Since high water losses occur in these systems due to evaporation and deep percolation of irrigation water, sprinkler and trickle irrigation systems are gaining popularity day by day. Sprinklers or drip irrigation systems can be used and must run for considerable time to insure that water penetrates the soil three to four feet deep with mature trees. To determine the quantity of water required in trickle irrigation, Daniell (1989) suggested the use of class A pan and derived an equation for weekly water requirements to irrigate pecan trees. Using weather, soil and crop data Miyamoto (1984) developed a computer programme adaptable to microcomputers for predicting soil water storage and irrigation needs.

Water stress in pecan is correlated with soil moisture from budbreak through the end of nut sizing. Pecan trees bearing a moderate to heavy crop load may undergo water stress during the kernel-filling stage regardless of soil moisture level. Therefore, it is suggested that water stress during the kernel-filling period is a function of nut development, crop load, or both in addition to soil moisture. The reduced early season irrigation schedule provided a 38% reduction in irrigation water use with no significant effect on pecan tree water stress, yield, or quality, suggesting that pecan trees can tolerate moderate early season water stress with no effect on pecan yield or quality under southeastern U.S. environmental conditions (Wells, 2015).

11.6. Weed Control

Weeds must be controlled in young and mature pecan orchards to prevent reduced growth (Smith *et al.*, 2002), kernel development and quality (Daniell, 1974) and to increase yield (Foshee *et al.*, 1997). Foshee *et al.* (1997) conducted weed control experiment for 9 seasons in young pecan orchard cv. Desirable, showed that total weed control with herbicides resulted in a cumulative yield through the ninth growing season of up to 4584 kg/ha, compared with 1279 kg/ha with no weed control. In the humid environment, irrigation did not increase crop value obtained from the young trees, except for one year. At the end of the ninth season, total weed control with herbicides was the only treatment to have a positive net present value. These data indicate that establishment costs for young Desirable pecan trees can be recovered as early as the eighth growing season if competition from weeds is totally eliminated.

Bermudagrass and Johnsongrass are especially severe competitors for water and nutrients but numerous other grasses and weeds can result in poor tree growth. A "sod and strip" system that combines herbicide use with mowing of grass is the most practical and widely used system of orchard floor management. Herbicides must be used effectively and properly to kill the weeds, to reduce chemical costs, and to protect the applicator and the environment.

Pre-emergence and contact herbicide are generally used for controlling the weeds. Diuron and simazine (2 to 5 kg/ha) are widely adopted to control grass and broadleaf weeds as pre-emergence weedicides but should not be used before the pecan plants have established for three years. When a single herbicide is used, perennial weeds can flourish which need to be checked by the use of contact herbicides, such as, paraquat or glyphosate. McEachern and Storey (1984) got satisfactory control of pecan weeds with glyphosate (roundup) in a

10-year trial. Paraquat (1 kg/ha) has long been a reliable weedicide of pecan orchards which should not be allowed to come in direct contact with green stem and foliage. Application of a single heribicide may lead to a build up of tolerant weeds. It is, therefore, advisable to use a mixture of herbicides and similarity two herbicides can be used in rotation.

Pecan trees will make the best growth in the first few years after planting if a circle 6 feet in diameter is kept weed free. In view of many limitations of cultivation, the sod-herbicide system has been suggested giving the advantage of both cultivation and sod culture (Daniell, 1989). The weeds are controlled by using herbicides 2 m each side of the row while sod system is maintained between the rows. The weeds competing with the trees are eliminated by the weedicides without mowing, thus the mechanical injury to trunks and feeder roots does not occur. The weed-free strip facilitates the mechanical harvesting and the sod reduces soil erosion. Results from a four year study revealed that application of glyphosate to the trunk of established trees did not significantly affect mortality, growth, yield or nut quality. However, repeated application of glyphosate to foliage can have an adverse effect (Foshee *et al.*, 2008).

A survey was conducted in Georgia to determine the most troublesome weeds in pecan orchards and document common herbicide weed control practices. Weed control practices and infestations in pecan were divided between winter and summer seasons. The most troublesome pecan winter weed species were wild radish and Italian ryegrass, whereas the most troublesome summer season weeds were Palmer amaranth and bermudagrass. Other weeds included crabgrass species, bahiagrass, Florida pusley, purslane species, morningglory species, curly dock, and cutleaf evening-primrose. The most widely used POST herbicide in both the winter and summer season was glyphosate. The most commonly used year-round herbicides with soil persistence were pendimethalin, diuron, flumioxazin, halosulfuron, simazine, indaziflam, and oryzalin (Grey *et al.*, 2014).

11.7. Mulching

The tree basins are mulched with dry grass and black polythene sheets as this conserves soil moisture, check weed population and add organic matter to the soil. In an experiment, Giles pecan seedlings were either not mulched or mulched with wood chips arranged in a 1- or 2-m-wide square that was 30 cm deep. Tree height was positively related to mulch width each year of the 3-year study, and trunk diameter was positively related to mulch width during the second and third years of the experiment. Leaf P and K concentration during 2 years and leaf N during 1 year of the study were positively related to mulch width (Smith *et al.*, 2000). Foshee *et al.* (1996) showed that the common yard-waste like hardwood leaves, pine [Pinus] nuggets, pine straw, grass clippings, and chipped limbs, etc., used as mulching at depths of 10, 20 or 30 cm in a 3 x 3-m area around young pecan cv. Desirable trees effectively increased growth of young pecan trees.

11.8. Intercropping

Rice *et al.* (1993) noted that the legume cover like crimson clover/hairy vetch (*Trifolium incarnatum/Vicia villosa*) supplied up to 186 kg N/ha to the trees and increased the populations of certain beneficial insects (lacewings and ladybirds). There were also fewer aphids infesting the trees with this cover, particularly at the peak times in May and September.

Kremer and Kussman (2011) stated that intercropping alleys in agroforestry provides an income source until the tree crop produces harvestable yields. However, cultivation of annual crops decreases soil organic matter and increases soil erosion potential, especially on sloping landscapes. Perennial crops maintain a continuous soil cover, increase water infiltration, reduce soil erosion, and improve overall soil quality. They made a study to assess the effects of a perennial legume, kura clover (*Trifolium ambiguum* M. Bieb.), on soil quality in a recently established pecan (*Carya illinoinensis* Wangenh. C. Koch) orchard. The pecan-kura clover agroforestry practice was established on deep loess soils of the Missouri River hills landscape. Results illustrate that kura clover as the alley-cropped component improved soil fertility and biological activity through increased organic matter and improved soil structure, and yielded high quality forage valuable for the cattle-feeding operation. Kura clover maintained or improved soil quality, reduced soil erosion potential, and benefited pecan growth by providing a source of soil nitrogen and improving soil structure for adequate water infiltration and aeration.

11.9. Mineral nutrition

Maintaining healthy, well nourished trees is critical to producing good yields of high quality nuts every year. Investment in a good nutrition program affects this year's crop as well as the crop potential for next year. Fertilization is one of the most important practices for bearing trees. If the trees are to produce a good crop, terminal growth should be 6 inches each year. Young plants or soon after transplanting (upto one year) pecan plants are not fertilized, the will injured the roots. They must first become well established and start growing vigorously. The first application should only be made one year after planting. Some aspects of pecan nutrition should clearly be understood before taking decisions about fertilization programmes. Pecan requires nutrients for growth and fruit production. Pecan responds very slowly to fertilizers, i.e. the effects of nutrients applied in any one year will be reflected in growth and yield of the trees after two or three years. For the same reason, tissue analysis may not very helpful in depicting the nutritional condition of the tree.

The yield does not decrease even at leaf N levels lower than the established threshold levels (Malstrom *et al.*, 1983). Filling of the nuts in negatively correlated with the number of nuts on a tree. A fertilizer Programme which increases the number of nuts will adversely affect their filling and thus quality. The nutrients should not only be available for nut set but throughout the period of their development as well (Woodroof, 1979). Hence pecan nutrition should be managed for a number of years rather than only for a few years.

11.9.1. Role of nutrients and deficiency symptoms

11.9.1.1. Nitrogen

Nitrogen is the element that most commonly limits pecan growth and ultimately orchard profitability. It provides better tree growth, a higher kernel percentage, and a healthier tree. When properly maintained, nitrogen can help to provide optimal year to year production. Nitrogen deficiencies result in poor growth and poor tree health. The key to nitrogen management is to balance applications with the needs of the tree. The management of nitrogen fertilization in the pecan orchard will vary for irrigated versus non-irrigated orchards and from one year to the next within a given orchard, depending upon crop load. Nitrogen uptake in the pecan tree is driven by demand. There are several factors to consider when deciding on the best timing, material, and method to apply N fertilizer. Trees have two high-demand periods, approximately 3 weeks after budbreak when N storage reserves are depleted by the rapidly growing shoots and expanding leaves, and when fruit are beginning the kernel filling process during late summer (Smith *et al.*, 2007). The early spring foliage flush is nourished primarily from reserves held within the tree, while the nitrogen demand during the kernel fill stage is satisfied from an earlier application made during the current season. In the absence of leaf sample recommendations, dry land orchards should be fertilized with 75 lbs of N/ acre in late February or early March so that winter rains will help to improve soil moisture and ultimately nitrogen uptake (Smith, 2005). Application of liquid N via a drip or sprinkler irrigation system may increase N availability to the tree, thus reducing the amount of N required to attain the same response (Smith *et al.*, 1979). Typically, N is applied through the irrigation system in four equal applications beginning about budbreak and then at 2-week intervals (Smith, 2005).

Nitrogen deficiency symptoms in pecan have been described by many workers. Light green to yellowish green foliage, weak shoot growth and earlier fall defoliation are the main symptoms of nitrogen deficiency. Nitrogen applications also affected K, Ca, Mg, Mn, Cu and Zn concentration in the leaflet (Sparks and Baker, 1975). In a four year trial of nitrogen fertilizer application on sandy loam and clayey soils, the deficiency symptoms appeared on unfertilized sandy soil when tissue nitrogen level was lower than 2 per cent (Malstrom *et al.*, 1983).

Smith *et al.* (1985) applied 0, 56,112 and 224kg N/ha for six consecutive years to the mature trees of Western cultivar and got a rapid response. The yield was increased by 56 kg N/ha. They, however, justified the higher nitrogen rates on the basis of better tree appearance, shoot growth and production of new shoots than 56 kg nitrogen rate. Leaf nitrogen concentration was not a good indication of the total nitrogen absorption because the difference was relatively small between no nitrogen and highest nitrogen rate (2.13 % to 2.35 %). Smith *et al.* (2007) applied nitrogen (^{15}N) @ of 1.7 g/cm^2 trunk cross sectional area to evaluate the effect of nitrogen on pre-bud break (PBB) and during rapid fruit development (RFD). In PBB treatment 100 per cent of nitrogen was applied in March while in RFD 70 per cent was applied in March and 30 per cent in July. In the second year of study nitrogen was applied in the same way. At harvest in November fruit contained 41 per cent and 36 per cent of the total nitrogen absorbed during the first year from the March and July applications, respectively.

11.9.1.1.1. Nitrogen use efficiency

Due to the increasing cost of fertilizers and the negative effect of their excessive use on the environment, a more rational way of nutrient application is needed. Sanchez *et al.* (2009) evaluated the nitrogen use efficiency (NUE), productivity and quality of pecan. Three N doses were evaluated (160, 320, and 480 kg ha^{-1}). Ammonium sulfate (20.5% N) was used as N source. N was split into seven applications during the following phenological phases: (1) pre-sprouting (late February to early March, with 12.5% of total N dose); (2) beginning of fruit set (middle to end of April, 25% of total N); (3) fruit growth (early June, 12.5% total N); (4) aqueous stage (mid July, 12.5% of N); (5) milky stage (mid August, 12.5% of N); (6) fruit ripening (middle to end of September, 12.5% of N); and (7) postharvest N application (mid December, 12.5%). At harvest, average yield per tree, number of nuts kg^{-1}, kernel percentage and NUE (kg of produced fruit kg^{-1} of applied N) were evaluated. Results indicate that the maximum NUE was obtained with 160 kg ha^{-1} which yielded 16.91 kg of nuts per kg^{-1} of N applied. The optimum N dose for efficient use of nitrogen, pecan productivity and quality was the treatment with 160 kg ha^{-1}.

11.9.1.2. Phosphorous

Phosphorous is important for energy storage as well as the production of wood and nuts. Symptoms of phosphorous deficiencies are unthrifty growth and change of colour from dull green to yellowish green in severe cases. The shoot growth is short, slender and older leaflets become necrotic as the deficiency accentuates. But phosphorous deficiencies have been rarely observed under field conditions. In heavy bearing varieties, phosphorous deficiency can be expressed as a marginal leaf scorch, which may begin 7-10 days before shuck split and premature defoliation. High concentrations of phosphorous can inhibit the uptake of nitrogen, as well as iron, zinc and copper, by the pecan tree.

Sparks (1986b) compared the effect of root-applied phosphorous with foliar-applied phosphorous under greenhouse conditions and found that root application was better than foliar application in preventing phosphorous deficiency symptoms and increasing its concentration in the leaf, trunk and root. Phosphorous sprays also caused leaf scorching. Sparks (1988) obtained increased nut growth and decreased leaf scorch and premature defoliation when 2.2 kg H_3PO_4 per tree was applied in the field. Leaf phosphorous value for maximum nut growth and minimum leaf scorch were greater than 0.14 and 0.16 per cent, respectively.

Phosphorous is relatively immobile in the soil, so a single application of 60-100 lbs/ acre of P_2O_5 incorporated at planting can last for several years (Smith, 2005). Subsequent applications as needed should also be incorporated because surface applications may require several years to be of benefit to the orchard. Applications should be made any time leaf levels fall below 0.12

11.9.1.3. Potassium

Potassium is essential for the movement of carbohydrates, regulation of osmosis, and the activation of enzymes within the pecan tree. Maintaining an appropriate balance of nitrogen and potassium within the tree is of vital importance. If leaf potassium content is near minimum levels, heavy nitrogen applications will induce a visible potassium shortage termed nitrogen scorch which can lead to serious defoliation. Potassium deficiency may result from destruction of the organic layer along with increased nut production, which enhances potassium demand, especially in the K-rich shuck (Diver *et al.*, 1984).

Deficiency symptoms of this element begin with the loss of green colour from the margins towards the mid vein of the leaves which in acute cases appear as marginal chlorosis or leaf scorch (Alben *et al.*, 1942). Necrotic spots may develop on the surface of the leaf. Severe deficiencies may induce premature defoliation, shoot die-back, and small, poorly filled nuts. Plants low in potassium, even without showing any symptoms, get severely injured or killed by low winter temperatures (-7.2°C) and the yield from the damaged trees also declines (Sharpe *et al.*, 1952). Sparks (1977) found that heavy fruiting induced leaf scorch and defoliation.

Excessive Stage II fruit drop (i.e., June drop) often linked to K deficiency in pecan. Wood *et al.* (2010) found that elevating leaf and fruit K concentration through soil banding of potash over drip irrigation emitters: (1) increased fruit retention by reducing Stage II fruit drop; (2) increased in-shell nut yield; and (3) increased nut quality by increasing percentage kernel. A comparison of the K concentration of retained fruit versus abscised fruit during the Stage II fruit drop window found that retained fruit possessed endogenous K concentrations of 1.2% to 1.7% (dry weight basis) in one orchard and 1.45% to 1.9% in a second orchard, whereas aborted fruit possessed K at 0.65% to 1.2% in one orchard and 0.75% to 1.2% in a second orchard, respectively, thus establishing ~1.25% K as a "drop threshold" under conditions of the study.

The most common method of supplying potassium is by soil application of muriate of potash. Rates should be based on leaf potassium and the amount of nitrogen applied. The N/K ratio should be maintained at 2:1 for most varieties in order to prevent leaf scorch. Applying K as a banded application rather than a broadcast application has been more effective in increasing K absorption. Apply 100 lb./acre of K_2O in a band about midway between the trunk and canopy on one side of the tree (Smith *et al.*, 2012).

Pecan trees are very inefficient in absorbing soil-applied potassium taking up to 5 years. Surface-applied potassium usually increases residual soil potassium by moving up to a depth of 70 cm (Smith *et al.*, 1985; Worley *et al.*, 1974). Smith *et al.* (1985) found that potassium application at 0, 93 or 186 kg/ha significantly increased the yield of mature Western cultivar. Potassium application also increased significantly leaf potassium concentration but no relationship existed between this concentration and yield.

11.9.1.4. Calcium

Yellowing, bronzing and reduction in leaflet size were noticed under calcium deficiency. As the deficiency persists, the leaflets come closer and overlap due to shortening of rachis (Alben *et al.*, 1942). Cole and Hunter (1962) reported that leaves from calcium deficient trees were abnormally high in zinc and the soil around the trees was acidic (pH 4.7 to 4.9). Liming improved the growth of the trees and reduced zinc content of plants. Lime induces the leaching of phosphorous and potassium by increasing the soil pH which is conducive to plant growth and fixation of nitrogen by leguminous cover crops. Lime is generally surface applied to raise the soil pH and it along with NPK fertilizer increases top and root growth. On raising the pH of the subsoil from 5.1 to 6.5, tap root growth increased by 64 per cent (White, 1982).

11.9.1.5. Sulphar

Wells *et al.* (2014) reported that foliar Sulphur sprays increased pecan nut weight over the control. Foliar Sulphur application enhanced leaf chlorophyll index. Results suggested that foliar Sulphur sprays may provide pecan producers with a tool with which to maximize nut size and increase the profitability of their crop.

11.9.1.6. Magnesium

Magnesium deficiency is relatively rare in pecan; however, it can occur in trees growing on acid or sandy soils, especially in orchards with high soil potassium levels. Magnesium deficiencies are characterized by an intervienal chlorosis, which forms a "Christmas tree" pattern on the leaf. In very severe cases, a marginal leaf scorch may follow chlorosis. Magnesium deficiency has often been observed on sandy soils and is attributed to leaching (Alben *et al.*, 1942).

Magnesium deficiency is best prevented by maintaining soil pH at 6.0-6.3 and by the use of Dolomitic limestone as a liming material (Smith *et al.*, 2012). Application of magnesium sulphate at the rate of 34 kg/ha annually for five years or as single dose of 224 kg/ha increased leaf magnesium level after five year of initial application (Worley *et al.*, 1974). Dolomite raised soil pH and soil magnesium but not magnesium level in leaves.

11.9.1.7. Zinc

Zinc (Zn) has an important metabolically role in plants growth and development and is therefore called an essential trace element or a micronutrient. Zinc is uptake and transfers the form of Zn_{2+} in plants and is an essential nutrient that has particular physiological functions in all living systems, such as the maintenance of structural and functional integrity of biological membranes and facilitation of protein synthesis and gene expression, enzymes structure, energy production and Krebs cycle; also has a positive impact on crops yield; therefore crops quantitative and qualitative yield is strongly dependent on zinc (Zn) in the soil. Calcareous soils with high intake of phosphorus (P) and soils with high pH are confronted with zinc deficiency. Zinc is an active element in biochemical processes and there is chemical and biological interaction between it and some other elements such as phosphorus, iron and nitrogen in plants. Phosphorus and copper have an antagonistic impact on zinc (Mousavi *et al.,* 2013).

Zinc has a major influence on the economic return of a pecan operation due to its effect on flowering, fruit size, leaf efficiency, and nut yield (Sparks, 1987). It is particularly important to leaf expansion and shoot elongation. Zinc deficiency may be induced by either burning of the pecan forest floor or acceleration of decomposition of the humus layer by high soil temperatures from excessive tree thinning (Rosborough *et al.*, 1946). Similarly, zinc deficiency may occur from disruption of the humus-topsoil layer by cultivation of cultivation in native groves or in orchards (Sparks, 2004). Zinc is the most critical trace element as most soils are low in zinc or the zinc is not available, this element must be added every year.

The most familiar characteristic of zinc deficiency is pecan rosette, which begins as chlorosis and curling of young leaves, resulting in a wavy leaf margin. Additional symptoms may be a rosette pattern, narrow leaves, and terminal die-back. Rosetting is caused when shoots do not elongate and leaves are bunched at the terminal of the shoot, analogous to rose petals. The most common and noticeable symptoms of rosette are bronzing and mottling of leaves; early defoliation; dead twigs in tops of trees; abnormally small nuts; small yellowish, chlorotic leaves; and short, thin twigs growing on older scaffold branches with rosettes of small yellowish-green leaves at the tips. Zinc (Zn) deficiency is a typical nutritional disorder in pecan trees grown under field conditions in calcareous soils. The effects of Zn deficiency on the leaf characteristics of pecan trees include not only decreases in leaf chlorophyll and Zn concentrations, but also a reduction in the thickness of the palisade parenchyma, an increase in stomatal density and pore size and the practical disappearance of Zn leaf pools (Ojeda-Barrios *et al.*, 2012).

Sparks and Payne (1982) tested the reliability of the satisfactory range of 50-100 ppm zinc in the leaflets and found the occurrence of the deficiency symptoms only when zinc levels in the leaflets had fallen below 40 ppm. Zinc concentration in leaf changed from 60 μg/g in the beginning of the season to 120 μg/g in the middle of the season to 100 μg/g at the end of the season (Diver *et al.*, 1984). Shell, kernel and shuck in normal fruits have respective zinc concentrations of 9.2, 46.0 and 71.0 μg/g (Sparks, 1975). The zinc concentration in the fruit varied from 20 to 42 μg/g during the season, on a dry weight basis, with the highest concentration occurring at the beginning of fruit growth (Singanuson *et al.*, 2003).

Zinc deficiency has been mostly corrected either by applying zinc to the soil or its spray on the foliage. It is commonly applied as zinc sulphate which gives quicker response than other compounds. Since zinc is mobile, it is applied directly to the soil surface singly or combined with regular fertilizer programme. Soil application of zinc is helpful only on acidic soils and a higher dose is required on heavier than on light soils. Payne and Sparks (1982) compared the broadcast and band application methods and found more rapid tree response from the former than the latter.

Apply 2 lb/100 gal $ZnSO_4$ (36% Zn) in water to runoff beginning as the first leaf unfurls during budbreak and then at 2-week intervals until shoot growth ceases (Smith, 2005). Foliar zinc should be applied anytime particularly when leaf concentrations fall below 50 ppm or when visible symptoms of zinc deficiency are present. When Zn is root absorbed, leaf Zn concentrations are more uniform, thus 40 µg/g is adequate (Sparks, 1982). However, soil-applied zinc is relatively ineffective in supplying the Zn requirement of pecan trees grown in the high pH, calcareous soils (Smith *et al.*, 1980).

11.9.1.8. Manganese

Manganese is essential to the light reactions of photosynthesis. A low leaf Mn concentration was detected in bearing pecan (*Carya illinoinensis*) trees growing in an alluvial soil with an alkaline pH. Trees lacked vigour and leaves were pale, but there was no discernible leaf chlorosis or necrosis. Southwestern U.S. pecan (*Carya illinoinensis* (Wangenh.) K. Koch] orchard soils are typically alkaline and calcareous, making micronutrients such as manganese (Mn) poorly available for root uptake.

Manganese deficiency has been characterized as 'mouse ear' or 'little leaf' which usually occurs at high soil pH in coastal areas and dooryard plantings (Gammon and Sharpe, 1956). The shortening of mid-vein of the leaflets makes them round wrinkled, turning into a cup (Shear and Faust, 1980; Smith *et al.*, 2001).

The deficiency usually occurs when the leaf manganese level falls below 100 ppm in comparison with the normal range of over 300 ppm. To correct this condition, manganese sulphate (1 to 7 kg/plant) can be applied to the soil by evenly broadcasting under the tree canopy. The quantity of these compounds will vary with tree size and severity of symptoms (Woodroof, 1979). Manganese leaf concentrations in high yielding pecan orchards ranges from 85 to 294 µg/g (Pond *et al.*, 2006). Medina (2004) reported that manganese leaf level for high yielding pecans in North-Central Mexico is 108 µg/g. It was found that foliar application of Mn at 0.68 g Mn/L was the best. Optimum concentration of foliar Mn in immature pecan trees was estimated to be 151.64 $\mu g.g^{-1}$ (Sherman *et al.*, 2017).

Three foliar applications of MnSO4 beginning at budbreak, then twice at 3-week intervals at 0 to 3.3 kg/ha increased leaf Mn concentration curvilinear and alleviated leaf symptoms. Results indicated that three foliar applications of MnSO4 at 2.15 kg/ha plus a surfactant were adequate to correct the deficiency (Smith *et al.*, 2001).

11.9.1.9. Boron

In case of boron deficiency the leaves become small and uniformly chlorotic and the shoots short slender and die eventually (Alben *et al.*, 1942). The symptoms of boron deficiency resemble to the rosette of pecan. Generally, boron deficiency is not observed in the orchard. The normal leaf boron content ranges from 10 to 90 ppm. Sharpe and Gammon (1958) suggested application of 10 to 15 kg borax/ha in the soil along with the fertilizer to encounter the deficiency of boron.

11.9.1.10. Iron

As a component of chlorophyll, iron is essential to the process of photosynthesis. Iron deficiency can be induced by cool, wet environmental conditions in the spring that inhibit Fe translocation from the root system to the aerial portions of the tree (Brown *et al.*, 1959). Symptoms of iron deficiency look similar to nitrogen deficiency, exhibiting chlorosis of the leaf. The terminal young leaves are the first to be affected and show interveinal chlorosis with pronounced green veination. Pecan's hypersensitivity to water logging near budbreak is reflected in the widespread occurrence of iron deficiency in young trees following excessive irrigation of calcareous soils (Sparks, 2000). Depending upon the cause of iron deficiency, correction may take varying routes, including foliar applications of iron, changing the amount of lime applied to the orchard, or foregoing phosphorous applications.

11.9.2. Soil/leaf analysis

Adequate, inadequate or excessive amounts of nutrients could be present in the plant. It is therefore possible to determine the nutrient status of an orchard and to fertilize accordingly. The single most reliable indicator to the nutritional needs of the pecan tree is the soil and leaf analysis. When trying to represent nutrient availability with soil samples is that fertilizers may not be distributed equally over the entire orchard (Smith *et al.*, 2012). But soil sampling may be quite useful for monitoring soil pH and lime requirements. However, before transplanting (a month before) take a soil test and, based on its results, adjust soil fertility. Adjust the pH to 6.5 and bring phosphorus and potassium levels into the high soil test range. If the soil pH has not been determined within the past five years, a soil test may be useful.

The most reliable indicator of pecan tree fertility requirements is leaf analysis. Leaf analysis determines the concentrations of specific elements in the plant. Leaf analysis may be used to diagnose or confirm a particular nutrient problem in an orchard after symptoms are present. More importantly, leaf analysis determines mineral condition of plants is supported by the relationship between nutrient concentration and relative yield, growth or quality (Jones *et al.*, 1991). Frequently, it reveals that certain fertilizers being used are not needed, resulting in a more economical fertilizer program. Leaves for analysis must only be taken from healthy trees and must be free of sunburn, deficiency symptoms, insect damage and disease. Leaves should be sampled for analysis during 15th July to 15th August as this is the period in which nutrient concentrations in the leaves are most stable. Yuan *et al.* (2014) concluded that the appropriate period for analyzing the standards of nutrition diagnosis in pecan leaves is from July 27th to August 13th. Nutrient Sufficiency levels are given in Table 2.

Table 2: Sufficiency levels for essential elements.

Element	Sufficiency range (%)	Element	Sufficiency range (ppm)
N	2.30-3.00	Zn	60-150
P	0.14-0.30	Fe	50-300
K	1.00-2.50	Mn	100-2000
Mg	0.30-0.60	B	15-50
Ca	0.70-1.75	Cu	6-30
S	0.20-0.35	Ni	>2.5

Smith *et al.*, 2012

Nitrogen concentration in the leaf is affected by leaf age, leaf position, genotype and presence of fruit on the shoot. In 15 year old 'Stuart' pecans nitrogen concentration from full bloom to leaf abscission changed from 1.39 to 1.42 per cent, whereas nitrogen content went from 100 to 170 mg/leaf (Kim and Wetzstein, 2005). In Arizona, leaf nitrogen concentrations decreased relatively constantly @ 0.003 per cent per day, from flowering to fruit maturity, overall change was from 2.97 to 1.80 per cent (Walworth *et al.*, 2008). In New Mexico total leaf nitrogen percentage decreased from early May (3.45 %) to the end of June, increased and remain steady from July 23 to August 6 (2.50%), then decreased rapidly until fruit maturation i.e. 15 October (1.75 %) and in November 15, close to defoliation leaf nitrogen dropped to 1.20 per cent (Kraimer *et al.*, 2001). Smith (2010) stated that minimum pecan leaf standards in July for fertility recommendations of 1.0% K and 0.14% P. Pecan [*Carya illinoinensis* (Wangenh.) K. Koch] trees developed necrotic areas in the leaves late in the growing season if there is low value of P and K.

11.9.3. Fertilizer recommendations and their method of application

Pecans require nutrients for growth and fruit production. Substantial nutrient recycling in pecan is expected based on several factors including; production of prodigious leaves creating abundant leaf litter (Acuna-Malconado *et al.*, 2003); formation of organic matter from foliage of deciduous trees, typically nutrient rich in N, S, B, Cu, Zn, and bases (Tisdale *et al.*, 1985); minor nutrient loss from nut predation due to the nut's low mineral content and the relatively low nut production per tree, and; return of nutrient rich shuck to the soil (Diver *et al.*, 1984). Application of fertilizers resulted in supply of nutrients to soils and helps to correct nutrient deficiencies. Nitrogen is essential for good pecan growth. The first application should be at bud break in April with additional applications in May and June. When extremely heavy crops are set, nitrogen should also be applied in July and August. As a rule, 10 pounds of actual nitrogen are needed for 100 pounds of pecans set per acre in June. The number of nitrogen applications should be consistent with crop size. Young trees should receive very small but frequent applications of nitrogen in response to growth. When fast growth occurs, add additional applications until early June. Smith *et al.*, (2004) studied the effect of nitrogen application (75 or 159 kg/ha) on adult Mohawk trees over a six year period and reported that nitrogen concentrations were not affected or else responses were inconsistent. Trunk diameter of transplanted adult trees was increased in fertilized treatments (1.38 to 4.14 kg/tree) relative to trees receiving no nitrogen (Herrera and Duarte, 2000), but no change was found in nitrogen concentration of leaves with rates ranging from 0 to 4.14 kg/plant. Yuan *et al.* (2014) concluded that N, P, K and Ca, Mg, respectively as primary and secondary base fertilizer should be added in the pollination flowering period and the flushing period, while organic fertilizer should be deep fertilized before the harvest of autumn fruit.

Trace elements are very important in nut production. Zinc is the most critical trace element as most soils are low in zinc or the zinc is not available, this element must be added every year. When zinc is lacking or present in insufficient quantities, nutlets fail to set and yields are very low. Zinc is best applied as a foliar spray just after pollination at a rate of 1 to 2 lb of zinc sulfate/100 gallon of water for young trees and 5 to 10 lb/100 gallon for mature trees. Apply two more times at two to three week intervals. For soil application, zinc chelate is recommended over zinc sulfate. Zinc chelate 10 per cent can be applied with liquid fertilizers as a broadcast or banded treatment. Use at a maximum rate of 1 part zinc chelate (10 %) to 50 parts fertilizer in ammonium phosphate solutions or 1 part zinc chelate (10 %) to 10 parts fertilizer in nitrogen solutions.

11.10. Use of plant growth regulators

Growth regulators are required to shorten the internodes or to induce more lateral growth. Use of growth regulators to control pecan tree size had variable effects on nut quality. Alar (daminozide) sprayed @ 4000 and 5000 ppm increased yield, bud forcing, precociousness, growth density, green color intensity, shuck disease, percentage kernel, nuts/kg and reduced shoot growth (Storey *et al.*, 1970, Hooks and Storey, 1971), trunk growth and delayed maturity (Storey *et al.*, 1970).

The use of either Atrinal (dikegulac) or Alar did not reduce terminal growth or induce earlier nut production on old dehorned trees (Arnold *et al.* 1981). Dikegulac has induced lateral shoot development (Malstrom and McMeans, 1977a, Worley, 1980). Foliar applications of dikegulac made three weeks after bud break retarded shoot growth (Malstrom and McMeans, 1977a). Paclobutrazol has reduced internode growth and induced lateral buds to break, but this material has been discontinued by the manufacturer. Gibberellic acid (GA_3) and NAA inhibited the pistillate flowering of the subsequent season, whereas prohexadione-calcium, ethephon and BA + TIBA promoted floral initiation when topically applied to canopies before the kernel filling stage of seed development. These bioregulators exhibit potential for integration into a bioregulator-based strategy to mitigate pecan alternate bearing by selective and timely use in 'off' or 'on' cycle years, depending on the bioregulator (Wood, 2011).

Cottrell *et al.* (2010) sprayed three plant growth regulators viz. forchlorfenuron, gibberellic acid and aviglycine to control the chlorotic injury to pecan foliage by Black pecan aphid. Treatments were applied to orchard foliage and observed that the Foliage pretreated with forchlorfenuron + gibberellic acid resulted in significantly less aphid-elicited chlorosis. Foliar application of ReTain was applied on 'Desirable', 'Oconee' and 'Success' for improving pecan nut retention. The plants were sprayed at 7 and 21 days after the cessation of pistillate receptivity and reports of two year study provides positive results for nut yield, nut size, and nut quality (Graham, 2015).

Four plant growth regulators (Biozyme TF (BZ) @ 1 L /ha, Biofrut® (BF) @ 300 g /ha, thidiazuron (TDZ) @ 10 g/ha and 5 g/ha in 1st and 2nd application, respectively and paclobutrazol (PBZ) @ 200 g/ha) were applied the last week of March and the first week of April during four consecutive years in order to evaluate their effect on nut yield and quality (Nunez, 2016). First three growth regulators were sprayed with 1800 L/ha of water whereas PBZ was soil applied. Marketable nut yield was affected significantly among the treatments as compared to control. TDZ treatment decreased alternate bearing intensity from 0.27 showed in the control treatment, to 0.05. These results suggested that application of plant growth regulators could help to manipulate yield and alternate bearing in warm regions where chilling is not enough to induce and break dormancy period (Nunez, 2016).

Pecan buds passed through 11 growth stages and the 2 types of flowers (male and female) did not mature at the same time in both seasons (dichogamy). To synchronize the time of full bloom of male and female flowers in different pecan cultivars, hydrogen cyanamide (Dormex, 3%), potassium nitrate (5%), urea (10%) and water (control) were applied to one-year-old shoots of different cultivars at 4 weeks before normal bud break (1 February). Trees treated with Dormex or potassium nitrate resulted in the lowest percentage of dormant and opened vegetative buds, followed by urea, whereas the control trees had the highest percentage. Potassium nitrate and hydrogen cyanamide resulted in the highest percentage of female flowers; both treatments induced the earlier opening of female flowers (by 2 weeks) compared to the control. Potassium nitrate was more effective in stimulating the formation of male flowers, followed by hydrogen cyanamide and urea. The application of Dormex or potassium nitrate could synchronize the time of full bloom of male and female flowers for each cultivar in both seasons. Dormex and potassium nitrate were highly effective in increasing the initial and final fruit set percentages, whereas urea was the least effective with regard to both parameters (Fayek *et al.,* 2008).

Insufficient fruit retention limits profitability of certain pecan [*Carya illinoinensis* (Wangenh.) K. Koch] cultivars. Wood *et al.* (2009) examined efficacy of aminoethoxyvinylglycine (formulated as ReTainReg.; Valent BioSciences, Libertyville, IL), a natural ethylene inhibitor, for increasing crop-load through increased fruit retention in pecan trees grown at three distinct locations within the U.S. pecan belt. Several years of field studies found that timely postpollination ReTainReg. sprays [132 mg L^{-1} a.i. (11.7 oz./acre)] to canopies could increase fruit retention of 'Desirable' and increase crop yield by 16% to 38% in trees carrying a "moderate to heavy" crop. ReTainReg. did not detectably increase fruit retention on trees carrying a "light" crop-load. The ReTainReg.-associated increase in yield of "heavy" crop-load trees did not necessarily decrease subsequent year yield. ReTainReg. appears to offer commercial potential as a crop-load management tool for 'Desirable' through regulation of Stage II drop (i.e., June-drop), but may not be efficacious for all cultivars.

Sharkawy and Osman (2009) concluded that pecan fruits can be harvested earlier (about two weeks) when the trees were sprayed with Dormex (2%) at the end of January and followed by Ethrel spraying at 750 ppm on first of September without any effect on pecan fruit quality. Results showed that the treatments increased bud opening percentage and advanced flowering and harvesting time as compared with control. Dormex at 2% treatment was the most effective. Percentage of pre-harvest nut germination decreased. Fruit set, number of fruits/tree, yield (kg/tree), fruit physical characteristics expressed as fruit weight, kernel %, fruit dimensions and oil content were improved by all treatments as compared with control.

Mitigation of alternate bearing (AB) through regulation of floral initiation of pistillate flowers is central to improving crop load management of pecan [*Carya illinoinensis* (Wangenh.) K. Koch] trees and orchards. Wood (2011) examined the influence of key bioregulators {i.e., an auxin [as B-napththaleneacetic acid

(NAA)], a cytokinin [6-benylamino purine (6-BA)], an ethylene generator (ethephon), and an auxin transport inhibitor [2,3,5-triiodobenzoic acid (TIBA)]} on subsequent season pistillate flowering. Gibberellic acid (i.e., GA3) and NAA inhibited, whereas prohexadione-calcium (P-Ca; calcium 3-oxido-5-oxo-4-propionylcyclohex-3-enecarboxylate), ethephon, and BA+TIBA promoted floral initiation when topically applied to canopies before the kernel filling stage of seed development. These bioregulators exhibit potential for integration into a bioregulator-based strategy to mitigate pecan AB by selective and timely use in "off" or "on" cycle years, depending on the bioregulator. Field studies provided evidence that a "cytokinin-gibberellin balance," with partial modulation by auxin and ethylene, acts in the endogenous primordial environment of floral meristems as a "second-level signal" regulating a key step in a three-step process for initiation of pistillate flowers in pecan.

Soil application of paclobutrazol at 1.25 g/m^{-2} inhibited pecan branch growth and stimulated short bearing branches, which promoted fruit yield (Luo *et al.*, 2016).

11.11. Physiological disorder and their management

11.11.1. Alternate bearing

Alternate bearing occurs in almost all tree fruit crops, but is especially severe in pecan. This severity is linked to three traits inherent to pecan: i) the time of fruit maturity, ii) the nature of fruit growth and iii) the chemical composition of the kernel (Sparks, 1974; Conner and Worley, 2000). In contrast with many fruit crops, pecan fruit matures late in the growing season leaving little time prior to leaf fall for carbohydrate storage to support the next season's developing flowers and fruit. In addition, most of the dry matter of the nut accumulates at the end of the season. Finally, kernels contain about 70 per cent lipids, which require considerable energy to produce. These traits, when combined with a heavy fruit set, can leave trees depleted of carbohydrate reserves, especially in the roots, at the end of the season and contribute to poor fruit set the following year (Barnett and Mielke, 1981).

In an experiment Reid and Smith (1993) showed that thinning pecans can reduces the alternate bearing. In a comparison of fruit thinning dates, it was shown that removal before the dough stage of nut enlargement (June- late July) stimulated flowering in the following spring. A trunk shaker (Nuthustler model 2138) was used successfully to remove a portion of the crop in late July, detaching single fruits rather than entire clusters. In late July, trees of Mohawk in Oklahoma and Giles in Kansas, which had set heavy crops, were shaker-thinned to achieve a wide range of retained loads (10-90% of terminals bearing fruits in Mohawk and 65-95% in Giles). Thinning greatly influenced Mohawk nut quality at harvest, the ideal balance between % kernel and yield being achieved with about 50% of fruiting shoots. In contrast, nut quality appeared to be unaffected in Giles over the range of thinning. Return cropping in Giles trees was greatly enhanced by fruit thinning. Rohla *et al.* (2007) also indicated that fruit thinning of overloaded trees improved return bloom. Removal of entire fruit clusters appears as effective as thinning fruit within a cluster to maintain adequate nut quality and promote return bloom.

Khalil *et al.* (2015) concluded that Alternate bearing is a serious problem in pecan *Carya illinoensis* (Wangenh.) C. Koch] production which result in excessive fruits of low quality in one year, usually followed by very low fruit next year and the growers end up with low economic return in both years. Application of foliar ethylene has been used successfully as an alternative method for fruit thinning. Five ethephon treatments (0, 200, 400, 600 ppm, water spray) were applied to 27 year old pecan cultivar Wichita at Leyendecker Plant Science Center, College of Agricultural, Consumer and Environmental Sciences, New Mexico State University, Las Cruces, USA. Foliar ethephon was applied as a source of ethylene. Foliar ethephon significantly increased leaf chlorosis, leaf defoliation and kernel quality and reduced total kernel weight, but showed non significant effect on number of nuts at harvest. It was concluded that 600 ppm foliar ethephon could be used to obtain good quality nuts during year of excessive fruit load and may increase fruit set in the following year of low crop load enabling the growers to get maximum profit on regular basis from pecan orchards.

11.11.2. Fruit drop

Wood (1993) explained the factors responsible for fruit-drop of pecan. Pecans possess a high degree of dichogamy. This is generally divided into protandrous and protogynous types but many cultivars show major overlaps in flowering, resulting in the potential for considerable selfing and a subsequent decline in nut quality and increased levels of fruit abortion. Three distinct phases of fruit drop occur. The first takes place during and shortly after flowering, the second 2-4 weeks after pollination, and the third from about mid-July to late October. Poor pollination is the primary reason for fruit drop, weather conditions often being responsible. Severe Zn deficiency, pesticides, pests and diseases may also be causal factors.

11.11.3. Water-stage fruit-split

Water-stage fruit-split (WSFS) is a relatively common and often major problem of certain pecan [*Carya illinoinensis* (Wangenh.) K. Koch] cultivars. Wells and Wood (2008) demonstrated that insufficient availability of B or Ni to developing ovary tissues potentially predisposes developing fruit to WSFS when environmental triggers occur. They also advised for repeated foliar spraying of nickel (Ni) and boron (B) single or both to reduce crop loss resulting from WSFS in such orchard situations.

11.11.4. Mouse-ear or little-leaf disorder

Wood *et al.* (2006) reported that the mouse-ear or little-leaf disorder and the associated replant disease of pecan [*Carya illinoinensis* (Wangenh.) K. Koch] is a Ni deficiency and is cured by timely foliar application of Ni (at 100 mg L^{-1}); thus documenting the first known example of Ni deficiency in orchard crops. Deficiency is inducible on soils with adequate Ni content by (a) excessively high soil Zn, Cu, Mn, Fe, Ca, or Mg; (b) root damage by root-knot nematodes; or (c) dry and or cool soils at time of bud break. Symptoms associated with Ni deficiency, but varying with severity of deficiency, include early-season leaf chlorosis, dwarfing of foliage, blunting of leaf/leaflet tips, necrosis of leaf or leaflet tips, curled leaf/leaflet margins, dwarfed internodes, distorted bud shape, brittle shoots, cold-injury-like death of over-wintering shoots, diminished root system with dead fibrous roots, failure of foliar lamina to develop, rosetting and loss of apical dominance, dwarfed trees, and tree death. Observations implicate excessive soil accumulation of light metals from long-term fertilizer usage as a primary cause of Ni deficiency. These results raise the possibility that Ni is a contributing factor in certain recalcitrant maladies and replant disorders of certain woody perennial crops and support the role of Ni as an essential nutrient element for higher plants.

11.11.5. Vivipary in pecan.

Crop loss to vivipary (i.e., germination of nuts on the tree) is potentially a major profit-limiting problem affecting yield and quality of nutmeats from certain pecan [*Carya illinoinensis* (Wangenh.) K. Koch] cultivars at certain locations. Improved orchard management tools and strategies are needed to avert major economic loss. Field research on 'Cheyenne' pecan assessed the possibility of controlling vivipary via supplemental irrigation and nitrogen (N) management. The incidence of vivipary on irrigated trees was 25-fold greater than that on non-irrigated trees, and incidence on trees receiving excessive N fertilization was 21-fold greater than that of trees not receiving N. There was a strong 'Irrigation × N' interaction, with trees receiving 'Irrigation+N' exhibiting 41-fold more vivipary than did 'Non-irrigated-N' trees. The incidence of vivipary in orchards therefore appears amenable to management via an appropriate irrigation and N fertilization strategy. Treatment of developing 'Oconee' fruit with (±)ABA [(2-cis, 4-trans)-5-(1-hydroxy-2,6,6-trimethyl-4-oxo-2-cyclohexen-1-yl)-(3-methyl-2,4-pentadienoic acid)] or fluridone [(1-methyl-3-phenyl)-5-3-(trifluoromethyl) (phenyl-4(1H)-pyridinone)], an inhibitor of ABA biosynthesis, reduced and increased the incidence of vivipary, respectively. Additionally, treatment of developing 'Sumner' fruit with fluridone reduces the germination time of harvested pecan nuts. This indicates that the influence of supplemental irrigation and N on incidence of vivipary is likely via alteration in ABA metabolism within developing fruit (Wood *et al.*, 2015).

11.11.6. Shuck die back

This disorder is due to the hormonal imbalance. This condition is commonly associated with Success and its hybrids. Nuts affected by this disorder drop from one to two weeks early. They do not fill properly due to the peduncle being girdled and restricts movement of nutrients into the pecan. The shuck turns black and opens at the tip in a normal manner. Still no control is recommended.

11.12. Diseases and their management

11.12.1. Scab

It is caused by fungus *Cladosporium caryigenum.* Scab is the most serious disease of pecans throughout most of its growing range. The cultivars differ in their susceptibility to the disease and northern cultivars are more tolerant than southern cultivars of pecan. Early symptoms are the appearance of numerous small, irregular, olive-green to black spots on leaves or young shoots and small dark brown to black circular lesions on the nuts (Goff *et al.*, 1993). These small spots may fuse to form large blackened areas. Immature leaves and severely affected nuts fall off or fail to fully develop. On nuts, the scab lesions are small black dots which become sunken as they mature and have misshapen nuts, which have no commercial value. If it invades the nut kernel, it can cause off flavors and pink rot. The shells of infected nuts may have an oily appearance. In the spring when temperature and moisture conditions become favourable, the fungus becomes active and within a few days spores are formed at the surface (Goff *et al.*, 2003). They are spread by wind and rain to developing leaves and nutlets. New spring growth on the trees is infected when the leaf surfaces are wet, especially after rain. Susceptibility for the disease varies in different cultivars. Ukulinga, Shoshoni, Moore and Barton are regarded as highly tolerant, while Mohawk, Wichita and Chocktaw are susceptible.

Pecan scab is readily controlled with properly applied sprays of effective recommended fungicides, in conjunction with strict adherence to recommended sanitary practices, or by use of resistant cultivars. Protective measures such as burning of all diseased leaves, twigs and shucks are important. Keep drainage channels open so that water moves quickly out of orchards. Improve air circulation by removing low-hanging limbs and crowded trees. Fungicides are an important part of the pecan disease management program. Sprays of Bordeaux mixture or zineb during the middle of April and continued at three week intervals are much effective (Goff *et al.*, 2003). Plant only tolerant or resistant cultivars such as Chickasaw, Choctaw, Gormley, Graking, Mohawk, Mount, Oakla, Stuart, and Peruque.

11.12.2. Powdery mildew

Causal fungus of the powdery mildew is *Microsphaera alni*. Powdery mildew occurs sporadically and does not cause serious damage to most pecan varieties. It is occasionally a problem on highly susceptible seedling trees and can be identified by the white powdery growth on the surface of nuts and young shoots (KenKnight, 1968a). The fungus most often develops between mid-July and mid-August. Due to rainfall washing the spores from the nut or leaves, the fungus seldom is a problem during months of frequent rain. Early-infected nuts may remain small and have an early shuck split and shriveled kernels. Infected leaves are somewhat distorted in shape and are covered with a faint gray colored dusty material. Infected nuts or leaves are not significantly damaged by the fungus. Its feeding is limited to the outer cells of the shuck. This does not restrict the movement of nutrients into the developing kernels. The fungus is sometimes associated with nursery trees or sucker growth around the base of mature trees. Scab sprays usually provide adequate control of powdery mildew.

11.12.3. Brown Leaf Spot

It is caused by *Cercospora fusca* which affects only mature leaves and does not appear until the latter part of May or mid-June (KenKnight, 1968a). The disease is limited to humid regions and is of minor significance. Primary lesions develop on the lower leaf surface as small dots which gradually enlarge and become reddish brown with a gray cast. Later, these lesions attain circular or irregular shape and in favourable season pecan trees may be completely defoliated if the fungus is not controlled. The cultivar Stuart is comparatively more susceptible than others (Woodroof, 1979). Repeated sprays of Bordeaux mixture or zineb or dodine are quite effective in preventing brown leaf spot.

11.12.4. Vein Spot

Vein spot is a common disease of pecan leaves which is caused by *Gnomonia nerviseda* fungus (KenKnight, 1968a). It is a foliage disease and is more serious on several cultivars in some regions. Foliage symptoms of this disease are similar to the lesions formed by the scab fungus. The fungus infects the vascular tissue at the junction of petiole to the rachis and base of rachis causing premature leaf fall (Sanderlin *et al.*, 1983). Vein spot lesions, however, tend to be linear rather than round and also tend to be restricted to veins. Lesions develop on veins of leaflets and are characteristically dark brown to black. Severely infected leaflets shed soon after symptoms are observed. The fungus overwinters in fallen leaves beneath tree. Fungicides and sanitary procedures will control this fungus. It is easily controlled by most scab control fungicides.

11.12.5. Leaf blotch

It is caused by the fungus *Mycospharella dendroide.* The disease is only serious on neglected, declining trees and nursery plants and the latter are usually the source of infection (KenKnight, 1968a). However, low vigor trees and nursery plants are also susceptible. It overwinters in infected leaves on the orchard floor. The symptoms include the appearance of olive-green, velvety tufts on the under surface in early summer and yellow spots on the upper surface of leaflets in late season. The black pimple like structures of perfect stage gives a black shiny blotched look to the leaflets in the mid summer. The leaves may fall on severe infection. Premature shedding of leaves adds to the reduced plant vigor. Fungicides applied following the pecan scab spray will also control this fungus.

11.12.6. Crown gall

The crown gall is a bacterial disease caused by *Agrobacterium tumefaciens* and which invades many plants and is worldwide in distribution. It is characterized by round to irregular swollen tumors or galls, usually found at or near the soil line on the trunk or roots. Infected trees show a lack of vigor, foliage lacks normal green color, and occasionally the tree may die. To prevent crown gall, do not plant infected trees or plant trees in an area known to be infested, as this practice can spread the causal bacteria. It is important not to locate a nursery in an area where crown gall has been observed previously. The injury to plants by cultivation should be avoided. As a preventative control measure, it would be advisable to treat the seed with Galltrol-A before planting.

11.12.7. Bunch disease

Although the cause of the bunch disease is not completely understood, evidence indicates it is a *mycoplasma* or it may be *phytoplasma.* It is caused by virus which is supposed to be transmitted by sucking insects (KenKnight, 1968a). The characteristic symptom of bunch disease is bushy growths of slender willowy shoots, resulting from an abnormal forcing of lateral buds. Normally only the most distant bud will break dormancy and develop into a twig and finally a limb. A large number of buds on bunch infected trees will break dormancy at one time. This results in a dense growth of small twigs. Bunch disease is very conspicuous in the spring and early summer because the diseased shoots leaf out earlier than non-infected shoots. It is believed that the pathogen is insect transmitted, probably by leafhoppers. Nut production is very limited and size is restricted on the infected trees. Severely-infected trees should be destroyed. If only a few branches are infected, they should be removed much beyond the infected area. Highly susceptible cultivars should not be grown in the bunch disease prone regions. The cultivar Stuart is resistant to this virus.

11.12.8. Fungal leaf scorch

It is caused by fungus. Infected leaves turn brown along the leaf margin or tip. The necrotic areas may extend from only a small area at the leaf margin to completely encircle the margin of the leaf (Hunter *et al.*, 1986). As the lesion matures it becomes gray. Infection can result in leaf shedding from mid-July to mid August. Fungicide applications applied for scab are effective when applied in early to mid September. The varieties Western, Cheyenne, Chickasaw, Shoshoni, Stuart and Cape Fear are some of the more susceptible varieties to this group of fungi.

11.12.9. Pink mold

It is caused by the fungus *Trichothecium roseum* which occurs on scab infected nuts. Infected nuts develop a white mold growth in old scab lesions. In areas not exposed to air and sunlight, the white area become pink in color. The fungus may invade the kernel. The kernel becomes oily in appearance and leaks onto nearby nuts. The kernel is rancid and not eatable. In areas where scab disease control is regularly practiced, pink mold is not a problem. If pink mold is a severe problem, it may be necessary to remove the oily nuts or to separate the problem load from disease free nuts.

11.12.10. Twig Dieback

A twig dieback of pecan trees is associated with a fungus namely *Botryosphaeria berengeriana*. Some limbs may die back as much as 24-30 inches. It is most often found on limbs that are under stress. The stress may be caused by over-production, defoliation by scab and other foliar pathogens, drought or excessive shading. Twig death is usually general over a tree except under shady conditions, where it will occur only in the shaded areas. To control dieback, prune out dead wood and destroy it. Determine the cause of stress and correct it.

11.12.11. Kernel rot

It is caused by the fungus of *Phytophthora* spp. It is most often associated with very wet conditions in the later stages and of nut development. The shucks turn black. The kernel becomes discoloured and is not eatable. To control the kernel rot, copper based fungicides are applied from the water stage to shuck split on a 10 day schedule.

11.12.12. Stem canker

Initial symptoms appeared as small, elliptically-shaped, protruding lesions on the trunk bark and branches, progressing to cankers ranging from 6 to 60 cm and eventually leading to plant death. Necrosis and dark streaks were observed in longitudinal sections of the trunk, near the vascular tissue. The causal agent was identified as *Lasiodiplodia subglobosa* (Poletto *et al.*, 2016).

11.12.13. Black spot

A new threat to pecan production was observed in June 2011 when black spots were found on fruits that adversely affected the yield and quality in many pecan production areas in Jiande and Xinchang counties in Zhejiang, China. Initially, the symptoms were small, black spots on young fruits, with sunken, oval to irregular lesions later gradually developing. On heavily infected fruits, lesions coalesced to form large necrotic regions. The causal pathogen was identified as *Pestalotiopsis microspora* (Shi *et al.*, 2015).

11.12.14. Stem end blight

This disorder is associated with a fungus which attacks the nuts in the latter part of July and August. Shucks turn black rapidly and drop soon after. Lesions are black, sunken and shiny. When infected nuts are cut open, the liquid in the kernel has turned brown. This can be controlled with foliar sprays of a fungicide applied when the pecans begin to fill.

11.13. Insect pests and their management

11.13.1. Aphids

Many aphids cause injury to pecans. Black pecan aphids (*Myzocallis caryaefolia*) feed on both sides of leaves and appearance of large yellow angular blotches is the first sign of its injury. The blotches later turn brown and further damage causes premature leaf fall adversely affecting the current as well as subsequent crops (Wood and Reilly, 1998). Tedders and Wood (1985) estimated the damage of mixed aphid populations (*Monellia caryella, M. pecansis* and *M. nigropuncta*) on mature Stuart pecan trees. Early season feeding by aphids had a major adverse effect on tree productivity, growth and energy reserves. Nut volume and weight, shoot starch and reducing sugars were reduced during early fruit development. Likewise, nut volume and weight, kernel weight and lipid content, shoot starch and reducing sugars and nut yield were reduced at fruit maturity.

Wash off leaves with a high pressure hose sprayer. Wash as high as possible in tree. Repeat at least twice per week. A single application of aldicarb (28 kg/ha) by placement near the drip emitters during the spring resulted in rapid control of aphids (Alverson, 1985). No aphid was found six weeks after application and the efficacy was maintained for the rest of the season. Aldicarb also controlled mites for the entire season (Dutcher and Harrison, 1984). Pesticide residues in the nuts were very low (Villiers and Viljohn, 1987).

11.13.2. Stem borer (*Xylosandrus germanus)*

The stem borer is sporadically observed on pecan-nut trees. The first sign of infestation is red-brown granular excretions around the base of the trunk. This discharge comes from the pink coloured larvae which have burrowed into the trunk and branches of the trees. The tunnels vary in size according to the age of the larvae and can be as much as pencil thickness. Numerous tunnels occur in a single tree trunk. Young larvae hatch from the middle of December until the end of February. The larvae reach their maximum size of approximately 40 mm during spring and early summer. At this stage the larvae become inactive and change into pupae in the tunnels. The pupal stage lasts approximately six weeks and it appears that only one generation occurs per year. The larvae remain in the trunk for about eleven months.

Remove larval excreta around the stem just after spraying. In young trees a piece of soft wire can be used to kill the larvae in the tunnels. This method, although primitive, is very effective and must be undertaken during winter when the tunnels and the excreta are more noticeable around the stem of the tree.

11.13.3. Mites (*Tetranychus hicoriae*)

A heavy infestation of this mite may cause serious injury to foliage. They feed mainly on the lower side of leaflets along the midrib. Initially, a brown discolouration appears at the midrib which subsequently covers the entire leaflets (Osborn *et al*., 1966). The insecticides applied for the control of aphids will also check the mites. Ball (1982) found sulphur and cyhexatin are the most effective acaricides.

11.13.4. Hickory Shuckworm (*Laspeyresia caryana*)

This is a serious pest of pecan and other hickories which lays eggs on foliage and green nuts. Upon hatching, the cream coloured small caterpillars attack the immature nuts which prematurely fall to the ground. When the shell becomes hard, the caterpillars tunnel in the shuck hampering the kernel development. Overwintering of fully grows caterpillars occur in shucks on the ground. All the shucks after harvest should be burnt after collection. It is difficult to control the pest by insecticidal sprays because of overlapping generations and re-infestation from adjoining unsprayed trees. Remove and clean up all old shucks and debris. Also, the use of a blacklight insect trap placed in the trees can help. First application of guthion should be made at the beginning of shell hardening followed by another two applications at two week interval (Woodroof, 1979).

11.13.5. Bark borer (*Xylouops hasilare* (Say)

Larvae of bark borer feed on the living bark of pecan-nut trees, especially in young plantings. They later bore into the hard wood. Penetration is usually where branches are formed and can occur in branches of any thickness. The holes in the branches that serve as shelter for the larvae, are about 70 mm long and 5 mm in diameter when the larvae reach maximum size. Feeding marks on the bark are covered with larval excreta spun together with threads in such a way that the larvae can move freely underneath the threads. As the larva feeds, this "house" of spun threads and excreta becomes bigger and could be found around a branch. Although infested trees do not die, the branch is ring-barked and it could die back. Good control can be achieved with a registered chemical, even if only the lesions on the branches are treated. It is not necessary to remove the excreta from the branches before spraying. Spraying of the entire tree is not recommended.

11.13.6. Pecan Weevil (*Curculio caryae*)

The pecan weevil is a late-season pest of pecan and hickory that may sometimes destroy a large part of the crop both in the orchards and storage. The type of damage includes premature nut drop, black spots on the kernel, shriveled kernel and destruction of kernel by larvae (Hall *et al.*, 1981). After pupating in the soil, the adults move to the tree and damage the nuts before shell hardening during August and September and lay eggs after shell hardening. As soon as grubs hatch, they start feeding on kernels. The adult or weevil appears in late August and early September, about when the nut kernels harden. Grub leaves the nut after cutting a hole and enters the soil between September and December. It remains there for one or two years. Early maturing cultivars are more attacked than late maturing ones.

Sprays of insecticide before the female adults lays her eggs directly kills them, because once eggs have been deposited in the nut they are safe from insecticides. Five foliar applications of carbaryl at 2.7 kg/ha at interval of 7 to 10 days show significantly less damage by the weevil than 8 to 9 soil application of carbaryl (Payne *et al.*, 1985). The first spray should be applied as soon as the nuts begin to harden in the early cultivar.

11.13.7. Pecan Phylloxera (*Phylloxera devastatrix*)

Pecan phylloxera produces galls or bumps on new growth. Leaves, twigs, or nuts may be affected. Galls may range from 1D 10 to 1 inch in diameter. They are produced by small, soft bodied insects closely related to aphids. Phylloxera overwinters as eggs in bark crevices. Tiny nymphs hatch in the spring and move to feed on tender tissue. Several generations may occur during the growing season and infestations can continue as long as the tree has fresh, young growth.

Sprays may reduce continued buildup over the season but will not affect galls that have already developed. Dormant oil spray, as recommended for scales, will help with control if applied thoroughly to the tree trunks. This spray is targeted for overwintering eggs. During the growing season, control should be targeted at the "crawler" stage before the galls form. The crawler stage is active just before or at bud-break.

11.13.8. Pecan leaf casebearer (*Acrobasis juglandis*)

As the name indicates, the caterpillar keeps itself closed in a leaf case. The caterpillers emerging from overwintering cases in early spring feed on buds, leaves and flowers and later enter in the shoot. The broken shoots are symptoms of their attack. The month lays eggs on the underside of a leaf along midvein. It has only one generation in a year (Johnson, 1969).

The use of a blacklight insect trap or the use of a pheromone trap can help in reducing the number of insects and damage. The pest should be controlled by pre-pollination spray of an insecticide (parathion or malathion). If the insecticidal sprays are carried out to control other pests, a separate application is not required.

11.13.9. Fall Webworm (*Hyphantria cunea*)

The caterpillars attack limbs and can cause defoliation, occasionally. During summer, a loose silken web is formed covering leaves, twigs and small branches. The egg masses are deposited on leaves and hatching takes place in about one week. Depending on the region, the month has one or two generations in a year. It passes the winter in pupal stage in cocoons on the ground. Remove webbing when first seen. Insecticidal sprays to control shuckworm are enough to eradicate this as well.

11.13.10. Borer insect / Ambrosia Borer (*Euplatypus segnis* Chapuis)

The borer insect *Euplatypus segnis* Chapuis is associated to *Fusarium solani, Fusarium oxysporum, Alternaria alternata* and *Botryodiplodia theobromae* fungi which produce regressive death in pecan (*Carya illinoensis*) trees. It also has been observed that some trees can survive with or without chemical treatment. The results indicated that content of terpenes, hydrolyzables tannins, cellulose, lignin and silicium increased significantly in the damaged trees in comparison to healthy trees. The observations allowed inferring that these components increased as a chemical defense answer to insect invasion and to the enzymatic action of the associated phyto-pathogenic fungi (Alvidrez Villarreal *et al.,* 2011).

11.13.11. Nematodes

Pathogenic root-knot nematodes (*Meloidogyne* spp.) invade small roots of the pecan tree and cause small galls to form on the roots. Pecan roots are also susceptible to infection by a ring nematode (*Criconomella rusium*). Infected nursery trees should not be planted in an orchard. They should be destroyed. Infected trees sometimes show stunting and resetting of the foliage, which applications of zinc compounds and fertilizers will not correct. Use the same control measures as described for the crown gall disease.

11.13.12. Birds

Crows and blue jays are well known to cause great losses to pecans. Crows are particularly a menace to pecan as they start damaging the nuts soon after kernel filling either on the tree or by carrying them away. The damage is more serious to small, well-filled and thin-shelled nuts. Crows attack the nuts mostly early in the morning. After killing a crow, it should be hanged high on the top of a pecan tree to scare away the crows. Automatic (scare-away machine) explosion devices can also be used as they are more economical than employing watchmen for bird scaring.

12. Maturity

Pecans are mature on the tree when the shucks around the nut split and expose the nut. Under North Indian conditions pecan nuts usually ripen in September-October. As soon as the nut is physiologically ripe, the green husk becomes dry, cracks open and the nut drops out. If the shucks remain tight and do not split at maturity those nuts are considered worthless. However, Early harvest of pecan has been recommended in order to capture early market and to reduce bird damage but shucks do not dehisce and easily separate from the shell in immature fruits. The development of nut colour is also hampered. The development of nut colour markings as dots and streaks on the shell is an indication of nut maturation. With the onset of shuck dehiscence along the sutures, the shell colour changes from white to brown (Kays and Wilson, 1977). Shuck dehiscence, colour markings on shell and clean separation of packing tissues from kernel indicates that the fruit is ready for picking (Eddy and Storey, 1988). In case of delay or absence of shuck dehiscence in some geographical regions, the nuts are harvested and shuck is removed mechanically.

If cattle graze in the orchard, they should be removed at least two months before harvest to allow time for the manure to decompose to prevent contamination of the nuts. Any low hanging branches that could interfere with harvest should be removed. Remove any sticks that litter the orchard floor. The orchard floor (ground cover) should be mowed at frequent intervals late in the season to prevent the accumulation of plant residue which could interfere with nut harvest.

13. Harvesting

The nuts should be harvested at proper time as the quality of nuts is influenced by the time of harvest. When 70 to 80 percent of the nut clusters have open shucks and kernel moisture is below 10 percent, harvesting can begin. Some pecans will be on the ground at this time. Start by shaking a test tree. If 80 percent of the nuts drop after 10 seconds or less, proceed to harvest other trees of the same variety. Harvesting trees too early may damage the bark because sap is still flowing. Harvesting too late increases loss due to wildlife depredation and allows nutmeats to darken and lose quality.

Mature nuts do not fall at one time, so the nuts either thrashed down from the trees with bamboo poles or allowed to fall naturally on the ground where from these are gathered manually. To aid with the harvesting, a tarp may be spread under the tree before the trees are thrashed. Another harvest method is to wait for the nuts to fall and pick them up by hand. Hand shaking and gathering are possible only when plantings are small and labour is cheap and easily available but the method is slow, tedious and expensive. Some varieties will drop almost all of their nuts with just one shaking, while other varieties require a second shaking a couple of weeks after the first. The length and intensity a tree is shaken will depend on crop load and tree size. Large trees can handle more severe shaking and require longer shaking time, compared to small trees. In mechanical harvesting, a hydraulic shaker is used to shake each tree to dislodge nuts. Shakers typically have padded arms which clamp to tree trunks. Nuts are mechanically swept into windrows, where they are picked up by a mechanical harvester. The hulls are then separated from nuts at a hulling facility. This mechanical harvest method is similar to walnut harvesting.

14. Yield

The commercial production in pecan begins by 7th to 10th year of planting depending on the precocity of a cultivar and growth of the tree (Madden *et al.*, 1969). Furthermore, yield varies greatly with the cultivar, age of the trees, 'off' and 'on' year and planting density. The yield ranging from 20 to 100 kg per tree is expected from fully grown trees. Haulik and Holtzhausen (1987) reported an average annual yield of 25.1 kg per tree over a 22-year period and the extrapolated average yield of 2642 kg/ha for 10 years which compares favourably with the yields reported from the USA.

15. Drying, grading and storage

Harvested or collected nuts are then subjected through a series of hullers to remove any loose shucks sticking to some nuts. Remove cracked nuts, sticktights, nuts with weevil holes, irregularly colored nuts or other damaged nuts. For getting higher prices from early season pecans and reduce loss due to wildlife depredation, the moisture content needs to be lowered to 4 per cent to develop good eating quality and prevent molding. Nuts are generally dried to 4 per cent kernel moisture at 75°F, with relative humidity below 60 per cent and moderate air flow. When pecans lie on the wet soil after harvest for few days, they discolour and get infected with mould and eating quality deteriorates. In order to prevent moulding, discoulouration and rancidity, pecan nuts are dried soon after harvest to bring the moisture level of kernels to around 4 per cent. The drying also imparts satisfactory colour, texture, flavour and stability to the nuts which occurs naturally under field conditions if the weather and soil remain dry at harvest. The nuts should be dried at room temperature by constant circulation of low moisture air for about three days. Artificial heat drying of nuts is undesirable. The kernel moisture level was brought down to about 4 per cent within 24 hours by dehumidified air at 30°C with a velocity of 18 to 31 metres/minute.

Before marketing and storage of pecans, they need to be graded. Knowing ahead of time the quality of the nuts will allow to anticipate the price they will bring, whether they are marketed wholesale or direct to the public. Grading consists of determining the percentage of edible kernel and percentage of kernel grades.

Growers who choose not to market their pecans during the harvest season will need to place the crop in cold storage as pecan is consumed over a long period after harvest necessitating proper storage because of its semi-perishable nature. If nuts are not stored properly, they become inedible due to mould, insects and rancidity along with discolouration and staleness. In general, the lower the temperature, the longer the storage life of in-shell and shelled pecans. Freezing does not damage pecans since their moisture content is low. Only high quality pecans, i.e., having good size, colour, flavour and oil content should be stored to fetch high price in the off-season market compensating the extra cost involved in refrigeration. The relative humidity requirements of unshelled and shelled pecans for storage for more than three months are 75-80 per cent and 65-70 per cent, respectively. The refrigerated storage preserves natural flavour, colour and texture as long as eight years, retards rancidity and staleness and prevents moulding and insect damage (Woodroof, 1979). Unshelled and shelled pecan can be kept at room temperature (21.1°C) for 6 and 3 months, respectively, but the chances of discolouration, rancidity and moulding are more. El-Wahab *et al.*, (1984) found the cold storage of unshelled nuts for 20 month at 5°C to be the best which delayed decay and rancidity. At -6.7° to -3.3°C unshelled nuts can be preserved for 30 month and shelled halves for 18 to 24 months without deteriorating the quality (Woodroof, 1979).

Long term storage should be for 24 months at a temperature of – 18°C with 70 per cent relative humidity. In-shell pecans can be stored for 6 months at 22°C, 12 months at 0°C or 24 months at – 18°C, whereas shelled pecans can only be stored for 3 months at 22°C, 9 months at 0°C or 18 months at – 18°C. Storage at < 2 per cent O_2 for 52 days can cause a fruity flavour to develop (Santerre *et al.*, 1990). O_2 transmission rates for packaging materials should be > 0.08 ml/100 cm per 24 h (Dull and Kays, 1988). Vaccum packaging can offer a further benefit of protection from breakage.

16. Marketing

Many pecan growers, particularly smaller producers, market their pecans directly to the consumer in a retail store through fund-raising organizations etc. For these growers, pecan prices are set based on consumer demand for a fresh product. Pecans can also be marketed wholesale or directly to the consumer. Selling wholesale is the standard method for moving the largest volume of nuts in the shortest time. Growers who plan on wholesaling their pecans should contact a buyer before the pecans are ready to sell. Marketing of pecans directly to the consumer offers the opportunity to skip the middleman and obtain retail instead of wholesale prices. The key to selling pecans retail is to offer a quality product. Never sell poor or low quality pecans to the consumer. If the quality is there, the consumer will pay the price and return for more. If the quality is low, however, forget about repeat buyers. Types of direct to the consumer sales include orchard sales, roadside stands, farmers' markets and Internet sales.

Price is affected by a number of global economic factors as well as the alternate bearing habit of pecan trees, supply and demand, supply and cost of other competing nuts (walnuts, almonds) etc. Wholesale prices are often lowest when the national crop is large. Wholesale market price is additionally affected by variety being sold and the sizing of the nuts, quality of the lot being sold, and presence of defects.

17. Economics of cultivation

An estimation cost for growing of pecan nut plants per kanal (1 Kanal=1.5 bigha or 20 Kanal= 1 hectare) for six year has been presemted in Table 3.

Table 3: An estimation cost for growing of pecan nut plants per kanal for six year

S.No.	Components	Quantity	Cost of item	1st year	2nd year	3rd year	4th year	5th year	6th year	Total cost
1.	Grafted Plants	08	Rs. 250	2000/-	-	-	-	-	-	2000
2.	Measuring Tape	01	Rs. 150	-	-	-	-	-	-	150
3.	Ploughing, Levelling and digging of pits	Rs. 800/kanal		800	-	-	-	-	-	800
4.	Transportation cost	Rs. 50		-	-	-	-	-	-	50
5.	Number of labors (hoeing & weeding)	02	Rs. 300	600	600	600	600	600	600	3600
6.	Chemical fertilizer	Urea	Rs. 15/kg	15	15	15	30	30	45	150
		DAP	Rs. 30/kg	30	30	30	30	30	30	180
		MOP	Rs. 20/kg	20	20	20	40	40	40	180
7.	Organic manure	FYM*	10 kg	280	280	560	560	840	840	3360
8.	Weedicides		100	100	100	100	100	100	100	600
9.	Fungicides & Pesticides		200	200	200	200	200	200	200	1200
									Total investment	12,270
10.	Number of pecan nut produced		-	-	-	-	-	50	125	
11.	Total number of nuts		-	-	-	-	-	400	1000	
12.	Total weight of nuts		-	-	-	-	-	3.2 kg	8 kg	
13.	Total money generated		-	-	-	-	-	3,840	9,600	13,440
		Total money generated	-		Total investment		=			Total income
		13,440.00	-		12,270.00		=			1,170.00

*FYM was applied @ 10 kg/plant in the first year and @ 10 kg increased every alternate year

*Average size of nut produced 8 g

Rate of one kg of Pecan nut 1200/-

18. Organic cultivation

Nowadays increased emphasis around the world is on food safety and nutrition, with the use of less or no chemicals. Since up to 90% of the world pecan is produced in U.S., and the most important thing is that pecan virgin stands still exist on undisturbed sites in its native habitat. Native pecan groves are often managed both for nuts and for grazing of livestock. With the grazing cattle in pecan groves, can generate additional income from the land as well as enhance nutrient cycling. Although some new varieties developed are being grown conventionally, but there is a tremendous scope for growing pecan organically for the grower as well as consumer. New ARS (Agricultural Research Service, USDA) studies in Weslaco, show that it is possible for growers to boost their revenue further by growing pecans organically (Bradford, 2006). The quality advantage of organic pecan directly translates into much larger profits for organic growers. It has been observed that pecans grown organically contain ellagic acid and catechin four times and two times respectively as compared to inorganically grown pecans. Oil content and oleic acid were also found significantly higher in organically grown pecans (Malik *et al.*, 2009).

Evidences support the hypothesis that the humus-topsoil-root association is key to the overall nutrition of pecan in nature via recycling of nutrients, mainly from decaying leaves and shucks (Smith *et al.*, 2012). Further for nutrient supplementing in organic pecan orchards, in a study to evaluate legumes for their ability to supply nitrogen and provide habitat to beneficial insects, leaf nitrogen concentrations were typically maintained above the minimum sufficiency level (2.25%) by either crimson clover plus hairy vetch, or red clover plus white clover. A crimson clover-hairy vetch ground cover supplied the equivalent of 90 – 142 lbs. N/acre, while as, a red clover-white clover mixture supplied up to 118 lbs. N/acre. White clover or red clover alone were less effective in supplying N than when grown together (Smith *et al.*, 1996). Also the accumulated effect of all aphid predators and parasitoids attracted to the legume aphids may reduce the density of the pecan aphids (Rice *et al.*, 1998).

19. Effect of climate change

Climate change is one of the greatest environmental, economic and social challenges in the history of mankind and nowadays is considered as the biggest environmental problem of the world. Climate change has a significant global. Agriculture has been unfavourably affected in recent years, as the current and anticipated conditions in many cases seem to be rather prohibitive for the prosperity of the cultivated crops. On the contrary, these new conditions have made it possible for new plant species previously cultivated only in subtropical regions, to thrive in some country like Greece. Moreover, economic reasons would make it rather necessary for the agricultural industry to cultivate alternative crops, which are thoroughly analysed in the present study. Based on Heating Degree Days (HDD) Greece is divided into four climatic zones. The variations in the mean maximum, mean minimum and mean temperature in each climate zone as well as the rainfall over the last 50 years (from 1964 to 2013) are reviewed. The outcome of this research is that it is not feasible for the studied alternative crops to thrive in all climate zones or vice versa. However, some of them and particularly crops such as quinoa, maca, psyllium, chia, cassava and pecan can be cultivated in all climate zones. It has also to be noted that adequate water is necessary for almost all the examined crops in order to achieve optimal growth and yield and therefore irrigations are rather necessary for specific species and climate zones (Georgakopoulos *et al.*, 2016).

Climate change is expected to alter the global industry in fruits and nuts dramatically as tree crops struggle in rising temperatures. Even if polluters take greater action to cut carbon emissions, the impact of climate change will likely be severe enough that the nearly $100 billion-a-year fruit and nut industry needs to reassess planning (Luedeling *et al.*, 2011). The principal impacts of projected climate change are more likely from changes in the frequency of threshold events and extremes, such as the date of last spring freeze, the length of the growing season, and heat accumulation, than from changes in mean climatic states (Winkler *et al.*, 2011). In other words, 1 to 4°C average increase in global temperature will not itself be the problem. The problem will be extreme events – such as storms, droughts, killing forests, and changes in the length of the growing season.

Pecan is resistant to severe droughts of moderate duration in spite of its high water requirement. However, if the drought persists, dieback is initiated from the tree top branches and will continuously progresses downward into the crown with the duration of the drought (Sparks, 2005). Although pecans have sufficient freeze tolerance to withstand winters and a growing season sufficiently long for fruit development, but they usually fail to mature nuts. This failure has been attributed to the cooler springs during the season (Sparks, 1989). The dry and hot weather during the spring are advantageous for early maturing and sale of nuts. In contrast late maturity of nuts is supposed to be not profitable (Sparks, 1992).

Bud break in pecan is under the interactive control of heating and of chilling. Heat required for bud break varies inversely with chill accumulation, and bud break may occur with no chilling once sufficient heat accumulates. Variability in bud break may increase dramatically when there are fewer than ~100 chilling degree days during the winters (Sparks, 1993). It has been observed that freezing temperatures following bud break has about two times more effect on reducing the production than freezing temperature preceding nut maturity (Sparks, 1997). High relative humidity and low temperatures during flowering may decrease the pollen release from anthers (Yates, 1993).

20. Future thrust

Among temperate nut crops, pecan nut stands second position after walnut and has great potential for its cultivation even in lower elevations (500 -1500 m amsl) which are considered to be hotter for walnut. It can also be easily grown in drought prone areas due to its long tap root system. At present very little produce comes from the seedling or unknown plantation. However introduction of few varieties having good nut and kernel quality by the state department of hilly areas are helpful in the production of the quality nuts and also in the breeding programs. For more production of quality nuts, area expansion under pecan nut cultivation is the major priority. Other strategic points which need to be emphasized includes:

- Identification of potential growing areas especially in the northern states of the country.
- Introduction and selection of promising, precocious, regular bearing and high yielding planting material.
- Collection, characterization, conservation, evaluation and utilization of germplasm.
- Development of precocious and cross compatible varieties having better in shell nut as well as kernel characteristics.
- Establishment of bud wood / gene bank for ensured quality of planting materials and for further mass multiplication is urgently needed.
- Dwarf clonal rootstocks need to be introduced for adaptive research and high-density plantation culture.
- Effective research programs on cultural techniques, propagation, pollination, nutrition and postharvest handling and storage of pecan nut need to be strengthened and initiated for commercialization
- Establishing/strengthening the mother blocks.
- Assessment of cost effective water and nutrient management technologies under different agro-climatic conditions.
- The existing marketing system for pecan nut needs to be strengthened.

References

Acuna-Malconado, L.E.; Smith, M.W.; Maness, N.O.; Cheary, B.S.; Carroll, B.L. and Johnson, G.V. 2003. Influence of nitrogen application time on nitrogen absorption, partitioning, and yield of pecan. *J. Amer. Soc. Hort. Sci.,* **128**: 155-162.

Adams, J.C. and Thielges, B.A. 1977. Research underway on pecan timber improvement. *Louisiana Agriculture,* **20**(2): 14-15.

Ajamgard, F.; Rahemi, M. and Vahdati, K. 2016. Development of improved techniques for grafting of pecan. *Scientia Horticulturae,* **204**: 65-69.

Alben, A.O., Hammer, H.E. and Sitton, B.G. 1942. *Proceedings to American Society for Horticultural Science,* **41**: 53-60.

Allan, P. Brutsch, M.O., Smith, I.E., and Wolstenholme, B.N. 1980. *Pecan Quart.,* **14**: 7-14.

Alverson, D.R. 1985. Influence of edaphic factors on emergence of the pecan weevil, Curculio caryae (Horn.) Colcoptera: Curculionidae) In: Pecan Weevil: Research persepective (Neel, W.W. (ed.). pp 11-26, Quail Ridge Press, Brandon, MS

Alvidrez Villarreal, R.; Hernandez Castillo, F. D; Garcia Martinez, O.; Mendoza Villarreal, R.; Rodriguez Herrera, R.; Aguilar Gonzalez, C. N. 2011. Secondary metabolite changes in pecan (*Carya illinoensis*) tissue damaged by *Euplatypus segnis* Chapuis and associated fungi. *American Journal of Agricultural and Biological Sciences,* **6**(4): 553-559.

Amling, H.J. and Amling, K.A. 1983. Physiological differentiation of pistillate flowers of pecan and cold requirement for their initiation. *J. Amer. Soc. Hort. Sci.,* **108**: 196-198.

Anon. 2016a. Maxico increased its nut production by 83%. *Fresh Plaza.,* 8455.

Anon. 2016b. Directorate of Horticulture Kashmir. http://hortikashmir.gov.in.

Anon. 2017. USDA (NASS) Pecan crop production report. Pecan report. http://www.pecanreport.com/news-blog.

Arnold, C.E., Crocker, T.E. and Aldrich, J.H.1981. Rejuvenation of a mature pecan orchard by dehorning and subsequent chemical applications. *Pecan Quart.,* **15**(3): 11-18.

Assaf, R. 1977. A new method of vegetative propagation of physiologically adult shoots of pistachio, walnut and pecan. *Fruits,* **32**(5): 309-319.

Avila Trevino, J. A; Arreola Avila, J.G; Rodriguez de la O, J L; Trejo Calzada, R; Valdez Cepeda, D. and Rosa, A B de la 2013. Morphogenic responses in the in vitro propagation of pecan (*Carya illinoinensis* [Wangenh] K. Koch). *Revista Chapingo Serie Ciencias Forestales y del Ambiente,* **19**(3): 469-481.

Ball, J.C. 1982. Impact of fungicides and miticides on predatory and phytophagous mites associated with pecan foliage. *Environ. Entom.,* **11**: 1001-1004.

Barnett, J. and Mielke, E.A. 1981. Alternate bearing: A re-evaluation. *Pecan South,* **8**: 20–23.

Bhardwaj, K.N. 1983. Standardization of vegetative propagation techniques in pecan nut (*Carya illinoensis*).*Thesis Abstracts*, Haryana Agricultural University, **9**: 65.

Blayney, D. and Gutierrez, P. 2017. Economic importance of the Pecan Industry. Cooperative Extension Service. New Mexico State University. Guide Z-501

Bolt, L.C. 1983. Bench-grafting of pecan root cuttings. *Pecan Quart.,* **17**: 20-21.

Bonner, F.T. 1976b. Storage and stratification recommendations for pecan and shagbark hickory. *Tree planters Notes,* **27**(4): 3-5.

Bonner, F.T. and Maisenhelder, L.C. 1974. Carya nut., hickory. In: Schopmeyer, C.S. Seeds of woody plants of United States. Agric. Handbk. Washington DC. 269-272.

Bottari, N. B.; Lopes, L. Q. S.; Pizzuti, K.; Alves, C. F. dos S.; Corrêa, M. S.; Bolzan, L. P.; Zago, A.; Vaucher, R. de A.; Boligon, A. A.; Giongo, J. L.; Baldissera, M. D. and Santos, R. C. V. 2017. Antimicrobial activity and phytochemical characterization of *Carya illinoensis*. *Microbial Pathogenesis*, **104**: 190-196.

Bradford, J. 2006. Organic production systems for pecans. *18th world congress of soil science-Philaddelphis*, Pennsylvania, USA. 162: 24.

Brown, J.C.; Lunt, O.R.; Holmes, R.S. and Tiffin, L.O. 1959. The bicarbonate ion as an indirect cause of chlorosis. *Soil Sci*., **88**: 260-266.

Camarena, D.M.; Nunez, H. and Puebla, M.A. 2012. The pecan nut in Mexico: characteristics of commercial supply and consumers preferences. *Acta Hortic*., **930**: 77-82.

Caminada, P. 1979. Germination of the pecan nut. *Zimbabwe Rhodesia Agricultural Journal* **76**(6): 237-238.

Cole, J R. and Hunter, J H. 1962. Unusual case of delayed foliation and dieback of moneymaker Pecan. *Plant Disease Reporter,* **46**: 894-895.

Conner, P. and R. Worley, R. 2000. Alternate bearing intensity of pecan cultivars. *Hort. Sci.,* **35**: 1067–1069.

Cortes-Olivares, J., Philips, G.C. and Butler-Nance, S.A. 1990. Micro-propagation of pecan. *Hort. Sci.,* **25**: 1308.

Cottrell, T.E., Wood, B.W. and Ni, X. 2010. Application of plant growth regulators mitigates chlorotic foliar injury by the black pecan aphid. *Pest Management Science,* **66**(11): 1236-42.

Daniel, J.W. and Moss, R. 1985. The cumulative production through the 12th year (overgrown at that time) of pecan trees in a high density planting with and without irrigation. *Proc. Ga. Pecan Grow. Assoc*., **16**:11-18.

Daniell, J.W. 1974. The effects of chemical weed control on yield and quality of pecans. *Proc. S. E. Pecan Growers Association,* **67**:35–38.

Daniell, J.W. 1989. Irrigation. In: Pecan production in the southeast: A guide for growers. Goff W.D., McVay J.R., Gazaway W.S. (eds.) Ala. Coop. Ext. Serv., Auburn Univ, AL, pp 73–86.

Diver, S.G., Smith, M.W. and McNew, R.W. 1984. Influence of fruit-development on seasonal elemental concentrations and distribution in fruit and leaves of pecan. *Communications in Soil Science and Plant Analysis,* **15**: 619-637.

Dull, G.G., and Kays, S.J. 1988. Quality and mechanical stability of pecan kernels with different packaging protocols. *J. Food Sci.,* **53**:565-567.

Dutcher, J.D. and Harrison, I.C. 1984. Application of reduced rates of sptemic insecticides for control of foliar pen arthropods. *J. Econ. Entomol.,* **77**: 1037-1040.

Eddy, M. and Storey, J.B. 1988. The influence of harvest date on oil and flavour development in 'Desirable' and 'Wichita' pecans. *Hort. Sci.,* **23**(3): 782.

El Sharkawy, S.M.M. and Osman, I.M.S. 2009. Effect of foliar applications of some growth regulators on flowering, fruit production and quality of Wichita and Western Schley pecan cultivars. *Egyptian Journal of Horticulture,* **36**(1): 29-46.

Elmore, C.D. and Polles, S.G. 1980. Nitrogen fertilization effects on amino acid composition of pecan (*Carya illinoensis*) nutmeats. *J. Agric. Food Chem.,* **28**: 902.

El-Nabawy, S., Rawash, M.A., El-Hamady, A., Desouky, I. and Khalil, F. 1980. Effect of stratification and GA_3 on germination of pecan seeds and subsequent seedling growth. *Ann. Agric. Sci.,* **25**: 323-338.

El-Wahab, F.K., El-Hamady, A.M., El-Nabawy, S.M., Rawash, M.A. and Hagagg, L.F. 1984. Effect of storage and prestorage treatments on pecan fruit decay and oil properties. *Gartenbauwissenschaft,* **49**: 61-64.

FAS. 2015. Foreign Agricultural Services. Global Agricultural Trade System Online (GATS). USDA.

Fayek, M. A; Fayed, T. A; El Said, E. H. and El Hamed, E.E.A. 2008. Utilization of some chemicals for synchronizing time of male and female flowers in pecan (*Carya illionensis* Koch). *Research Journal of Agriculture and Biological Sciences,* **4**(4): 310-320.

Foshee, W. G, III; Goodman, R. W; Patterson, M.G; Goff, W.D. and Dozier, W. A. Jr. 1997. Weed control increases yield and economic returns from young 'Desirable' pecan trees. *Journal of the American Society for Horticultural Science*, **122**(4): 588-593.

Foshee, W. G; Goff, W. D; Tilt, K. M; Williams, J. D; Bannon, J. S. and Witt, J. B. 1996. Organic mulches increase growth of young pecan trees. *Hort. Science,* **31**(5): 811-812.

Foshee, W.G., Blythe, E.K., Goff, W.D., Faircloth, W.H. and Patterson, M.G. 2008. Response of young Pecan trees to trunk and foliar applications of Glyphosate. *Hort. Sci.,* **43**(2): 399-402.

Foshee, W.G., Goodman, R.W., Patterson, M.G., Goff, W.D. and Dozier W.A. 1997. Weed control increases yield and economic returns from young 'Desirable' pecan trees. *J. Amer. Soc. Hort. Sci.*, **122**:588–593.

French, R.B. 1962. Analyses of pecan, peanut and other oils by gas liquid chromatography and ultra violet spectrophotometry. *J. Amer. Oil Chemists Society,* **39**: 176-178.

Gammon, N. Jr. and Sharpe, R.H. 1956. Mouse-ear: A manganese deficiency of pecans. *Proc. Amer. Soc. Hort. Sci.,* **68**: 195-200.

Georgakopoulos, P.; Travlos, I. S.; Kakabouki, I.; Kontopoulou, C. K.; Pantelia, A. and Bilalis, D. J. 2016. Climate change and chances for the cultivation of new crops. *Notulae Botanicae, Horti Agrobotanici, Cluj-Napoca,* **44** (2) 347-353.

Goff, W.D., Brasher, L.R. and McGuire, J.A. 1992. Germination of unstratified pecans is affected by exposure to high temperature and by soaking. *Scientia Horticulturae,* **50**(1): 159-163.

Goff, W.D., Campbell, L. Thompson, T.E. Bannon, J.S. and Latham, A.J. 1993. Scab occurrence on pecan clones in Alabama in a year of high disease incidence. *Fruit Var. J.,* **47**: 47–51.

Goff, W.D., Nesbitt, M.L. and Browne, C.L. 2003. Incidence of scab and foliage condition on pecan cultivars grown without fungicide or insecticide sprays in a humid region. *Hort. Technology,* **13**: 381–384.

Graham, C.J. 2015. Evaluation of ReTain® plant growth regulator on pecan production. *Acta Hortic.,* **1070**: 183-192.

Grauke, L. J. and Thompson, T.E. 1996. Pecans and hickories. In: Janick, J. A. and Moore, J. N. (ed.). Fruit Breeding—III Nuts. pp. 185–239. John Wiley and Sons, Inc.

Grey, T.L; Turpin, F. S, II; Wells, L. and Webster, T.M. 2014. A survey of weeds and herbicides in Georgia pecan. *Weed Technology*, **28**(3): 552-559.

Gustafson, W.A., Storey, J.B. and Shreve, L.W. 1982. Bench-grafting of pecan. *Hort. Sci.,* **17**(6): 969-970.

Hall, M. J., Hedger, G.H., Eikenbary, R.D. and McNew, R.W. 1981. Impact of pecan weevil on pecan production in a pest-managed commercial orchard. *Environ. Entomol.,* **10**: 668 – 672.

Hanna, J.D. 1987. Pecan rootstocks. In: Rom, R.C. and Carlson, R.F. (eds) *Rootstocks for Fruit Crops*. John Wiley & Sons, New York, pp. 401–410.

Hansen, K.C. and Lazarte, J.E. 1984. *In-vitro* propagation of pecan seedlings. *Hort. Sci.,* **19**: 237–239.

Hansman, D. and Novoa, O.C. 1986. Micro-propagation of temperate nut trees. *Horticultural Abstracts,* **56**(6): 403-416.

Harlow, W.H., Harrar, E.S., Hardin, J.H. and White, F.D. 1991. Textbook of dendrology covering the Important Forest Trees of the United States and Canada.7th edn. McGraw-Hill, New York, pp. 269–271.

Hassanen, S.A. and Gabr, M.F. 2013. *In-vitro* rooting of pecan (*Carya illinoensis* Wang Koch). *World Applied Sciences Journal,* **21**(3): 315-319.

Haulik, T.K. and Holtzhausen, L.C. 1988. Anatomy of staminate flower ontogeny of the pecan as determined by scanning electron microscopy. *S. Africa J. Plant Soil,* **5**: 205-208.

Haulik, T.K. and Holtzhausen, L.C. 1987. Evaluation of five pecan cultivars (*Carya illinoensis* (Wang.) K. Koch) for yield, precocity, alternate bearing, leaf:nut ratio and length of growing season in the Transvaal mid-level. *S. Afr. J. Plant Soil*, **4**: 70-74.

Heation, E.K., Daniell, J.W. and Moon, L.C. 1982. Effect of drip irrigation on pecan quality and relationship of selected quality parameters. *J. Food Sci.*, **47**(4): 1272.

Herrera, E. and Martinez, D. 1983. Effect of seed size on vigor of pecan seedlings. *Pecan Quart.*, **17**(1): 18-19.

Herrera, E.A. and Duarte, E. 2000. Nitrogen rates and time of application effect on growth and nut production of mature transplanted pecan trees [*Carya illinoinensis* (Wang.) K. Koch]. *Phyton-International Journal of Experimental Botany,* **66**: 129-135.

Hinrichs, H.A. 1961. The relationship of native pecan tree spacing to yield. *Okla. Agric. Exp. Sta. Bull.,* B-574. 11p.

Hooks, R.F. and Storey, J.B. 1971. Effect of succinic acid-2, 2-dimethyl-hydrazide (SADH) and heading-back on pecans. *Hort. Sci.,* **6**: 237-238.

Hunter, R. E., Thompson, T.E. and Sanderlin, R.S. 1986. Control of pecan diseases through genetic resistance. *Proc. S. E. Pecan Growers Association,* **79**:51–54.

Isbell, C.L. 1928. Growth studies on the pecan. *Ala. Agric. Exp. Stn. Bull.*, **266**: 1-63.

Johnson, W.T. 1969. Plant pests, pp. 67–156. *In:* Jaynes, R.A. (ed.), Handbook of North American nut trees. W.F. Humphrey Press, Inc., Geneva, New York

Jones, J.B., Wolf, B. and Mills, H.A. 1991. Plant Analysis Handbook. Micro-Macro Publishing, Inc., Athens, GA.

Joolka, N. K. and Rindhe, A. B. 2000. Standardization of method and time of budding in pecan. *Indian Journal of Horticulture*, **57**(1): 51-53.

Joolka, N. K; Rindhe, A. B; Sharma, M.K. 2001 Standardization of method and time of grafting in pecan. *Indian Journal of Horticulture*, **58**(3): 212-214.

Kays, S.J. and Wilson, D.M. 1977. Alteration of pecan kernel color. *Journal of Food Science,* **42**: 982-988.

KenKnight, G. E. 1968a. Resistance of pecan to disease. *Proc. S. E. Pecan Growers Association,* **61**:168–173.

Khalil, S. K.; Mexal, J.; Abdur Rehman; Said Wahab; Amanullah; Khalil, I. H.; Zahid Hussain; Khan, A. Z. Direk, M. 2015. Management of alternate bearing dilemma in pecan through foliar ethephon thinning in New Mexico, USA. International Conference on Sustainable Agriculture and Environment, September 30 - October 03, 2015, Konya, Turkey. *Proceedings book*, Volume I & II: 834-842.

Kim, T. and Wetzstein, H.Y. 2005. Seasonal fluctuations in nutrients and carbohydrates in pecan leaves and stems. *Journal of Horticultural Science and Biotechnology,* **80**: 681-688.

Knox, C.A. and Smith, R.H. 1981. Progress in tissue culture methods for production of Riverside stocks. *Pecan Quart.,* **15**: 27–34.

Kraimer, R.A., Lindemann, W.C. and Herrera, E.A. 2001. Distribution of N-15-labeled fertilizer applied to pecan: A case study. *Hort. Sci.,* **36**: 308-312.

Kremer, R. J. and Kussman, R. D. 2011. Soil quality in a pecan-kura clover alley cropping system in the Midwestern USA. *Agroforestry Systems,* **83**(2): 213-223.

Kuden, A. B; Tuzcu, O; Bayazit, S; Yildirim, B. and Imrak, B. 2013. Studies on the chilling requirements of pecan nut (*Carya illionensis* Koch) cultivars. *African Journal of Agricultural Research*, **8**(24): 3159-3165.

Lazarte, J.E. 1984. Propagating pecan rootstocks by tissue culture. *Am Nurseryman,* **19**(11): 79-82.

Liu GuangQin; Zhu HaiJun; Zang Xu; Sheng JingYa; Zhou BeiBei. 2011. Maoshan 1, a new pecan cultivar. *Journal of Fruit Science,* **28**(6): 1132-1133.

Luedeling, E.; Girvetz, E.H.; Semenov, M.A. and Brown, P.H. 2011. Climatic changes lead to declining winter chilling during 1950-2099. *PLoS One*. 6(5): 20155.

Luo, X.; Li, Z.; Sun, Z. and Wan, X. 2016. Analysis of pecan cultivars Mahan and Western in East China. *Genetics and Molecular Research,* **15**(3) 15038732.

Madden, G.D. and Brown, E.J. 1973. Blossom dates of selected pecans. *Pecan Quart.,* **7**: 17-19.

Madden, G.F., Brison, F.R. and McDaniel, J.C. 1969. Pecans. In: Jaynes, R.A. (ed.), Handbook of North American nut trees. W. I. Humphrey Press, Geneva, New York. Pp. 163-189

Malik, N.S.A.; Perez, J.L.; Lombardini, L.; Cornacchia, R.; Cisneros-Zevallos, L. and Braford, J. 2009. Phenolic compounds and fatty acid composition of organic and conventional grown pecan kernels. *Journal of Science of Food and Agriculture*, **89**: 2207-2213.

Malstorm, H.L. and McMeans, J.L. 1982. Shoot length and previous fruiting affect subsequent growth and nut production of 'Money Maker'pecan. *Hort. Sci*., **17**: 970-72.

Malstrom, H.L. and McMeans, J.L. 1977a. A chemical method of pruning young pecan, *Carya illinoensis*, trees. *Hort. Sci.,* **12**(1): 68-69.

Malstrom, H.L., Fenn, L.B. and Riley, T. 1983. Nitrogen fertilization of pecans in Far West Texas. *Pecan Quart.,* **17**: 5-13.

Malstrom, H.L., Riley, T.D. and Jones, J.R. 1982. Continuous hedge pruning affects light penetration and nut production of "Western" pecan trees. *Pecan Quart.,* **16**(3):4-15.

Manning, W.E. 1978. The classification within the Juglandaceae. *Ann. Missouri Bot. Gard.,* **65**: 1058-1087.

Marquard, R.D. 1992. Fruit set of pecan requires a low percentage of live pollen in controlled pollination. *Hort. Sci*., **27**: 473.

Marquard, R.D. 1992a. Pollen tube growth in Carya and temporal influence of pollen deposition on fertilization success in pecan. *J. Amer. Soc. Hort. Sci*., **117**: 328-331.

Maximos, S.E., Bondok, A.Z., Salama, M.A. and Youssuf, N.F. 1984a. Effect of daminozide on shoot vigour in relation to fruiting in Curtis pecan trees. *Annals Agri. Sci.,* **29**: 1027-43.

Maximos, S.E., Bondok, A.Z., Salama, M.A., Habib, S.M. and Youssuf, N.F. 1984b. Histological studies on pistillate and staminate floral bud's differentiation of 'Curtis' pecan. *Annals Agri. Sci.,* **29**: 1015-25.

Maximos, S.E., Rawash, M.A., Dcssokey, A.N. and Youssef, N. 1979. *Egyptian Jr. Hort.,* **6**: 157-167.

McEachern, G.R., Wolstenholme, N.B and Storey, J.B. 1978. Chilling requirement of three pecan cultivars. *Hort. Sci.,* **13**: 694.

McEachern, G.R. and Storey, J.B. 1984. Low volume, high concentrate, Roundup herbicide control of pecan weeds. *Pecan Quart.,* **18**:27–30.

Medina, C. 2004. Normas DRIS preliminarespara nogal pecanero. *Terra,* **22**: 445-450.

Mehta, K.A. and Thakur, B.S. 2004. Pecan (Carya illinoensis (Wang) K. Koch) cultivation in India-problems, prospects and future strategies to enhance productivity. Proc. Compendium of summer school on enhancement of temperate fruit production- in changing climate, 2001. Pp 231-238.

Melanson, R. A.; Sanderlin, R. S.Grauke, L. J.. and Heerema, R. 2015. Hot-water treatment of pecan scions as a means of phytosanitation to reduce the potential introduction of *Xylella fastidiosa*, the causal agent of pecan bacterial leaf scorch, into orchards and new geographic regions. *Acta Horticulturae,* **1070:**201-209.

Melanson, R. A.; Sanderlin, R. S.Grauke, L. J.; Heerema, R. 2015. Hot-water treatment of pecan scions as a means of phytosanitation to reduce the potential introduction of *Xylella fastidiosa*, the causal agent of pecan bacterial leaf scorch, into orchards and new geographic regions. *Acta Horticulturae,* **1070:** 201-209

Miyamoto, S. 1984. A model for scheduling pecan irrigation with micro-computers. *Trans Amer. Soc. of Agr. Eng.,* **27**: 456-463.

Mousavi, S. R.; Galavi, M. and Rezaei, M. 2013. Zinc (Zn) importance for crop production - a review. *International Journal of Agronomy and Plant Production,* **4**(1): 64-68.

Muhammad Sajid; Ijaz Hussain; Abdur Rab; Khan, S. M.; Zubair Aman; Shah, S. T.; Naveed Ahmad and Ali, Q. S. 2017. Sulphate of potash and dipping time influence the seed germination and growth of pecan nuts. *Pure and Applied Biology*, **6** (1): 310-316.

Mullenax, R.H. and Young, W.A. 1972. Female flower development of the pecan. *Proc. S. E. Pecan Growers Association,* **65**: 83-92.

NASS. 2015. http://www.nass.usda.gov

Nunez, M.H., Cervantes, M.A., Diaz, M.G., Grageda, G.J. 2016. Effect of plant growth regulators applied during bud break of pecan nut yield. *Acta Horti.,* **1109**: 95-100.

Obeidy, A.A. and Smith, M.A.L. 1990. Establishing axenic cultures from mature pecan embryo explants on media with low water availability. *Plant Cell Reports,* **9**: 463–465.

Obeidy, A.A. and Smith, M.A.L. 1993. Organogenesis from mature pecan cotyledons and embryonic axes. *Hort. Sci.,***28**: 213–215.

Ojeda Barrios, D; Abadia, J; Lombardini, L; Abadia, A. and Vazquez, S. 2012. Zinc deficiency in field-grown pecan trees: changes in leaf nutrient concentrations and structure. *Journal of the Science of Food and Agriculture,* **92**(8): 1672-1678.

Osburn, M.R., Pierce, W.C., Phillips, A.M., Cole, J.R. and Kenbright, G.E. 1966. Controlling insects and diseases of the pecan. *USDA Agric. Handbook*, **240**: Rev. 1–55

Payne, J.A. and D. Sparks. 1982. Zinc levels in pecan leaflets from broadcast and band applications over a 6-year period. *Hort. Sci.,* **17**: 235-236.

Payne, J.A., Dutcher, J.D. and Ellis, H.C. 1985. Soil application of carbaryl for control of the pecan weevil. *Proc. S.E. Pecan Grow. Assoc.,* **78**: 65-68.

Peterson, J.K. 1990. Silvics of North America-2. Hardwoods. Agriculture Handbook 654, USDA Forest Service, Washington, DC, pp. 205–210.

Phillips, G.C. and Ramirez, J.J. 1983. Pecan tissue culture. In: Proceedings of the 17th Western Pecan Conference. Cooperative Extension Service and Western Irrigated Pecan Growers Association, New Mexico State University, Las Cruces, New Mexico, pp. 101–109.

Pichkhaya, N.S. 1987. *Subtrop Kul'tury*, No. pp. 106-110

Poletto, I.; Muniz, M. F. B.; Poletto, T.; Stefenon, V. M.; Baggiotto, C. and Ceconi, D. E. 2015.Germination and development of pecan cultivar seedlings by seed stratification. *Pesquisa Agropecuária Brasileira,* 50 (12): 1232-1235.

Poletto, T.; Maciel, C. G.; Muniz, M.; Blume, E.; Poletto, I. and Brioso, P. 2016. First report of stem canker caused by *Lasiodiplodia subglobosa* on *Carya illinoinensis* in Brazil. *Plant Disease*, **100** (5) 1016-1017.

Pond, A., Walworth, J., Kilby, W.M., Gibson, R., Call, R. and Nunez, H. 2006. Leaf nutrient levels for pecans. *Hort. Sci.,* **41**: 1339-1341.

Preece, J.E. and Read, P.E. 2003. Novel methods in micro-propagation. *Acta Horticulture,* **616**: 71-76.

Ramirez-Martinez, J.J. 1983. *In-vitro* establishment and proliferation of apical and axillary buds of pecan (*Carya illinoinensis* Koch). Effects of media, culture date and physiological age. MS thesis, New Mexico State University, Las Cruces.

Randall, J. J.; Rascon, A.; Heerema, R. J.; Potter, M. T. Grauke, L. J. and Heerema, R. 2015. Molecular mechanisms of pecan flower induction. *Acta Horticulturae,* **1070:** 89-99.

Rathore, D.S. 1985. Effect of Indole-butyric acid on the rooting of pecan cuttings under mist *Indian Journal of Horticulture,* **42**: 61-62.

Ravindran, C.; Sharma, M. and Kher, R. 2008. Present status and problems of Pecan nut (*Carya illinoinensis* Wang.) cultivation in Jammu and Kashmir. *Acta Hortic*., **772**: 179-182.

Reid, W. and Smith, M. W. 1993. Thinning pecans reduces alternate bearing. 83rd Annual report, Northern Nut Growers' Association. Pp. 91-94.

Reid, W. R. ; Grauke, L. J. and Heerema, R. 2015. The phenology of early fruit ripening in pecan. *Acta Horticulturae,* **1070**: 27-32.

Rice, N. R; Smith, M. W; Eikenbary, R. D; Arnold, D. C; Tedders, W.L; Wood, B.W; Taylor, G.G; Landgraf, B.S. and Barlow, G.E.1993. Evaluation of low input pecan orchard floor management systems. *Annual Report Northern Nut Growers' Association,* **84**: 95-103.

Rice, N.R.; Smith, M.W.; Eikenbary, R.D. and Arnold, D. 1998. Assessment of legume and nonlegume ground covers on Coleoptera: Coccinellidae density for low-in-put pecan management. *American Journal of Alternative Agriculture*, **13**(3): 11-123.

Rohla, C. T; Smith, M. W. and Maness, N. O. 2007. Influence of cluster thinning on return bloom, nut quality, and concentrations of potassium, nitrogen, and non-structural carbohydrates. *Journal of the American Society for Horticultural Science*, **132**(2): 158-165.

Romberg, L.D. 1931. The present status of the pecan pollination problem. *Proc. Texas Pecan Growers Assn.,* **11**: 56-61.

Romberg, L.D. 1968. Pecan varieties in 1968. *Proc. S. E. Pecan Growers Association,* **61**: 56-65.

Romberg, L.D. and Smith, C.L. 1968. Progressive report on the breeding of new pecan varieties. *Proc. Texas Pecan Grower Assn.,* **29**: 12-21.

Rosborough, J.F.; Smith, C.L. and Romberg, L.D. 1946. The pecan in Texas. *Texas Agr. Ext. Bul.*, 120.

Sampaio, V.R. and Barbin, D. 1980. A comparison between bench and field grafting in pecans. *An. Esc. Super. Agric. Luiz de Queiroz,* **37**(1): 329-334.

Sampaio, V.R. and Barbin, D. 1981. Influence of the time and size of the rootstock root on pecan bench grafting (*Carya illinoensis*).

Sanchez, E; Soto, J. M; Sosa Cerecedo, M; Yanez, R M; Munoz, E; Anchondo, A. 2009. Nitrogen use efficiency in pecan. *Terra Latinoamericana*, **27**(4): 311-317.

Sanderlin, R.S., Hunt, A.S. and Babcock, D.K. 1983. Epidemiology and control of vein spot disease of pecan caused by *Gnomonia nerviseda. Plant Disease,* **67**: 1209-1213.

Santerre, C.R., Scouten, A.J. and Chinnan, M.S. 1990. Room temperature of shelled pecans: Control of oxygen. *Proc. S. E. Pecan Growers Association,* **83**: 113-121.

Sargent, C.S. 1965. Manual of the tree of North America (exclusive of Mexico) 2nd ed. New York, Dover Publications. Pp: 934

Sharma, S.D. and Rosanghura, S. 1987. Rooting pattern of primary and secondary roots in air and mound layered pecans. *Fruit Sci. Rpt.,* **14**: 119-123.

Sharpe, R.H. and *Gammon*, N. 1958. Observations on time of bark grafting with wax or polyethylene covers. *Proc. S. E. Pecan Growers Association,* **51**: 54-55.

Sharpe, R.H., Blackmon, G.H. and Gammon, N. 1952. Relation of potash and phosphate fertilization to cold injury of Moore pecans. *Proc. S. E. Pecan Growers Association,* **45**: 81-85.

Shear, C.B. and Faust, M. 1980. Nutritional ranges in deciduous tree fruits and nuts. *Hort. Rev.,* **2**: 142-163.

Sherman, J.; Heerema, R. J.; Vanleeuwen, D. and St. Hilaire, R. 2017. Optimal manganese nutrition increases photosynthesis of immature pecan trees. *Hort. Science,* **52**(4): 634-640.

Shi, H. J.; Zhang, C. Q.; Shan, L. Y.; Xu, K. Y.; Xu, J. P.; Qi, Q. Q. and Xu, Z. H. 2015. First report of *Pestalotiopsis microspora* as a causal agent of black spot of pecan (*Carya illinoinensis*) in China. *Plant Disease*, **99** (9): 1276.

Singanuson, R., R.L. Mason, and B.R. D'Arcy. 2003. Compositional changes of Australia-grown Western Schley pecans [*Carya illinoinensis* (Wangenh.) K. Koch) during maturation. *J. Agric. Food Chem.,* **51**: 406-412.

Singh, S., Krishnamurthi, S and Katyal, S.L. 1967. Fruit culture in India. 2nd ed. ICAR, New Delhi, pp. 338-341.

Smith, M. W. 2010. Relationship of leaf necrosis and defoliation to phosphorus and potassium concentrations in selected tissue and to certain fruit quality parameters of pecan. *Scientia Horticulturae*, **125**(2): 117-122.

Smith, M. W; Carroll, B. L. and Cheary, B. S. 2000. Mulch improves pecan tree growth during orchard establishment. *Hort. Science,* **35**(2): 192-195.

Smith, M. W; Cheary, B. S. and Landgraf, B. S. 2001. Manganese deficiency in pecan. *Hort. Science,* **36**(6): 1075-1076.

Smith, M.W. 1977. Shoot meristem and callus tissue culture of pecans, *Carya illinoensis* (Wang) IC Koch. Ph.D. Diss., T'exas A&M Univ., College Station.

Smith, M.W. 2005. Understanding alternate bearing. *Pecan South.*, **38**(10): 32-37.

Smith, M.W. and Hinriches, H.A. 1980. High density pecan planting in Oklahoma. *Pecan Quart.,* **14**: 44.

Smith, M.W., Ager, P.L. and Endicott, D.S.W. 1985. Effect of nitrogen and potassium on yield, growth, and leaf elemental concentration of pecan. *J. Amer. Soc. Hort. Sci.*, **110**: 446-450.

Smith, M.W., Cheary, B.S. and Carroll, B.L. 2004. Response of pecan to nitrogen rate and nitrogen application time. *Hort. Sci.*, **39**: 1412-1415.

Smith, M.W., Cheary, B.S. and Landgraf, B.S. 2001. Manganese deficiency in pecan. *Hort. Sci.,* **36**: 1075-1076.

Smith, M.W., Wazir, F.K. and Akers, S.W. 1989. The influence of soil aeration on growth and elemental absorption of greenhouse grown seedling pecan trees. *Communications in Soil Science and Plant Analysis,* **20**: 335-344.

Smith, M.W., Wood, B. and Raun, W.R. 2007. Recovery and partitioning of nitrogen from early spring and midsummer applications to pecan trees. *J. Amer. Soc. Hort. Sci.,* **132**: 758- 763.

Smith, M.W.; Asrat, S. and Natasha, R.R. 1996. Legume cover crops as nitrogen source for pecan. *Journal of Plant Nutrition*, **19**(7): 1117-1130.

Smith, M.W.; Charles, T.R. and William, D.G. 2012. Pecan leaf elemental sufficiency ranges and fertilizer recommendations. *Hort. Technology*, **22**(5): 594- 599.

Smith, M.W.; Kenworthy, A.L. and Bedford, C.L. 1979. The response of fruit trees to injection of nitrogen through a trickle irrigation system. *J. Amer. Soc. Hort. Sci.,* **104**: 311-313.

Smith, M.W.; Storey, J.B.; Westfall, P.N. and Anderson, W.B. 1980. Zinc and sulphur in pecan leaflets as affected by application of sulphur and zinc to calcareous soils. *Hort. Science,* **15**: 77-78.

Smith, M.W.; Wood, B.W. and Raun, W.R. 2007. Recovery and partitioning of nitrogen from early spring and mid-summer applications to pecan trees. *J. Amer. Soc. Hort. Sci.,* **132:** 758-763.

Smith, M.W., Cheary, B.S. and Carroll, B.L. 2002. Fescue sod suppresses young pecan tree growth. *Hort. Sci.,* **37**:1045–1048.

Sparks, D. 1967. Shoot and apical bud abortion on non-bearing mature Stuart pecan trees as a function of time. *Proc. S. E. Pecan Growers Association,* **60**: 135-43.

Sparks, D. 1974. The alternate fruit bearing problem in pecans. *Ann. Rep. Northern Nut Grow. Ass.,* **65**: 145-158.

Sparks, D. 1975. Concentration and content of 14 elements of fruit of pecan. *Hort. Sci.,* **10**: 517-519.

Sparks, D. 1977. Effects of fruiting on scorch, premature defoliation, and nutrient status of Chickasaw pecan leaves. *J. Amer. Soc. Hort. Sci.*, **102**: 669-673.

Sparks, D. 1982. Zinc concentration in pecan leaflets associated with zinc deficiency symptoms. *Hort. Science,* **17**: 670-671.

Sparks, D. 1986b. Growth and nutrition of pecan seedlings from potassium phosphate foliar sprays. *Hort. Sci.,* **21**: 451-453.

Sparks, D. 1987. Apparent effect of zinc treatment on the growth rate of pecan production and yield. *Hort. Science,* **22**:899-901.

Sparks, D. 1988. Growth and nutritional status of pecan in response to phosphorus. *J. Amer. Soc. Hort. Sci.*, **113**: 850-859.

Sparks, D. 1989. Predicting nut maturity of the pecan from heat units. *Hort. Science*, **24**: 454-455.

Sparks, D. 1992. Pecan cultivation: The Orchard's Foundation. Pecan production Innovations. Watkinsville, Ga.

Sparks, D. 1993. Chilling and heating model for pecan budbreak. *J. Amer. Soc. Hort. Sci.,* **118**(1): 29-35.

Sparks, D. 1997. 'Desirable' pecan. *Fruit Varieties Journal,* **51**(1): 2-7.

Sparks, D. 1997. A model for predicting pecan production under arid conditions at high elevations. *J. Amer. Soc. Hort. Sci.,* **122**(5): 648-652.

Sparks, D. 2000. pecan in warm climate. In: Temperate fruit crops in warm climates, by Eraz, A. (ed.). Kluwer Academic Publishers, Norwell, Mass, pp. 381-403.

Sparks, D. 2004. A letter from Romberg, L.D. *Pecan South*, **36**(12): 14-20.

Sparks, D. 2005. Adaptability of Pecan as a species. *Hort. Science,* **40**(5): 1175-1189.

Sparks, D. and Baker, D.H. 1975. Growth and nutrient response of pecan seedlings, *Carya-Illinoensis* Koch, to nitrogen levels in sand culture. *J. Amer. Soc. Hort. Sci.*, **100**: 392-399.

Sparks, D. and Brack, C.E. 1972. Return bloom and fruit set of pecan from leaf and fruit removal. *Hort. Sci.,* **7**: 131-132.

Sparks, D. and Payne, J.A. 1982. Zinc concentration in pecan leaflets associated with zinc deficiency symptoms. *Hort. Sci.,* **17**: 670-671.

Sparks, D. and Yates, I.E. 2002. Pecan pollen stored over a decade retains viability. *Hort. Sci.,* **37**(1): 176-177.

Srivastava, K. K; Sharma, A.K; Sharma, M.K. and Aroosa Khalil. 2007. Standardization of budding time and methods in pecan. *Haryana Journal of Horticultural Sciences*, **36**(1/2): 55-56.

Stein, L.A., McEachern, G.R. and Storey, J.B. 1989. Summer and fall moisture stress and irrigation scheduling influence pecan growth and production. *Hort. Sci.,* **24**: 607-611.

Storey, J.B., Madden, G. and Garza-Falcon, G. 1970. Pecan research 1965-69. Influence of pruning and growth regulator on pecans. *Ter Agric. Exp. Sta.,* PR 2709.

Stuckey, H.P. and Kyle, E.J. 1925. Pecan diseases and spraying outfits. pp 191-192. In: Pecan growing, The Macmillan Company, New York, 233 pp.

Tedders, W.L. and Wood, B.W. 1985. Estimate of the influence of feeding by *Monelliopsis* pecan is and *Monellia caryella* (Homoptera: Aphididae) on fruit foliage, carbohydrate reserves, and tree productivity of mature 'Stuart' pecans. *J. Econ. Entomol.,* **78**: 642-646.

Thompson, T.E. and Grauke, L. J. 2012. 'Lipan' pecan. *Hort. Science,* **47**(1): 121-123.

Thompson, T.E. and Hunter, R.E. 1985. 'Pawnee' pecan. *Hort. Sci.,* **20**(4):776.

Thompson, T.E. and Young, E.F. 1985. Pecan cultivars – past and present. Tex. Pecan Growers Assn., College Station.

Thompson, T.T. 1984. Pecan cultivars: current use and recommendations. *Pecan Q.* **18**: 20-26

Thompson,T. E. and Grauke, L. J. 2001. 'Nacono' pecan. *Hort. Science*, **36**(6): 1162-1163.

Tisdale, S.M.; Nelson, W.L. and Beaton, J.D. (eds.). 1985. Soil fertility and fertilizers, 4th ed. Macmillan, New York.

Torikata, H., Kobayashi, Y. and Suganuma, H. 1988. Investigation of flowering and bearing of pecan trees in Aichi Prefecture. *J. Japan Soc. Hort. Sci.,* **57**: 145-51.

Upadhyay, S. K; Badyal, J; Awasthi, R.P. 2011. Effect of pruning intensities on growth, nut quality, yield and leaf nutrient status of pecan. *Indian Journal of Horticulture,* **68**(2): 184-188.

Upson, S., Rohla, C., Locke, J. and Springer, J. 2012. Pecan production: Establishing and managing an improved variety pecan enterprise in the Southern Great Plains. The Samuel Roberts Noble Foundation. Pp: 40

USDA. 2000. www.ams.usda.gov.in

van Staden, J., Wolstenholme, B.N. and Dimalla, G.G. 1976. Effect of temperature on pecan seed germination. *Hort. Sci.,* **11**: 261-262.

Villiers, E.A. and Viljoen, H.M. 1987. Evaluation of aldicarb, endosulfan and fenthion for control of stink bugs on macadamia nuts. *Information – Bulletin. Citrus and Subtropical Fruit Research Institute, South Africa.* **174**: 13 – 16.

Walworth, J., Pond, A., Nunez, H. and Kilby, W.M. 2008. Nutrient balance for pecans. In: Grageda, J., Sabori, R., Maldonado, L. and Ortega, P. (eds.), Reunion tecnica internacional de nogal pecanero Vol. Memoria Tecnica 27, INIFAP, Hermosillo, Son., Mexico., pp. 26-35.

Wells, L. 2010. Budding and Grafting of Pecan. University of Georgia Cooperative Extension Bulletin 1376: 8 p

Wells, L. 2015. Irrigation water management for pecans in humid climates. *Hort. Science*, **50** (7): 1070-1074.

Wells, L.; Brock, J. and Brenneman, T. 2014. Effects of foliar sulfur sprays on pecan independent of pecan scab control. *Hort. Science,* **49**(4):434-437.

Wells, M. L. and Wood, B. W. 2008. Foliar boron and nickel applications reduce water-stage fruit-split of pecan. *Hort. Science,* **43**(5): 1437-1440.

Wells, M.L. 2011. Response of pecan orchard soil chemical and biological quality indicators to poultry litter application and clover cover crops. *Hort. Sci.,* **46**(2): 306-310.

Wells, L. and Conner, P. 2012. Pecan varieties for Georgia orchards. *University of Georgia Circular,* 898.

Wetzstein, H.Y. 1989. Pollination and development of the receptive stigma in pecan, *Carya illinoensis*. *Acta Horti.,* **240**: 193-196.

Wetzstein, H.Y. and Sparks, D. 1983. The morphology of pistillate flower differentiation in pecan. *J. Amer. Soc. Hort. Sci.,* **108**: 997-1003.

Wetzstein, H.Y. and Sparks, D. 1985. Structure and *in-vitro* germination of pollen of pecan. *J. Amer. Soc. Hort. Sci.,* **110**: 778-781.

Wetzstein, H.Y., Ault, J.R. and Merkle, S.A. 1989 Further characterization of somatic embryogenesis and plantlet regeneration in pecan *Carya illinoinensis*. *Plant Sci.,* **64**: 193–201.

White, A.W. 1982. Effects of topsoil-subsoil fertilizer and lime amendments on top and root growth of 'Stuart' pecan seedling in a simulated Norfolk soil profile. *Hort. Sci.,* **17**: 380-381.

Winkler, J.A.; Guentchev, G.S.; Perdinan, P.; Tan, N.; Zhong, S.; Liszewska, M.; Abraham, Z.; Niedzwiedz, T. and Ustrnul, Z. 2011. Climate scenario development and applications for local/regional climate change impact assessments: An overview for the non-climate scientist. Part I: Scenario development using downscaling methods. *Geography Compass*. 5/6: 275-300.

Witt, H.J., Allison, J.R. and Daniel, J.W. 1989. Economic analysis of space management practices in high-density pecan groves. *J. Am. Soc. Hort. Sci.,* **114**: 61-64.

Wolstenholme, B.N. 1979. The ecology of pecan trees. Part 1. Characteristics of the native habitat. *Pecan Quart.,* **13**(2): 32-35.

Wood, B. W. 1993. Factors influencing fruit-drop of pecan. 83rd Annual report, Northern Nut Growers' Association. Pp. 68-71.

Wood, B. W.; Grauke, L. J. and Heerema, R. 2015. Regulation of vivipary in pecan. *Acta Horticulturae,* **1070**: 33-42.

Wood, B. W; Lombardini, L; Heerema, R. J. 2009. Influence of aminoethoxyvinylglycine on pecan fruit retention. *Hort. Science,* **44**(7): 1884-1889.

Wood, B. W; Reilly, C. C. and Nyczepir, A. P. 2006. Field deficiency of nickel in trees: symptoms and causes. *Acta Horticulturae*, **721**: 83-97.

Wood, B. W; Wells, L. and Funderburke, F. 2010. Influence of elevating tree potassium on fruit drop and yield of pecan. *Hort. Science*, **45**(6): 911-917.

Wood, B.W. 1982. *In-vitro* proliferation of pecan shoots. *Hort. Sci.,* **17**: 890–891.

Wood, B.W. 1989. Pecan production responds to root carbohydrates and rootstock. *J. Amer. Soc. Hort. Sci.*, **114**:223–228.

Wood, B.W. 2011. Influence of plant bioregulators on pecan flowering and implications for regulation of pistillate flower initiation. *Hort. Science*, **46**(6): 870-877.

Wood, B.W. and Reilly, C.C. 1984. Pecan kernel proteins and their changes with kernel development. *Hort. Sci.,* **19**: 661-663.

Wood, B.W. and Reilly, C.C. 1998. Susceptibility of pecan to black pecan aphids. *Hort. Sci.,* **33**: 798–801.

Wood, B.W., Payne, J.A. and Grauke, L.J. 1990. The rise of the United States pecan industry. *Hort. Sci.,* **25**: 721–723.

Woodroof, J.G. 1930. Studies of staminate inflorescence and pollen of Hickoria pecan. *J. Agri. Res.*, **40**: 1059-1104.

Woodroof, J.G. 1979. Tree nuts. AVI Publishing company, INC., Westport, Connecticut.

Worley, R.E. 1980. Effect of dikegulac on pecan shuck dehiscence and number of spring shoots. *Hort. Sci.,* **15**: 180.

Worley, R.E. 1991. Pecan tree spacing and tree size. *In*: Pecan Husbandry: Challenges and Opportunities. *1st National Pecan Workshop Proceedings.* Pp: 10

Worley, R.E., Harmon, S.A. and Carter, R.L. 1974. Effect of repeated N,P,K and lime applications on soil pH, P and K under old and young pecan trees. *J. Amer. Soc. Hort. Sci.,* **99**: 57-62.

Wu WenHao; Cao Fan; Liu ZhuangZhuang; Peng FangRen; Liang YouWang; and Tan PengPeng. 2016. Effects of NAA treatment on the endogenous hormone changes in cuttings of *Carya illinoinensis* during rooting. *Journal of Nanjing Forestry University*, **40** (5): 191-196.

Xu KuiYuan; Yang XianYu; Yuan ZiQian; Ling Hua; Xia GuoHua; Huang JianQin; Wu LianHai and Wang ZhengJia. 2015. Biological characteristics and fruit drop of *Carya illinoensis* 'Mahan'. *Journal of Anhui Agricultural University*, **42** (2): 165-170.

Yates, I.E. 1993. Environmental regulation of anther dehiscence and pollen germination in pecan. *J. Amer. Soc. Hort. Sci*., **118**(6): 699-706.

Yates, I.E. and Wood, B.W. 1989. Organogenesis from immature pecan embryonic axes *in-vitro*. *J. Amer. Soc. Hort. Sci*., **114**: 1025–1029.

Yates, I.E., and Sparks, D. 1989. Hydration and temperature influence *in-vitro* germination of pecan pollen. . *J. Amer. Soc. Hort. Sci*., **114**: 559-605.

Yates, I.E., and Sparks, D. 1990. Three year old pecan pollen retains fertility. *J. Amer. Soc. Hort. Sci*., **115**: 359-63.

Yates, I.E., and Sparks, D. 1993. Environmental regulation of anther dehiscence and pollen germination in pecan. *J. Amer. Soc. Hort. Sci*., **118**: 699-706.

Yates, I.E., Sparks, D., Conner, K. and Towill, L.1991. Reducing pollen moisture simplifies long-term storage of pecan pollen. *J. Amer. Soc. Hort. Sci.,* **116**: 430-434.

Yuan ZiQian; Yang XianYu; Ling Hua; Shen Chen; Zhao WeiMing; Huang JianQin and Wang ZhengJia. 2014. Dynamic changes of leaves' main mineral elements in pecan 'Mahan's growth period. *Acta Botanica Boreali Occidentalia Sinica*, **34**(7): 1443-1449.

Zhang JianHong; Shi JuanJuan; Xia GuoHua; Huang YouJun; Huang JianQin and Wang ZhengJia. 2014. Research of lignified cuttings and rooting mechanism of *Carya illinoensis. Journal of Anhui Agricultural University,* **41**(2): 203-208.

Zhang Rui; Li Hui; Peng FangRen; Hao MingZhuo and Zhai Min. 2014. Flowering morphological characteristics, pollen viability and stigma receptivity of *Carya illinoinensis. Journal of Nanjing Forestry University Natural Sciences Edition*, **38**(3): 50-54.

Zhang Rui; Peng FangRen; Le DongLiang; Liu ZhuangZhuang; He HaiYang; Liang YouWang; Tan PengPeng; Hao MingZhuo and Li YongRong 2015. Evaluation of epicotyl grafting on 25- to 55-day-old pecan seedlings. *Hort. Technology*, **25** (3): 392-396.

Zhang Rui; Peng FangRen; Yan Pan; Cao Fan; Liu ZhuangZhuang; Le DongLiang and Tan PengPeng. 2015. Effects of root pruning on germinated pecan seedlings. *Hort. Science,* **50** (10): 1549-1552.

Zhang WeiMei; Yang XianYu; Wu LianHai; Zhu JunHua; Shen Chao; Huang JianQin; Wang ZhengJia 2016. Male flower development and pollen germination of 'Mahan' pecan. *Journal of Anhui Agricultural University,* **43** (6): 908-914.

Zhang Xiang; Zhai Min; Xu YingChun; Li YongRong. 2014. Effect of different pruning measures on branch growth and carbon-nitrogen metabolite accumulation in branch and leaf of *Carya illinoensis. Journal of Plant Resources and Environment*, **23**(3): 86-93.

11

Pistachio (*Pistacia vera* L.)

M.U. Rehman, G.H. Rather, M. Maqbool Mir, Umar Iqbal
Yasmeen Gull, Ejaz Ahmad Parray, Shahid Qayoom

1. Introduction

The pistachio nut tree, *Pistacia vera* L. has been extensively cultivated since ancient times in the Middle Eastern and Mediterranean countries; in India, Afghanistan. The climatic conditions favorable to pistachio culture are similar to those for almond and olives, and prevail in extensive regions of the western United States also. Pistachio is a nut which comes originally from Iran and has the history of 4000 to 5000 years. Pistachio is a drought-tolerant, slow-growing, deciduous tree that reaches to a height of 30 ft. The tree survives in arid regions where summers are hot and the winters cold. Rocky, poor terrain with little rainfall is no problem as long as there is good drainage. The fruits grow in clusters. Although called a nut, botanically, pistachios are actually drupe. The pistachio consists of a bone ivory shell that encases the edible meat portion, called the seed.

1.1. Nutritional value and uses

Pistachio figures prominently among other nuts due to its nutritional value (Table 1). Because of its high value and favorable taste, planting pistachio trees has become common in other parts of the world. Pistachios are utilized mostly for fresh consumption. The processed nuts are used for preparation of candy, baked goods, and ice cream. It also have traditional medicinal uses such as toothache relief. The resin is used as a blood-clotting agent in Europe and the Middle East, while husks are used in India for dying and tanning. Pistachios have been reported as a remedy for scirrhus and sclerosis of the liver, abscesses, poor circulation, and other medical problems (Tous and Ferguson, 1996).

Table 1: Nutritional Value of Pistachio per 100g.

Edible portion		Minerals & Vitamins		Fatty Acids	
	Grams		Grams		
Protein	20.8	Potassium	1.020	Palmitic Acid	6.270 g
Fat	51.6	Phosphorus	5000.000	Myristic Acid	0.100 g
Carbohydrates	16.4	Magnesium	158.000	Lauric Acid	0.050 g
Others	11.2	Calcium	136.000	Oleic Acid	36.000 g
		Iron	7.300	Linoleic Acid	9.500 g
		Selenium	0.450	Stearic Acid	0.710 g
		Nickel	0.080	Palmitoleic Acid	0.290 g
		Vitamin C	7.000	Linolenic Acid	0.280 g
		Vitamin E	5.200	Arachidic Acid	0.280 g
		Nicotinamide	1.450	Total sterol	108.000 mg
		Vitamin B1	0.690	Beta-Sitosterol	90.000 mg
		Vitamin B2	0.200	Campesterol	6.000 mg
		Carotene	0.150	Stigmasterol	2.000 mg
		Folic acid	0.058		

Source: USDA Nutrient Database for Standard Reference, 2014.

2. Origin and distribution

The origin of pistachio was not known up to the beginning of the present century. According to Dioscurides Pedanius (historian/botanist of 1st century AD), Pistachio is a Latin word derived from 'pissa'= resin and 'aklomai' = to heal. That is to say, it is a plant with wholesome resin. Another explanation given by Davatchi (1958), its origin is in the Persian word 'peste'. However, it is generally acknowledged that the nuts of Jacob were pistachio nuts, called Gatoum = batam by the Arabs (Moldenke and Alma, 1952).

Syria and Mesopotamia were considered as the natural habitat of pistachio (Bailey, 1947). According to Vavilov (1951) the origin of pistachio is the near-east which covers Asia Minor, Caucasus, Iran and the mountain region of Turkmenistan. The tree was introduced into Europe at the beginning of the Christian era (Moldenke and Alma, 1952). The cultivated varieties of pistachio were introduced into Syria from Turkey and into Italy towards the year 800. Pistachio was introduced into Spain from Italy and France (Lemaister, 1959). It was later introduced into the USA in 1852-1854 (Lemaister, 1959).

The multiple species of the genus *Pistacia* comprise both the male and female commercial cultivars, as well as the rootstocks they are propagated on. *Pistacia* species are native to latitude between 40° and 70° in western Asia and Asia Minor. *Pistacia vera*, the species used to produce commercial nuts, and multiple other *Pistacia* species are still growing wild in numerous desert locations in

Lebanon, Israel, Syria, Iran, Iraq, India, Kyrgyzstan, Afghanistan, southern Europ, and the desert countries of Asia and Africa (Ferguson and Kallsen, 2016). Today it is grown as an orchard tree mainly in Iran, USA, Turkey, Afghanistan, Greece and Syria. Also, it seems that the first commercial plantations were mainly in Iran, Turkey, US, China and Afghanistan, and other countries adjacent to the wild pistachio nut stands.

3. Acreage, production, export value

According to the latest statistics (FAOSTAT. 2016) released by the World Food and Agriculture Organization (FAO) affiliated to the United Nations, the pistachio production in the world stood at 857,878 MT in 2014, on an area of 826,523 ha. Iran ranking first with a production of 415,531 MT, followed by the U.S. with a production of 233,146 MT and Turkey with 80,000 MT. Area occupied in Iran under this crop is 316,780 ha. while as in U.S. it is 89,436 ha. Although, Iran ranks first in terms of production and area, it does not enjoy such a high position in global market due to exporting challenges. In India, pistachio production is very less, with few plantation in Jammu and Kashmir and Himachal Pradesh. Despite the crop being third fastest growing product with an annual growth rate of 8.5 during the period 2000-2013 (Pietro, 2015), pistachio cultivation in India is minimal, though having a great potential to adapt it on commercial scale. India alone imported 10,450MT of pistachio during 2013 (FAOSTAT, 2016), which clearly shows the demand in the country, and need for exploring the field.

4. Taxonomy

Pistacia L. belongs to the family Anacardiaceae (cashew family), order Sapindales (Stevens, 2008). It contains nine species and five subspecies according to Al-Saghir and Porter (2012). *Pistacia vera* L. (cultivated pistachio) is by far the most economically important species in the genus. Linnaeus (1753) was the first to establish the genus and recognized six species of *Pistacia: P. lentiscus, P. narbonensis, P. simaruba, P. terebinthus, P. trifolia,* and *P. vera. Pistacia narbonensis* and *P. trifolia* are synonyms of *P. vera; P. simaruba* is *Bursera simaruba* (L.) Sarg. of the closely related Burseraceae.

Prior to Linnaeus, Tournefort (1700) considered *P. lentiscus* a distinct genus, *Lentiscus,* while he classified *P. terebinthus* and *P. vera* under the genus *Terebinthus.* At present the most complete taxonomic study is that of Zohary (1952), who included 11 species in the genus and divided them into four sections:

- *Lentiscella* Zohary - containing *P.mexicana* Humb., Bonpl. and Kunth and *P.texana* Swingle.
- *Eu Lentiscus* Zohary - containing *P.lentiscus* L., *P.saportae* Burnat, and *P.weinmannifolia* Poiss.ex Franch.
- *Buttmela* Zohary - containing *P.atlantica* Desf.; and
- *Eu Terebinthus* Zohary - containing *P.chinensis* Bunge, *P.khinjuk* Stocks., *P.palaestina* Boiss., *P.terebinthus* L., and *P.vera* L.

Since Zohary's (1952) treatment, several new species have been added to *Pistacia*. Yaltirik (1967a, 1967b) classified *Pistacia* species in Turkey and described a new species, *P. eurycarpa*. This had been called *P. atlantica* var. *kurdica* by Zohary (1952), a synonym. Zohary considered *P. palaestina* as a separate species, whereas Yaltirik (1976a, 1976b) retained it as a variety of *P. terebinthus*. Kokwaro and Gillett (1980).described a new species from East Africa, *P. aethiopica*.

Yi *et al.* (2008) assessed the phylogeny of *Pistacia* and recognized the following twelve species: *P. atlantica, P. chinensis, P. integerrima, P. khinjuk, P. lentiscus, P. mexicana, P. palaestina, P. saportae, P. terebinthus, P. texana, P. vera,* and *P. weinmannifolia.* However, on the basis of morphology, Al-Saghir and Porter (2012), recognized these texa (*P. integerrima* as a subspecies of *P. chinensis*), plus *P. chinensis* subsp. *falcate, P. eurycarpa,* and *P. lentiscus* subsp. *emarginata.*

5. Cytogenetic

All species of genus *Pistacia* are diploid with chromosome numbers of 2n = 24, 28 and 30. The basic chromosome number in this genus is x = 12, 14 and 15. *Pistacia khinjuk* with a chromosome complement of 2n = 24 has a symmetric karyotype with a median and submedian centromere position. The ancestral species have a symmetric karyotype hence, it seems that x = 12 is the initial basic chromosome number in this genus and the x = 14 and x = 15 derived from x = 12. No polyploid races have been reported in *Pistacia*. It seems that diploidy played an important role in evolution and speciation. The diploid chromosome number for cultivated pistachio, *Pistacia vera* is 2n = 30. Fasihi Harandi and Shahsavan Behboodi (1997) evaluated the species and subspecies of Iranian pistachio to find their karyotype and gave chromosomal complement of *Pistacia vera* as 2n = 30, *Pistacia khinjuk* as 2n = 24 and other subspecies of *Pistacia atlantic* as 2n = 28.

6. Botanical description, growth habit, flowering and fruiting phenology

Pistachio is deciduous and capable of growing moderately tall (25 – 35 ft), and, if not trained to a trunk, it will grow as a bush with multiple trunks. Once the tree is mature both vegetative and reproductive growth is apically dominant, growing or setting fruit primarily at shoot and flower apices. The leaves are large (called imparipinnate compound leaves), grayish green, and have three to five roundish leaflets that are 2 to 3 inches long (Ferguson and Kallsen, 2016). The fruit bud differentiation occurring during the year prior to blossoming (Ayfer, 1963). Shoot extension begins at the end of March and terminates between the end of April and the middle of May. Generally 1 or 2 axillary buds located distally on the new growth are vegetative. They are considerably smaller than the inflorescent buds, and may give rise to lateral branches in the following year, or they may remain dormant. Inflorescent buds begin expansion at the end of the following March, and anthesis occurs generally in the later part of May and for about three weeks their growth and differentiation is rapid (Ayfer, 1963). Thus pistachio bears its fruit laterally on wood produced during the previous season (Crane, 1984). Approximately 80% of the fruiting occurs on the wood that is < 10cm long, and the shoot >30cm in length may be important in establishing vegetative buds in a position above the main tree crown for canopy expansion during the following 'off year' (Stevenson *et al.*, 2000).

Both staminate and pistillate inflorescences are panicles that may consist of 100 to several hundred individual flowers enclosed in small scales. Both types of flowers are apetalous, and the pollinating agent is wind. Staminate and pistillate cultivars having similar flowering times must be provided to ensure adequate pollination of the latter. The fruit is a semidry drupe and consist of three parts: a combined exocarp and fleshy mesocarp (which form the fleshy hull), and a hard endocarp (called the shell, which encloses the seed, i.e., the edible kernel). Its maturity is manifested by a change in the epicarp (skin) from translucent to opaque, and a softening and loosening of the epicarp and mesocarp (hull) from the endocarp (shell) which encloses the embryo (kernel) (Crane and Iwakiri, 1985). The endocarp has a thin red-violet fleck and the seeds range in colour from light to dark-green. Endocarp dehiscence is first noted along the ventral suture in late July, about the time ultimate kernel size is attained and progress along both sutures until physiological maturity about the middle of September (Ozbek, 1978). Physiological maturity is signaled by easy separation of the epicarp (hull) from the shell (Crane, 1984), the equivalent of flesh separation from the pit in a free-stone peach.

6.1. Flowering and pollination

The female flowers have a pistil topped with a large three-lobed stigma on a short style, and an ovary that has three carpels, two of which abort. The male flowers have three to five large anthers and release large amounts of pollen. However, the timing of pollen release is variable and more than one variety of male trees is included in an orchard to ensure pollination. The ratio of male to female trees in orchards varies from 1:8 to 1:24 depending on the variety, location and size of orchard. While large orchards (>25 ha) use a higher ratio (1:30) (Rafati *et al.* 2014), small orchards of a few hectares should use a low ratio to ensure pollination. For successful pollination, female flowers must receive pollen within two days of opening. There are significant differences in the success of pollination between the different male varieties (Shuraki and Sedgley, 1994).

The fruits grow from the pistil of the female flower. The pistil has a single ovary at its base. The ovary forms the fruit wall that includes the shell and the hull. Within the ovary is a single ovule. The ovule, which contains the female germ cell (egg), will develop into the edible kernel of the nut. Extending from the ovary is a three-part style. Each of the three parts of the style terminates in a stigma. As the flower opens, these stigma surfaces become receptive to pollen.

Fruit set typically follows successful pollination. Pollen is released from staminate flowers on male trees and is carried by air currents to stigmas of the pistillate flowers. When a pollen grain lands on a receptive stigma surface, it germinates to form a pollen tube. The pollen tube is an elongated cell that grows through stigma and style tissue to the basal ovary and into the ovule. As it does so it carries male germ cells to the egg cell. Many pollen tubes germinate and grow through the style, but only one successfully reaches the ovule.

Martinez and Herrero (1995) studied the path of the pollen tube in pistachio (cv. White Batoury), a chalazogamous species where the pollen tube penetrates the ovule via the chalaza. Special attention was paid to the way in which the pollen tube gained access to the ovule. A single anatropous ovule with a big funiculus occupied the entire ovary cavity. At anthesis, a physical gap existed between the ovule and the base of the style. However, upon pollen tube arrival a protuberance, the ponticulus, developed in the uppermost area of the funiculus between the style and the ovule. This structure appeared to facilitate access to the ovule by the pollen tube. The pollen tube penetrated the ovule via this ponticulus. Upon penetration, callose developed in the ponticulus cells surrounding the pollen tube. After pollen tube passage, the upper layer of the ponticulus lignified and isolated the ovule from the style. This separation was further enlarged 2 weeks later when the ovary started to develop without expansion of the ovule and a large gap developed separating the ovule from the style. Except for the induction of callose formation by the pollen tube in the funiculus, this process was independent of pollination and appeared to be developmentally regulated since it occurred in the same way and at the same time in pollinated and unpollinated flowers. The ponticulus, although by a different mechanism, appeared to be playing the role of an obturator regulating access of the pollen tube to the ovule. Furthermore, this access was restricted to a particular time during the development of the ovule.

Successful pollen tube on reaching enters ovule and releases its contents. Fertilization involves the fusion of a male germ cell with the egg cell. The fusion of the male and female germ cells produces a single-celled zygote that eventually grows to form an embryonic plant. Thus, the reproductive process leads to fruit set. Weather conditions can greatly affect pollination and frost, rainy and/ or windy weather at flowering can cause poor fruit set. Fruit set is also affected by mild winter temperatures that cause an uneven bud burst in spring. Pistachios have a great chilling requirement than most other nuts (Ferguson *et al.*, 2005).

6.2. Fruit development

Pistachio nuts have an unusual growth pattern and the process begins slowly. In the first half of the season, the shell and hull develop; in the latter half of the season, the kernel grows to fill the shell. The whole thing occurs in three growth stages, identified as Stage I, II, and II. Stage I, is a period of rapid increase in the size and weight of the shell and hull, and begins from mid-April through mid-May. The first division of the zygote does not occur for several weeks after flowering during which time the ovary grows to its final size. During this period the pistachio shell achieves full size but is empty, since the kernel has not yet begun to grow. Pericarp growth takes place up to 4 weeks after pollination and preceded embryo, endosperm and funicle proliferation between 8 – 16 weeks (Shuraki and Sedgley, 1996). In stage II, from mid-May to late August (depending on the cultivar), the shell does not increase in size but thickens and hardens by lignification, and little growth of the kernel occurs. The shell is high in hemicelluloses (38.5% in mature shell) and xylan (33.1%), around which lignin (29.4%) provides a rigid structure (Salgado *et al.* 2012). Initially the cells in this layer are flexible and lack these structural elements. This stage of growth is the result of lignification of only these endocarp cells. During a lignification stage, the cells begin to produce smaller amounts of the hemicellulose and lignin components, transporting these elements to the cell wall, where they are polymerized by laccases, peroxidases, and other chemical couplers (Li and Chapple, 2010).

In mid- to late-June, the beginning of stage III, the kernel begins to grow, achieving nearly full size by late July or early August. The weight gains from late June through August are from kernel growth and maturation. This unusual growth pattern allows flexibility in irrigation, as there are times (during bloom, nutshell expansion, and kernel growth) when decreasing irrigation could decrease nut quality and yields through effects on size, weight, and embryo growth. Decreased irrigation during Stage I and III has been demonstrated to decrease nut size and increase non-split nuts, respectively, while decreased irrigation during Stage II has little effect on final yield or quality (Goldhamer *et al.*, 2004).

The shell begins to split in late July. Full maturity (in late August or during September, depending on the heat unit accumulation and cultivar) is indicated by the easy separation of the hull from the shell. When the hull begins to separate/slip from the shell on the majority of the nuts, the shell simultaneously begins splitting from the tip to the base, along the lateral sutures of the shell (Zarei *et al.*, 2014).

As the pistachio nut develops through the season, simultaneous changes occur in the chemical composition of the kernel. Pistachio kernels are initially composed of sugars, predominantly sucrose, followed by much lower levels of glucose, fructose, and inositol, a sugar alcohol. Sucrose increases sharply during the first month of kernel growth, peaking at about 35 percent dry weight in early July and decreases rapidly to about 6% in mid-August, where it remains through maturity. Crude fats comprise about 4% of kernel dry weight in mid-July. As sugar content decreases in late July, crude fats sharply increase to 40% by the end of August (Zarei *et al.*, 2014).

7. The physiology of the tree

Pistachio trees are temperate deciduous trees with physiological adaptations to desert and saline environmental conditions. Areas suitable for pistachio production have long, hot, dry summers with no rainfall, humidity below 50% during the growing season, and moderate winters with sustained temperatures in the range of 40°F and no hard freezes. Overnight freezes down to 4°F to 11°F can damage 1 year-old shoot growth but not major scaffold or trunk wood on mature, dormant trees. Frost damage also appears to be increasing in young, vigorous orchards established in saline conditions. The soils in these orchards are often more saturated late in the season, delaying young, vigorous trees from entering dormancy and making them more susceptible to the late-fall freezes. *Arbuscular mycorrhiza* colonization improves the drought tolerance of seedlings by increasing the accumulation of osmotic adjustment compounds, nutritional and antioxidant enzyme activity, and thus increases host biomass and plant growth (Abbaspour *et al.*, 2012). Humic acid, the major component of organic matter, tend to moderate salinity stress and decrease leaf proline accumulation, also it decreases levles of ABA (a stress-induced hormone) by their stimulatory effects on plant growth and development (Moghaddam and Soleimani, 2012).

The growth differs among trees on different rootstocks, with greater stimulation of neoformed (newly-formed) growth in trees on the more vigorous rootstocks. Most annual extension growth in mature tree crowns is preformed, contrasting with the relatively high rate of neoformation found in young tree crowns. Large amounts of neoformed growth in young trees allow the trees to become established quickly and secure resources, whereas, predominantly preformed growth in mature trees allow for continued crown expansion without outgrowing available resources. Understanding the source of variation in shoot extension growth on different rootstocks has important implications for orchard management practices (Spann *et al.,* 2007).

Pistachio trees bear heavy (on-year) and light (off-year) fruit crops in alternate years. Nutrient uptake occurs primarily during nut filling in both on-year and off-year trees. The sink demand regulates the uptake and distribution of N, P, and K in Pistachio trees. In on-year trees, N and K uptake increases by 35% and 1112% respectively, during nut filling compared with off-year trees. During this period, nutrients are allocated largely to embryo development in on-year trees and to storage in perennial tissues in off-year trees. Nutrient uptake is negligible between harvest and leaf senescence. Although root growth is reduced during nut filling in on-year trees compared with off-year trees, there is no relationship between root growth and the uptake of N, P or K from soil (Rosecrance *et al.* 1996).

The tree simultaneously bears the current year's crop on shoots grown the previous year, as well as the buds for the following year's crop on the new apical growth of the same shoot. Nuts are born in panicles on a rachis, referred to as cluster on one-year old wood. The development of a cluster begins three seasons before it is harvested. At cessation of vegetative growth in season 1, usually in June, shoots terminate in a large, apical, terminal bud. In trees of reproductive age, these large terminal buds contain the primordial that will develop, next season, into a new shoot, complete with leaves and their axillary vegetative and reproductive buds. As the new shoot grows, some buds in the axils of the expanding leaves on this new shoot are vegetative, and some rapidly begin differentiating into male or female flower buds, depending on the sex of the tree. Differentiation continues through mid-June of season 2, and by early July, further differentiation within these axillary flower buds ceases, and they enter a quiescent period. Prior to the off-year of the alternate-bearing cycle, many of these differentiating flower buds will not make it past this stage. They fall from leaf axils beginning in July and continuing through October. Female flower buds resume growth briefly with pistil initiation and then remain quiescent until March of season 3, when differentiation of functional female flower parts is completed. Male florets remain quiescent from July of season 2 until the following March, at which time they complete differentiation and mature by mid-spring (Ferguson and Kallsen, 2016).

The proteolytic activity is important in the mobilization of nitrogen reserves stored in the bud storage proteins during dormancy to support the fast developing inflorescence in spring after bud dormancy break. The activity reaches a peak at the beginning of bud opening and flowering and is directed against inflorescence bud dehydrin-like proteins (IBP32 and IBP27) (Peri *et al.* 1999). Respiration rates of pistachio reproductive and vegetative organs are related to their developmental stage. Both for leaves and for infructescence, the highest rate can be observed during the earlier stages of growth corresponding to the phases of most intense organ growth (Marra *et al.*,2009).

Climatic factors influence the reproductive physiology of the crop in the aspects related to the commercial quality of the yield (percentage of dehiscent and blank fruits). Heat units above 30°C during the growing season (from May to September) have the best significances with the percentage of split fruits. High temperatures during the vegetative growth of the tree increase the percentage of split fruits, with maximum temperature in June being those of major influence. Moreover, the percentage of blank fruits show better relationship with the temperatures of April, when flowering and pollination occurs; generally it is observed that the minimum temperatures of this month affect the final percentage of blank fruits (Gijon *et al.,* 2014).

8. Climate

Determination of chilling and heat requirements of pistachio (*Pistacia vera* L.) cultivars is important for satisfactory growth and development, particularly when large-scale commercial production is desired. Adequate winter chilling is essential but this varies with variety. Chill units of 1000 hours below 7°C is said to be required to produce an even bud break and good set, but it should not be less than 670 h (Jiang-JianGuo, 1999). Less units of chilling usually result in irregular flowering and delayed leaf-out. Under such mild winter conditions, where chilling requirement is not met, application of hudrogen cynamide (H2CN2) at 5.0% ~30 days before expected budbreak can increase budbreak as substitute for chilling (Arreola-Avila *et al.*, 2011). Pistachio requires extended dry summer for its proper development and productivity. A frost-free period in excess of 10700 GDH (Growing Degree Hours) is considered necessary to ensure the flower development and fruit ripening (Kuden *et al.*,1995). Alpine areas and regions too far north or too close to moderating coastal influences may therefore be unsuitable.

While pistachios will tolerate winds, however, when trees are young, strong winds during flowering and heavy cropping can cause reduced yields and broken branches. In most situations, secure staking of young trees is essential. The incidence of fungal and bacterial diseases increases with an increase in humidity, particularly in the growing season, and infections are likely to over-winter and recur the following season. Therefore, coastal areas and regions receiving > 20mm of rain per month during summer are unlikely to be suitable for pistachio production.

9. Soil

Pistachio are similar to other nut crops in that high yields and pistachio quality are best achieved on deep, uniform loam soils with a neutral pH. These soils provide the optimal combination of permeability with sufficient water-holding capacity and root zone aeration. Orchards with soils that vary from sands to clay texture are subject to variable water stress across the orchard because of different water holding capacities of the soils. Layers in rootzone can cause water logging by slowing water movement through the root zone and creating temporarily saturated soil layers. These injure roots by depriving them of oxygen and enhancing conditions that favor root diseases. Some subsoil layers can form physical barriers to roots that simply cannot grow through hard, dense, or compacted layers. Pistachio yield can be considerably affected by salinity and clay content of the soils (Salehi and Hosseinifard, 2012). Saline soils have been found to decrease phosphorus concentrations in leaves, stem, roots; as well as leaf potassium concentration, and increased sodium concentrations in these plant organs (Shahriaripour *et al*., 2011). The deleterious effect of salts can get depleted by adding gypsum in conjunction with organic amendments (Yazdanpanah *et al*., 2011; Mahmoodabadi *et al*., 2013).

10. Land /pit preparation

Holes are usually hand-dug and the stock inserted with the top of the root-ball slightly above the soil level; this compensates for settling after watering and reduces the potential for excessive moisture around the 'crown' of the seedling. Tamping of the backfilled soil by foot is done sparingly to avoid damage to the small roots extending from the crown. The bamboo stick supporting the nursery stock is also removed at this time and the resulting small hole is covered with soil to prevent drying of the crown roots. The trees are re-staked using six-foot stakes to provide additional support.

11. Raising of rootstock

Propagation of rootstock is done from seeds or using softwood cuttings. Seeds can be obtained from relatively large fruits when the hull turns blue-green in the mid-fall; fruits with pinkish hull may comprise blank seeds and must be rejected. The hull contains a germination inhibitor and should be removed by soaking the nuts in water followed by rubbing over a screen. Seeds that float in water have aborted embryos and should be discarded (Hartmann *et al*. 2002). Seeds are then stratified in moist and cool environment (2-4°C) for 4 weeks, and at times for 6 weeks, before germination at 20°C. For better root production of germinated seeds, IBA (100mg/l) or Agrobacterium rubi strain A1 have been found promising in increasing the number of lateral roots (Orhan *et al*., 2007). Use of containers that promoting horizontal root growth in the pistachio nursery is very important as because pistachios grown on arid desert soils develop a vertical rather than a horizontal root system that increase seedling mortality at the time of transplanting (Popov *et al*., 1994). With early planting of seeds, the more-vigorous seedlings may reach budding size by fall. Because of their long tap root, best results are obtained by planting seedling rootstocks in their permanent orchard location, then grafting or budding *in-situ*.

12. Propagation

12.1. Cuttings

Currently, rooting of cuttings is not the most common practice of vegetative propagation in pistachio. Clonal propagation through softwood cuttings can be performed by collecting the softwood cuttings from etiolated mother trees and rooted under mist (Aleta *et al*., 1997). Softwood cuttings of one-year old *P. chinensis* seedlings can be treated with 5000 mg/l IBA and result in 92% rooting (Pair and Khatamian, 1980). Explants taken from *P. vera* seedlings can be rooted *in vivo* following a quick dip (5 seconds) in concentrated IBA solution 500 mg/l, with 100% rooting (Abousalim, 1990). As far as rooting of adult material is concerned, rooting attempted *in vitro* using the quick dip method, achieved 50% success when the cut ends of 4-year-old micro-cuttings were dipped for 10 s in 1000mg/l IBA (Abousalim, 1990). Softwood *P. vera* cuttings were successfully rooted under a mist system after very high auxin concentrations were used; 35000 mg/l IBA found to be optimal and 88% rooting was obtained after six weeks from planting (Al Barazi and Swabe, 1982).

12.2. Grafting

The scion- or bud-rootstock compatibility is decisive in the success of grafting pistachio. Climate conditions and appropriate rootstock choice are the two key factors influencing grafting success. Bud-stock compatibility can be enhanced by treating rootstocks with mycorrhiza, which helps by efficient uptake of nutrients (Aznarte-Mellado *et al.*, 2014).

For winter grafting, dormant wood of the female cultivar is collected from the mother tree in early winter and stored at 2°C in sealed bags. Propagators usually graft trees using a wedge graft. To do this a piece of scion stick containing two good vegetative buds is wedged into a slit in the trunk of the rootstock ensuring that there is good cambium contact. The graft is taped firmly and the tree is placed in a greenhouse until the buds grows. Selection of scion wood from semi-hardened water sprouts or non-fruiting shoots is generally the best.

12.3. Cleft grafting

It is the most common method and may be done either in the greenhouse or field. Scion and rootstock diameters should be similar in size. Rootstock of diameter 0.7 cm or more than 2cm can be cleft grafted as long as a good cambial match between scion and rootstock is maintained. While cleft grafting, do not line up the cambia precisely, but slant the scion into the cleft so that the cambia cross at one point. This method of grafting is usually more successful if performed in late summer.

12.4. Saddle grafting

Generally, rootstock between ¼ to ½ inches is useful for saddle grafting. The scion is prepared by cutting upward through the bark and into the wood on opposite sides of the scion, and the saddle shape is formed by removing pieces of wood. The rootstock is cut transversely and receives two upward cuts on either side. This should be done to expose the vascular cambium of the rootstock to match vascular cambium in the saddle of the scion. The apex of the rootstock is carved to fit the saddle. The graft needs to be tied, and all exposed cut surfaces sealed, till the graft union has formed. This method of grafting is usually more successful if performed in late summer.

12.5. Budding

Careful selection of ideal scion buds is essential for successful budding. Since most dormant buds on mature pistachio trees are flower buds, propagator should select vegetative buds based on their size and location. Vegetative buds are almost always smaller than flower buds and tend to occur at the base of current-year growth on flowering shoots. Vegetative buds tend to be more pointed than flower or fruit buds that are more rounded. Young, rapidly growing shoots with greater than 20 inches of current growth will tend to have more vegetative buds. Generally, younger trees less than six years old are the best source of budwood.

12.5.1. T-Budding

T-budding is the most efficient and most commonly used method in commercial pistachio production. T-budding is performed when the rootstock bark is slipping during the early summer and the scion buds have matured and hardened. After the seedlings are well established, the bud can be inserted 0.6m or more above the ground. Generally, rootstock between ¼ to ½ inches are useful for T-budding. The best results can be obtained during June with T- and inverted T-budding (Arpaci *et al.*, 1998). Liu-ZhenLiang *et al.* (2004) reported 80% success, when 1 year-old twigs with vigorous buds from 2- to 4-year-old mother trees were used as scions and plants with more than 1 cm diameter were used as rootstocks. Budding was best in August when the air temperature at 100cm height was 25-30°C, with a relative humidity of 60-70%. Also the highest affinity between scion and rootstock was recorded with Kerman on UCB-1 rootstock.

12.5.2. Chip budding

Generally, rootstock between ¼ to ½ inches are useful for chip budding. Accurate matching of cambia is required for chip buds. The best results can be obtained by chip budding in May-June using dormant buds from budsticks kept under refrigerated conditions (Aleta *et al.*, 1997). Chip budding onto the *P. vera* during early autumn is found to be most successful (Arpaci *et al.*, 1998).

12.5.3. Patch budding

Patch budding requires a larger caliper rootstock and is therefore more likely to be done in the field in the spring or fall. Rootstocks sometimes produce excess sap (gumming), which can be a problem when budding in early summer. Larger rootstock between ½ to ¾ inch in diameter are needed for patch budding.

12.6. Micropropagation

Micropropagation of *Pistacia vera* L. using different explant sources and suitable medium has been presented in Table 2 and 3. The initiation of cultures from adult material has usually involved pruning, grafting and application of benzyladenine (BA) and gibberellic acid (GA3) as spray treatments to stimulate new growth of shoots (Gonzales and Frutos, 1990).

Mature pistachio cultivars have been extremely difficult to establish due to the high incidence of contamination (Abousalim, 1990) and to oxidation of phenolic compounds (Barghchi and Alderson, 1989). Oney (1996) reported that small meristem tips (approx. 1mm) from actively growing shoot of mature *P. vera* plants can be used to reduce the contamination problem. Multiple shoot production has been obtained from shoot tips cultured on semi-solid MS medium with B5 (Gamborg *et al.*, 1968) vitamins and 4.4 μM BA (Oney, 2000a), with a multiplication rate of 20 shoots for each explant for each 30-day cycle.

Table.2: *in vitro* shoot tip/ nodal culture studies of pistachio, *Pistacia vera* L.

Species and cultivars	Explant source	Medium (μM)	Morphogenic response	References
P. vera L. rootstocks	tb/ab	MS + BA (18)	Multiple shoot proliferation, Plantlet regeneration	Barghchi, 1982
P. vera L. rootstocks	tb/ab	MS + BA + NAA ½ MS	Plantlet regeneration,Callus formation	Alderson and Barghchi, 1982
P. vera L. rootstocks	st/ab	MS + kin (9.3-18)MS + kin (9.3-18) + NAA (1.4-1.5) +PhG Modified MS + IBA (12.3)	Single shoot growth,Reduced multiple shoot,Rooted plantlet	Barghchi and Alderson, 1983a,b
P. vera L.	ab	EM + BA (13.3) + GA3 (1.4) + IAA (0.006)	Shoot proliferation, Plantlet regeneration	Bustamante-Garcia, 1984
P. vera L. rootstock	st/nb	MS + BA (18) Modified MS + IBA (12.3)	Multiple shoot proliferation, Plantlet regeneration	Barghchi and Alderson, 1985
P. vera L. rootstocks	at/nb	MS + kin (9.3-18.6) MS + BA (18) Modified MS + IBA (12.3)	Single shoot proliferation, Rooted plantlets	Barghchi, 1986b
P. vera L. 'Kerman' and 'Peter's'	at/nb	MS + BA (1.3) + IBA (0.3) MS + 2,4-D + BA	Shoot growth, Callus formation	Barghchi, 1986a
P. vera L.	st	MS + BA	Shoot growth and necrosis	Barghchi, 1986c
P. vera	st/ab	MS + BA	Shoot growth	Martinelli, 1988
P. vera L. 'Mateur'	st	MS + BA (17.8) MS + IBA (4.9-9.8)	Multiple shoot formation, Rooting	Abousalim, 1990
P. vera L. 'Antep'	nb	MS + kin (4.7) + NAA (5.4)	Plantlet regeneration	Yucel *et al.*, 1991
P. vera L.	st/nb	MS + BA (0.4) + IBA (0.5)½ MS + IBA (7.4-12.3)	Multiple shoot formation, Rooted plantlets	Yang and Ludders, 1993
P. vera L. 'Mateur'	st	MS + BA (13.3)	Necrotic shoots	Abousalim and Mantell, 1994
P. vera L. 'Kerman' and 'Stewart'	st	MS + BA (4.4) + IBA (1.0) + MeJa (1.3-4.4 μM)	Shoot multiplication, Rooted plantlets	Dolcet Sanjuan and Claveria, 1995
P. vera L.	st	DKW + BA + IBADKW + IBA	Multiple shoot formation, Rooted plantlets	Parfitt and Almehdi, 1994
P. vera L. 'Antep'	nb	MS + BA (2.2-35.5)	Multiple shoot formation, Plantlet regeneration	Onay, 2000a

ABA – abscisic acid; ab – apical bud; alf – aseptic leaf; at – apical tip; Au – auxin; BA – 6-benzylaminopurine; Cyt – cytokinin; 2,4-D – 2,4-dichlorophenoxy acetic acid; DKW – Driver-Kuniyuki-Walnut medium; ER – Eriksson medium; GA3 – gibberellic acid; IAA – indole-3-acetic acid; IBA – indole-3-butyric acid; if – immature fruits; ike – immature kernels; inf – inflorescence; ke – ker*nel; kin – kinetin; lf – leaf; MeJa – methyl jasmonate; MS – Murashige and Skoog medium; NAA – á-naphthalenacetic acid; nd – nodal bud; PEMs – proliferative embryogenic mass; PhG – phloroglucinol; st – shoot tip; tb – terminal buds; TDZ – thidiazuron.*

Table.3: *in vitro* organogenesis and embryogenesis studies of pistachio, *Pistacia vera* L.

Species and cultivars	Explant source	Medium (µM)	Morphogenic response	References
P. vera L. rootstocks	at/nb/lf	Mod MS + Cyt + Aux	Callus formation,Organogenesis	Barghchi 1982; Barghchi and Alderson, 1983b
P. vera 'Lambertin'	at/nb/nf	MS + BA (1.3) + IBA (0.25) MS + 2,4-D + BA	Shoot growth,Callus formation	Barghchi and Martinelli, 1984
P. vera L.	lf	MS + 2,4-D (18) + NAA (10.7) + kin (9.3)	Callus formation	Zaheer *et al.*, 1989
P. vera L. 'Antep'	ke	MS + BA (8.9)	Somatic embryogenesis	Onay *et al.*, 1995
P. vera L. 'Antep'	st/nb	MS + NAA	Somatic embryogenesis	Taskin *et al.*, 1996
P. vera L. 'Antep'	PEMs	MS + BA (8.9)	Synthetic seeds, Encapsulated PEMs	Onay *et al.*, 1996
P. vera L. 'Antep'	at/nb	MS + BA MS + BA or TDZ	Organogenesis,Embryogenesis	Onay, 1996
P. vera L. 'Antep'	ke	Liquid MS + BA or ABA	Maturation,Germination, Embryo development	Onay *et al.*, 1997
P. vera L. 'Antep'	lf	MS + TDZ (4.5-9.1)	Somatic embryogenesis	Onay and Namli, 1998
P. vera L. 'Antep'	ike	MS + BA	Somatic embryogenesis, Germination	Firat and Onay, 1999
P. vera L. 'Antep'	if/lf	MS + BA or TDZ	Somatic embryogenesis	Onay and Jaffree, 2000
P. vera L. 'Antep'	alf	MS + BA	Development of somatic embryos from single epidermal cells	Onay, 2000c
P. vera L. 'Antep'	ke	MS + BA	Somatic embryogenesis	Onay, 2000b
P. vera L. 'Antep'	at/nb	MS + BA or ABA	Somatic embryogenesis,Maturation	Onay *et al.*, 2000

ABA – abscisic acid; ab – apical bud; alf – aseptic leaf; at – apical tip; Au – auxin; BA – 6-benzylaminopurine; Cyt – cytokinin; 2,4-D – 2,4-dichlorophenoxy acetic acid; DKW – Driver-Kuniyuki-Walnut medium; ER – Eriksson medium; GA3 – gibberellic acid; IAA – indole-3-acetic acid; IBA – indole-3-butyric acid; if – immature fruits; ike – immature kernels; inf – inflorescence; ke – kernel; kin – kinetin; lf – *leaf; MeJa – methyl jasmonate; MS – Murashige and Skoog medium; NAA – á-naphthalenacetic acid; nd – nodal bud; PEMs – proliferative embryogenic mass; PhG – phloroglucinol; st – shoot tip; tb – terminal buds; TDZ – thidiazuron.*

Rooting of the shoots has been induced on MS medium supplemented with 9.8µM IBA. Rooted plantlets can survive in soil after a short acclimatization period. Parfitt and Almehdi (1994) obtained 88% rooting using DKW medium containing 1.2×10^{-5} M IBA. In this method improved shoot growth, survival when rooting, and acclimatization was observed as compared to sucrose-based medium. Contamination was controlled using 2% CO_2 atmosphere.

The use of somatic embryogenesis to overcome problems of red pigmentation and bud necrosis in in vitro propagation of pistachio was investigated by Chatibi *et al.* (1998). Cotyledons, leaves and apical internodes were cultured in inducing medium supplemented with various combinations of 2,4-D and kinetin. Explants were transferred to a second medium for callus proliferation and embryo maturation. Callogenesis levels were similar for all types of explant but somatic embryogenesis on the inducing medium varied with explant source. Somatic embryos were obtained from cotyledons in the presence of 1.0 mg 2,4-D + 1.0 mg kinetin/litre (60%); from leaves in the presence of 2.0 mg 2,4-D + 1.5 mg kinetin/litre (63%); and from internodes in the presence of 5.0 mg 2,4-D + 2.0 mg kinetin/litre (68%).

Ghoraishi and Mantell (1998) developed a modified protocol for the micropropagation of *P. mutica* and *P. vera* in order to control shoot-tip necrosis (STN) and improve shoot growth. Shoot explants were cultured on MS medium supplemented with 100 micro M IBA. Rooting occurred after 5 days. Under saturated or near-saturated water vapour (99% RH), microplants of both species showed 100% incidence of STN and this was completely alleviated by reducing RH to 89%. This was achieved by using Suncap closures to improve air exchange in the culture vessels, and covering the media surfaces to prevent evaporation inside the culture vessel. Also, relative shoot length increase was significantly improved from 1.92 to 3.6 mm/week in *P. mutica* and from 3.2 to 6.04 mm/week in *P. vera*. Scanning electron microscopy of the stomatal structure and surface leaf wax showed that microplants grown under the improved culture system had characteristics similar to those of plants grown in the greenhouse.

Tilkat *et al.* (2009) developed a protocol for direct shoot and plantlet regeneration from in vitro regenerated leaf explants of male *Pistacia vera* L. cv. 'Atl'. Leaves excised from axenic shoot cultures of pistachio were used to induce organogenesis on a Murashige and Skoog (MS) medium with Gamborg vitamins supplemented with combinations of different concentrations of BAP and IAA. The highest adventitious shoot regeneration in 35% of the explants, with the number of shoots ranging from 2 to 3 per explant, occurred in the explants cultured during the establishment phase in the medium with 1 mg l^{-1} IAA and 2 mg l^{-1} BAP. For shoot multiplication, the highest number of new microshoot/explants (5.76) was obtained in a culture medium supplemented with 1 mg l^{-1}

BAP, but it was not significantly different from the number obtained at 2 mg l^{-1} BAP. A high rooting frequency (84%) for microshoots was recorded on a medium supplemented with 2 mg l^{-1} IBA. In vitro rooted plantlets were transferred to pots filled with a mixture of soil, sand and peat (1:1:1). They were weaned in a growth room and finally moved to a greenhouse.

Tilkat *et al.* (2011) developed a method for direct plant regeneration from in vitro-derived leaf explants of *Pistacia vera* L. cvs. "Siirt" and "Atl". The *in-vitro* adventive organogenesis involved six steps that included (1) explant disinfection, (2) the addition of cytokinins (benzyladenine; BA, Kinetin; Kin) for the in vitro installation of mature shoot tips, and the successful rapid proliferation of shoot initials, (3) induction of shoot initials from the regenerated mature leaves, (4) regeneration and elongation of shoots from the shoot initials, (5) rooting of the shoots and (6) acclimatization of the plantlets. Exudation of phenolics was not observed when the rinsed shoot tips were washed twice, in sterile distilled water, on a shaker at 150 rpm for 1 hour. Explants on the Murashige & Skoog (MS) medium with Gamborg vitamins supplemented with 1 mg L-1 BA produced the highest number of shoots and longest shoots in the second step. Leaves excised from axenic shoot cultures of pistachio were used to induce adventive organogenesis on a MS medium with Gamborg vitamins supplemented with combinations of different concentrations and combination of BA and indole-3-acetic acid (IAA). In the third stage, the best medium for shoot induction in both cultivars was MS medium with 1 mg L-1 IAA and 2 mg L-1 BA. For shoot multiplication, the highest number of new microshoot/explants for male and female was obtained in a culture medium supplemented with 1 mg L-1 BA, but it was not significantly different from the number obtained at 2 mg L-1 BA in the fourth step. The elongated shoots (>4 cm long) were rooted on a full-strength MS basal medium supplemented with 2 mg L-1 of indole-3-butyric acid (IBA) in the fifth step. Finally, the rooted plants were transferred to soil.

An effective pistachio (*Pistacia vera* L.) micropropagation system was developed by Benmahioul *et al.* (2012) involving rapid axillary bud proliferation and ex vitro rooting. The highest shoot proliferation frequency was obtained from nodal explants cultured on Murashige and Skoog (MS) basal salts containing Gamborg (B5) vitamins and supplemented with 4 mg l^{-1} 6-benzyladenine (BA). The addition of 2 mg l^{-1} meta-topolin (mT) generated an optimal number of shoots with suitable morphological features, while kinetin (KIN) was found to be unsuitable for pistachio shoot proliferation. Microcuttings were rooted ex vitro after being dipped in rooting powder. The peak ex vitro rooting response was achieved after shoot explants were treated with RhizoponReg. 2% indole-3-butyric acid (IBA). Rooted plantlets were transplanted in plastic pots containing a peat-perlite-vermiculite (1:1:1) mixture and then transferred to the greenhouse. After 2 months, 81.5% survival of rooted microshoots was achieved.

12.6.1. Somatic embryogenesis

Somatic embryogenesis of pistachio was first achieved using immature kernels as explants (Oney *et al.*, 1995). Subsequently, somatic embryogenesis of several genotypes was reported using immature and mature zygotic embryos, cotyledons and juvenile leaf tissues as explants (Onay, 2000b; Onay and Jeffree, 2000). Kernels from immature fruits (12 weeks after pollination) have been used as explants. Following removal of the outer pericarp and shells, material is surface-steriliized, the testa is removed from each seed and the kernels are washed with sterile distilled water before culturing in liquid MS medium, supplemented with B5 vitamins, 200mg/l casein hydrolysate, 50mg/l ascorbic acid, 8.9 µM BA and 30 g/l sucrose. Cultures are incubated in continuous light at 25±2°C. embryogenic cultures have also been induced from juvenile leaves. Onay and jaffree (2000) reported induction of embryogenic cultures from juvenile leaves of pistachio on medium containing thidiazuron (TDZ). Embryogenic cultures consist of proembryonal masses (PEMs), which are evident after 14 days. The induction of somatic embryos from leaf explants of pistachio involves single epidermal or subepidermal cells (Onay, 2000c).

Embryogenic cultures consisting of PEMs can be maintained on induction medium supplemented with 4.4 µM BA and on induction medium without growth regulators with subcultures at 122-14 day intervals. Cultures have survived for >2 years on plant growth regulator-free medium. When somatic embryos are transferred on to medium for development, abnormal embryos are recultured on induction medium in order to produce more PEMs. Proembryonal masses are transferred on to semi-solid MS medium containing 17.9 µM BA and 60 g/l sucrose in order to initiate somatic embryo development. Approximately 49 somatic embryos develop from each 250 mg fresh weight (FW) of PEMs. A week after transfer of PEMs on to maturation medium, cotyledonary somatic embryos are evident. Germination occurs on semi-solid or in liquid MS medium with B5 vitamins without phytohormones, and with 40 g/l sucrose, 500mg/ l case in hydrolysate and 50 mg/l ascorbic acid in continuous light. Onay *et al.* (1997) reported that liquid MS medium is optional for recovery of plantlets; however, the rate of conversion in low. Onay *et al.* (2000) also observed that abscisic acid (ABA) is superior to BA for somatic embryo germination although BA is superior to ABA for promoting subsequent development of pllantlets. Somatic embryo-derived plantlets have been successfully transferred to soil mixture. Plantlets are to be covered with a beaker to maintain 90±5% relative humidity (RH) for 4-5 weeks before transfer to a glasshouse (25°C day, 20°C night; 18 h photoperiod).

12.7. Micrografting

Chatibi et al. (1998) reported micrografting in pistachio. In *in-vitro* and *ex-vitro* micrografting experiments with pistachio cv. Mateur, the highest success rate (>70%) was obtained when scions were taken from *in-vitro* shoots originating from meristematic apices which had been cultured on Zygotic(Ra-Z) medium containing 1.0 mg BAP [benzyladenine]/litre, 0.2 mg IBA/litre and 0.05 mg NAA/litre. The scions were cleft-grafted on to seedling rootstocks with 0.2-0.3 cm diameter.

The success of various *in-vitro* micrografting methods of shoot tips of pistachio (*Pistacia vera* L. var. Siirt) have been examined by Onay *et al.* (2004). Excised zygotic embryos that germinated in vitro were used as rootstocks. Current year shoot tips from mature trees of pistachio micrografted onto *in-vitro* juvenile rootstocks, resulted in the restoration of shoot-bud proliferation. The results indicate that the easiest and most successful method for grafting was slit micrografting. High levels of micrograft take were achieved with 2-4 mm (56.75%) and 4-6 mm (79.25%) long scions obtained from the regenerated shoot tips. The survival rate of the shoot tips was directly related to time of the year. The best growth of micro-scion was obtained with the *in-vitro* forced shoot tips rather than with shoot tips excised from tree. Slow growth and lack of axillary shoot development on the micro-grafts was noticeable when the micro-grafts were cultured on hormone-free and germination medium. In vitro micro-grafted plantlets were successfully weaned and no problems were encountered with the establishment of micro-grafted plants in vivo.

13. Rootstocks

Choosing the rootstock is one of the most important decisions in orchard development. The rootstock diameter should be large enough to accommodate pistachio buds which are broad and large as compared with most fruit tree buds (Joley, 1979; Woodroof, 1979). The nut producing species *Pistacia vera* is grown on seedling rootstocks of different *Pistacia* species or interspecific hybrids. *Pistacia vera* itself has not proved to be a good rootstock due to low vigour (Monastra *et al.*, 1997). However some selection of *P. vera* like; Sarakhsi, Daneshmandi, Badami-e-Zarand A, Badami-e-Zarand B (salinity tolerant), Kalle-ghouchi are common in some pistachio growing regions (Karimi *et al.*, 2011).

The best pistachio rootstock-cultivar combinations for irrigated conditions in Sanlurfa in Turkey were investigated by Atl *et al.* (2011). Four different Pistacia species (*Pistacia khinjuk* Stocks, *Pistacia atlantica* Desf., *Pistacia vera* L. and *Pistacia terebinthus* L.) were used as rootstocks and five standard pistachio cultivars (Siirt, Ohadi, Kellekoci, Kerman and Selection 14) were grafted on them. On the basis of selected criteria, such as, tree vigor, bearing, yield and some quality characteristic, after a long trial, they concluded that the best highlighted combinations were determined. *P. khinjuk* x Siirt cultivar and *P. atlantica* x Siirt combinations gave the best results for irrigated conditions in Sanlurfa.

In addition to *P. vera*, there are nine species that have been used as rootstocks for pistachio: *P. chinensis, P. terebinthus, P. atlantica, P. integerrima, P. khinjuk, P. lentiscus, P. mutica, P. palaestina* and *P. eurycarpa* (Barghchi and Alderson, 1989). Some inter specific hybrid rootstocks have also been developed such as: Pioneer Gold II (PG II) and UC Berkeley I (UCB I) from *P. atlantica* x *P. integerrima*; Badami-Riz-e-Zarand from *P. atlantica x P. vera* (Karimi *et al*., 2014); or *P. palaestina x P. terebinthus* (Monastra *et al.*, 1997). All these rootstocks have characteristics which make them more or less suitable for specific planting situations; some of these are described below.

13.1. *Pistacia vera:* In some Middle East countries, generally *P. vera* is used as a rootstock. Seedlings of these species make more lateral roots and thicker stems than the others and they can reach budding size in a shorter time (Ayfer *et al.* 1990). However, there are differences in growth among seedlings, so between rootstocks and cultivars, stock-scion incompatibility may be seen that needs intergrafting. The initial pistachio cultivar evaluations demonstrated that the *Pistacia vera* seedling tree that produces the edible pistachio nut was susceptible to the soil borne pathogens *Verticillium dahliae* and *Phytophthora* spp., and to nematodes (Woodruff, 1979).

13.2. *Pistacia terebinthus*: Earlier *P. terebinthus* was common, but it is susceptible to *Verticillium* wilt. Besides, it is more difficult to bud than *P. atlantica* or *P. integerrima*. Its use often resulted in less uniformity than the other rootstocks. However, it is less vigorous than the other commonly used rootstocks. Currently, it is seldom used. However, it has some characteristics which may make it useful in specific situations, and it may be valuable in future breeding programmes (Monastra *et al.*, 1997). The rootstock is most cold hardy and pistachios growing on this rootstock can withstand temperatures as low as 15°F with no damage. Also the most efficient for absorption of nutrients especially zinc and copper. Because of its very slow growth habit, *P. terebinthus* when grafted at the place where it grows, makes dwarf trees and begins to produce early, large and quality fruits. It is generally accepted that *P. terebinthus* is one of the most valuable rootstock and together with *P. vera* can be a suitable rootstock for intensive cultivation under irrigated conditions (Ayfer *et al.*, 1990).

13.3. *Pistacia atlantica*: Until *P. integerrima* was discovered to be tolerant of *Verticillium, P. atlantica* was the most common rootstock. *P. atlantica* is more cold hardy than *P. integerrima* and the two interspecific hybrids, but less cold hardy than *P. terebinthus*. It is less vigorous than *P. integerrima* but more vigorous than *P. terebinthus*, and slightly slower to come into production (Alberghina, 1996). However, it is found to be superior in enhancing leaf concentrations of B, Cu, Zn, and P via uptake from range of soil types (Brown *et al.*, 1994). Top worked cultivars on this rootstock quickly outgrow and out yield those grafted onto *P. vera* in spite of the characteristically slow growth in the nursery of the rootstock (Joley, 1979).

13.4. *Pistacia integerrima*: After the discovery of its tolerance to *Verticillium* wilt, it became the most widely planted rootstock. It is vigorous, buds easily set and results in a uniform tree stand (Monastra *et al.*, 1997). It is the least cold tolerant of the commonly used rootstocks and can be damaged by winter freezes and early frosts, particularly when the tree is young. The root stock is less efficient than *P. atlantica* for boron, zinc and copper uptake.

13.5. *Pistacia khinjuk*: It show a rapid increase in length, but make thinner seedlings than the others so they reach budding size later. However, there is no stock-scion incompatibility between cultivated varieties and these rootstocks (Ayfer *et al.* 1990).

13.6. PG II: It is the result of open pollination in an isolated situation with multiple *P. atlantica* trees as the female parent and multiple *P. integerrima* trees as the pollen parents. The rootstock has increased vigor compared to *P. atlantica*. PG II is moderately susceptible to Verticillium. It is relatively efficient in zinc uptake, compared to *P. terebinthus*.

13.7. UCB I: This is the result of a closed greenhouse pollination of *P. atlantica* tree with the pollen of *P. integerrima*. The rootstock has increased vigor compared to *P. atlantica* and generally equal or greater vigor compared to *P. integerrima*. UCB I is tolerant to moderately tolerant to *Verticillium*. The rootstock is least efficient in zinc uptake and at par with *P. integerrima*.

14. Variety/hybrid

14.1. Kerman: This is a primary and major female cultivar commercially grown and was released during 1957. It is very good cultivar, producing heavy yield of large and attractive nuts. However, it has a strong alternate-bearing tendency, a high percentage of blank nuts in some years, and has relatively high level of non-split nuts (Parfitt *et al.*, 2016).

14.2. Sirora: This is the main cultivar grown in Australia. It constitutes about 99% of female pistachios grown in Australia, where as Kerman 1%. Sirora was developed from a selection of Red Aleppo. It is similar to Kerman but is slightly smaller and with curled leaves. It is late flowering, usually synchronizing with Group 4 males, and ripens about 8 days before Kerman. Fruits have a red colouring similar to Red Aleppo. Air dried dehulled nuts weigh 110g/100 nuts (those of Kerman weigh 120-140 g) and measure 18 mm long, 12 mm back to front and 11 mm wide. The suture is only slightly thickened. Mature kernels are green, a character associated with superior flavour. Sirora has higher yields than Kerman (Maggs, 1990). This female cultivar appears to have lower chilling requirements, slightly less biennial bearing and produces fewer no-splits than Kerman.

14.3. Golden Hills: It is a female pistachio.The cultivar is a cross between '2-35' as female and 'ES#2', released by the California Agriculture Experimental Station during 2005. The nut matures uniformly, can be harvested 2-4 weeks earlier than 'Kerman' and also flowers earlier than Kerman. In-shell nut yields in the first 4 years are equal to those of Kerman, but the percentage of edible split nuts is higher. It appears to be a good early maturing cultivar, with uniform harvest period and little navel orangeworm infestation (Parfitt *et al*. 2007).

14.4. Lost Hills: This is a new female cultivar developed by Parfitt *et al.* (2008). It flowers 4 to 7 days before Kerman and matures about 2 weeks earlier than Kerman. Nut size and weight are larger than Kerman. The percentage of splits is about 26% higher than for Kerman. Lost Hills appears to be a less alternate bearing cultivar than do Kerman or Golden Hills. However, a potentially undesirable characteristic is its lost and loose kernels.

14.5. Peters: It is a good pollinator (male cultivar) and was found in the early 1900s. It produces abundant pollen that is shed over a relatively long period (2 weeks), and its pollen viability over time is also very good.

14.6. Chico: This was introduced from the Chico USDA Plant Introduction Station in 1962, and is a male selection from seed introduced from Aleppo, Syria, probably of species hybrid origin and tested as Chico 23. It is prolific pollen producer and blooms early. Chico sheds pollen prior to and during the earliest part of the Kerman bloom period. The cultivar is not advised in case xenia (male contribution) effects are present in pistachio for nut size.

14.7. Famoso: This male pistachio variety is a cross between '2-35' as female and 'ES#$" as male. Famoso flowering is found to be more consistent than 'Peters', produces a large amount of viable pollen, and provides a better overlap in flowering period with 'Kerman' (Kallsen *et al*., 2016). Other male cultivars being used are Ask, Gazvin, and Nazareth.

14.8. Randy: It is a male pistachio. The use of this cultivar as a pollen source for the pistachio cultivars Golden Hills and Lost Hills (Parfitt *et al.,* 2010).

Table.4: List of female cultivars grown in various countries.

European	Iranian	Turkish & Syrian	Californian
Napoletana	Momtaz	Uzun	Red Aleppo
Trabonella	Ohadi	Kirmizi	Bronte
Bronte	Kalehghouchi	Red Aleppo	Trabonella
Agostana	Ghermeza	Achoury	Sfax
Girasola	Tbeahimi	Alemi	Kastle
Notaloro	Ogah	El Bataury	Rashti
Cappuccia	Wahidi	Obiad	
Femminello	Rafsanjani	Ayimi	

In India, very few pistachio plants (Kerman as female and Peters as male) can be seen in Jammu and Kashmir, and Himachal Pradesh.

15. Cultivation

15.1. Planting

Planting of unbudded nursery-grown rootstock seedlings begins in January or February while they are still dormant or in the early leafing stage. Planting dormant trees allows for a much greater margin of error than planting after leafing, but the potential for freezing temperature exists. Before leaving the nursery, the one-year-old seedlings are topped to approximately 24 inches (60 cm), irrigated, and placed in standard fruit bins for shipment.

Pistachio trees are long-lived. Time spent properly designing an orchard saves years of farming frustrations and prevents poor economic returns. The well-designed orchard ensures: the earliest economic returns, maximum production and least management cost. A pistachio orchard reaches its maximum bearing potential when space over the soil is completely occupied by nut-bearing surface with sufficient light penetration. Appropriate row spacing is particularly important because pistachio is relatively slow to begin bearing. Five to six years are required between planting and the first nut harvest. The initial number of trees planted per acre is directly related to the orchard's onset of economic production and the time required reaching its maximum bearing potential. The closer the trees are planted, the more likely tree thinning required latter to maintain its maximum bearing potential. For such orchards, where plants are to be thinned latter on, spacing can be maintained at 4 x 8m. However the new orchards are not being planted with the assumption, where tree thinning will be necessary to ensure maximum production at maturity. Canopy development, nut size, nut split percentage and productivity are found to be best with the plant to plant and row to row spacing of 8m x 8m (Arpaci and Atli, 1995; Arpac *et al.* 2000).

15.2. Pollinizer placement

Pistachio trees are dioecious, with male and female flowers born on separate trees. Since pistachio pollen is wind blown, males are interspersed around the border of and within the orchard in a regular pattern. the male trees can be planted after every fifth tree in every fifth row with a ratio of 1:24. Additional male trees are some times planted in border rows of the orchard. However, in areas, where few or no other pistachios are grown, the ratio of male to female can be 1:14 to ensure adequate pollination.

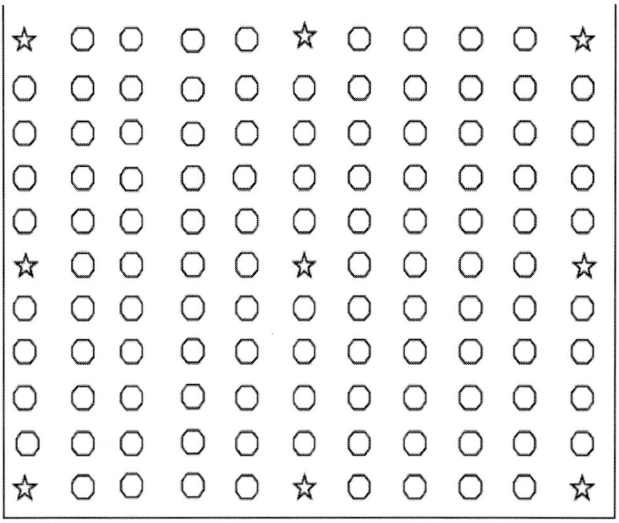

Fig. 1: A rectangular planting system with 1 male tree to 24 female trees

15.3. Training

Training of young pistachio is essential to establish the desired tree shape to facilitate harvest and cultural operations. Trees are trained to open center system, leaving much of the interior wood for early fruiting. During the first growing season of the transplanted budded trees in the orchard, low laterals developing from the shoots are pinched rather than removing to provide the greatest possible leaf area. The remaining rootstock leaves are similarly pinched for the same reason. The first winter training consists primarily of restricting rootstock growth on successfully budded trees. Taller shoots are headed at 1.1m to force primary scaffold development. Near this heading cut several laterals will be pushed out, which grow vigorously and require tying into a vase-shaped position in May-June. Failures to do so resulted in flat scaffolds and too open a canopy. Moderately vigorous laterals can also develop along the distance of the trunk and are better to be removed, as they can quickly develop size and compete with primary scaffold growth. Leaves and short spurs along the trunk can be left, as they aid in trunk diameter growth. Primary scaffolds are allowed to grow 30 – 35 cm before they are pinched to 28 – 33 cm by early June.

Young pistachios tend to be of insufficient girth and primary limbs are often unable to support the weight of new secondary growth. Tying of limbs is, therefore, often practiced during this development stage. It is generally not advisable to attempt development of secondary and tertiary branching in the same year. Secondary branches would require tipping at 28 – 33 cm by mid-July to have sufficient vigor to produce tertiary shoots of 76 cm in length by mid-September.

Trees with secondary branches are thinned to 2-3 branches arising from each primary. Limbs which are directly opposite should be removed. Branches should be spaced at 8-10 cm down the primary. Following selection, each is headed to 28-33 cm and tied up. The practice continues to increase total canopy size and limb diameter so that the tree acquires sufficient structural strength.

15.4. Pruning

Pistachio bears crop laterally on one year-old wood. An individual flower bud (inflorescence) botanically is a panicle consisting of an enlarged central axis with lateral branches that produces as many as 200 individual flowers. These structure supporting the flowers are known as the 'rachis'; and the rachis eventually supporting nuts are termed as 'cluster'.

Dormant, one year-old fruit wood is represented by both preformed (spur) and neoformed (whip) growth. Spurs can grow from large diameter, one year-old structural branches in young trees where apical dominance is not as great. Spurs often develop in the first flush (preformed growth) of whips, if those buds did not differentiate into flower buds of the previous season. Long, whip-like growth usually develops flower buds in the preformed growth (first flush). If a one year-old whip is left untipped, the lateral buds are all capable or producing spur growth, which could then develop fruit buds for the following year. Pruning of mature deciduous trees have several major objectives:

i. **One is to confine the tree to its allotted space:** Apical dominance in pistachio promotes tree crowding, loss of tree shape, and limits flower bud formation on lower fruit wood due to lack of light. The eventual reduction in growth on the sides of the canopy promotes added growth in the top of the tree, where light is not limiting. Although high in fruitfulness, the upper canopy eventually grows beyond the height manageable by hand pruning crews. Therefore, pruning bearing pistachio is to confine the tree to the allotted space and promote light penetration for viable fruit wood and nut production throughout the tree.

ii. **Secondly pruning is performed to renew fruiting wood and distribute light throughout the canopy:** Pistachio expresses strong apical dominance in their vegetative growth habit, meaning that vigorous current season shoots develop only from buds closest to the terminal of one year-old branches. Vegetative buds further down the branch are capable of producing long, vigorous shoots, but develop only short spur growth because of hormonal control exhibited by the terminal bud. This characteristic becomes more pronounced with tree age. Thus branches may elongate for years without lateral branching, giving it a spindly appearance and causing the tree to bear fruit further from its central axis. This increases the tendency for main fruiting limbs to become less upright with each successive crop. Early pruning studies indicated apical dominance could be overcome in older trees by cutting the entire canopy back into three and four year old wood (Crane and Iwakiri, 1980). This treatment produced new shoots on 14-17 year-old wood.

iii. **Finally, pruning can be used to mitigate strong alternate bearing:** The degree of alternate bearing is largely determined by the previous year's crop load. A large crop born on one year-old wood, reduces the fruiting potential for the following year by triggering the abscission of flower buds produced at the base of current season's growth. Pruning can mitigate the phenomenon by regulating crop load (Elloumi *et al.*, 2014).

Growers have the option to go for either hand pruning or use mechanical hedging and topping equipments to accomplish the pruning task. Hand pruning allows each tree canopy to be modified individually. Heading and thinning cuts can be applied as needed to optimize light interception, stimulate new growth in weak areas, and control tree size.

Yield data collected in one pruning trial showed that hand-pruned pistachio trees produced more pounds of split nuts over six years than tress single or double side hedged by machine. The difference between hand—pruned and single-side hedged annually was 1,348 pounds of split nut over six years; double sided hedging every other year differed significantly from hand pruning by 1,923 pounds of split nuts (Beede *et al.*, 2002). However, this considerably escalates the cost to \$200/acre compared to \$50 for mechanical pruning; hence the complications associated with hand pruning of large acreage may not be justifiable.

The goal of hand pruning can be accomplished in four steps. The first is to remove branches that are broken, too low, crossing over the center, or overlapping one another. These are the relatively obvious cuts that are to be performed. Second, thinning cuts must be made to the outside of the canopy with the objective of 'pushing' the bearing branches upright. This is accomplished by thinning off branches that are at less than 45 degrees. Branches must be cut back sufficiently to have those remaining be pointed upward and not out. Few thinning cuts are needed in the center of pistachios, since apical dominance causes most of the new growth to occur on the canopy periphery. The third step is to perform selected heading cuts in areas closer to the tree axis that are weak. These will hopefully stimulate latent buds to develop into new shoots that can be used to renew fruitwood. The final step is to top the one year-old whips produced primarily in the top of the canopy. Shorter growth can also be tipped elsewhere with the goal of creating as much new spur growth as possible. It is essential that the pruner knows the difference between flower and vegetative buds so that at least one of the latter is retained on the tipped one year-old wood.

Machine pruning is fast, inexpensive and a large acreage can be pruned with a single machine. Machine utilizes equipment with a gang of rotating circular saws set at pre-determined heights or widths. All wood coming in the way is removed by indiscriminant heading cuts, which certainly requires supplemental hand pruning for better yields. Research conducted prior to both 'on' and 'off' years showed no difference in yield between single and double-sided moderate side hedging (cuts made into one and two year-old wood) and hand-pruned controls (Ferguson *et al.*, 1991). Zhang (2014), also didn't found any significant difference of yield between hand pruned and machine side hedged pistachio trees and observed that pistachio compensates for the flower buds removed during machine hedging by increasing the number of nut set/ rachis

15.5. Influence of stress factors

Salt stress is considered as one of the most important abiotic factors limiting plant growth and yield in many areas of the world (Abbaspour *et al.*, 2006). Plant biomass is strongly reduced by high salinity levels and the soil adverse effects are accentuated with increasing level of salinity (Tavallali *et al.*, 2009). It also enhances the trees susceptibility to *V. dahliae* (Mohammadi *et al.*, 2007).

NaCl-induced strong oxidative stresses as indicated by high levels of lipid peroxidation and electrolytes leakage can be efficiently reduced by Zn treatments, which act on antioxidant enzyme activity to scavenge the reactive oxygen species produced in response to salt stress (Tavallali *et al.*, 2009). Salicylic acid is a common plant-produced signal molecule that is responsible for inducing tolerance to a number of biotic and abiotic stresses. Salicylic acid improves growth of pistachio seedlings under salt stress by increasing relative chlorophyll content, water content, chlorophyll fluorescence ratio, and photosynthetic capacity, when applied at the rate of 0.5-1mM (Bastam *et al.*, 2013).

15.6. Irrigation

While pistachio trees are less thirsty than other nut trees and able to survive periods of drought; irrigation is essential for commercial production. Soil moisture monitoring devices should be used to ascertain frequency and duration of irrigation schedules. Pistachio trees are more tolerant to saline conditions than other nut trees. For best quality yields and tree health, trees should never be subjected to moisture stress during the growing season. This is particularly important during nut fill in late summer because inadequate soil moisture at this stage can cause a high percentage of non-splits (nuts with shells that do not split open). All eco-physiological parameters like stomatal conductance, transpiration, net CO_2 uptake are influenced by irrigation and must be restored to 50% ETc (crop evapo-transpiration) (Palma *et al.*, 1997). Higher levels of irrigation lower the photosynthetic rates and decrease fruit/inflorescence, which may be due to root sensitivity to soil oxygen depletion in saturated soils (Monastra *et al.*, 1995; Palma and Novello, 1998). Deficit irrigation of pistachio trees in April and May result in substantial increase in the formation of early-split nuts in late summer (Doster *et al.*, 2001).

Selection and design of the irrigation system is influenced by site topography and soil type. The choice of irrigation system, in turn, influences final land leveling and tillage requirements. Properly designed and managed micro-sprinkler, surface drip, subsurface drip, and sprinkler system help minimize physical soil limitations more than surface irrigation systems. The advantage of choosing a flood or furrow irrigation system over a micro-irrigation system is reduced initial capital cost. Irrigating pistachio trees with 120 mm water at 20-30 days interval has been found to double the yields during both on- and off-years (Kanber *et al.*, 1993). From an another experiment Kanber *et al.* (1993) showed that with irrigation at 20-day intervals, fruit yield in on-yielding years was 57.1 kg/tree, compared with 34.5 kg/tree for controls (no irrigation). Irrigation did not affect fruit quality characteristics, such as nut weight and split nut percentage. With frequent irrigation, root activity occurred nearer the soil surface than with less frequent irrigation. Currently irrigation water usage ranges from 900 mm/ha in some regions down to as 200 mm/ha in other orchards.

15.7. Weed control

A plan to control weeds should be in place before planting trees and carried through the life of the orchard. Weed control is especially critical during the first years of establishment, since heavy weed competition can reduce early tree vigor and productivity. In addition to direct competition for water and nutrients, certain weeds host several important insects, diseases, nematodes and rodents. Weeds have differing characteristics, making them unique from each other. They have different life cycles (annual, biennial, or perennial), have differing general plant structures (broadleaf, grass, or sedge), grow flat or erect, reproduce by seed and/or other reproductive structures and compete differently for valuable resources. The weeds most commonly found in pistachio orchards are Nettle, Mustard spp., chickweed, Bermuda grass etc.

There are many local conditions that influence management decisions, including topography, soil structure and texture, permeability, type of irrigation, availability of equipment, labour, herbicide and cost. The specific method(s) of weed control will vary from site to site, but usually include the combination of mechanical, cultural, and chemical control options. In-row tillage can be practiced in new plantings and disking in established plantings to eliminate weeds growing in between the rows, when low-volume irrigation tubing has not been laid out. Complete non-tillage is an option where weeds are maintained in the middles as green cover. This is an excellent way of reducing soil erosion, especially on sloping lands. A flail mover can be used to help maintain the weeds in a low-growing state. Herbicides, when properly used, can be a cost effective tool for controlling weeds in most pistachio orchards. Mohassel *et al.* (2001), in a study reported that two herbicides in combination give better results than application of single herbicide alone. The herbicide combinations which were suggested to be best to control weeds like bermuda grass, Russian knapweed, field bindweed, Russian thistle and camel thorn are: glyphosate (3kg/ha) plus 2,4-D (1.5kg/ha); glyphosate (3kg/ha) plus DCPA (0.9kg/ha); and glyphosate (3kg/ha) plus ammonium sulphate (6kg/ha). Combination treatment of 24% urea and 6% yolk can also be an excellent control to lambsquarters, hoary cress, orchard grass, downy brome, and bind weed (Davarynejad, 2001). However, glyphosate (3.28kg/ha) plus 4% ammonium sulphate has been reported to be more effective to control the weeds in pistachio orchards (Sadri *et al.*, 2010).

15.8. Intercropping

Relatively early slow growth of pistachio trees extends the time period before middles are shaded, and the economic bearing starts. To provide income before the pistachio begins producing; growers can plant crops such as, watermelons, beans or other crops between the rows of young trees. Surucu and Demirkiran (2013) reported that barley, vetch and mixture of both can be adapted as intercrops in pistachio growing fields. They showed that intercropping did not yield any significant differences on nutrient contents, except Mn of pistachio leaves. Care must be taken in choosing the intercrop to ensure that chemicals used on the intercrop and pistachios are compatible for both crops.

15.9. Mineral nutrition

Pistachio trees require 14 elements for normal growth and reproduction. These essential elements are classified as either macronutrients (N, P, K, Ca, Mg, S) or micronutrients (Fe, Mn, Cl, B, Cu, Zn, Ni, Mo) based on the concentration normally present in plants. Each is essential for particular functions in the plant (Duan-YunFeng *et al.* 2009). Plant nutrients are also important in disease resistance and fruit quality, and the balance between the various elements can affect plant health and productivity. In addition to the essential nutrients, several elements (Cl, B, Na) may be toxic to the tree if present at excessive levels in the soil or irrigation water (Ferguson *et al.,* 2002). Optimization of pistachio productivity and quality requires an understanding of the nutrient requirements of the tree, the factors that influence nutrient availability and the methods used to diagnose and correct deficiencies.

Pistachio tree roots spread through a large volume of soil, and soil type is usually variable within the orchard. Soil chemistry differs with depth. Surface soil can be quite different from the soil just a foot down. Soil sampling should be taken from the site of most active root growth, which is commonly the upper four feet of the profile. For a thorough analysis, soil profile should be divided into three depths (0-40 cm, 40-80 cm, and 80-120 cm), and the samples should be taken from three to ten different spots around the tree and composited. Replicate samples should then be taken across the orchard (Kermani and Hokmabadi, 2012).

It is also important to consider the effect irrigation method has on root distribution and soil fertility within the root zone. Flood or basin irrigation applies water over a large area relatively uniformly and results in wider distribution of roots and area of nutrient uptake. Hence, sampling near the edge of the tree canopy but to one side of where fertilizer applications are made, provides a reasonable assessment of soil nutrient status. With mini-sprinkler system, sampling should be performed within the wetted pattern, but avoiding its edge where salts may accumulate. Orchards under drip irrigation require sampling approximately half-way between the emitter source and the edge of the wetted area. Due to the large difference in soil water content with distance from the emitter source, sampling too close to the emitter can lead to enormously low soil nutrient assessment of some elements, particularly nitrogen because it exists as a leachable form in soil solution.

Leaf analysis is more useful in diagnosing mineral deficiencies and toxicities in tree crops than soil analysis. The mineral composition of a leaf is dependent on many factors, such as its stage of development, climatic conditions, availability of mineral elements in the soil, root distribution and activity, irrigation, etc. Fully expanded sub-terminal leaflets are randomly collected from the non-fruiting branches at about six feet from the ground. Four to ten leaves are typically collected per tree, and 10—20 trees are sampled in each orchard block. Leaves sprayed with micronutrients typically should not be analyzed for that nutrient since the surface contamination cannot be removed. Hence, no leaves have received in-season nutrient sprays for the elements of interest should be sampled. Samples should be kept in labeled paper bags and submitted to the analytical service within 24 hours of collection (Assimakopoulou and Elena, 2010).

Leaf samples should be done from late July through August. The critical levels of different elements are established through experimentation and observations (Table 5). However the comparison of good tree against poor ones, can be done at any time. Samples at times other than from late July through August may have nutrient concentrations different than those recommended in the critical values tables and must be interpreted with care. Although valuable as a tool to assess orchard nutritional status, critical values are not absolute. They are based on general tree health and not yield or crop quality. Some nutrients, such as boron during bloom and potassium and nitrogen during kernel filling, may require supplementation for short periods to optimize production (Weinbaum *et al.*, 1995; Brown, 2001).

Table 5: Critical and suggested levels for August leaf samples.

Element	Critical value	Suggested Range	Reference
Nitroge (N)	1.8%	2.2-2.5%	Weinbaum *et al.*, 1995
Phosphorus (P)	0.14%	0.14-0.17%	
Potassium (K)	1.6%	1.8-2.0%	Brown *et al.*, 1999
Calcium (Ca)	1.3%	1.3-4.0%	
Magnesium (Mg)	0.6%	0.6-1.2%	
Manganese (Mn)	30ppm	30-80 ppm	
Boron (B)	90 ppm	150-250 ppm	Brown *et al.* 1993
Zinc (Zn)	7 ppm	10-15 ppm	Uriu and Pearson, 1986
Copper (Cu)	4 ppm	6-10 ppm	Uriu *et al.* 1989

15.9.1. Nutrient deficiency and management

15.9.1.1. Nitrogen

Nitrogen is the most widely needed fertilizer element in pistachio. Nitrogen is used by plants to synthesize amino acids and nucleic acids that are necessary for all functions of the plant. Nitrogen deficiency resulted in reduced shoot growth in pistachio. Shoots are thinner, shorter and in more severe deficiencies have reddish bark. Nitrogen is mobile and new leaf production is at the expense of older leaves if N is deficient. Young leaves pale as older leaves turn yellow and drop from the tree early. Excessive leaf drop results in a tree with sparse foliage. The petioles and midribs of N deficient leaves become red (Gonzales and Uriu, 1985). Nut yield will decrease with increasing N deficiency.

April is a good time for nitrogen fertilization. The efficiency of nitrogen uptake from the soil before early leaf out is essentially zero, because nitrogen uptake occurs simultaneously with water use. Consequently, applications made in the winter are subject to leaching beyond the major root zone at three to four feet, especially in production areas averaging 15 or more inches of rainfall. Due to exhaustion of the N in the on-year of pistachio tree, the stored N in the beginning of the off-year is lower than on-year trees (Weinbaum *et al.*, 1995). Nitrogen uptake in pistachio occurs primarily between mid-May to late August. Nitrogen uptake efficiency improves with multiple applications at lower rates. The demand for N and the capacity to take N up from the soil is greater in the on-year. Nitrogen uptake is largely driven by crop load since nut fill accounts for 90% of the accumulated seasonal nitrogen. Since off-year trees are lower in stored N, applying half the season's N prior to shell hardening and the remainder in July and August might be considered (Weinbaum *et al.* 1995).

15.9.1.2. Potassium

The potassium is an activator of many plant enzymes. Potassium has important functions in plant water relations where it regulates ionic balances within cells. Potassium regulates the leaf stomata opening and subsequently the rate of transpiration and gas exchange. Plants also need potassium for the formation of sugars and starch, for the synthesis of proteins, and for cell division. It increases the oil content of pistachio and contributes to its cold hardiness. Studies indicate that K uptake and distribution are driven by fruit development, with the crop serving as a sink. K uptake occurs mainly during the nut-fill period, with little uptake during the spring vegetative flush and postharvest periods. K fertilization is found to be effective in increasing leaf K status, nut yield and quality (Zeng *et al.,* 1998).

Mature, heavy bearing pistachio trees have been shown to take up approximately100 kg/ acre of K during the on-year and 50kg/ acre during the off-year. Much of this K is in nuts and is removed from the orchard at harvest (Brown *et al.*, 1999). Since potassium is used in large quantities by the developing crop, there is considerable potential for K deficiency to develop if fertilizer is not applied for many years. Deficiencies are more likely to appear in lighter soils where reserves are less.

Potassium-deficient leaves begin showing symptoms in early to mid-summer. Leaves become pale (resembling N deficiency) and the edges of the leaflets later fold upward and curl in revealing a grayish cast. The symptoms are most pronounced on older leaves of current shoots. Potassium-deficient trees slow in growth and produce smaller leaves. More severe deficiency symptoms include yellowing of the leaf, which start at the tip and progress down the curled leaf margins. The affected area widens with time. The yellow leaf tissue eventually turns brown and dies resembling sodium and/or chlorine toxicity. Trees moderately deficient in K may look normal but produce yields below their potential.

Soil banding K fertilizers in neutral, loam soils at rates of 50 to 100kg per acre of K_2O was found to increase soil and leaf K concentrations. Increases in nut weight and split nut percentage along with reduced blanking resulted in higher yield (Brown *et al.*, 1999). The positive response to K fertilization was found regardless of whether potassium chloride (KCl), potassium sulphate (K_2SO4) or potassium nitrate was the source of K. some soils will reduce the plant availability of applied K fertilizers through soil chemical processes, hence soils high in certain clay minerals may not respond as positively to K fertilization. For optimizing the response to K fertilization, inject it into the irrigation water during periods of greatest tree demand (nut fill). Potassium fertilizer rates can be adjusted based on a suggested leaf tissue concentration of 2.0% by weight. Excessive K fertilization can interfere with the uptake of other nutrient cations like Ca and Mg.

15.9.1.3. Magnesium

Magnesium is an activator for many enzymes involved in energy transfer and growth processes. It is a component of chlorophyll and thus is essential for photosynthesis. Magnesium deficiency has not been widely reported in pistachio. Deficiencies are more likely to occur in sandy and acid soils. Alkaline, highly calcareous and boric soils, may fail to provide sufficient available magnesium for optimal tree growth due to antagonistic competition for uptake by excessive calcium and other cations present on the soil colloids. In a similar manner, high rates of gypsum may induce magnesium deficiency in some soils.

Deficiency symptoms appear mid-season on the lower leaves of shoots as tip and lateral margin yellowing or as interveinal yellowing. The leaf margins may later become scorched. The scorching progresses inward, leaving a green, inverted 'V' at the base of the leaf. In severe deficiency, the interveinal yellowing may turn to scorching. Scorched leaves will then drop. Magnesium deficiency may be confused with potassium deficiency. Suspected Mg deficiency should be confirmed by leaf tissue analysis.

15.9.1.4. Zinc

The micronutrient, Zinc, is the third most common deficiency in deciduous trees. It is needed in a number of enzymes as well as in membrane integrity and hormone synthesis. Zinc is required to make auxin, the plant hormone responsible for cell elongation and growth. Zinc also is needed to make chloroplasts, the leaf component that contains chlorophyll.

Zinc deficiency is widespread throughout all pistachio growing areas. It is prevalent in sandy and older, oxidized soils. Orchards with regular applications of animal manure are prone to zinc deficiency due to their high phosphorus levels. Zinc is rendered less available for plant uptake in soils with pH values of 8.5 and greater. Zinc deficiency prevails in younger tissue due to its immobility. Symptoms of Zn deficiency may be restricted to terminal leaves on individual branches of a tree. A tree may have several Zn-deficient limbs while the rest of the tree appears healthy. This pattern of Zn deficiency further demonstrates the immobile nature of Zn within the tree.

Zinc deficiency symptoms appear early in the season, especially when levels are quite low. Newly planted and early bearing trees are particularly prone to zinc deficiency. Spring symptoms are delayed opening of vegetative and flower buds that can be delayed as much as a month. This gives the appearance of cold injury to the one year-old wood in the upper canopy. When vegetative buds do open, the terminal leaves are small, chlorotic and appear in tufts (little leaf). Severe deficiency also causes dieback of young terminal shoots. Marginally deficient leaves are only slightly reduced in size and show chlorotic areas between the lateral veins. These symptoms can be masked by growth later in the season. However, deficient leaves remain chlorotic during entire season. Zinc deficient leaves have a distinct wavy leaf margin. Nuts on zinc-deficient shoots are markedly smaller in size and much redder than normal. The number of nuts per cluster is also dramatically reduced and most of them can be blanks.

Most soils are not truly low in zinc. Rather, the chemistry of the soils prevents Zn from being sufficiently available to the plant. Correcting Zn deficiency through soil applications can therefore be excessively slow and expensive. Soil texture and pH are the primary factors affecting the efficacy of Zn fertilization. In sandy soils, trenching 1 to 3 kg zinc sulfate 36% per tree, depending on size, has produced good response. Soil treatment becomes less effective in fine-textured soils or soils with high pH. Fertigating 35 kg per acre of zinc sulfate annually for two years in a clay loam soil (pH 7.5), increased soil Zn levels but not zinc concentrations in leaf tissue. Fertigating zinc sulfate in sandy loam soils (pH 6.5) did result in some uptake by the tree (Kallsen, *et al.*, 1998). Treating alkaline soil with acidic or acid forming chemicals may release substantial amounts of

Zn chemically bound in the soil and thus, increase tree uptake. Nitrogen fertilizers like ammonium sulfate, urea-ammonium nitrate, ammonium phosphate, urea have an acidifying affect on the soil but their free hydrogen content is insufficient to make large changes quickly. Sulphuric acid, urea-sulphuric acid, and phosphoric acid provide a more rapid response and they can be applied uniformly through most low-volume irrigation systems.

Foliar sprays applied post-bloom when leaves are 50% expanded (end of April to early May) are effective at increasing tissue Zn concentrations. Absorption decreases markedly as leaves mature and acquire a fully developed waxy cuticle that binds the applied Zn. Spring applications are made with 1kg zinc sulphate (36% Zn) per 375 lt of spray solution per acre adjusted to a pH of 4.5 to improve solubility and absorption. A single application is usually sufficient for the growing season. Foliar sprays of Zn provide only temporary correction, however, and must be repeated to prevent deficiency (Brown *et al.* 1996).

Zinc deficiency and the risk of fall cold injury is greatly reduced in young trees by inducing dormancy with high zinc rates in late October. Zinc sulphate (36% Zn) at the rate of 20 kg per acre are commonly applied between late October and early November without tree damage. Unlike the spring timing, fall application provides zinc nutrition in advance of bloom, when deficiency effects fruit set and nut development. Another efficacious timing for this high rate is in late February and early March during swelling of the bud scales (Beede *et al.*, 1991).

15.9.1.5. Boron

Boron is required for cell wall structure and may also have a role in maintaining call membrane integrity. Reproductive structures have been shown to be particularly sensitive to B deficiencies. Their demand is higher than that of vegetative growth (Brown *et al.*, 1993). The boron requirement in pistachio is the highest known for any tree crop. Boron appears to be extremely important in pistachio flowering, pollen viability and nut set. Moderate deficiencies are common in orchards on sandy and moderately acidic soils. Avanzato *et al.* (2011) opined that boron is an essential element of plant growth, involved in various physiological phases, from bud break to the fruit ripens. It influences flowering (lack can cause the drop of entire flower clusters), pollen viability and production.

Boron deficiency symptoms appear early in the season as tip dieback of growing points and deformity in young leaves. Shoot tips die back or the terminal buds remain dormant. Lateral, normally dormant buds, sprout into growth with short internodes. This new growth may then also dieback. In some shoots brown-gray areas of bark distributed irregularly over the shoot may precede this dieback.

These shoot-growth abnormalities produce a tree that is stunted with a bushy appearance.

Pistachio leaves deficient in boron are yellow and crinkled. The leaf tips curl upward and sometimes become so misshapen that they appear bract-like in form. Mature leaves are thick and feel leathery to the touch. They also become brittle and drop prematurely. The petioles and stems may also be thickened, corky, cracked and shortened. Entire flower clusters on young, boron-deficient pistachio often drop during bloom.

Toxicity resulting from excessive boron is occurring more frequently as growers are increasingly forced to use lower quality water. Symptoms begin showing in mid to late summer as boron accumulates in the leaf tissue. First the tip, and then the margin of the leaves become brown and necrotic. In severs cases, the necrotic tissue progresses into the interveinal areas, and the margins of the leaflets curl upward. Every leaf on the tree may be so severely scorched that the tree appears as if it were injured by fire.

Soil or foliar application can correct boron deficiency. Similar to other micronutrients, boron is most available in lighter, acidic soils. There are a number of B fertilizer materials containing different amounts of actual B. soil application of Borax (11%B) by the first week in September at the rate of 25 to 35 kg per acre can correct the deficiency, lasting for 3-5 years. Soils lighter in texture and low in buffering capacity require the addition of less born for correction. Applications through drip systems may be reduced by 25-30% due to higher efficiency from concentrated treatment. A foliar spray of soluble B (e.g. Solubor having 20.5% B) at the rate of 1-1.5 kg in 375 litres of spray solution per acre can be applied anytime as the B deficiency is detected. Spring applications during shoot extension are more effective in correcting current season deficiency than treatments in June or July. The best long-term correction can be expected from a combination of soil and foliar treatments applied as yearly maintenance. Avanzato *et al.* (2011) found that boron treatment applied in July caused a notable increase of production and significant increase in dry leaf weight.

15.9.1.6. Copper

Like zinc, copper is a component of many enzymes in the plant and plays a role in energy metabolism. Deficiencies of copper are relatively common in pistachio. Like zinc, copper deficiency is more likely in newly planted and early bearing orchards. Commonly zinc and copper deficiency symptoms occur together. Copper deficiency can be quickly differentiated from zinc because its symptoms appear in mid-summer, not during early spring. The immature leaves near the shoot terminal first develop tip burn and become somewhat heart-shaped. Slight shriveling of the shoots then occurs, and small dark lesions develop near the shoot tip. Scorched leaves defoliate. Terminal dieback subsequently occurs in late summer. Some affected shoot terminals curl downward, resembling a Sheppard's crook. Kernels are often badly shriveled.

Copper sulphate ($CuSO_4$ 25%) injected through low-volume irrigation systems at a rate of 20 kg per acre annually for two years increased soil copper concentration but did not increase Cu concentration in pistachio leaf tissue (Kallsen *et al.*, 1998). Soil treatments with copper sulphate and CuEDTA over several years provided highly variable responses (Uriu *et al.* 1989). Some showed no correction whatsoever, some provided correction, and still others showed phytotoxicity. This variability is most likely linked to soil pH and texture. Lighter, slightly acidic soils might be more responsive. The risks associated with soil treatment are also not justified since foliar treatment has proven highly effective. Post-bloom foliar application of chleated copper (CuEDTA 10-15%) applied at ¼ to 3/8 of a kg in 375 litres of water per acre when leaves are 50% expanded (end of April to early May) are extremely effective in correcting copper deficiency before symptoms appear in mid-summer (Uriu *et al.* 1989).

Application of copper sulphate at ¼ to ½ kg per 375 litres of spray solution per acre during this same time period has been successful, but they occasionally cause phytotoxicity. In many cases, treatment for copper deficiency is only required during orchard and canopy development. Once the trees are mature, the incidence of copper diminishes substantially, presumably because of increases in root density.

15.9.2. Fertilizer application

The average NPK removal per 1000 lbs of marketable yield is 28lbs of N, 3 lbs of P, and 24 lbs of K. This value includes all nutrients removed in hull, shell, kernels, blank nuts and other non-marketable yield per 1000 lbs. Evidence form research suggest that an average of 25 lbs N and 22 lbs of K are utilized to support tree growth requirements in a mature tree greater than 10 year old. Similarly a crop that results in 4000 lbs of marketable yield, would remove from the orchard 112 lbs N, 96 lbs K, and 12 lbs P (Brown and Siddiqui, 2013). The fertilizers are applied through the drip system beginning in late April during the establishment year, and continue into nut development. Nitrogen (UAN32) is applied alone in April and in liquid blends (10-0-10 and 15-0-05) combined with potassium. A total of 200 pounds of N and 125 pounds of K are applied annually due to the high potassium requirement of pistachio (Table 6). Zinc (Zinc sulfate 36%) and boron are foliar applied at 50% leaf expansion (mid-April). Boron is applied through the drip system in June or July in orchards with chronically low tissue levels (Brar *et al.*, 2015).

Table 6: Nitrogen & Potassium Fertilization Programme in Pistachio

Application time	Fertilizer source	Gallons (lbs.)Per acre	N lbs/acre	K lbs/acre
April	UAN32	7.0 (77)	25	0
May	10-0-10	51.5 (500)	50	50
June	10-0-10	51.5 (500)	50	50
July	15-0-5	51.5 (500)	75	25
Total			200	125

Fertilizer trials were conducted by Tekin *et al.* (1995) in a 37-year-old pistachio cv. Uzun orchard at Gaziantep, Turkey and recorded highest yield with 800 g N/tree + 600 g P_2O_5/tree + 60 kg sheep manure /tree + 3-times foliar sprays of micro-nutrients.

15.9.3. Nutrient demand during ON and OFF year:

The influence of alternate bearing on nutrient utilization and total tree nutrient content was investigated in mature pistachio cv. Kerman trees by Brown *et al.* (1995). Tree nutrient contents and, presumably, the size of nutrient storage pools in dormant trees varied between 'on' and 'off' years. There was 22 and 14% more N and P, respectively, in dormant trees following 'off' than 'on' years. The greater N and P accumulation in 'off' year trees presumably supports the large fruit demand for N and P during the 'on' years. In contrast to N and P, there was greater K and Ca accumulation in perennial tree parts during 'on' years than during 'off' years. The greater K accumulation in perennial tree parts and ~30% greater removal of K in annual organs during 'on' than 'off' years suggest that K uptake could be 4 times higher in 'on' year trees than in non-cropping, 'off' year trees.

Interrelationships between crop load, nitrogen (N), phosphorus (P), and potassium (K) uptake, and root growth were examined by Rosecrance *et al.(*1996) in mature, alternate-bearing pistachio (*Pistacia vera*) trees. Pistachio trees bear heavy (on-year) and light (off-year) fruit crops in alternate years. Uptake and partitioning of N, P, and K among tree parts were determined during spring flush (mid-March to late May), nut filling (late May to early September), and postharvest-leaf senescence (late September to early December). Nutrient uptake occurred primarily during nut filling in both on-year and off-year trees. In on-year trees, N and K uptake increased by 35 and 112%, respectively, during nut filling compared with off-year trees. During this period, nutrients were allocated largely to embryo development in on-year trees and to storage in perennial tissues in off-year trees. Nutrient uptake was negligible between harvest and leaf senescence. Although root growth was reduced during nut filling in on-year trees compared with off-year trees, there was no relationship between root growth and the uptake of N, P or K from the soil. The data support the hypothesis that sink demand regulates the uptake and distribution of N, P, and K in pistachio trees.

15.9.4. Foliar fertilization

The results showed that nitrogen significantly increased the flower bud retention, nitrogen level of leaf, current shoot and bud, photosynthesis rate, chlorophyll content and leaf area. Nitrogen was more effective in increasing flower bud retention of pistachio trees when applied at 0.3% on June 17. Abscission of flower buds decreased progressively as nitrogen content in leaf, current shoot and flower bud and also leaf area increased. From the results, Baninasab *et al.* (2007) concluded that application of nitrogen to the pistachio trees might be a useful method in decreasing flower bud abscission and increasing crop in the following year.

Zinc (Zn2+) deficiency in pistachio is frequently reported. Zn2+ fertilisation can be an effective way of correcting Zn2+ deficiency thus enhance yield of pistachio grown in Zn deficient soils. The results suggest that 2 g L^{-1} Zn application at 15 day-intervals (before, during and after blossom) would provide adequate Zn nutrition in pistachio grown under rainfed conditions (Kizilgoz *et al.*, 2010).

Karimi *et al.* (2012) observed that foliar application of K_2SO_4 at 0.1 or 0.2% during endospermic growth stage of seed and cotyledons appearance resulted in increase yield, splitting percentage; nut fresh mass and kernel dry mass and decrease blank percentage in pistachio trees cv. Owhadi.

Fruit dropping in pistachio orchards is one of the numerous problems. Results revealed that the highest level of primary fruit set was detected in a combination of Fe (as iron sulfate at 300 mg L^{-1}) and Zn (1000 mg L^{-1}). The maximum final fruit set percentage was measured in a combination of Fe and Cu (200 mg L^{-1} as copper sulfate). The highest of Splitting rate was detected in a combination of Cu and Fe and the lowest one in control. But spray of Zn, Cu and Fe did not affect the blankness compared with the control. Highest yield was produced in two levels of Zn and maximum vegetative growth took place in Cu treatment. Foliar application with Zn, Cu and Fe increased the Zn, Cu and Fe concentration of leaves, respectively (Soliemanzadeh *et al.* (2013).

15.10. Use of plant growth regulators

Alternate bearing in pistachio is due to the abscission of immature inflorescence buds during a heavy crop year. Pontikis (1990) found that spraying 5 times with 5000 p.p.m. NAA at intervals of 30 days beginning in mid-Jun. resulted in retention of 90.1% of inflorescence buds until end March, only 8.7% of the buds retained produced inflorescences and fruits. Fewer applications gave less bud retention.

Spraying of paclobutrazol at or above 500 ppm delayed flowering when sprayed on 25th June (when current shoot growth was completed) and the delay increased as the dose increased. Trees sprayed with paclobutrazol at 500 ppm flowered at the same time as the female cultivar Siirt. Percentage pollen germination was highest in controls (85%), and was generally lower from paclobutrazol-treated trees, although the inhibitory effect of paclobutrazol declined as the rate increased. Average shoot length decreased from 10.16 cm in controls to 1.0 cm in trees sprayed with paclobutrazol at 4000 ppm. Pistachios growing in unirrigated conditions suffer from reduced shoot growth due to water deficit and the additional decrease in shoot growth caused by paclobutrazol was considered undesirable (Ak, 1998).

An experiment was conducted by Alizadeh and Rahemi (2003) to study the effects of urea and benzyladenine (BA) on the incidence of flower bud abscission in pistachio cv. Owhadi. Application of BA at 100 mg litre^{-1} singly or in combination with 0.6% Nitrogen significantly increased floral bud retention. A positive correlation was observed between N content of flower-bearing bud branches and floral bud retention; however, such relationship was not observed between N content of leaf and flower bud retention. The N content of leaves might have been transferred to the other organs during the growing season. BA reduced kernel weight and increased blankness. A negative correlation was found between kernel weight and blankness.

The effects of gibberellic acid and a cytokinin on hull-splitting and aflatoxin levels in pistachio (*Pistacia vera* L. cv. 'Owhadi') were studied by Rahemi and Pakkish (2005) in a commercial orchard over 2 years. Gibberellin (GA_3) at 25, 50 and 100 mg $^{l-1}$ and benzyladenine (BA) at 50 and 100 mg l^{-1} reduced hull-splitting and infection by *Aspergillus spp.* with a strong correlation between splitting and infection (R2=0.88). GA3 and BA at high concentrations reduced hull-splitting to 0.6% and 0.3%, respectively. Kernels from trees treated with GA3 and BA at 50 or 100 mg l^{-1} had no aflatoxins compared with 4.7 ng g^{-1} in non-treated trees. BA also increased crude fat and chlorophyll levels in the kernels of pistachio fruit. Applications of GA_3 or BA are recommended for commercial pistachio production.

An effect of plant growth regulator applications on alternate bearing in pistachios was investigated by Acar *et al.* (2006). In the study, 10, 25, and 50 ppm NAA, 25 ppm benzyladenine (BA) and 0.25% urea were used, and 0.25% urea was added to all doses of plant growth regulators. Plant growth regulators were sprayed to the fruits, buds and leaves to the point of drip. Applications were done two times in year. First application was done at starting kernel development, and second application was done 20 days after the first application. According to the results, the highest yield was obtained from 25 ppm BA+0.25% urea in cv. Uzun (4.98 kg per tree) and 47.29% of inflorescence buds retained in this application. The lowest yield was obtained from the control as 4.05 kg per tree and 47.56% of inflorescence buds retained in the control. At the cultivar Siirt, the highest yield (8.26 kg per tree) was obtained from 0.25% urea application and 51.58% of inflorescence buds retained in this application. The lowest yield was obtained from 50 ppm NAA application (5.16 kg per tree) and 58.49% of inflorescence buds retained in this application. At the Siirt, 5.53 kg yield per tree was obtained from the control and 6.01 kg per tree obtained from 25 ppm BA+0.25% urea.

Alternate bearing, the occurrence of high yield 'on' year followed by low yield 'off' year, is striking in pistachio (*Pistacia vera* L.). Floral buds of pistachio are formed a year before bloom, but abscise during the years with heavy crop ('on' year). Abscission of floral buds is due to competition between growing seeds on 1-year-old shoots and developing buds on current season growth. Rahemi and Ramezanian (2007) studied the effects of chemical fruit thinning on alternate bearing and nut characteristics in a commercial orchard of 'Owhadi' pistachio cultivar in Rafsanjan, Iran. Ethephon (100 and 200 mg l^{-1}), naphthaleneacetic acid (NAA) (125 and 250 mg l^{-1}) and urea (2.5 and 5%) were applied to the branch units of each individual 'on'-year trees. The results showed that ethephon at both concentrations significantly increased fruit thinning and floral bud retention for the subsequent year. Other treatments also increased fruit thinning and floral bud retention but were inferior to ethephon..

Pistachio trees exhibit some physiological disorders such as abscission of inflorescence buds, fruit abscission, the production of blank, non-split, early-split and deformed nuts. Since the occurrence of these disorders is closely related to the yield, the effect of free polyamine treatments on these disorders and the yield was investigated by Khezri *et al.* (2010) in two consecutive years ("on" and "off" years). Free polyamine treatments were applied one week before and also two weeks after full bloom. Results showed that Spermine (Spm), unlike Putrescine (Put), decreased the physiological disorders of pistachio and also increased the yield. Free polyamine treatments did not affect the percentage of early-split nuts. Spermidine (Spd) decreased the percentage of both fruit abscission and deformed nuts but did not affect the yield. In both the "on" and "off" years the effect of Spm and Spd was greatly dependent on both the time of application and the concentrations used. Results of free polyamine analysis showed that exogenous application of these compounds, especially two weeks after full bloom, increased the free polyamine contents of shoots in both "on" and "off" years confirming the absorption of free polyamines via spray application in pistachio crop. Exogenous application of Spm two weeks after full bloom decreased the abscission of inflorescence buds in "on" year supporting the idea that free polyamines could trigger the abscission process of inflorescence buds. Since Spm decreased the percentage of blank, non-split and deformed nuts, it seems that Spm dominantly plays a key role in the growth and development of pistachio nuts.

Nut size and split nut yield are key economic factors in commercial pistachio (*Pistacia vera* 'Kerman') production. Nut size tends to decrease as trees age and can be significantly reduced during the on-crop year. Plant bioregulators (PBRs) are powerful tools for solving production problems in the field. Gibberellic acid (GA_3), 3,5,6-trichloropyridyloxyacetic acid (3,5,6-TPA) and 1-(2-chloro-4-pyridyl)-3-phenylurea (CPPU) were tested for their capacity to stimulate embryo growth and increase nut size. Abscisic acid (ABA), methyljasmonic acid (MJA), 2-chloroethylphosphonic acid (CEPA) and 2,3,5-tri-iodobenzoic acid (TIBA) were tested for their capacity to loosen mature fruit and increase the number of fruit (nuts) harvested. Branch studies testing multiple application times, rates, and combinations of PBRs were undertaken with the goal of increasing split nut (dry weight) yield. Application of GA_3 (170 mg L^{-1}) during late bloom-early fruit set (20 May), CPPU (10 mg L^{-1}) at the initiation of the period of exponential embryo growth (initiation of nut fill, INF) (10 June), and the combination of GA_3 and CPPU 30 days after INF numerically, but not statistically, increased split nut (dry weight) yield per branch. TIBA (125 mg L^{-1}) applied 4 days before harvest was the most effective fruit-loosening treatment. MJA (10 mg L^{-1}) applied 14 days before harvest was nearly as effective as TIBA. Both treatments reduced adherence of fruit tissues to the seed (nut), thereby reducing shell staining. Neither treatment caused significant leaf drop (Lovatt and Zheng, 2010).

Fresh pistachios undergo biochemical and physiological changes after harvest and consequently it has a short shelf life. Mirdehghan *et al.* (2013) studied the effects of foliar spraying of putrescine (Put) and spermine (Spm) before harvest and evaluating the postharvest behavior of fresh pistachio. Put (0, 1, 2 mM) and Spm (0, 1, 2 mM) were applied to pistachio tree, 'Kalleh Ghochi' and 'Fandoghi'. Treatments were performed 20 and 35 days before harvest. Following Put and Spm treatments, fruits were stored at 1.5+or-1 degrees C and 90-95% RH for 0, 20 and 45 days. The results revealed that fresh pistachio storability could be extended by Put and Spm treatment due to their effect on delaying the softening and decreasing weight loss, fungal infection. In addition both polyamines at different concentrations increased the firmness of fresh pistachio nuts significantly in comparison to the control.

16. Diseases and their management

16.1. Verticillium Wilt

The Verticillium wilt pathogen (*Verticillium dahlia*) is a wide spread soil-borne fungus that attacks hundreds of plant species. They can survive in the soil for many years with or without the presence of susceptible hosts. It invades the small roots, just behind the root tip, and then colonizes the xylem elements, thereby disrupting the flow of water through the plant. The disease assumes two forms, one form causes a rapid collapse and death of part or the entire tree and the other form is a slow decline that eventually leads to tree death. Salt stress before infection by *V. dahliae* can increase the severity of Verticillium wilt of pistachio, especially in salt-sensitive rootstocks (Mohammadi and Banihashemi, 2008). Mohammadi *et al.* (2006) reported a higher number of *V. dahlia* isolates, obtained in areas deficient in potassium and phosphorus, and were present in soil with pH of 7.3-8.6.

Control of the disease is difficult due to soil borne nature of the causal agent. Verticillium wilt has been biologically controlled by *Trichoderma* spp. (Jamdar *et al.*, 2013). Most rootstocks are susceptible, but some (*P. integerrima, Pistacia x saportae*) are functionally resistant and use of these rootstocks is the best control available (Werner *et al.*, 2001). Soil fumigation or solarization has not been successful in controlling this disease in pistachio.

16.2. Fruit rot

It is caused by *Aspergillus flavus*; a pathogen infecting pistachios, previously been damaged by insects, animals, early splits and mechanical harvest (Hejri *et al.*, 2006). Shell staining can indicate early split nuts where the kernel is exposed to infection (Doster and Michailides, 1999). Risk of infection is taken seriously because the fungus is associated with the presence of a carcinogenic toxin called Aflatoxin (Zahra *et al.*, 2014). Early split nuts are more susceptible to insect infestation and *Aspergillus flavus* and aflatoxin contamination (Pearson, 1996).

Plant resistance to aflatoxin contamination is not commercially available and fungicides are not effective. The use of antagonistic microorganisms against mycotoxin fungi is a promising approach to manage aflotoxin contamination via biological control. Certain yeast strains *Rhodotorula glutinis*, *Candida and Cryptococcus* spp.) and *Trichoderma harzianum* antagonistic isolates have provided good protection against the *A. flavus* and confined the spreading, sporulation and reduced the amount of aflatoxin production (Zahra *et al.*, 2014). The presence of fatty acid hydroperoxide, which can form in plant material either pre-harvest under stress or postharvest under improper storage conditions is correlated with high levels of aflatoxin production. Exposure to plant growth regulator methyl jasmonate (volatile methyl ester of jasmonic acid) can inhibit aflatoxin production by as much as 96%. Treatment can delay spore germination and inhibit the production of a mycelial pigment (Goodrich-Tanrikulu *et al.*, 1995). GA3 or BA at 50-100mg/l reduces the infection by reducing the hull splitting. BA also increased crude fat and chlorophyll levels in the kernels of pistachio fruit (Rahemi and Pakkish, 2005).

16.3. Panicle and shoot blight

The causal agent for causing blight in pistachio trees is supposed to be *Botryosphaeria dothidea.* The fungus attacks all current season growth and kills buds, shoots and fruit clusters. The pathogen survives in infested buds and in old infections where it forms cankers. The cankers can produce inoculum for several years. It is readily spread by rain and to some extent by certain insects and birds. Although the tree is not killed, severe outbreak of this disease renders the tree completely unproductive and at times cause 100% yield losses (Michailides *et al.*, 2014). The disease grows throughout the season, with a particular fondness for killing the nut-bearing clusters just before harvest. Infected nuts remain on the tree and can become a host to overwintering Navel orange worm.

Control is difficult and often inadequate, allowing the disease to build up to unmanageable levels. Early detection and immediate control measures are critical to prevent the disease from becoming endemic in an orchard. Pruning is very useful for removing sources of inoculum (Holtz, 2002). Ma-ZhongHua *et al.* (2001) reported that water stress adds to the severity of the blight. The chemicals which are used against the disease include: Benomyl @ 0.75-1 lbs a.i./acre applied once at bloom; Azoxystrobin @ 0.013lbs a.i./acre applied 6 times in a season. Other pre-mix fungicides which have proved to be most effective in controlling the disease include: fluxapyroxad + pyraclostrobin; boscalid + pyraclostrobin or tebuconazole + trifloxystrobin (Michailides *et al.*, 2014).

16.4. Alternaria late blight

The pathogen (*Alternaria alternate, A. tenuissimma, A. arborescens*) attacks leaves and fruits but the primary damage is to leaves, which shed early causing debilitating defoliation (Ma and Michailides, 2004). Symptoms on the nuts consist of black lesions with red halos, resulting in yield losses often exceeding 20% (Krid *et al.*, 2011). The disease develops in mid to late summer and continues into fall. It is associated with areas of high humidity and repeated dew events. Monthly average temperature above 30°C and air relative humidity 64%-70% accelerates the occurrence and development of this disease (Li-ShuRong *et al.*, 2007). It survives on any dead plant organic matter and its conidia are spread by wind. Orchards sustaining repeated premature defoliation may decline in production. The disease can be controlled by applying Azoxystrobin @ 0.013lbs a.i./acre (Luo *et al.*, 2007). Copper hydroxide applied at 6.16 lbs a.i./acre gives partial control with repeated applications.

17. Insect pests and their management

17.1. Navel Orangeworm (NOW)

Navel Orangeworm (*Amyelois transitella*) is considered the most important insect pest problem in pistachio (Kuenen and Siegel, 2010). The moth deposit eggs on the outer hull of the developing pistachio. Upon hatching from eggs, the larvae of NOW will infest the nutmeat if there is any damage to the outer hull protecting the shell and the nuts. The larva produce excrement (frass) and webbing in the nut. Often more than one worm can often be found eating the nut. Infested nuts are prime sites for *Aspergillus* mold infestations, which produce aflatoxin contaminants (potent carcinogens). Crop damage from this pest can reach levels of 5% or more, which combined with the strict limits on aflatoxin contamination in nuts, require management of this pest.

The primary management technique for NOW is winter removal of un-harvested nuts from the trees and the ground cover. The moths require nuts to survive the winter. The overall density of mummy nuts serving as potential oviposition sites prior to the next year's crop has greater impact on the abundance of Navel Orangeworm during the growing season and subsequent harvest (Burks *et al.*, 2008).

Use of *Steinernema carpocapsae* an entomopathogenic nematode @ 100000 infective juveniles IJs/m^2 in 500ml of water has been found to play a role in post harvest control of Navel Orangeworm (overwintering larvae) on fallen pistachio with a great potential (Siegel *et al.*, 2004), and also reduces pesticide reliance. Chemical control includes Azinphos-methyl @ 2lbs a.i./acre once per season at hull split or Carbaryl used periodically @ 4-5lbs a.i./acre. Degree-day based timing of treatments increase insecticide efficacy for the control of Navel Orangeworm (Holtz, 2002).

17.2. Oblique banded leaf roller (OBLR)

The oblique banded leaf roller (*Choristoneura rosaceana*) attacks a wide range of fruits and nuts. The overwintering larvae will feed on newly formed leaves and clusters. The primary damage is through defoliation. The two months generation time of OBLR combined with its habit of webbing leaves around itself for protection, has made chemical control difficult. Treatment may occur in May to control overwintering larvae and July through August to control the second generation (Holtz, 2002).

Permethrin is most commonly used to control OBLR. It is applied at 0.2-0.4 lbs a.i./acre. Azinphos-methyl @ 2lbs a.i./acre is infrequently used against OBLR, because its use is limited to once per season. Moreover, *Bacillus thuringensis* sprays, liquid-lime-sulphur, show promise in controlling oblique banded leafroller (Holtz, 2002).

17.3. Scale insects

Frosted scale (*Parthenolecanium pruinosum*) and European fruit lecanium (*Parthenolecanium corni*) are the two species of unarmored scale, which have become the dominant scale insect pests in pistachio. Both species have numerous host plants. These scales have only one generation per year and overwinter on twigs and limbs as nymphs. With the warmer temperatures of spring the nymphs start to feed and grow again. Scales are most common on vigorously growing trees and are normally kept under control by native parasites and high summer temperatures. Moderate to high scale populations can retard shoot growth and shell splitting. When populations build up, heavy amounts of honeydew can affect photosynthesis and provide a substrate for sooty mold (Yang-Sen *et al.*, 2008).

Soft scales molt twice before they reach maturity. The stage just before adult molting is frequently referred to as the rubber stage. In this stage, the scales remain soft, somewhat translucent, and are still susceptible to parasites and insecticides. Upon molting to adults, the shell hardens and becomes opaque reducing the effectiveness of insecticides and parasites. An effective parasite of soft scale is a tiny parasitic wasp (*Metaphycus luteolus*), which destroys the scale in its early instars before it can reproduce or cause substantial injury. In addition, other parasites such as the predacious Lady beetle, Lindorus, prey on these scales. Chemical control includes carbaryl @ 4-5 lbs a.i. plus 6 gallons of mineral oil/ acre, applied in late dormant period (mid-February). In growing season carbaryl alone can provide an effective control.

17.4. Plant Bugs

The large plant bugs (leaf footed bug – *Leptoglossus clypealis*; red shouldered stink bug – *Thyanta pallidovirens*; flat green stink bug – *Acrosternum hilare*), are sporadic pests of pistachio and the damage to pistachio nuts varies by season, as well as among insect species and development stages, with larger bugs typically causing more damage than smaller bugs. Feeding early at the nut development stage results in nut abortion and thus most damaged nuts are dropped from the cluster without reducing fruit load. The midseason period is the most critical because the damaged nuts remain in the cluster, causing a disorder known as kernel nut necrosis. If yeast or bacteria are present on the mouthparts of feeding insects, a condition known as stigma tomycosis may develop, which causes the meat to soften and become rancid. After shell hardening, the kernel is largely protected from bug feeding (Daane *et al.*, 2005).

Permethrin is the most effective insecticide for control of the large plant bugs. carbaryl is occasionally used at a rate of 5 lbs a.i./acre, particularly if NOW is a concern shortly before harvest. Biological control of the bugs can periodically regulate populations. Naturally occurring egg parasitoids are primarily responsible. The most common are *Trissolcus* spp. and *Gryonpenn sylvanicum*.

17.5. Pistachio twig borer moth (*Kermania pistaciella* Amsel)

The pistachio twig borer moth has been a major and important pest in pistachio orchards in Iran. Some quantity and quality damages of pistachio twig borer moth were studied by Tezerji (2011). The adult moth appears in late March-early April, lays its eggs in flower and fruit clusters, as well as on new shoots and petioles during high pest populations. The newly hatched larvae directly penetrate into the cluster tissues from which they bore into the rachis fruit cluster. Larvae tunneled in the pedicel of fruit cluster and branches and began to feed. There are four larval instars in pistachio woodborer life cycle. Results showed that, the pistachio twig borer moth causes severe economic damage through fruit dropping, decreasing quality and quantity of crop yield and rachis weakening of fruit clusters. Number of nuts per cluster, and split nut percentage significantly decreased in fruit clusters with larval tunneling in comparison with intact fruit clusters. In contract, irregular hull cracking pistachios, blank and non full growth kernel nuts significantly increased in fruit clusters with larval tunneling in comparison with intact fruit clusters.

17.6. Phloem feeding psyllid (*Agonoscena pistaciae*)

It is an important key pest of *Pistacia vera* in several Asian countries. Plant response to this pest is regulated in part by the salicylate and/or jasmonate signaling pathways. The effects of exogenous methyl jasmonate (MeJa) and salicylic acid (SA) application on *P. vera* leaves were assessed by Shahabinejad *et al.* (2014) comparing the number of nymphs and eggs per leaflets. The effects of MeJa and SA were concentration dependent, with higher concentration of SA failing to induce a more active response. SA was found more efficient than MeJa, as 3 days after hormonal spraying the egg number decreased by 79 and 56%, after SA and MeJa treatments, respectively, and the nymph density by 73 and 64%. Both MeJa and SA increased antioxidant enzyme activities, whereas they decreased total protein content. SA affected negatively also the total phenolic content. The levels of hydrogen peroxide were not changed by hormonal treatments.

The control of pistachio pests relies almost exclusively on pesticides. However, for most pistachio pests, sanitation procedures are the primary means of control. Cultural and mechanical control such as: clearing up leaf litter and debris as pruned branches and twigs both inside and around the orchards; removal of unharvested nuts; destruction of any nuts left on the soil surface, and tilling and disking the orchard floor, particularly under the tree canopy, may help reduce overwintering populations of many pest species. Furthermore, appropriate and regular irrigation and fertilization regimes are considered efficient treatments against wood boring species. Pruning and removal of weak, damaged and dead pistachio twigs, branches and trunks are necessary to disrupt the egg laying sites for scolytid beetles (Mehrnejad, 2014).

18. Physiological disorder and their management

18.1. Flower bud abscission

Reduction of water, increased salinity and unfavourable environmental conditions in most pistachio orchards, especially during flowering time has been associated with low fruit set and enhanced flower abscission. Numerous studies have shown that ascorbic acid and polyamines can reduce the bad effects of salinity in pistachio (Bastam *et al.*, 2013; Kamiab *et al.*, 2013). Foliar application of 550mg/l of ascorbic acid at flowering time increased pistachio fruit set (Molamhamadi *et al.*, 2014). Ascorbic acid is one of the most important and abundant water soluble anti-oxidants in plants and is responsible for increasing tolerance to a number of biotic and abiotic stresses (Bastam *et al.*, 2013).

18.2. Fruit abscission

Pistachio fruit abscission after the end of flowering is a physiological phenomenon, the main cause of which is lack of fertilization (Acar and Eti, 2007). The abscission of fruits can approach upto 80%. Pollen type, pollination and fertilization problems, unfavourable environmental conditions, impaired absorption of carbohydrates, minerals and growth regulators are reported as some reasons for this phenomenon (Crane and Iwakiri, 1985). In order to reduce fruit abscission and increase yield; Spermine (1mM), ascorbic acid (0.5g/l) and amino acids (33g/l) in combination can be successfully (Kamiab *et al.*, 2015)

18.3. Blank Nut

Blank nuts and non-split nuts are a concern for pistachio growers. Blank nuts result when the embryo fails to grow inside the shell and this can occur during both nut setting and nut filling. Blanking is thought to be influenced by poor tree nutrition and moisture stress. Blank and semi-blank fruits were identified by slower growth and lack of exocarp coloration during development stage II. Blank fruits result from one of 5 patterns of ovule degeneration caused by funicle degeneration, embryo sac absence, embryo sac degeneration, lack of pollen tube penetration of the embryo sac or failure of endosperm cellularization. Funicle degeneration is found to be the most common cause of blanking and was observed either at the site of pollen tube penetration or at the chalazal end of both pollinated and unpollinated ovules. Degeneration of funicular tissue extends towards the vascular bundles. Small-seeded or semi-blank fruits result from degeneration of the cotyledons during the later stages of embryo development (Shuraki and Sedgley, 1996).

18.4. Non-Splits

Non-splits are nuts that have a completely closed shell. They are rejected as the split allows consumers to open the nut easily. Non-split however, can be marketable if they are cracked for kernel. Splitting of nuts in response to kernel growth begins about one month prior to nut maturity. It would therefore seem that factors that promote kernel growth also promote shell splitting. These factors include irrigation management and crop load. It has also been suggested that the non-split characteristic may be transmitted via the pollen. Reports indicate that the percentage of non-split nuts increases and the percentage of blanks decreases with an increase in crop load (Polito and Pinney, 1999). Ferguson *et al.* (2005) suggested that to overcome this disorder, trees should not be water stressed from mid-August through September, and boron levels should be maintained above 120ppm by dry weight of July leaf sample.

18.5. Shell discoloration

Shell discoloration of pistachio (*Pistacia vera*) nuts was related to fungal decay, caused by *Aspergillus spp., Ulocladium spp.* and *Alternaria spp*., and insect infestation of the kernel (Doster and Michailides, 1999). Nuts with ruptured hulls (early split nuts and nuts with cracked hulls) varied considerably in the amount of shell discoloration, ranging from none to extensive. For both types of hull rupture, as shell discoloration increased, kernel decay also increased. Nuts with no discoloration had little or no fungal decay and navel orangeworm (*Amyelois transitella*) infestation. Processed nuts with an oily-shell appearance had the highest incidences of fungal decay and navel orangeworm infestation; nuts with a crinkled shell, nuts with extensive dark brown discoloration, and nuts with moderate dark brown discoloration along the suture had relatively high levels of decayed and infested kernels; while nuts with yellow discoloration, nuts with moderate dark brown discoloration not along the suture, and nuts with no discoloration had little or no decay and infestation. Also, as shell discoloration of processed nuts increased, kernel decay increased. It is suggested that shell characteristics may be used by processors to identify poor quality nuts and to improve the quality of pistachio nuts sold to consumers ((Doster and Michailides, 1999).

19. Harvesting

At maturity the pistachio hull slips easily off the shell and the colour generally turns a pastel shade of crimson. However, some nuts that are shaded inside the canopy can be yellow and blank nuts tend not to soften in color. Change in hull color is closely connected with shell splitting and so it is important to harvest nuts when they are fully mature to ensure maximum shell split. Nut maturity can also be determined by the ease at which the nuts detach from the tree. If the nuts fall when the branch is shaken, they are ready to harvest. Early-splits are not desirable because the shell splits open and the exposed kernel may deteriorate. In early-split nuts, the shell splits open earlier, with the hull still attached. As it is rare for the whole crop to mature at exactly the same time, more than one harvest is necessary, usually a week or two apart. Harvest of course, is dependent on fine weather and the availability of harvesting equipment.

Fine weather during harvest ensures greatest nut quality and harvest efficiency. Pistachios are small nuts with an open shell so it is important that they do not fall to the ground. Nuts that do fall to the ground may become contaminated if the hull is damaged. The fungus, *Aspergillus flavus*, inhabits damp soil and has potential to produce a carcinogen called aflatoxin. To avoid contact with the soil, harvesting machinery is therefore based on the shake and catch principle where the tree is shaken and the nuts fall onto a catching frame and are conveyed into a bin. The most popular harvester is the side by side shakes and catch machine that is commonly used in large orchards and by harvest contractors. Contract harvesting is an option where the harvesting machine can move freely within the tree rows, and where the orchard is of a size for contract harvesting to be cost-effective.

Smaller shake and catch harvesters mostly operate with a catching frame not unlike an inverted umbrella that can be opened under the tree canopy when the tree is shaken, and closed for transport to the next tree. For young trees and smaller orchards where the expense of a shake and catch machine is not an option, a combination of non-mechanical shaking and catching devices have been used. The simplest is a tarpaulin or length of shade cloth material spread under the canopy and the branches are knocked with long poles.

20. Yield

Pistachio starts bearing after 6 years and typically reaches full production by the 12th or 13th year. Pistachio yield data is not meaningful unless averaged over some years, due to substantial alternate bearing. Well-managed mature orchards on an average yield around 4000 kg of dried nuts per hectare (Kanber *et al.*, 1993).

21. Post harvest management

Nuts that are destined for the dried pistachio market are quickly transported to the hulling and drying plant. If hulls are not removed within 24 hours, the nuts generate heat and the shell can become stained. The sooner the nuts are hulled and dried, the better quality the nut will be. The nuts after drying should be stored in appropriate conditions to avoid mold growth and undesirable chemical reactions such as oxidative rancidity (Kashaninejad and Tabil, 2011). If there is a delay between harvest and hulling, the bins of nuts should be spread out in a cool-room.

The harvested nuts must be quickly transported for processing. The fresh nuts are cooled immediately after harvest in a cool room. Fresh pistachios are sold in the hull and so do not need hulling. Pistachio destined for the fresh market is cooled straight after harvest. All sticks, leaves and other debris and reject nuts are removed before they are graded, hand sorted and packed into polystyrene fruit boxes. The nuts should be stored in a cool-room at all times. Pistachios destined for sale as dried nuts are hulled (hulling means removing the hull, leaving behind the shelled kernel, whereas on shelling only kernel is left) as soon as possible. The hulling machines remove the hulls and the nuts are then washed to remove any remaining hulls. Floating nuts that are blank are removed and the good nuts are elevated to the drier. Slow drying produces the best quality kernel but this can be difficult when a huge volume of nuts has to be dried at one time. To overcome this problem, a three stage drying process has been developed. From the float tanks, the wet nuts are surface dried with fans and tumblers. They are then aggressively dried for a few hours until 'shell dry'. The nuts are then slowly dried using ambient air to retain kernel quality. Heat during drying can damage the kernel. Freshly harvested nuts have a moisture content of around 45%. The nuts are stable in storage when the moisture content is about 6% (Payne, 2004).

The next step in pistachio handling is removing non-splits via a needle picker machine. The nuts then proceed to the grader that separate nuts into appropriate sizes. To achieve the most uniform grades, the size graders operate via a combination of round and slotted holes. The standard nut sizes are:

- Extra Large – 72 or less nuts / 100g
- Large – 73 to 90 nuts/ 100g
- Medium – 91 to 107 nuts/ 100g
- Small – 108 or more nuts/ 100g.

Kernenls should be green in color with sweet flavor. On the basis of split and color of the hull, the nuts can further be graded as:

First grade wide split: the nuts are light in color and easy to open;

Natural splits: the nuts are light in color with narrow split.

The discoloured nuts can be removed by electronic sorting equipment and the product is given final quality sort by hand before packaging as unroasted nuts or proceeding for further processing. The nuts that are to be roasted and salted are soaked in brine before processing to the roasting ovens. Reject nuts and non-splits can be cracked for their kernel.

22. Industrial and export potential

Economic development is one of the important goals of each society in the world and economic growth is one of the basic factors in order to gain this goal. Many factors cause economic growth but in economic studies, export is the most efficient too to achieve economic growth (Francis *et al.*, 2007). Global exports of Pistachios are forecasted to increase by 3% to 335,000 MT of that of exports during 2014/15, primarily due to continued growth in Turkey's imports, with top markets in China and the European Union expected flat (Anonymous, 2015). Pistachio producing nations are not typically thought to be importers, yet statistics shows that the U.S. and Iran also imported. Reverse trade is typically a result of re-imported domestic product that was exported for the purpose of separating the pistachio nut its shell. This process allows producers to take advantage of lower labor rates in other nations, such as China. Other instances for re-imported product can be attributed to rejected shipments based on quality, sanitary or food safety standards (Anonymous, 2014).

The largest pistachio import markets are Hong Kong, Canada, Germany, Belgium, and India with the imports of 8,000MT in-shell during 2014/15 (Anonymous, 2015). Not all countries import directly from pistachio producing countries. Transshipments may occur for the following reasons: strategic trading purposes, quality issues, or evasion of domestic and foreign laws and regulations.

23. Utilization/processing

Drying is a critical step in the processing of pistachio. Temperature and thickness of pistachio on the dryer plays an important role in the quality of the product. Though there is no or little effect of temperature on protein content, however, the oil content increases with the increase in temperature (Mirmohseni *et al.*, 2012). The nuts are dried either by sun drying or in heated air ovens. The pistachio nuts should be spread on trays which can then be placed in the sun to dry. It usually takes 3 to 4 days for drying depending on ambient temperature; the higher the temperature the faster the drying. The drying trays are covered with hardware cloth to protect the nuts from birds and rodents. Heated-air drying method is much faster than sun drying. A forced air draft oven or a rotary-screen drier can be used. The time required to dry the nuts to about 5% moisture depends on the temperature used. At 160°F it takes about 10 hours. Higher temperatures (upto 200°F) will reduce drying time (to about 7 hours at 180°F and 5 hours to 200°F), but will also increase the extent to which the shells separate. This may become undesirable if the shell opens to a point where the kernel falls out during subsequent handling.

After drying, good quality splits are graded, followed by roasting, salting, and packaging. Around 90% of consumer-packed nuts from California are roasted and salted in the shell. Salting before roasting is done by soaking the nuts in a 15% salt solution for about 20 – 30 minutes. After soaking in the brine salt solutions, the nuts can be dried at 160°F to 200°F for about 10 – 15 minutes to remove excess water before roasting.

Roasting is accomplished by heating the dried pistachio nuts in a rotating drum of a roaster and gradually raising the temperature to 250°F or other roasting temperatures until the desired degree of roasting is obtained. Various combinations of roasting temperatures and times can be used: 250°F for 25 minutes, 300°F for 20 minutes, or 350° for 15 minutes.

23.1. Flow diagram for pistachio Processing

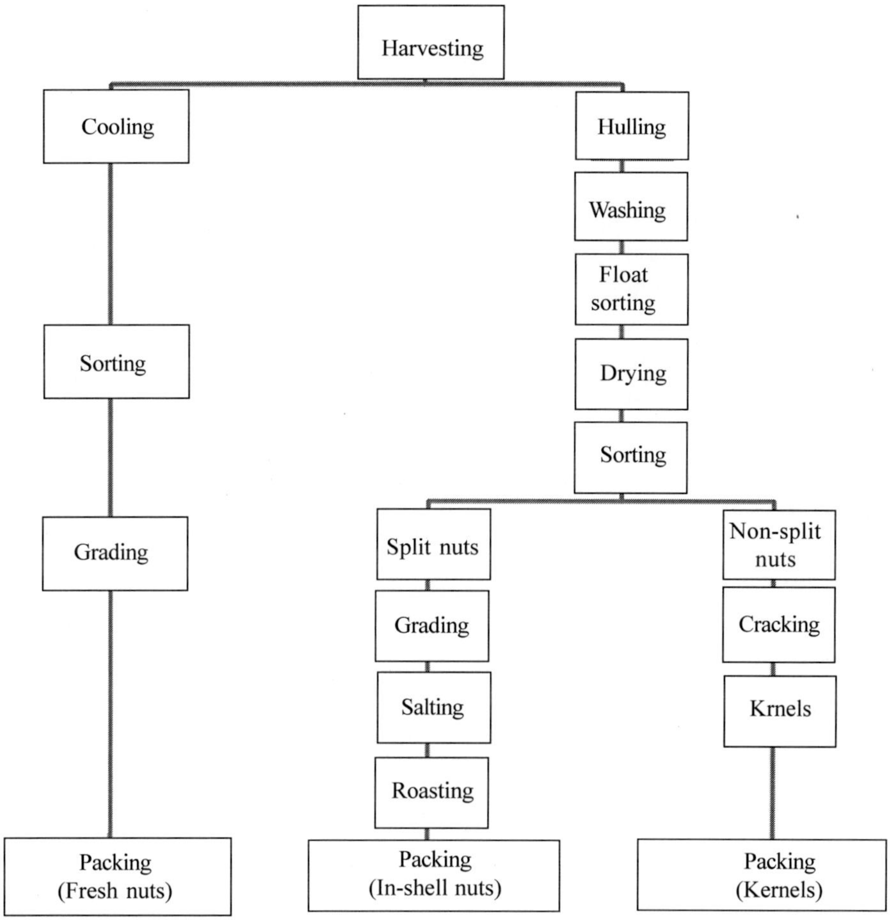

24. Storage

Pistachios may be stored at 38°F at a relative humidity of 60% to maintain good quality for about 2 years. Deterioration of roasted nuts begins in 12 to 15 months. Recommendations for shelled pistachios are a temperature of 5°C with moisture content of 4 to 5% for long storage (Payne, 2004).

25. Packaging

Proper packaging is must and serves several functions like protecting, facilitating use of, and promoting the product, as well as providing information about the product and its use. The protection function is critical in both transport and storage, protecting a product under a variety of temperatures and moisture conditions and against being crushed or dropped during handling is no small undertaking. Shape, colour, material and cost of the packaging can influence on consumer perceptions that are very important. By using vacuum method, the quality of pistachio doesn't change for a long time, and can protect the product against damages by hot or cold weather, and even direct sun shine for about two years (Aghdaie, 2009).

Raei and Jafari (2011) investigated the influence of different packaging materials and storage conditions on the quality attributes of Pistachio. Results revealed that pistachio samples stored under ambient conditions had better shelf life compared with 40 degrees C treatments. Among packaging materials, two layer plastic pouches resulted in higher quality attributes for the stored pistachios, especially regarding their fat quality. This confirms that two layer pouches had less permeability to oxygen compared with other packaging types. Regarding moisture loss, three layer plastic pouches were the best ones which can be very important from economic point of view.

The food manufacturing trade uses shelled pistachios in confections, baked goods, ice creams and as a flavoring and topping. Nuts can be processed into beverages (Pistachio milk). The roasted kernels are milled, and blending this pistachio paste with water at pH 8.5 for 30 min. the milk thus formed can be flavoured by adding 5% sugar and 0.02% vanilla (Ahmad-Shakerardekani *et al.*, 2013).

26. Economics of cultivation

The profitability of an operation can be projected by using the estimated prices and costs. Projected profit is the difference between expected income and expected costs. Expected income per hectare is obtained by multiplying the expected price and the expected yield. One simple and effective way to analyze orchard performance over time is to compare industry production figures with the individual orchard's production over the same period. If the individual orchard is falling further below average, it may be time for changes in management practices or for replacing the orchard. A sample cultivation economics of a pistachio orchard at San Joaquin valley (south) during 2015 for the first six years and the ninth year as reported by Brar *et al.* (2015) is given below (Table 7). The production practices in the study were those considered typical for pistachio in the valley, but may not apply to every situation.

Table 7: Sample costs ($/acre) to establish a Pistachio orchard in San Joaquin Valley, 2015.

Conditions/particulars	Year						
	1st	2nd	3rd	4th	5th	6th	9th
Planting costs (land preparation, disking, trees etc.0	1932	53					
Cultural costs (irrigation, pruning, pest/ weed control etc.)	657	734	905	1000	1292	1449	2198
Harvesting costs (shaking, picking, hulling etc.)						271	390
Overhead costs (office exp., leaf analysis, equipments etc.)	695	695	695	695	704	712	820
Interest on operating capital @ 5.75%	126	22	24	26	39	28	53
Total cash cost	3411	1503	1624	1722	2035	2460	3461
Income from prod.						1548	7224
Net cash costs	3411	1503	1624	1722	2035	912	
Profit above cash costs							3763

27. Trade and marketing

Iran is the leading pistachio producing and exporting country. It has 39.4% of global harvested area, accounts for only about 47.6% of global pistachio production, and has about 30.75% of world pistachio export. The USA, Iran's main competitor in the world market, with only 10% of global pistachio harvested area, contributed 24.6% of global pistachio production in 2013 (FAOSTAT, 2016). The main importers and consumers of world pistachio are European countries. Iran share in European markets has reduced due to strong competition from USA in the European Union markets and WTO decisions.

Despite good capacity for production and marketing of pistachio, Iran faces various internal and external problems in order to retain its better position of pistachio trade. Internal challenges are mainly management process problems such as: traditional methods in production and harvesting process, misuse of modern technology and pest control, lack of advanced irrigation and preservation systems, lack of capital and investment, irregular supply of product in market during the year, inadequacy of packaging industries and poor quality of product in international market. While external challenges are: lack of marketing advertisement and defective packing system, unfamiliarity with international markets and principles, indirect relations with overseas importers, improper planning for marketing and exporting of surplus pistachio, low accessibility to global markets and non-accession in WTO (Sedaghat, 2011).

28. Organic cultivation

The success in organic production depends upon the systematic approach of soil and plant. Therefore soil and plant should be planned together and a closed system should be formed by selecting resistant and demand varieties with agro-ecosystem. In organic farming, solutions are based on rational, experimental and experimental ecological knowledge (Lammerts *et al.*, 2004). Various organic based fertilizers were extensively used in agriculture to compensate nutrient deficiency in soil. Recently, the variety of fertilizers used in organic farming has increased and in addition to organic materials such as compost, humic and fulvic acids and leonardite, the production of fertilizers which contain extracts have been started commercially. It has been reported that soil microbial activity is affected in different ways by seaweed extracts (Blunden, 1991), and leonardite (Demirkiran *et al.*, 2008). Studies on feeding of small pistachio trees (1 year-old) with organic and inorganic fertilizers are very limited in the scientific world (Aaliabadi, 2006). In a study, the effect of peat on the growth and nutrition of *Pistacia lentiscus* L. was tested. For growing media each compost was added at a rate of 40%, fresh pine bark at 20 or 40% and peat at 20, 40 or 60%. In relation to plants growing in peat-based substrate, nutrition and P uptake was notably enhanced (Ostos *et al.*, 2008).

Soils of pistachio trees in Turkey, significantly lack many plant nutrients and have generally low level of organic matter content, and organic materials addition in soil improves the soil structure and supply plant with nutrients (Tamer and Karaca, 2004; Aaliabadi, 2006; Demirkiran *et al.*, 2008). Demirkiran and Cengiz (2010) in their study observed that micro nutrients especially Zn and Cu content of plant leaves are negatively affected by application of chemical fertilizers (macro), and some organic materials (alsil, humic acid and straw). Moss had positive effect on Mn uptake by pistachio trees, whereas peat, straw and humic acid positively affected P uptake. In general organic materials especially peat, gyttja and sea moss increased nutrient contents of pistachio leaves in areas where pistachio lacks nutrients.

Ensuring a vigorous healthy crop is the best insurance against pests and diseases. Fungicide/ bactericide preparations and certain insecticide products permitted for pest control include diatomaceous earth, quassia, soft soap, white oil emulsion and herbal preparations including natural pyrethrum. Products approved for disease control include sulphur, lime sulphur, copper sulphate, sodium silicate, herbal preparations and seaweed extracts. Lipophilic extracts obtained from different parts (leaf, branch, stem, kernel, shell, skin, seeds) of *Pistacia vera* in the range of 128-256 micro g/ml concentration have got noticeable antifungal activity, also the kernel and seed extracts as antiviral activity at the same concentration (Ozcelik *et al.* 2005). Numerous organisms, including bacteria, yeasts and nontoxigenic fungal strains of *Aspergillus flavus* and *Aspergillus parasiticus*, have been proved to be able to control the aflatoxins produced primarily by two closely related fungi, *Aspergillus flavus* and *Aspergillus parasiticus* (Yin-YanNi *et al.* 2008). To be most effective they should be applied after taking into account the level of disease, existing weather conditions, most effective concentration of preparation, and the most effective method of application and timing of application.

29. Effect of climate change

Temperate orchards are in urgent need of climate change adaptation strategies because of the high investments incurred in orchard development and the long productive life span of trees. Due to their deciduous nature, temperate trees must fulfill their chilling and heat requirements in order to break dormancy (Campoy *et al.*, 2011). Both requirements are attained to a certain climate regime. They must work together to ensure that dormancy is broken late enough to keep trees from starting to grow in winter. On the other hand, growth must be resumed early enough to allow trees to complete their annual reproductive cycle before the onset of the following winter season. Given these climatic requirements, productive cultivation of each tree cultivar is confined to a certain agroclimatic zone, and choosing the right tree cultivar for given climate regime is crucial for orchard productivity.

Climate change is likely to affect chilling (Baldocchi and Wong, 2008; Luedeling *et al.*, 2011). With global temperatures expected to rise by up to 6°C by the end of the 21st century, compared to pre-industrial levels (IPCC, 2007), it is unlikely that this agroclimatic metric will remain stable (Else and Atkinson, 2010; Luedeling *et al.*, 2011). Advancing trends in bloom dates of many trees indicate that dormancy breaking processes are indeed changing, most likely in response to climate change (Guedon and Legave, 2008; Legave *et al.*, 2009). This is true for a wide range of species, such as apples, pears, apricots, peaches, plums, walnuts, almonds and pistachios. Within each species, different cultivars have different chilling requirements (Guerriero *et al.*, 2010), and identifying an appropriate cultivar is essential for anyone planting trees for commercial production.

The pistachio (*Pistacia vera* L.) requires extended dry summers for its proper development and productivity. These conditions are characteristic of arid and semiarid areas, as are mild winters which are more constant due in part to climate change. Under these conditions chilling requirements for the optimum budbreak of this species are not met. Therefore, the use of chilling compensators is one of the ways to solve this problem. Hydrogen cyanamide applied at 5.0% ~30 days before expected budbreak can increase budbreak and substitute for chilling when pistachio is grown in environments with mild winters (Arreola Avila *et al.*, 2011).

On the other hand frost injury in recent years has exposed Pistachio to serious risks. Highest sensitivity to early spring cold is in the fully-opened flower stage and maximum damages are inflicted at -6°C. Pistachio nut size is an important parameter of production. Some factors influencing on nut size such as variety, pollination, irrigation and crop load are known. However effects due to climatic factors are considered too. The key period of maximum temperatures influencing nut size has been found to be in spring during hull enlargement. When maximum temperatures increase during hull enlargement, nut size increases. Under the same maximum temperature during that period, trees with good winter chill have been found to produce large nuts than poor chill years (Zhang and Joyce, 2009).

Shahsavari *et al.* (2011) suggested a measure against spring frost damage on inflorescence bud formation in pistachio orchards when temperature dropped to -6 degrees C and caused damage to flower buds and newly grown shoots. They found that a combination of irrigation and nitrogen fertilizer and 30% pruning had maximum percent of flower inflorescence buds in the trees for future crops.

30. Future research thrust

Fruit tree improvement is hampered by factors that make progress slower and more expensive than other crops. The large generation time of fruit trees make traditional breeding including crossing, evaluation and selection a lengthy process. Due to cross-pollination, pistachio is very heterogeneous. Breeding strategies exploit this genetic variation using seed orchards and controlled pollination. Because pistachio is dioecious, it is impossible to simply combine the best characteristics of the two cultivars, female and male, by hybridization. Few exceptional monoecious plants of *P. atlantica* have been reported in Turkey, which in future can be used as starting point for breeding monocious cultivated pistachio (Kafkas *et al.*, 2001).

Several soil fungi attack the pistachio. The most serious fungal disease is *Verticillium* wilt, which can quickly kill trees of varying ages (Moral *et al.*, 2010). Trees are also sensitive to the oak root fungus, *Armillaria mellea* (Michailides *et al.*, 1995). So, to tackle this soil-related problem, the breeding objectives of rootstocks include: resistance to upon nursery performance (Kafkas and Perl-Treves, 2001).

Pistachio nuts are classified according to nut characteristics, i.e. size, pericarp splitting rate and kernel colour (Atl *et al.*, 2011; Acar and Eti, 2011). In addition to nut characteristics, trees are selected for pollen production, flowering time, yield, quality and potential, uniformity of cropping (to avoid alternate bearing and empty seeds), early cropping and tree form. The main thrust includes tree vigour, age of bearing, growth habit and height and maturity, together with characteristics such as flowering time, early cropping, disease resistance, salt tolerance, tree form for mechanical harvesting and higher nut yield and quality.

The application of classical breeding to the pistachio tree has had only success due to the heterozygous nature of pistachio trees. The integration of molecular tools with conventional methods will be a qualitative advance in pistachio programmes (Aznarte-Mellado *et al.*, 2014). The establishment of sound protocols for regeneration from elite material is an essential prerequisite for the production of improved cultivars using modern techniques.

References

Aaliabadi, H.M. 2006. Study on the effects of different treatments on vegetative factors of *Pistacia vera* seedling. *Acta Horticulturae*, **726**: 205-208.

Abbaspour, H.; Afshari, H.; and Abdel Wahhab, M.A. 2012. Influence of salt stress on growth, pigments, soluble sugars and ion accumulation in three pistachio cultivars. *Journal of Medicinal Plants Research,* **6**(12): 2468-2473.

Abbaspour, H.; Fallahyan, F.; Fahimi, H.; and Afshari, H. 2006. Response of *Pistacia vera* L. in salt tolerance to inoculation with arbusscular mycorrhizal fungi under salt stress. *Acta Horticulturae,* **726**: 383-389.

Abousalim, A. 1990. Micropropagation and micrografting of pistachio (*P. vera* L. and *P. atlantica* Desf.). PhD thesis, University of London.

Abousalim, A. and Mantell, S.H. 1992. Micrografting of pistachio (*Pistacia vera* L. cv. Mateur). *Plant cell, Tissue and Organ Culture,* **29**: 231-234.

Abousalim,A. and Mantell, S.H. 1994. A practical method for alleviating shoot-tip necrosis symptoms in *in vitro* cultures of *Pistacia vera* cv. Mateur. *Journal of Horticultural Science*, **69**(2): 375-365.

Acar, I., and Eti, S. 2007. Abscission of pistachio flowers and fruits as affected by different pollinators Pakistan. *Journal of Biological Science*, **10**: 2920-2924.

Acar, I., and Eti, S. 2011. Nut quality of 'Kirmizi', 'Siirt' and 'Ohadi' pistachio cultivars as affected by different pollinators. *Acta Horticulturae,* **912**: 81-86.

Acar, I; Tahtaci, S. A; Arpaci, S; Aydin, Y. and Karadag, S. 2006. Determination of effects of plant growth regulator applications on alternate bearing in pistachios under suitable growing conditions. *Acta Horticulturae*, **726:** 539-544.

Aghdaie, S.F.A. 2009. Investigating effective factors on Iran's pistachio exportation. *International Journal of Marketing Studies*, **1**(2): 35-40.

Ak, B. E. 1998. The effects of paclobutrazol (PP-333) applications on inflorescence in male pistachio trees. *Cahiers Options Mediterraneennes*, **33**: 57-61.

Al Barazi, Z.; and Schwaba, W.W. 1982. Rooting softwood cutting of adult *Pistacia vera*. *J. Hort. Sci,* **57**(2): 247-252.

Alberghina, O. 1996. Comparison of two pistachio rootstocks. *Informatore Agrario*, **52**(15): 55-57.

Alderson, P.G.; and Barghchi, M. 1982. The potential for Pistacia propagation through shoot, leaf and callus culture. In: Fujiwara, A. (Ed.). *Plant tissue culture.* Maruzen, Tokyo. pp: 739-740.

Aleta, N.; Ninot, A.; Rouskas, D.; Zakinthinos, G.; Avanzato, D., and Gaspar, A.M. 1997. Pistachio propagation. *Options Mediterraneennes Serie-B, Etudes et Recherches*, **16**: 121-132.

Alizadeh, M. and Rahemi, M. 2003. Influence of foliar application of urea combined with 6-benzyladenine to decrease flower bud abscission in pistachio nut. *Iranian Journal of Agricultural Sciences,* **34**(3): 659-665.

Al-Saghir, M.G. and Porter, D.M. 2012. Taxonomic Revision of the Genus *Pistacia* L. (Anacardiaceae). *American Journal of Plant Sciences*, **3**: 12-32.

Anonymous. 2014. World Pistachio Trade. Schramm, Williams & Associates, Inc. 3-5.

Anonymous. 2015. Pistachios: World markets and trade. 2014/15 forecast overview. Foreign Agricultural Services. United States Department of Agriculture.

Arpac, S.; Atl, H.S.; Tekin, H., and Uygur, N. 2000. Effect of planting distance on tree growth, yield and some nut quality properties of the pistachio trees grown in arid conditions. *Bahce*, **29**(1/2): 3-12.

Arpaci, S., and Atli, H.S. 1995. The effect of different in-row spacings on yield, growth and some qualifications under arid conditions. *Acta Horticulturae*, **419**: 259-264.

Arpaci, S.; Aksu, O., and Tekin, H. 1998. Determination of the best suitable grafting method on different pistachio rootstocks. *Acta Horticulturae,* **470**: 443-446.

Arreola Avila, J. G; Nunez Barrios, A; Vargas Piedra, G; Valdez Cepeda, R; Borja de-la Rosa, A. and Trejo Calzada, R. 2011. Hydrogen cyanamide increases pistachio (*Pistacia vera* L.) bud break in arid regions with insufficient chilling. *Revista Chapingo Serie Ciencias Forestales y del Ambiente,* **17**(Special Issue): 259-265

Arreola-Avila, J.G.; Nunez-Barrios, A.; Vargas-Piedra, G.; Valdez-Cepeda, R.; Borja-de-la-Rosa, A., and Trejo-Calzada, R. 2011. Hydrogen cyanamide increases pistachio (*Pistacia vera*) bud break in arid regions with insufficient chilling. *Revista Chapingo Serie Ciencias Forestales-y-del-Ambiente*.17(Special issue): 259-265.

Assimakopoulou, A.; and Elena, K. 2010. Is there an influence of inorganic nutrition on the susceptibility of the pistachio to Camarosporium pistacia. *Options Mediterraneennes Serie-A, Seminaires Mediterraneens*, **94**: 181-185.

Atl, H. S; Aydn, Y; Arpac, S; Acar, I; Karadag, S; Bilgel, L; Sarpkaya, K; Kaska, N; Kafkas, S. and Ak, B. E. 2011. Determination of growth, bearing, yield and some quality characteristics of pistachio cultivars grafted on different rootstocks under irrigated conditions. *Acta Horticulturae*, **912** : 289-294.

Atl, H.S.; Aydn, Y.; Arpac, S.; Acar, I.; Karadag, S.; Bilgel, L.; Sarpkaya, K.; Kaska, N.; Kafkas, S., and Ak, B.E. 2011. Determination of growth, bearing, yield and some quality characteristics of pistachio cultivars grafted on different rootstocks under irrigated conditions. *Acta Horticulturae*, **912**: 289-294.

Avanzato, D; Vaccaro, A; Meli, M; Delfini, M; Capuani, G; Cocco, M. E-di; Sciubba, F; Tzareva, I. and Terziev, I. 2011. Effect of Boron treatment on 'Bianca' and 'Gloria' pistachio cultivars. *Acta Horticulturae,* **912:** 143-149.

Ayfer, M. 1963. Pistachio nut and its problems with special reference to Turkey. Univ. of Ankara Faculty of Agriculture Yearbook.pp. 189-217.

Ayfer, M.; Okay, Y.; and Erdogan, V. 1990. Antep Fistigi Anaclari ve Cogaltilmalari. In: Turkiye 1. Antepfistigi Simpozyumu Bildirileri, 11-12 Eylul. Gaziantep. pp: 38-48.

Aznarte-Mellado, C.; Sola-Company, P.J.; Robles, F.; Rejon, C.R.; Herran, R. de-la, and Navajas-Perez, R. 2014. Molecular characterization of the interspecific hybrid *Pistacia vigtos* (*P. vera* L. x *P. atlantica* Desf.). *Scientia Horticulturae*, **179**: 180-183.

Bailey, L.H. 1947. Pistacia. The standard Encyclopedia of Horticulture, III. 2648-2650.

Baldocchi, D., and Wong, S. 2008. Accumulated winter chill is decreasing in the fruit growing regions of California. *Clim. Cgange*, **87**: 513-5166.

Baninasab, B; Rahemi, M. and Javanshah, A. 2007. Effects of time of foliar application of nitrogen and its concentrations on the flower bud retention in pistachio trees. *International Journal of Soil Science,* **2**(1): 40-47.

Barghchi, M and Alderson, P.G. 1983a. *In vitro* propagation of *Pistacia vera* L. from seedling tissues. *Journal of Horticultural Science*, **58**: 435-445.

Barghchi, M. 1982. *in vitro* propagation of *Pistacia* species. Ph.D. thesis. Nottingham University.

Barghchi, M. 1986a. *in vitro* culture of mature commercial varieties of *Pistacia vera* L. *Proceedings of International Plant Propagators Society*, **35**: 331-333.

Barghchi, M. 1986b. *in vitro* micropropagation of Pistacia rootstocks. *Proceedings of the International Plant Propagators Society*, **35**: 334-337.

Barghchi, M. 1986c. Control of *in vitro* shoot tip necrosis in pistachio (*Pistacia vera* L.). In: *Plant Physiology Division Biennial Report*. Department of Scientific and Industrial Research, Palmerston North. pp.52.

Barghchi, M. and Alderson, P.G. 1983b. *in vitro* propagation of *Pistacia* species. *Acta Horticulturae,* **131**: 49-60.

Barghchi, M. and Alderson, P.G. 1989. Pistachio (*Pistacia vera* L.). In: Bajaj, Y.P.S. (ed.) *Biotechnology in agriculture and Forestry*. Vol. 5. *Tree II.* Springer-Verlag, Berlin.pp. 68-98.

Barghchi, M. and Martinelli, A. 1984. *in vitro* propagation of mature *Pistacia vera* varieties of Kerman (female) and Peter's (male) pistachio. In: *41st Easter School Symposium on Plant Tissue Culture and its Agricultural Applications.* University of Nottingham, Nottingham, Abstract 75.

Barghchi, M.; and Alderson, P.G. 1985. *in vitro* propagation of *P. vera* L. and commercial varieties of Ohadi and Kalleghoc. *J. Hort. Sci.,* **60**: 423-440.

Barghchi, M.; and Alderson, P.G. 1989. Pistachio (*P. vera* L.). In: Bajaj, Y.P.S. (ed.). Trees II (Vol. 3). Biotechnology in agriculture and forestry, pp. 68-98.

Bastam, N., Baninasab, B., Ghobadi, C. 2013. Interactive effects of ascorbic acid and salinity stress on the growth and photosynthetic capacity of pistachio (*Pistacia vera* L.) seedlings. *Journal of Horticultural Science and Biotechnology,* **88**: 610-616.

Beede, R.H.; Ferguson, L.; Munger, K.; Munger, D.; and Rose, D. 2002. Pistachio canopy management by moderate mechanical hedging and topping: Sixth year progress report. *Calif. Pistachio Ind. Ann. Rpt.,* 109-112.

Beede, R.H.; Ferguson, L.; Rose, D.; and Padilla, J. 1991. Effect of pruning on the yield components of pistachio. *Calif. Pistachio. Ind. Ann. Rpt.*, 90-92.

Benmahioul, B; Dorion, N; Kaid-Harche, M. and Daguin, F. 2012. Micropropagation and ex vitro rooting of pistachio (*Pistacia vera* L.). *Plant Cell, Tissue and Organ Culture,* **108**(2): 353-358.

Blunden, G. 1991. Agricultural uses of seaweeds and seaweed extracts. In: Seaweed Resources in Europe, Uses and Potential, John Wiley and Sons, Chichester.pp. 65-81.

Brar, G.S.; Doll, D.; Ferguson, L.; Fichtner, E.; Kallsen, C.E.; Beede, R.H.; Klonsky, K.; Tumber, K.P.; Anderson, N.; and Stewart, D. 2015. Sample costs to establish and produce pistachios. University of California Cooperative Extension. UCDavis.

Brown, P.; and Siddiqui, I. 2013. Managing pistachio nutrition: Overview, Findings and Recommendations. http://ucanr.edu/sites/scri/

Brown, P.; Ferguson, L.; and Picchioni, G. 1993. Boron nutrition of pistachio: Final report. *Calif. Pistachio Ind. Ann. Rpt.,* 57-59.

Brown, P.; Zhang, Q.; and Beede, B. 1996. Foliar spray applications at spring flush enhances zinc status of pistachio trees. *Calif. Pistachio Ind. Ann. Rpt.,* 101-106.

Brown, P.; Zhang, Q.; huang, Z.; Holtz, B.; and Hornung, C. 1999. Agronomic and economic responses of mature 'Kerman' pistachio trees to potassium applications in California. *Calif. Pistachio Ind. Ann. Rpt.,* 84-85.

Brown, P.H. 2001. Transient nutrient deficiencies and their impact on yield – a rationale for foliar fertilizers. *Acta Horticulturae*, **564**: 217-223.

Brown, P.H.; Zhang, Q.L., and Ferguson, L. 1994. Influence of rootstock on nutrient acquisition by pistachio. *Journal of Plant Nutrition*, **17**(7): 1137-1148.

Brown,-P. H; Weinbaum, S. A. and Picchioni, G.A.1995. Alternate bearing influences annual nutrient consumption and the total nutrient content of mature pistachio trees. *Trees Structure and Function,* **9**(3): 158-164.

Burks, C.S.; Higbee, B.S.; Brandl, D.G., and Mackey, B.E. 2008. Sampling and pheromone trapping for comparison of abundance of Amyelois transitella in almonds and pistachios. *Entomologia Experimentalis et Applicata*, **129**(1): 66-76.

Bustamante-Garcia, M.A. 1984. Micropropagation and rejuvenation of *Pistacia* species and the mechanism by which light influences root initiation. Ph.D. thesis. University of California, Davis, USA.

Campoy, J.A., Ruiz, D., and Egea, J. 2011. Dormancy in temperate fruit trees in a global warming context: a review. *Sci. Hortic.,* **130**: 357-372.

Chatibi, A; Kchouk, M. L; Mliki, A. and Ghorbel, A. 1998. Micrografting of pistachio (*Pistacia vera* L.) cv. Mateur. *Cahiers Options Mediterraneennes*, **33**: 121-130.

Chatibi, A; Kchouk, M. L; Thaminy, S. and Ghorbel, A. 1998. Somatic embryogenesis in pistachio (*Pistacia vera* L.) cv. Mateur. *Acta Horticulturae*, **470** : 460-466.

Crane, J. and Iwakiri, B.T. 1980. Preliminary notes on pruning mature pistachio trees. *Calif. Pistachio Ind. Ann. Rpt.,* 18-19.

Crane, J.C. 1984. Pistachio production problems. *Fruit Var. J.,* **38**(3): 74-85.

Crane, J.C., and Iwakiri, B.T. 1985. Vegetative and reproductive dominance in pistachio. *Hort Science*, **20**: 1092-1093.

Daane, K.M.; Yokota, G.Y.; Krugner, R.; Steffan, S.A.; Silva, P.G-da.; Beede, R.H.; Bentley, W.J., and Weinberger, G.B. 2005. Large bugs damage pistachio nuts most severely during midseason. *California Agriculture*, **59**(2): 95-102.

Davarynejad, G.H. 2001. Application of urea as a post-emergence herbicide for pistachio orchards. *Cahiers Options Mediterraneennes*, **56**: 265-266.

Davatchi, A.G. 1958. Etrude Biologuede la Fauna Entelogique des *Pistacia* sauvages et cultivars. *Extr. Rev. Pathol. Veg. Entomol. Agric. France*, **XXX**: 7.

Demirkiran, A.R., and Cengiz, M.C. 2010. Effects of different organic materials and chemical fertilizers on nutrition of pistachio (*Pistacia vera* L.) in organic arboriculture. *African Journal of Biotechnology*, **9**(38): 6320-6328.

Demirkiran, A.R..; Akkaya, A.; Turkmener, M.F.; Turkmener, M.C.; and Akkaya, S. 2008. Toprak verimliligini arttirmada kullanilabilecek alternative organic bir material: Gidya (Gyttja), Dunya Su Forumu, Sulama-Tuzlanma Toplantisi, Bildiri Sh. Sanliurfa, Turkiye (Turkish). **5**: 159-168.

Dolcet-Sanjuan, R. and Claveria, E. 1995. Improved shoot-tip micropropagation of *Pistacia vera* L. and the beneficial effects of methyl jasmonate. *Journal of the American Society for Horticultural Science*, **120**(6): 938-942.

Doster, M.A. and Michailides, T.J. 1999. Relationship between shell discoloration of pistachio nuts and incidence of fungal decay and insect infestation. *Plant disease,* **83**(3): 259-264.

Doster, M.A.; Michailides, T.J.; Goldhamer, D.A., and Morgan, D.P. 2001. Insufficient spring irrigation increases abnormal splitting of pistachio nuts. *California Agriculture*, **55**(3): 28-31.

Doster,-M-A; Michailides,-T-J. 1999. Relationship between shell discoloration of pistachio nuts and incidence of fungal decay and insect infestation. *Plant Disease*, **83**(3): 259-264.

Duan-YunFeng; Leng-PingSheng; Su-ShuChai; Xie-XiWen; and Yu-Fei. 2009. Seasonal changes of mineral nutrient element content in vegetative period of pistachio leaves. *Journal of fruit Science*, **26**(4): 502-506.

Elloumi, O.; Ghrab, M., and Mimoun, M.B. 2014. Effects of flower bud removal on seasonal starch storage and mobilization in fruiting and non-fruiting branches of pistachio trees cv. Mateur under dry and warm climate. *Scientia Horticulturae*, **172**: 19-25.

Else, M., and Atkiinson, C. 2010. Climate change impacts on UK top and soft fruit production. *Outlook Agric,* **39**: 257-262.

Fabbri, A., Ferguson, L., and Polito, V.S. 1998. Crop load related deformity of developing *Pistacia vera* cv Kerman nuts. *Scientia Horticulturae,* **77**: 219-234.

FAOSTAT. 2016. Food and Agriculture Organization Statistical Databases, United Nations. http://www.fao.org.

Fasihi Harandi, O. and Shahsavan Behboodi, B. (1997). Botany, distribution and ecology of Pistacia genus in Iran. *In: Proc. 6th Iranian Biology Congress*, College of Sciences, University of Kerman, Kerman (Iran), 25-27 August 1997.

Ferguson, L., and Kallsen, C.E. 2016. The Pistachio Tree: Physiology and Botany. In: Pistachio Production Manual, by Ferguson, L. and Hariland, D. pp. 19-26.

Ferguson, L., Polito, V., and Kallsen, C. 2005. The pistachio tree; Botany and physiology and factors that affect yield. Pistachio Production Manual. 4th ed.31-39.

Ferguson, L.; Maranto, J.; and Blackwell, B. 1991. Mechanical pruning of pistachios: Final progress report. *Calif. Pistachio Ind. Ann. Rpt.,* 140-143.

Ferguson, L.; Poss, J.A.; Grattan, S.R.; Grieve, C.M.; Wang, D.; Wilson, C.; Donovan, T.J.; and Chao, C.T. 2002. Pistachio rootstocks influence scion growth and ion relations under salinity and boron stress. *Journal of the American Society for Horticultural Science*, **127**(2): 194-199.

Firt, M.Z.; and Onay, A. 1999. Analysis of binomial data obtained from plant-tissue culture studies using generalized linear models. *Turkish Journal of Biology*, **23**: 261-267.

Francis, B., Iyare, S.O., and Lorde, T. 2007. Agricultural export diversification and economic growth in Caribbean countries: Co integration and Error-Correction Models. *The International Trade Journal,* **32**: 229-256.

Ghoraishi, S.R. and Mantell, S. H. 1998. Improved shoot growth and control of shoot-tip necrosis in *Pistacia mutica* and *P. vera* micropropagation systems. *Tree biotechnology towards the millennium*, pp. 163-172.

Gijon, M.C.; Memmi, H.; Perez-Lopez, D.; Couceiro, J.F.; Moriana, A.; Martinez, E.; and Guerrero, J. 2014. Influence of temperature on the production quality parameters in pistachio (*Pistacia vera* L.). *Acta Horticulturae*, **1028**: 249-254.

Goldhamer, D.A., Michailides, T.J., Salinas, M., Dorster, A., and Beede, R.H. 2004. Using early season water stress to increase shell splitting at harvest. In: California pistachio report, California Pistachio Commission. 115-117.

Gonzales, A. and Frutos, D. 1990. *In vitro* culture of *Pistacia vera* L. embryos and aged tree explants. *NATO ASI Series A.,* **186**: 335-338.

Gonzales, V.; and Uriu, K. 1985. Nutritional deficiency symptoms in pistachio. *Calif. Pistachio Ind. Ann. Rpt.,* 69-73.

Goodrich-Tanrikulu, M.; Mahoney, N.E., and Rodriguez, S.B. 1995. The plant growth regulator methyl jasmonate inhibits aflatoxin production by *Aspergillus flavus. Microbiology Reading*, **141**(11): 2831-2837.

Guedon, Y. and Legave, J.M. 2008. Analyzing the time-course variation of apple and pear tree dates of flowering stages in the global warming context. *Ecol. Model*, **219**: 189-199.

Guerriero, R., Viti, R., Iacona, C., and Bartolini, S. 2010. Is apricot germplasm capable of withstanding warmer winters? This is what we learned from last winter. *Acta Hortic.*, **862**: 265-272.

Hartmann, H.T., Kester, D.E., Davies, F.T., and Geneve, R.L. 2002. Propagation methods and rootstocks for fruit and nut Species. Plant propagation: Principles and Practices, 7th ed. 717-757.

Hejri, L.A.; Azar, M., and Javanshah, A. 2006. Isolation of saprophytic yeasts from pistachio fruits and leaves. *Acta Horticulturae*, **726**: 599-604.

Holtz, B.A. 2002. Plant protection for pistachio. *Hort Technology*, **12**(4): 626-632.

IPCC. 2007. Climate change 2007-synthesis report. Contributions of working groups I, II and III to the fourth assessment report of the Inter governmental panel on Climate change. Intergovernmental Panel on Climate Change. Geneva, Switzerland.

Jamdar, Z.; Mohammadi, A.H., and Mohammadi, S. 2013. Study of antagonistic effects of Trichoderma species on growth of Verticillium dahlia, the causal agent of verticillium wilt of pistachio under laboratory condition. *Journal of Nuts*, **4**(4): 53-56.

Jiang-JianGuo. 1999. The cultural techniques of Pistacia and the advances of its research. *South China Fruits*, **28**(5): 44-46.

Joley, L.E. 1979. Pistachio. In: Jaynes, R.A. (Ed.). Nut tree culture in North America. Northern Nut growers Assoc., Broken Arrow Road, Handen, Conn. pp. 163-174.

Kafkas, S., and Perl-Treves, R. 2001. Morphological and molecular phylogeny of Pistacia species in Turkey. *Theoretical and Applied Genetics*, **102**(6/7): 908-915.

Kafkas, S.; Cetiner, S., and Perl-Treves, R. 2001. Development of sex-associated RAPD markers in wild Pistacia species. *Journal of Horticultural Science and Biotechnology*, **76**(2): 242-246.

Kallsen, C.E., Parfitt, D.E., and Maranto, J. 2016. Male pistachio variety named 'FAMOSO'. Plant Patent Application Publication. United States. Pub. No.: US 20160309631P1.

Kallsen, C.E.; Holtz, B.A.; Wylie, C., and Villarux, L. 1998. Injecting zinc and copper fertilizers through low-volume irrigation systems to meet pistachio nutrient requirements. *Calif. Pistachio Ind. Ann. Rpt.,*105-106.

Kamiab, F., Salehabad, M.H., and Zamanibahramabadi, E. 2015. Evaluating the effects of foliar treatments of polyamines and some organic acids on quantitative and qualitative traits in some pistachio cultivars. *Journal of Nuts,* **6** (2): 131-142.

Kamiab, F., Talaie, A., Khezri, M., and Javanshah, A. 2013. Exogenous application of free polyamines enhance salt tolerance of pistachio (*Pistacia vera* L.). *Plant growth regulation,* **72**: 257-278.

Kanber, R.; Yazar, A.; Onder, S., and Koksal, H. 1993. Irrigation response of pistachio (*Pistacia vera* L.). *Irrigation Science*, **14**(1): 7-14.

Kanber, R; Krda, C; Yazar, A; Onder, S. and Koksal, H. 1993.Irrigation response of old pistachio trees (*Pistacia vera* L.). *Doga, Turk Tarm-ve-Ormanclk Dergisi*, **17**(3): 659-671.

Karimi, H. R; Sevandi Nasab, S. and Roosta, H.R. 2012. The effect of salicylic acid and potassium on some characteristics nut and physiological parameters of pistachio trees cv. Owhadi. *Journal of Nuts,* **3**(3): 21-26.

Karimi, H.R.; Ebadi, A.; Zamani, Z., and Fatahi, R. 2011. Effects of water salinity on growth indicies and physiological parameters in some pistachio rootstocks. *Journal of Plant Nutrition,* **34**(7): 935-944.

Karimi, H.R.; Maleki-Kuhbanani, and Roosta, H.R. 2014. Evaluation of inter-specific hybrid of *P.atlantica* and *P.vera* L. cv. 'Badami-Riz-e-Zarand' as pistachio rootstock to salinity stress according to some growth indices and eco-physiological and biochemical parameters. *Journal of Stress Physiology and Biochemistry*, **10**(3): 5-17.

Kashaninejad, M., and Tabil, L.G. 2011. Pistachio (*Pistacia vera* L.). *Postharvest Biology and Technology of Tropical and Subtropical Fruits*, **4**: 218-246.

Kermani, M.M.; and Hokmabadi, H. 2012. Trend of soil salinization and its effect on pistachio yield of Rafsanjan orchards. *Acta Horticulturae*, **940**: 287-292.

Khalilian, S. and Farhadi, A. 2002. Factors affecting agriculture sector exports in Iran. *Iranian Journal of Agricultural Economics and Development*, **39**: 71-84.

Khezri, M; Talaie, A; Javanshah, A. and Hadavi, F. 2010. Effect of exogenous application of free polyamines on physiological disorders and yield of 'Kaleh Ghoochi' pistachio shoots (*Pistacia vera* L.). *Scientia Horticulturae*, **125**(3): 270-276.

Kizilgoz, I; Sakin, E. and Aslan, N. 2010. The effects of zinc fertilisation on the yield of pistachio (Pistacia vera L.) grown under rainfed conditions. *African Journal of Agricultural Research*, **5**(24): 3427-3430.

Kokwaro, J.O. and Gillett, J.B. 1980. Notes on the Anacardiaceae of Eastern Africa. *Kew Bulletin*, **34**: 745-760.

Krid, S.; Triki, M.A.; Ghrab, M.; Gdoura, R., and Rhouma, A. 2011. Occurrence of Alternaria species involved in fruit blight of pistachio in Tunisia. *Journal of Plant Pathology*, **93**(4): S4.81.

Kuden, A.B.; Kuden, A.; Nikpeyma, Y., and Kaska, N. 1995. Effects of chemicals on bud break of pistachios under mild climate conditions. *Acta Horticulturae*, **419**: 91-96.

Kuenen, L. P.S., and Siegel, J.P. 2010. Protected emergence of overwintering Amyelois transitella (Lepidoptera: Pyralidae) from pistachios and almonds in California. *Environmental Entomology*, **39**(4): 1059-1067.

Lammerts, V.; Bueren, E.T.; and Struik, P.C. 2004. The consequences of the concept of naturalness for organic plant breeding and propagation. *Netherlands J. Agric. Sci.*, **52**(1): 85-95.

Legave, J.M., Giovannini, D., Christen, D., and Oger, R. 2009. Global warming in Europe and its impacts on floral bud phenology infruit species. *Acta Hortic.*, **838**: 21-26.

Lemaister, J. 1959. Le Pistacher (Etude bibliographique). *Fruits*, **14**: 57-77.

Li, X. and Chapple, C. 2010. Understanding lignification: Challenges beyond monolignol biosynthesis. *Plant Physiology*, **154**: 449-452.

Linnaeus, C. 1753. Species Plantarum. *Laurentii Salvii*, **2**: 1025-1026.

Li-ShuRong; Liu-ZhenLiang; Su-ShuChai; Wang-DaPeng; and Liu-KeFeng. 2007. Control of leaf brown spot of *Pistacia vera* introduced in Beijing area. *Scientia Silvae Sinicae*, **43**(11): 76-81.

Liu-ZhenLiang; Li-ShuRong; and Su-ShuChai. 2004. Grafting of *Pistacia vera* L. in Beijing. *Journal of Beijing Forestry University*, **26**(5): 85-88.

Lovatt, C. J. and Zheng, Y. 2010. Use of plant bioregulators to stimulate embryo growth and loosen fruit to increase split nut yield of pistachio. *Acta Horticulturae*, **884** : 549-557.

Luedeling, E., Girvetz, E.H., Semenov, M.A., and Brown, P.H. 2011. Climate change affects winter chill for temperate fruit and nut trees. *PLos ONE*. 6: e20155.

Luo, Y.; Ma-ZhongHua; Reyes, H.C.; Morgan, D.P., and Michailides, T.J. 2007. Using real-time PCR to survey frequency of azoxystrobin-resistant allele G143A in Alternaria populations from almond and pistachio orchards in California. *Pesticide Biochemistry and Physiology*, **88**(3): 328-336.

Ma, Z. H., and Michailides, T.J. 2004. A real-time PCR assay for the detection of azoxystrobin-resistant Alternaria populations from pistachio orchards in California. *Crop Protection*, **23**(12): 1259-1263.

Maggs, D.H. 1990. The Australian pistachio 'Sirora'. *Fruit Varieties Journal*, **44**(4): 178-179.

Mahmoodabadi, M.; Yazdanpanah, N.; Rodriguez-Sinobas, L.; Pazira, E., and Neshat, A. 2013. Reclamation of calcareous saline sodic soil with different amendments (I): redistribution of soluble cations within the soil profile. *Agricultural Water Management*, **120**: 30-38.

Marra, F.P.; Barone, E.; Mantia, M.La.; and Caruso, T. 2009. Toward the definition of a carbon budget model: seasonal variation and temperature effect on respiration rate of vegetative and reproductive organs of pistachio trees (*Pistacia vera*). *Tree Physiology*, **29**(9): 1095-1103.

Martinelli, A. 1988. Use of *in vitro* techniques for selection and cloning of different *Pistacia* species. *Acta Horticulturae,* **227**: 331-333.

Martinez Palle, E. and Herrero, M. 1995. The ponticulus: a structure bridging pollen tube access to the ovule in *Pistacia vera. Sexual Plant Reproduction*, **8**(4): 217-222.

Martinez-Palle, E; and Herrero, M. 1995. The ponticuluc: a structure bridging pollen tube access to the ovule in *Pistacia vera. Sexual Plant Reproduction,* **8**(4): 217-222.

Ma-ZhongHua; Morgan, D.P., and Michailides, T.J. 2001. Effects of water stress on Botryosphaeria blight of pistachio caused by Botryosphaeria dothidea. *Plant disease*, **85**(7): 745-749.

Mehrnejad, M.R. 2014. The psyllids of pistachio trees in Iran. *Acta Horticulturae*, **1028**: 191-194.

Michailides, T.J.; Morgan, D.P. and Doster, M.A. 1995. Diseases of pistachio in Calfornia and their significance. *Acta Horticulturae*, **419**: 337-343.

Michailides, T.J.; Morgan, D.P. and Puckett, R.D. 2014. Growers can reduce sprays during control of botryosphaeria panicle and shoot blight of pistachio by using a leaf wetness model. *Acta Horticulturae*, **1028**: 179-184.

Mirdehghan, S. H; Khanamani, Z; Shamshiri, M H; Hokmabadi, H. 2013. Preharvest foliar application of putrescine and spermine on postharvest quality of fresh pistachio. *Acta Horticulturae,* **1012**: 299-303.

Mirmohseni, S.R.; Moghadam, M.M.; Beheshti, B.; Almasi, M.; Borghei, A.M. and Ziaolhagh, H. 2012. The effects of some drying conditions on the oil and protein contents of pistachios. *Acta Horticulturae*, **963**: 147-154.

Moghaddam, A.R.L.; and Soleimani, A. 2012. Compensatory effects of humic acid on physiological characteristics of pistachio seedlings under salinity stress. *Acta Horticulturae*, **940**: 253-255.

Mohammadi, A.H. and Banihashemi, Z. 2008. Effect of different levels of NaCl on Verticillium wilt disease of pistachio in hydroponic culture. *Journal of Science and Technology of Agriculture and Natural Resources,* **12**(45(A)): 239-249.

Mohammadi, A.H.; Banihashemi, Z., and Maftoun, M. 2007. Interaction between salinity stress and Verticillium wilt disease in three pistachio rootstocks in a calcareous soil. *Journal of Plant Nutrition*, **30**(1-3): 241-252.

Mohammadi, A.H.; Haghdel, M.; Mohammadi-Moghaddam, M., Banihashemi, Z. 2006. Current status of Verticillium wilt disease of pistachio trees in Iran. *Acta Horticulturae*, **726**: 631-635.

Mohassel, M.H.R.; Nassiri, M. and Poorkazem, E. 2001. The effect of glyphosate and herbicide combinations on pistachio garden weeds in Kerman. *Cahiers Options Mediterraneennes*, **56**: 197-200.

Molamohamadi, A., Khezari, M., Safari, V.R., and Tavasolian, I. 2014. Evaluation effect of alphatochoferol and ascorbic acid on some growth traits and yield in pistachio. Master thesis, Shahid Bahonar University, faculty of Agriculture, Horticulture Department, Kerman, Iran. pp.200.

Moldenke, H.N.; and Alma, L. 1952. *Plants of Bible*. Chronica Botanica Co. Waltham.

Monastra, F.; Avanzato, D.; Martelli, S. and Dascanio, R. 1995. Pistachio trial under different volumes of irrigation in Italy. *Acta Horticulturae*, **419**: 249-252.

Monastra, F.; Rovira, M.; Vargs, F.J.; Romero, M.A.; Batlle, I.; Rouskas, D. and Gaspar, A.M. 1997. Isozyme characterization of different Pistacia species and their hybrids. Study of their performance as rootstocks for *Pistacia vera* L. *Options Mediterraneennes Serie-B, Etudes et Recherches*, **16**: 133-142.

Moral, J.; Lopez-Escudero, F.J.; Roca, L.F.; Blanco-Lopez, M.A. and Trapero, A. 2010. First report of Verticillium wilt of pistachio caused by Verticillium dahlia in Spain. *Plant Disease*, **94**(3): 382.

Murashige, T. and Skoog, F. 1962. A revised medium for rapid growth and bioassay with tobacco tissue cultures. *Physiologia Plantarum*, **15**: 473-497.

Onay, A. 2000a. Micropropagation of pistachio from mature trees. *Plant Cell, Tissue and Organ Culture*, **60**(2): 159-162.

Onay, A. 2000b. somatic embryogenesis in cultured kernels of pistachio, *Pistacia vera* L. cv Siirt. IN: *Proceedingss of the 2nd Balkan Botanical Congress,* Vol. II. Marmara Universittesi Publishing Section, Istanbul,pp. 109-115.

Onay, A. 2000c. Histology of somatic embryogenesis in cultured leaf explants of pistachio (*Pistacia vera* L.). *Turkish Journal of Biology*, **24**(2): 91-95.

Onay, A. and Jaffree, C.E. 2000. Somatic embryogenesis in pistachio. In: Jain, S.M., Gupta, P.K. and Newton, R.J. (eds) *Somatic Embryogenesis in Woody Plants.* Kluwer Academic Publishers, Dordrecht, The Netherlands pp. 361-390.

Onay, A., Pirinc, V., Yildirim and Basaran, D. 2004. *in vitro* micrografting of pistachio (*Pistacia vera* var. Siirt). *Plant Cell, Tissue and Organ Culture,* **77**: 215-219.

Onay, A.; and Namli, S. 1998. Antepfistiginin (*Pistacia vera* L.) yapraklarinda somatic embriyogenesis. In: *XIV. Ulusal Biyoloji Kongresi. Bitki Fizyolojisi-Bitki Anatomisi ve Hidrobiyoloji Seksiyonu, Samsun,* Vol.2. Ondokuz Mayis Universitesi Basimevi, Samsun, pp. 44-53.

Onay, A.; Firat, M.Z. and Namli, O. 1997. An improved method for embling production in pistachio, *Pistacia vera* L. using somatic embryos matured in a liquid medium. *Turkish Journal of Biology,* **21**(2): 159-174.

Onay, A.; Jaffree, C.E.; and Yeoman, M.M. 1995. Somatic embryogenesis in cultured immature kernels of pistachio, *Pistacia vera. Plant Cell Reports*, **15**(3/4): 192-195.

Onay, A.; Jeffree, C.E.; Theobald, C.; and Yeoman, M.M. 2000. Analysis of the effects of maturation treatments on the probabilities of somatic embryo germination and plantlet regeneration in pistachio using a linear logistic method. *Plant Cell, Tissue and Organ Culture*, **60**(2): 121-129.

Onay, A; Jaffree, C.E., and Yeoman, M.M. 1996. Plant regeneration from encapsulated embryoids and an embryogenic mass of pistachio, *Pistacia vera* L. *Plant Cell Reports,* **15**(9): 723-726.

Onay, A; Pirinc, V; Yldrm, H. and Basaran, D. 2004. In vitro micrografting of mature pistachio (*Pistacia vera* var. Siirt). Plant Cell, Tissue and Organ Culture, **77**(2): 215-219.

Oney, A. 1996. *In vitro* organogenesis and embryogenesis of pistachio, *Pistacia vera* L. PhD thesis, University of Edinburgh.

Orhan, E.; Esitken, A.; Ercisli, S.; and Sahin, F. 2007. Effects of indole-3-butyric acid (IBA), bacteria and radicle tip-cutting on lateral root induction in *Pistacia vera. Journal of Horticultural Science and Biotechnology*, **82**(1): 2-4.

Ostos, J.C.; Lopez-Garrido, R.; Murillo, J.M.; and Lopez, R. 2008. Substitution of peat for municipal solid waste- and sewage sludge-based composts in nursery growing media: Effects on growth and nutrition of the native shrub Pistacia lentiscus L., doi: 10.1016/j.biortech.2007.03.033. *Bioresour. Technol.*, **99**: 1793-1800.

Ozbek, S. 1978. Antep fistigi yetistriciligi. In: Ozel Meyvecilik. Cukurova Universitesi Ziraat Fakultesi Yayinlari, 128, Adana-Turkey. pp. 322-365.

Ozcelik, B.; Aslan, M.; Orhan, I.; and Karaoglu, T. 2005. Antibacterial, antifungal, and antiviral activities of the lipophilic extracts of *Pistacia vera*. *Microbiological Research*, **160**(2): 159-164.

Pair, J.C.; and Khatamian, H. 1980. Propagation and growing of the Chinese pistachio. *Proc. Inter. Plant Prop Soc.*, **32**: 497-503.

Palma, L-de.; and Novello, V. 1998. Effect of drip irrigation on leaf gas exchanges and stem water potential in pistachio cv. Larnaka. *Acta Horticulturae,* **470**: 317-323.

Palma, L-de.; Novello, V.; and Monastra, F. 1997. First observations on physiological effects of irrigation in pistachio. *Acta Horticulturae*, **449**: 545-550.

Parfitt, D. E; Kallsen, C.E; Holtz, B. and Maranto, J. 2010. 'Randy' male pistachio. *Hort. Science*, **45**(7): 1113-1115.

Parfitt, D.E. and Almehdi, A. 1994. Use of high CO2 atmosphere and medium modifications for the successful micropropagation of pistachio. *Scientia Horticulturae,* **56**: 321-329.

Parfitt, D.E., Kallsen, C., Maranto, J., and Holtz, B. 2007. 'Golden Hills' Pistachio. *Hort. Science,* **42**(3): 694-696.

Parfitt, D.E., Kallsen,C.E., Holtz, B. and Maranto, J. 2008. 'Lost Hills': A new pistachio cultivar for California. *Hort. Science*, **43** (1): 247.

Parfitt, D.E.; Kallsen, C.E.; Maranto, J. 2016. Pistachio cultivars. In: *Pistachio Production Manual,* by Ferguson, L. and Hariland, D.pp. 59-64.

Payne, T.J. 2004. Nuts. In: Processing fruits: Science and technology. 2nd ed. By: Barrett, D.M., Somogyi, L. and Ramaswamy, H.S.pp. 765-810.

Pearson, T. 1996. Machine vision system for automated detection of stained pistachio nuts. *Lebensmittel Wissenschaft and Technologie*, **29**(3): 203-209.

Peri, I.; Matusova, R.; Smirnoff, P.; Birk, Y.; and Golan-Goldhirsh, A. 1999. Inflorescence bud protease(s) against dehydrin-like inflorescence bud proteins of *Pistacia vera*. *Plant Physiology and Biochemistry Paris*, **37**(1): 51-56.

Pietro, G. 2015. Crop production. FAO Statistical Pocketbook. 26.

Polito, V.S., and Pinney, K. 1999. Endocarp dehiscence in pistachio (*Pistacia vera* L.). *International Journal of Plant Science,* **160**: 827-835.

Pontikis, C. A. 1990. Effects of 2-naphthaleneacetic acid on alternate bearing in pistachio. *Fruits Paris,* **45**(3): 281-285.

Popov, K. P; Lalymenko, N. K. and Lalymenko, L. A. 1994. New technology for raising pistachio seedlings. *Problems of Desert Development*, **2**: 77-81.

Raei, M. and Jafari, S. M. 2011. Influence of different packaging materials and storage conditions on the quality attributes of Pistachio (Pistacia vera L.) cv. Ohadi. *Annals Food Science and Technology*, **12**(2): 179-185.

Rafati, F.; Hajmohamadi, S.; and Hajmohamadi, S. 2014. Effect of seasonal winds on the pistachio trees productivity of sirjan in the last ten years. *International Journal of Advanced Biological and Biomedical Research*, **2**(2): 371-376.

Rahemi, M. and Pakkish, Z. 2005. Gibberellic acid and benzyladenine reduce hull-splitting and aflatoxin levels in pistachio kernels. *Journal of Horticultural Science and Biotechnology,* **80**(2): 229-232.

Rahemi, M. and Ramezanian, A. 2007. Potential of ethephon, NAA, NAD and urea for thinning pistachio fruitlets. *Scientia Horticulturae,* **111**(2): 160-163.

Rahemi, M.; and Pakkish, Z. 2005. Gibberellic acid and benzyladenine reduce hull-splitting and aflatoxin levels in pistachio kernels. *Journal of Horticultural Science and Biotechnology*, **80**(2): 229-232.

Rosecrance, R. C; Weinbaum, S. A. and Brown, P.H. 1996. Assessment of nitrogen, phosphorus, and potassium uptake capacity and root growth in mature alternate-bearing pistachio (Pistacia vera) trees. *Tree Physiology*, **16**(11/12): 949-956.

Rosecrance, R.C.; Weinbaum, S.A.; and Brown, P.H. 1996. Assessment of nitrogen, phosphorus, and potassium uptake capacity and root growth in mature alternate-bearing pistachio (*Pistacia vera*) trees. *Tree Physiology*, **16**(11/12): 949-956.

Sadri, Z.Z.; Montazeri, M.; and Dianat, M. 2010. Investigation on phenology of Cynanchum acutum in pistachio orchard in Aran & Bidgol (Esfehan province, Iran). *Proceedings of 3rd Iranian Weed Science Congress, Weed Biology and Ecophysiology,* Babolsar, Iran. 1: 78-80.

Salagado, J.M., Max, B., Rodriguez, R., Perez, N. and Dominguez, J.M. 2012. Extraction of ferulic acid from agro-industrial wastes and evaluation of bioconversion of ferulic acid to vanillin by *Streptomyces setonii.* Primer Congreso Iberoamericano sobre Biorrefinerias. 451-457.

Salehi, M.H.; and Hosseinifard, J. 2012. Soil and groundwater relationships with pistachio yield in the Rafsanjan Area, Iran. *Communications in Soil Science and Plant Analysis*, **43**(4): 660-671.

Sedaghat, R. 2011. Constraints in production and marketing of Iran's pistachio and the policies concerned: An application of the Garret Ranking Technique. *International Journal of Nuts and Related Sciences*, **2**(1): 27-30.

Shahabinejad, M; Shojaaddini, M; Maserti, B; Arvin, S. M. J. and Seyedi, S. M. 2014. Exogenous application of methyl jasmonate and salicylic acid increases antioxidant activity in the leaves of pistachio (*Pistacia vera* L. cv. Fandoughi) trees and reduces the performance of the phloem-feeding psyllid *Agonoscena pistaciae. Arthropod Plant Interactions,* **8**(6): 525-530.

Shahriaripour, R.; Pour, A.T.; and Mozaffari, V. 2011. Effects of salinity and soil phosphorus application on growth and chemical composition of pistachio seedlings. *Communications in the Soil Science and Plant Analysis*, **42**(2): 144-158.

Shahsavari, F; Hokmabadi, H; Zakerin, A; Tajabadi, A. and Shahsavari, N. 2011. Evaluation of some treatments after spring frost damage on inflorescence bud formation in pistachio orchards. *Acta Horticulturae,* **912**: 675-678.

Shuraki, Y.D.; and Sedgley, M. 1994. Effect of pistil age pollen parent on pollen tube growth and fruit production of pistachio. *Journal of Horticultural Science*, **69**(6): 1019-1027.

Shuraki, Y.D.; and Sedgley, M. 1996. Fruit development of *Pistacia vera* (Anacardiaceae) in relation to embryo abortion and abnormalities at maturity. *Australian Journal of Botany*, **44**(1): 35-45.

Siegel, J.; Lacey, L.A.; Fritts, R. Jr.; Higbee, B.S.; and Noble, P. 2004. Use of steinernematid nematodes for post harvest control of navel orangeworm (Lepidoptera: Pyralidae, Amyelois transitella) in fallen pistachios. *Biological Control*, **30**(2): 410-417.

Soliemanzadeh, A; Mozafari, V; Pour, A. T. and Akhgar, A. 2013. Effect of Zn, Cu and Fe foliar application on fruit set and some quality and quantity characteristics of pistachio trees. *South Western Journal of Horticulture Biology and Environment,* **4**(1): 19-34.

Spann, T.M.; Beede, R.H.; and DeJong, T.M. 2007. Preformation in vegetative buds of pistachio (*Pistacia vera*): relationship to shoot morphology, crown structure and rootstock vigor. *Tree Physiology*, **27**(8): 1189-1196.

Stevens, P.F. 2008. Angiosperm Phylogeny Website, version 9. http://www.mobot.org/MOBOT/research/Apweb/

Stevenson, M.T.; Shackel, K.A.; and Ferguson, L. 2000. Shoot length distribution and its relation to yield of alternate bearing pistachio trees. *Journal of the American Society for Horticultural Science*, **125**(2): 165-168.

Surucu, A.; and Demirkiran, A.R. 2013. Effects of intercropping on mineral content of pistachio in southeast of Turkey. *Journal of Food Agriculture and Environment*, **11**(3/4): 883-886.

Tamer, N.; and Karaca, A. 2004. Gidya'nin toprakta enzim aktiviteleri ile kadmiyum kapsami uzerine etkisi, A.U. Fen Bilimleri Enst. Toprak Anabilim Dali, Yusek Lisans Tezi, Ankaka (Turkish).

Taskin, T.; Cobanoglu, D.; Colak, G.; Namli, S.; Basaran, D.; and Gaspar, T. 1996. Plant regeneration through somatic embryogenesis in *Pistacia vera*. *Saussurea*, **27**: 59-65.

Tavallali, V.; Rahemi, M.; Kholdebarin, B. 2009. Ameliorative effects of zinc on pistachio (*Pistacia vera* L.) growth under salt-affected soil conditions. *Research Journal of Environmental Sciences*, **3**(6): 656-666.

Tekin, H; Guzel, N. and Ibrikci, H. 1995. Influence of manure and inorganic fertilizer on yield and quality of pistachio. *Journal of Plant Nutrition,* **18**(6): 1263-1272.

Tezerji, Z. S. 2011. Determination of damages of pistachio twig borer moth *Kermania pistaciella* Amsel, to fruit clusters of pistachio trees. *Acta Horticulturae*, **912**: 701-707.

Tilkat, E; Akdemir, H; Ozden Tokatli, Y; Yildirim, H; Suzerer, V. and Onay, A. 2011. Plant production through adventive organogenesis of mature pistachio, *Pistacia vera* L. cultivars 'Atli' and 'Siirt'. *Acta Horticulturae,* **912**: 567-573.

Tilkat, E; Onay, A; Yldrm, H. and Ayaz, E. 2009. Direct plant regeneration from mature leaf explants of pistachio, *Pistacia vera* L. *Scientia Horticulturae,* **121**(3): 361-365.

Tournefort, P. 1700. Institutions Rei Herbariae. *Typograhia Regia*, **2**: 579-580.

Tous, J. and Ferguson, L. 1996. 'Mediterranean Fruits'. In: J. Janick, ed, *Progress in New Crops,* ASHS Press, Arlington, pp. 416-430.

Uriu, K.; and Pearson, J. 1986. Zinc deficiency in pistachio-diagnosis and correction. *Calif. Pistachio Ind. Ann. Rpt.,* 71-72.

Uriu, K.; Teranishi, R.; Beede, R.; and Pearson, J. 1989. Copper deficiency in pistachio. *Calif. Pistachio Ind. Ann. Rpt.,*77.

USDA. 2014. Food Composition Database. United States Department of Agriculture. Agricultural Research Service. https://ndb.nal.usda.gov/ndb/

Vavilov, V.L. 1951. The origin, variation, immunity and cultivated plants. Tr. From Russian by Chester, K.S. *Chronica Botanica*, 1:6.

Weinbaum, S.; Brown, P.; and Rosecrance, R. 1995. Assessment of nitrogen and potassium uptake capacity during the alternate bearing cycle. *Calif. Pistachio Ind. Ann. Rpt.,* 56-60.

Werner, O.; Sanchez-Gomez, P.; Guerra, J.; Martinez, J.F. 2001. Identification of Pistacia x saportae Burnat (Anacardiaceae) by RAPD analysis and morphological characters. *Scientia Horticulturae*, **91**(1/2): 179-186.

Wollpert, J. 1986. Nitrogen and potassium nutrition and preliminary results on pruning of 13-year-old trees. *Calif. Pistachio Ind. Ann. Rpt.,* 80-85.

Woodroof, J.G. 1979. Tree nuts, production, processing, products. Vol. II, 2nd ed. AVI, West Port Connecticut.

Yaltirik, F. 1967a. Anacardiaceae. In: P.H. Davis, Ed., *Flora of Turkey and the East Aegean Islands*. Edinburgh University Press, Edinburgh. **2:** 544-548.

Yaltirik, F. 1967b. Anacardiaceae, Contributions to the Taxonomy of Woody Plants in Turkey. *Notes Royal Botanical Gardan.*, **28**: 11-12.

Yang, Z.H.; and Ludders, P. 1994. *in vitro* propagation of pistachio (*Pistacia vera* L.). *Gartenbauwissenschaft,* **59**(1): 30-34.

Yang-Sen; Li-Hong; Xu-BinQiang; Tuerxunnayi, T.; Tayier, A.; and Zhu-XiaoFeng. 2008. Study on species, field identification and biological characters of scale insects in orchards in Tarim Basin. *Xinjiang Agricultural Sciences*, **45**(2): 276-281.

Yazdanpanah, N.; Pazira, E.; Neshat, A.; Naghavi, H.; Moezi, A.A.; and Mohmoodabadi, M. 2011. Effect of some amendments on leachate properties of a calcareous saline-sodic soil. *International Agrophysics,* **25**(3): 307-310.

Yi, T., Wen, J., Golan-Goldhrish, A. and Parfitt, D.E. 2008. Phylogenetics and Reticulate Evolution in *Pistacia* (Anacardiaceae). *American Journal of Botany*, **95**(2): 241-251.

Yin-YanNi; Yan-LeiYan; Jiang-JinHua; and Ma-ZhongHua. 2008. Biological control of aflatoxin contamination of crops. *Journal of Zhejiang University Science B.*, **9**(10): 787-792.

Yucel, S.; Onay, A.; Colak, G.; and Basaran, D. 1991. The researches of obtaining from *Pistacia vera* L. apical bud and nodal bud by micropropagation. *Acta Horticulturae*, **289**: 167-168.

Zaheer, A.; Nasreen, A.; and Shah, F.H. 1989. Callus formation from the mesocarp tissues of *Pistacia vera* L. *Pakistan Journal of Scientific Industrial Research,* **32**: 549-550.

Zahra, S.; Ebrahim, R.; and Amir, S. 2014. Determination of aflatoxins in Iranian nuts using HPLC method. *Research Journal of Pharmaceutical, Biological and Chemical Science,* **5**(5): 27-31.

Zarei, M.; Davarynejad, G.; Abedi, B.; Kafi, M.; and Biabani, A. 2014. Changes inphysical properties, chemical composition and antioxidant activity of four pistachio cultivars at ten maturity stages. *Advances in Environmental Biology*, **8**(10): 106-115.

Zeng, D. Q; Brown, P. H. and Rosecrance, R. C. 1998. The effects of alternate bearing, soil moisture and gypsum on potassium nutrition of pistachio (*Pistacia vera* L.). *Acta Horticulturae*, **470**: 412-420.

Zhang, J. 2014. A preliminary report on mechanical pruning utilizing a '1-side mechanical slant topping' technique on pistachio trees. *Acta Horticulturae,* **1028**: 351-358.

Zhang, J.; and Joyce, C. 2009. A study on climate factors on 'Sirora' pistachio nut size. *5th International symposium on Pistachios and Almoonds. ISHS*. Sanliurfa, *Turkey. Book of Abstract,* pp. 21.

Zohary, M. 1952. A Monographical Study of the Genus *Pisacia. Palestine Journal of Botany Jerusalem*, **5**: 187-228.

12

Plum

D.P. Sharma, N. Sharma and Rimpika

1. Introduction

Plum is one of the important stone fruit crops of temperate region throughout the world. The Japanese group of plum is native to China but was domesticated in Japan and subsequently was introduced to different parts of the world. In India, Japanese plums were introduced by European settlers and missionaries around 1870. Plum grows well in areas ranging from 900 m to 1600 m above mean sea level.

1.1. Uses

Among deciduous fruits, plum is widely distributed in the world, because of wider adaptability among different species and cultivars of plum to varied agro-climates and soil conditions. Plum is consumed as dessert, dried and processed products. Plum fruit is considered best to overcome the jaundice. Its fruit use for a preparation of juice and squash. It ranks next to peach in commercial production. It is widely cultivated in several countries in Europe, Asia and in the America.

Nutritive value (Composition)

Nutrient	grams
Vitamin c	6.270
Thiamin	0.028
calcium	2.640
Iron	0.066
Potassium	113.520
Riboflavin	4.620
Carbohydrates	10.878

Source: USDA Fruit Nutrition Facts, 2001

Plum fruits are important sources of minerals, vitamins, sugars and organic acids in addition to protein fat and carbohydrates. Plum with high sugar content can be dried along with stone without causing any fermentation is known as prunes. It is reach source Vit. B1, Vit. A Riboflavin, sugars and carbohydrate.

2. Origin and distribution

P. domestica L. (European plum)

This species is native to western Asia, in the Caucasus mountains adjacent to the Caspian Sea. It was brought to North America by Spanish Missionaries (west coast) and English Colonists (east coast). Today, most production is in western states, with the vast majority in California's central valley, where climate disfavors disease and rain cracking.

Prunus salicina Lindl. (Japanese plum).

Contrary to the name, this species originated in China, where it was cultivated for thousands of years. It was brought to Japan 200-400 years ago, where it then spread around the world, being falsely called "Japanese plum". In the USA, it is grown primarily in California, and is the major fresh market plum seen in grocery stores.

3. Taxonomy and cytogenetic

Plum belongs to the family *Rosaceae* and genus *Prunus and basic chromosome number is 8.* It has numerous of cultivated species, but the commercially grown cultivars belong to two species i.e. *Prunus domestica* L. (European plum) and *Prunus salicina* Lindl. (Japanese plum).

i. **European plum *Prunus domestica* L:** It is relatively recent hybrid of wild parents. It is spontaneous chromosome doubling for a hybrid between P. *cerasifera* (diploid) and P. *spinosa* (tatraploid) may have produce haxapolid species. Trees are taller, more upright, longer lived, more cold hardy and late blooming than Japanese plum. Fruits are generally oval, smaller and more variable in colour than Japanese plums. Cvs. Reine Claude or greengage, Imperial gage, Hand Yellow egg, Golden drop and Lombard

ii. **Japanese plum *Prunus salicina* Lindl.:** It is one of the fresh market plum. These are most common fresh market plums. They are larger, rounder (or heart shaped), and firmer than European plums and are primarily grown for fresh consumption.

iii. **Damson, bullace, St. Julian and Mirabella plums:** *Prunus insititia* L. There are small wild plums natures to Europe. Fruits are small and oval, purple and clingstone for damsons and yellow and free stone for Mirabelles, with heavy bloom. They are used as rootstocks for plums and used primarily for jam, jellies and preserves.

iv. **American Plum: *Prunus Americana:*** It is native to North America. In the interior regions from Pennsylvania to the rocky- mountains where *P. domestica* have so largely failed to establish themselves; it is thought to have some desirable substitute. Of the innumerable number of varieties belonging to American species, vast majority are of limited local importance. Fruit have no commercial value. Most of these are self- unfruitful, so pollinizers should be placed in the orchards.

Section Prunocerasus (Plum cherries) : *Prunus americana* there are wild, small trees or shrubs found mostly in the eastern United States that produce small, round, edible fruits. They are used mostly for jam, jellies and preserves. Important in plum breeding programmes for incraseing disease resistance of Prunus *salicina.*

Lin *et al.* (1994) grouped the different species into 3 classes by similarities among their karyotypes. *P. salicina, P. simonii and P. ussuriensis were one group, P. domestica* and *P. cerasifera* another and the third group contained only *P. americana.* They studied the chromosome numbers of 226 plum cultivars from 6 species of Prunus and the karyotypes of the 6 species from 17 Chinese provinces by wall degradation hypotonic treatment. Triploid and chimaeric *P. salicina* types as well as diploid were observed, among which were cultivars Suilingwansuli and Suiling 73-81 from Heilongjiang which were triploid ($2n = 3x = 24$), and Shahezicuoli from Jilin which was a chimaera of tetraploid and diploid genomes ($2n = 4x = 32$, $2n = 2x = 16$). Wild *P. domestica* from Xinjiang was heptaploid ($2n = 6x = 48$) while cultivated forms were diploid ($2n = 2x = 16$).

4. Acreage, production and export value

Round the world during the year 2014, area under plum cultivation has been 2521100 ha, with the production of 11282527 MT, out of which China alone produces 6256906 MT on an area of 1829357 ha (FAOSTAT, 2017). Indian plum production during the period 2015-16 was estimated to be 82000 MT on an area of 22000 ha. In India high quality plum is being grown in hilly areas of Himachal Pradesh, Jammu and Kashmir, Uttrakhand and Uttar Pradesh. While as, low chilling plum cultivars are being cultivated in Punjab, Haryana and in some parts of Uttar Pradesh and Tamil Nadu (Table 1) (Anon, 2016).

Table 1: State wise share of Plum production in India (2014-15)

State	Production ('000 Tonnes)	Share (%)
Uttarakhand	37.89	52.51
Himachal Pradesh	10.83	15.01
Jammu and Kashmir	8.22	11.39
Tamil Nadu	5.33	7.39
Punjab	4.11	5.70
Nagaland	3.22	4.46
Mizoram	1.63	2.26
Haryana	0.93	1.29

Source: Anon, 2016

Annually, 700 million tonnes of plums are sold, with an export value of 650 million Euro. Chile and Spain are two major export countries, with an export of over 110 million tonnes. South Africa is the world's third exporter, but it is by far the most important supplier to the Netherlands. The Netherlands is the fourth importer of plums worldwide; the bulk of the import is re-exported again. Russia, United Kingdom and Germany are the main import countries (Fig. 1 & 2). The Dutch plum import reached a peak in the year 2007-2009. In 2014, Chilean export declined sharply due to a failed harvest, causing a temporary end to the steady growth of global plum trade. During the year 2013, Indian exports and imports were recorded as 120 MT and 1199MT respectively (FAOSTAT, 2017).

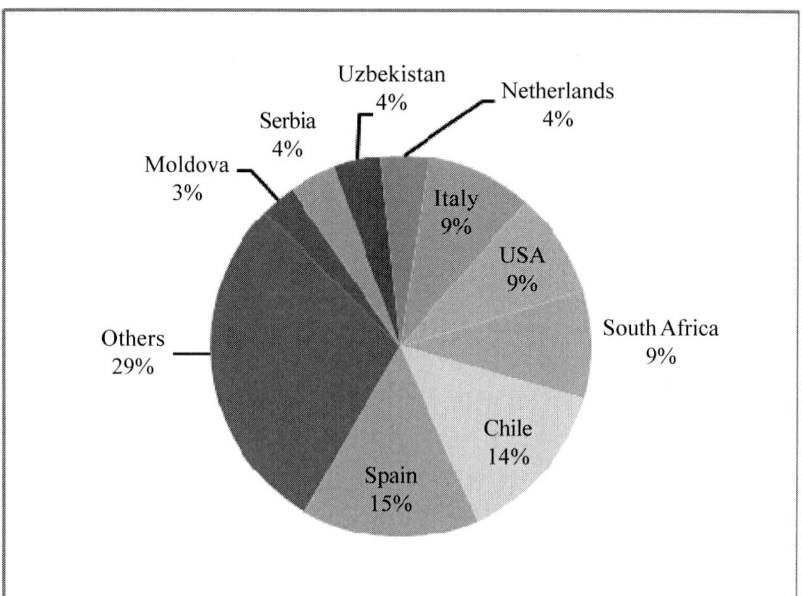

Fig. 1: Plum exporters (Average 2010-2104)

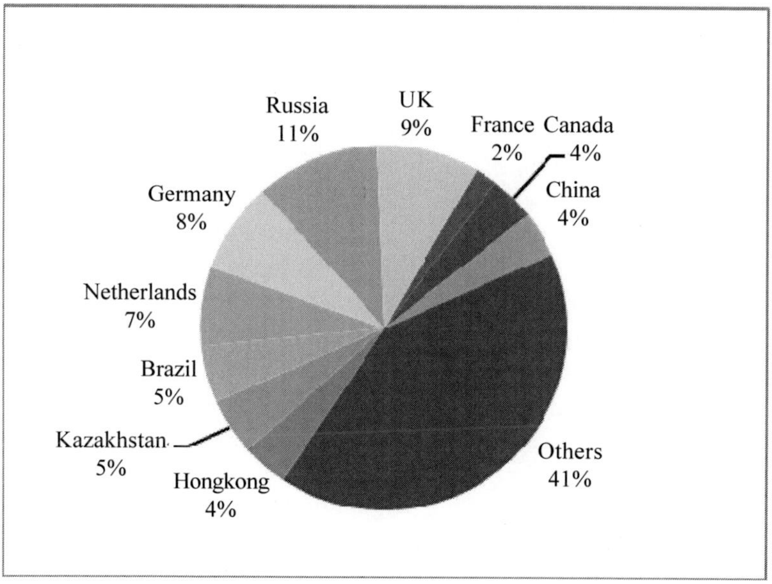

Fig. 2: Plum importers (Average 2010-2014)

5. Botanical description and growth habit

5.1. Plant

Plum trees are small to medium sized trees, similar to but more erect growing than peach. European plum trees are larger and more erect than Japanese plums. Leaves ovate or elliptic with acute or obtuse tips, short petioles, crenulate margins. Japanese plum trees have rougher bark, more persistent spurs, and more numerous flowers than European plums. They are also more precocious, disease resistant, and vigorous than European plums.

5.2. Fruit

Fruit is a drupe. Oval shaped in European types, round to conical in Japanese types. Bloom (epicuticular wax) is usually present on glabrous surface (thus, the fruit surface is termed "glaucous"). Plums require 2.5 to 6 months for fruit development, with most Japanese ripening in relatively short periods (3 months), and some prune & canning cultivars ripening in autumn. Thinning is necessary for proper size development for Japanese plums, but not always necessary for European plums, particularly prunes, since they are not as floriferous, and fruit set is generally lighter.

5.3. Flower bud development and flowering

Generally most deciduous fruit species initiate flowers during one season but blooming takes place the following year. Extensive studies suggest flower bud initiation in Prune and European plum starts in early June and ends in mid July to mid August while in Japanese plum, it takes place in mid July and terminates by early August. In early varieties the processes are more rapid than the late varieties. The reproductive organs develop throughout the winter and the development has been reported to cease only below 0°C temperature. Microsporogenesis starts only after the buds have 40-50 days exposure to 0.5 °C followed by a temperature above 4.5°C. After floral initiation, the development of flowers takes place rather very rapidly within the bud. A very slow development takes place in winter followed by rapid development in spring leading to opening of flowers and the shedding of pollen. The extent and quality of flower development in the post initiation period are governed by factors like age of the wood, position on the tree, temperature, water, carbohydrate, nitrogen and other nutrient elements, growth regulators and winter chilling. Flowering in plums takes place on bud of one year old shoot as well as on two year old spurs.

5.3.1. Flowers

Flowers are similar in morphology to peach trees, but white, smaller, and have longer pedicels. Flowers are borne mostly in umbel-like clusters of 2-3 individuals on short spurs, and solitary or 2-3 in axils of 1-yr-old wood. European plums bloom much later than Japanese types, and are therefore less frost prone.

5.3.2. Flowering

From the work it was observed that the flowering in the cultivars under study started in second week of February and cultivar Florida 1-2 was the earliest to flower and AU-Rosa was last to do so. Similar observations in variation in flowering period of Plum cultivar has been observed by Randhawa and Nair (1960) under Delhi conditions where flowering in plum cultivars started from first week of February and continued till third week of March.

5.3.3. Anthesis and stigma receptivity

The mean stigma receptivity two days prior to anthesis was 2.06 per cent which increased to 54.31 per cent one day before anthesis and on the day of anthesis it was maximum i.e. 94.67 per cent after which it declined gradually. The stigmas of plum cultivars were found receptive a day before anthesis, by Randhawa and Nair (1960) under Delhi conditions. Further during the present investigation it was observed that the stigmas turn brown and become dry and dead at the end of the receptive period i.e. 3 to 5 days.

5.3.4. Physiology of flowering

Moreno *et al.* (1992) studied flower bud growth and ovule longevity in the plum cultivars Italian and Brooks and the effects of autumn-applied ethephon (500 mg/litre) and of temperature. Fresh and dry weights of terminal flower buds were measured at 1-week intervals from 50 days to 1 day before flowering in 1988. Buds were also analysed for N, P, K, Ca and B. After flowering, ovule longevity was determined using a fluorescence method after staining with aniline blue. Ovule longevity was determined in 1990 using shoots excised at full bloom from untreated and ethephon-treated trees of both cultivars and held in growth chambers for 18 days at 5, 10, 15 or 20 degrees C. FW and DW accumulation was greater in flower buds of Brooks than in those of Italian, and ethephon reduced bud weights in both cultivars. Ethephon did not affect mineral content of flower buds of Brooks, but Italian flower buds contained a higher Ca concentration and a lower P concentration when treated with ethephon. B content was higher in the ethephon-treated buds of Italian trees than in untreated controls on some sampling dates. Ovule longevity was greater for Brooks than for Italian in both years. Ethephon treatment delayed ovule senescence in Italian flowers, but had little or no effect on Brooks flowers. Higher temperatures induced faster ovule senescence in both cultivars.

5.3.5. Pollination

Honey bees are the major pollinator in most plum cultivars (Mladenovic *et al.*, 2013). For Japanese plums, pollinizers are necessary for commercial production for most cultivars. 'Bruce', 'AU Producer', 'Beauty', 'Santa Rosa' (and its sports), 'Simka', 'Casselman' and 'Methley' do not require cross pollination. In *P. domestica,* about ½ of the major cultivars require pollinizers, but most of the major prunes produced in the USA do not. 'Stanley', 'French', 'Italian', and 'Sugar' prune types are self-fruitful.

Cross incompatibility is of frequent occurrence in Japanese plums (Nyeki and Szabo (1997). Efficient pollination is fundamental for guaranteeing abdicate productivity in Japanese plums. Total gametophytic self incompatibility distinguishes all cultivars of this pomological group and makes it necessary to plant suitable pollinizers so as to guarantee and adequate production of fertile pollen during anthesis. The main characteristics required for a good pollinizer include elevated production of flower, long anthesis period, abundant pollen production, high viability and fertility of pollen grains corresponding to high germinability.

The cultivars Santa Rosa, Queen Rosa and AU-Rosa cultivars showed maximum fruit set when self pollinated by bagging was done (11.39, 10.56 and 13.72 per cent respectively. This showed that these cultivars are self fruitful, while cultivars Florida 1-2, Frontier and Red Beaut showed very less fruit set i.e. 6.13, 5.11 and 5.12 per cent respectively, and were classed as partially self fruitful and therefore require suitable pollinizers for satisfactory fruit set and fruit production.

The investigation showed substantial variation under different modes of pollination. The fruit set under selfing by bagging varied from 5.11 to 13.72 per cent whereas under hand self pollination it was 8.38 to 16.54 per cent and with open pollination it ranged from 10.47 to 17.48 per cent. This clearly shows the inbuilt genetic variation of the cultivars as regards their potential to fruit set of their own or with the aid of a compatible pollinizer to produce commercial or fruits of higher quality.

MiZhen *et al.* (2001) observed that Meili and Dashizaosheng, promising cultivars for the Yangtze River valley, were pollinated using 5 pollinators. Furongli and Meili were good pollinators of Dashizaosheng while Beijing Wanhong was a good pollinator of Meili.

Fruit set lower than 5 per cent indicated self fruitful nature, but in some it was very high i.e. up to 67 per cent (self fruitful) producing higher yield with their own pollen (Yu *et al*., 2008) According to Bajwa *et al.* (1991) cultivar Florida 1-2 was self incompatible and needed mixed plantation of Kala Amritsari for successful pollination.

An increase in the fruit set was seen in all the cultivars under open pollination as compared to suiting by bagging, which showed that Japanese plums are cross pollinating and most of the cultivars are partially self fruitful. These findings are in line with the observations of Bajwa *et al.* (1991) who found that fruit set in cultivar Kala Amritsari increased to 23 per cent from 11.6 per cent when pollinated with 'Titron'. The fruit set in open pollination varied from 10.43 per cent (Red Beaut) to 17.48 per cent (Santa Rosa). One of the major reasons for low fruit set is flower emasculation by hand during self and cross pollination. Similar cause of low fruit set was observed by Guerra *et al.* (2010). Pollination intensity after open pollination differed significantly among cultivars and flowers within the cultivars. Most of the Japanese cultivars are auto incompatible and must be pollinated by an external pollinator and that choice of pollinator was determined by blossoming concordance and pollination compatibility. Plum cultivars and Japanese in particular are classified as self fertile, partially self fertile and sterile.

The cultivar Santa Rosa was found to be suitable compatible pollinizer for cultivars AU-Rosa, Florida 1-2, Fortune, Frontier, Queen Rosa and Ruby Sweet while a number of studies have shown that Santa Rosa is partly self fruitful (Costa and

Grandi, 1982) and is a potential source as pollinizer for a number of cultivars like Beauty (Ghanshyam *et al*., 1977) Oishiwase, Soldom and White Plum (Chung *et al*., 1998). The findings showed that Santa Rosa was better Pollinizer compared to other cultivars. In another study Nikolic *et al*.(2012) recommended the cultivars Stanley, and Italian Prune as good pollenizers for the cultivar Cacanska Najbolja.

Japanese plum (*Prunus salicina* Lindl.) which carries the S-RNase-mediated gametophytic self-incompatibility system, prevents self-fertilisation, and thus promotes out-crossing. The plum cultivar 'Black Diamond' Reg. has become, one of the most important cultivars in Israel in the last decade, yet its yield is low in comparison with its known potential. Honeybees (*Apis mellifera*) are the most important pollinators for plums and several studies have demonstrated an apparent relationship between the number of honeybees and the size of the fruit crop. It was found that 2.5 colonies ha^{-1}, resulted in the greatest level of fruit set (Sapir *et al.* (2007).

5.3.6 Number of days from peak flowering to fruit set

The time period from peak flowering to fruit set was minimum in the cultivars AU-Rosa and Santa Rosa (11 days) and cultivar Florida 1-2 took maximum of 20 days. Flowering duration is basically a cultivar phenomenon and bloom is also dependent on the prevailing climatic conditions. Bajwa *et al.* (1991) observed longest flowering time of 19 days in 'Kala Amritsari'.

5.4. The physiology of the tree

The root systems represents nothing more than the tree's anchorage in the soil or the water and nutrient transport pipeline to the stem and canopy. Research activities over the past few decades have established very clearly that the growing root is a very active site of complex metabolic processes and presents a huge interface for the exchange of chemical constituents from the tree to the soil and vice versa as well as a providing a site of high microbiological activity much greater than in the bulk soil. The soil-root interface or rhizosphere of a tree is much greater in area than the leaf-atmosphere interface. While the latter is in the range of several m^2 per unit area of soil surface, (the so-called Leaf Area Index, LAI) usually 5 to 10, for apple trees, the estimated root surface is some magnitudes larger. From the studies of Ebert and Lenz (1991) it can be calculated that a 5-year-old apple tree has a root surface area of more than 250 m^2. The root surface area varies during the growing period because of the development of short-lived root hairs, which significantly increase the contact area of the growing root to its environment. In comparison to leaves, the lifespan of active roots is much shorter. The fast turnover of roots requires a great deal of metabolic energy in the form of photosynthates from the leaf canopy for the growing process as well as for the synthesis of organic metabolites and uptake of mineral nutrients. The interrelationship between root and shoot growth during the season controls the development of the entire tree. The root, this very active part of the tree, serves the following important functions: Anchorage of the tree in the soil; uptake and transport of water; release and transport of organic and inorganic constituents; synthesis, decomposition and storage of metabolites. These functions also moderate the yield and fruit quality in relation to the activity of the root system. Root growth is favoured in well aerated light soils which are easily penetrated.The root systems of dwarfed trees also encounter strong competition from the roots of weeds and grasses for uptake of water and nutrients. In this respect the soil should not be compacted and any competition with weeds should be avoided by weed control in the tree rows. The soil should be maintained so that moisture, aeration, and organic matter components are conducive to the growth of fruit tree roots (Balal *et al.,* 2011).

5.5. Physiology of fruit ripening

Taking ethylene biosynthesis as a clue, changes of firmness, TSS and TA content, ethylene production, ACC and MACC content, the activity of ACC oxidase and ACC synthase were determined and the effect of ethylene on plum fruit ripening and softening were discussed by Kan *et al.* (2012). The results showed that firmness and TA content decreased and TSS increased during plum fruit ripening and softening. In the start-up stage of softening, ethylene production as well as ACC oxidase and ACC synthase activity increased slowly and linear correlations can be seen among them. The peak of ethylene appeared in the later stage of softening. The study proved that ethylene was not the promoter for plum softening, and other factors for promoting plum softening still existed.

Plum fruit exhibit varying types of ripening behaviour that is highly dependent on genotype, harvest maturity, and pre- or postharvest handling practices. Singh and Khan (2010) focused on recent advances in the physiology of plum fruit ripening. They reported that in plums: (1) harvest maturity is associated with the ability of fruit to ripen properly, which ultimately affects consumer acceptance; (2) harvest maturity, storage temperature and storage period affect the level of sugars and organic acids; (3) malonyl 1-amino-cyclopropane-1-carboxylic acid and malonyl transferase activities are the main regulatory factors controlling fruit ripening and some other possible ethylene-related disorders during cold storage; (4) development of chilling injury symptoms after cold storage are associated with ethylene biosynthesis and abnormalities of cell wall metabolism including reduction in pectin solubilisation and depolymerisation and decrease of galactose-rich pectin as well as postharvest oxidative stress; (5) great variations exist in the concentrations of bioactive compounds and antioxidant activities depending on various pre- and postharvest factors.

6. Climate

Plum can be grown in the areas where winters are cold and summers are hot or when it is dry or rainfall is high. Cold, wet and windy weather at bloom are harmful to crop. In general, European plums require 800-1500 chilling hours below 7°C whereas Japanese plums require 100-800 chilling hours for satisfactory bud break in spring. These chilling requirements may be influenced by environmental factors like occurrences of intermittent warm periods in winter, light, leaching of inhibiting substances by rain, cultural practices like bending of branches, withholding irrigation, reduction of apical dominance, summer pruning, chemical spray, rootstocks etc. Japanese plum can be grown from subtropical plane areas of Punjab, Haryana, parts of U.K. etc. upto an elevation of 1525 m above mean sea level in Himalayan hills or little above in other hills of north-eastern India including Nilgiri hills of south India. These are grown in low to mid hill areas of Himachal Pradesh (1000-1600 m amsl). European plums would thrive well in hills from 1525-2745 m above mean sea level. The blossoms of Japanese plum are sensitive to spring frost, not because they are less hardy but they bloom very early in the spring. A Northern slope is preferable particularly for these plums which tend to delay the bloom period and thus avoid early frost injury. Areas receiving 100- 125 cm rainfall, well distributed throughout the growing season may be considered suitable. Areas subjected to high wind velocity or cyclone and hail prone are not suitable and should be avoided.

7. Soil

The plum tree is not very fastidious in its soil requirement but well drained, deep loamy soils with pH 5.5 - 6.5 is considered best. European plum can tolerate heavier clay soils than Japanese plum .If plums are grown on light, very shallow, water logged, poorly drained or dry soils, the trees are short lived and fruits are of smaller size. The soil should be free from alkalinity and salinity.

8. Variety

8.1. Czar

The fruits are on the large size with dark purple plums. The flesh is yellow. Traditionally use for cooking, Czar is also a good eater if the fruit are allowed to fully ripen. It has distinct advantage over some other plum trees of being self-fertile. It resists frost damage well. Fruit is produced in August.

8.2. Early Laxton

It is a very early plum tree, producing ripe fruit from late July onwards. The plums have attractive yellow and light red skin. The flesh has a gage type flavour, yellow, full bodied and very juicy. It can be used for both eating and cooking. It was awarded an AGM from the Royal Horticultural Society in 1916. "Early Laxton" was first introduced in 1902 from Bedford, England. The plums ripen ready for eating in mid August. It does require another plum tree for pollination and falls in pollination group. It is partially self-fertile, but will crop better with other plum trees nearby. The pollination group is C.

8.3. Bluebyrd

Released by USDA in 1998, Bluebyrd is an excellent European type plum for the commercial orchard and home garden use. The fruit is blue with amber flesh, medium to large in size with excellent flavor and high sugar content. The tree is vigorous and productive, showing great resistance to black knot. Bluebyrd blooms before Stanley and requires cross pollination.

8.4. Wickson

A large, greenish-yellow heart-shaped plum with yellow flesh. The tree is upright and vigorous and tends to be a shy cropper. Any of the Japanese plums listed will pollinate Wickson, however Wickson is not considered a good pollinizer.

8.5. Vanier

A hybrid (Wickson x Burbank cross), that matures one week after Wickson. Fruit is red, medium size, yellow fleshed and clingstone. The trees are upright, vigorous and productive

8.6. Castleton Prune Plum

This early-maturing dual-purpose plum is a Valor x Iroquois cross released by Cornell University's New York State Agricultural Experiment Station. The fruit is blue, slightly ovate and similar in appearance to Stanley, ripening 2-3 weeks ahead of Stanley. The trees are spreading and consistently produce a heavy crop

8.7. Long John

A mid-season prune-type plum for the fresh market. The fruit is dark blue, very large, and of high quality. The trees are upright, medium in vigour and very productive

8.8. Stanley

A New York State Experiment Station introduction and leading cultivar in the Great Lakes region. A fine prune-type plum with excellent quality suited for both home use or processing. Fruit is large in size with a dark blue skin. Flesh is greenish-yellow, juicy and fine grained. The tree is early bearing and a good pollinizer for other European varieties.

8.9. Valor

It ripens just a few days after Stanley. It was developed at Vineland, Ontario. The fruit has purple-blue skin and yellow flesh, sizes medium-large. It is a semi-freestone and has great fresh market potential. The trees are healthy and productive.

8.10. Au Rosa

A dark red plum, medium to large in size with red flesh. Trees are very vigorous, spreading and moderately productive

8.11. Black amber

A mid-season plum with black-red skin and an amber flesh. The fruit is large and very firm. This variety is susceptible to bacterial diseases caused by extreme humid climates and is not recommended under these conditions

8.12. Black Ruby

It was released in the mid-90s by the USDA station in Byron, GA. It is a Queen Ann x Santa Rosa cross. The fruit is firm and colors to a red-black finish. Flesh is yellow and sweet so fruit is recommended for fresh-eating. This variety is tolerant to humid climates, making it a great choice for southern districts. Harvest in late August.

8.13. Early golden

A high quality early maturing plum with attractive red blush over golden yellow ground color, maturing 10 to 14 days before Shiro. Fruit is very sweet, small to medium in size. Trees are hardy, vigorous, and productive. It needs cross-pollination with another Japanese Plum to ensure heavy cropping.

8.14. Fortune

A Japanese-type plum; maturing in late August/early September. Fruit is medium to large, reddish-purple and very firm with yellow flesh. The trees are upright and vigorous. Cross pollination with other Japanese varieties is recommended

8.15. Friar

A large oval plum; deep black with light amber flesh. The fruit is freestone with a small pit, sweet and juicy when fully ripe. The tree is vigorous, upright and productive

8.16. Methley

An early plum with fine quality and appearance. The fruit is purple with red, sweet flesh, very juicy with a distinctive flavor. Ripening in early to mid-July, Methley is self-fruitful and a good pollinizer for Shiro.

8.17. Satsuma

A very distinctive, dark red plum with red flesh. The fruit is oval shaped, medium in size, and very sweet. The tree is upright and very productive. The fruit is susceptible to cracking in prolonged wet periods.

8.18. Shiro

A sweet, juicy yellow plum. Fruits are round, clingstone, and medium in size. The trees are spreading and very productive. Shiro is a good pollinizer for Methley, Santa Rosa and Satsuma.

8.19. Victoria

One of the best known and popular of plum trees; produces pale red fruit. The fruit is ripe for eating around mid-August. Victoria is versatile plum tree, produces fruit which can be eaten, used for cooking and for jams. It produces a heavy crop which often needs to be thinned if fruits are to be produced each year.The Victoria plums are produced in late autumn and it is self fertile. This plum has stood the test of time, originally bred in Sussex around 1840.

8.20. Bruce

One of the toughest Plum trees- a Chickasaw-Japanese plum hybrid. Flavor is tops; A frost-hardy tree, ideal for low areas. Semi-dwarf, weeping habit. Sunset orange fruit with a sweet mellow flavor. Needs a pollinator ; Ripens late May-June. It requires 500 chill hours.

8.21. Byrongold

Finally a yellow plum of high quality. To quote the United States Department of Agriculture, "Byrongold meets or exceeds all current expectations for tree vigor, productivity, precocity, and resistance to the bacterial and fungal diseases that are common in the Southeastern United States." Byrongold's fruit is a winner, too Large, over 2 inches in diameter, and round, it's bright yellow on the outside with a golden-yellow flesh that's deliciously crispy. Needs a pollinator. Ripens in late June to early July. It requires 450 chill hours.

8.22. Chickasaw

A true edible ornamental, thickets of native Chickasaw plum are a welcome sight to bears, raccoons, foxes, deer, man, and other wildlife. They provide abundant crops of small bright red plums that ripen to yellow and make delicious jelly. In the early spring, while other deciduous trees are still dormant, Chickasaws announce their presence with stunning clouds of tiny pure white fragrant flowers. Drought tolerant once established.

8.23. Flatwood

This native plum, also known as hog plum, is found from the Carolinas to central Florida to Texas and Arkansas. A hardy under story tree that has a natural crooked form, the Flatwood will produce 1/2 to 1 inch tart purplish-black or deep red fruit. Their natural pectin makes them a great source for jams, jellies and pies. Flatwoods can grow to 20 feet tall.

8.24. Gutherie chickasaw

An improved Chickasaw plum that doesn't sucker and is highly disease resistant. This new variety is yellow-skinned with a tangy, sweet yellow flesh. Fruit is 1 1/2 inches across and makes a fabulous jelly. A good choice for growers who have problems with pollination. Ripens mid-June.

8.25. Improved methley

A great improvement on a wonderful old standby. This new variety is a purple-skinned plum with a sweet, soft, melting amber flesh. One of the best pollinators for other varieties. Ripens mid-June and rewires 500-700 chill hours.

8.26. Robusto

This variety has a bronze-red skin and red flesh coupled with a great flavor and disease resistant. Clingstone; needs a pollinator. Ripens early June and rewires 400-500 chill hours.

8.27. Santa Rosa

Purple-red skin and yellow-red flesh coupled with a tangy, sweet flavor. Needs a pollinator. Ripens early June and rewires 400 chill hours.

8.28. Queensland Bellerosa

Queensland Bellerosa, derived from Burbank X Santa Rosa, is an early-ripening Japanese-type (*Prunus salicina*) hybrid with *P. simonii* and *P. americana* in its parentage. It produces fruits with an average weight of 63 g ripening 11 days before Santa Rosa. It flowers slightly later than Santa Rosa but overlaps sufficiently for pollination. The fruits are more attractive than those of Santa

Rosa or Wilson and the variety is resistant to *Xanthomonas campestris* pv. pruni. (Topp and Russell, 1990).

8.29. Segundo

This new variety has a red skin and orange-red flesh coupled with a great flavor and disease resistant. Clingstone; needs a pollinator. Ripens mid June and rewires 400-500 chill hours.

8.30. Honey Sweet

It is a plum variety developed through genetic engineering to be highly resistant to plum pox potyvirus (PPV) the causal agent of sharka disease that threatens stone-fruit industries world-wide, and most specifically in Europe. Field testing for over 15 years in Europe has demonstrated the stable and durable PPV resistance of "Honey Sweet". Resistance is based on gene silencing whereby the inserted gene induces a natural plant defense mechanism against viruses. This resistance has been transferred to seedlings through cross-hybridization as a single locus dominant trait making it useful as a parent for developing new plum varieties for specific growing areas and markets. "Honey Sweet" plums are of high quality and compare well to the quality and nutritional value of conventional plums (Ravelonandro *et al.* 2013).

8.31. Sui Li 5

It is a new cold-resistant and early-maturing plum cultivar which originated from a cross-combination 'Sui Li 3' with 'Moonlight'. The fruit color is bright red, and the shape is ellipse with a spike top. The average single fruit weight is 68.0 g, the maximum is 89.0 g. The clingstone fruit taste sour and sweet. The soluble solids content of fruit is 13.10%, the soluble sugar is 6.03%. The global consistency and storability of fruit are good. It is a early-maturing variety. The development period of fruit is about 80 days. It has a strong cold resistance (Sun *et al.*, 2010).

8.32. Tropical plums (Low Chill for Zones 8B-9)

Along with a low chill requirement, they have a high heat requirement to initiate flowering (a heat unit is sort of the opposite of chill units—just take our word for it). All the Gulf series plums pollinate each other (but cannot be pollinated with our other low chill plums).

8.33. Gulf beauty

Still another University of Florida patented release (USPP 11224). Richly sweet, small, red fruit. Heavy producer with excellent flavor. Needs a Gulf series pollinator. Ripens in early May. It requires 250 chill hours.

8.34. Gulf blaze

Patented University of Florida release (USPP 10880). Medium-sized, deep ruby-red fruit with yellow-red center. Excellent flavor. Needs a Gulf series pollinator. Ripens early to mid-May. It requires 250 chill hours.

8.35. Gulf rose

Another University of Florida patented release (USPP 12833). Heavy crops of medium-sized fruit. Rich flavor, rose center. About as beautiful as a plum can get; Needs a Gulf series pollinator. 275 chill hours. Ripens in early May.

8.36. Scarlet beauty

Developed in Wauchula, FL by plant breeder Ronald Lambert; the Scarlet Beauty is a medium sized red skinned, red streaked amber fleshed plum. Reported to be self fertile. It requires 150 chill hours. Ripens in May.

8.37. Satluj Purple

It performs well under subtropical condition of Punjab and has high yield, excellent fruit size, colour, taste and flavour (Mahajan *et al.,* 2008).

8.38. Tophit

The cultivar was bred by crossing 'Cacanska najbolja' x 'President' in Research Institute in Geisenheim (Germany). This cultivar belongs pomologically to group of semi-prunes. Trees are vigorous, medium vigorous in full bearing. Canopies are almost globular, medium dense. Flowers are formed on annual shoots and on older fruiting wood. Flowering is medium late to late. Fruits are very big (57.1 g). The colour of the fruit is dark blue, strongly glaucous. Fruit skin is firm, hardly detachable from flesh, slightly acidic. The stone is medium, oval. The flesh is yellow to golden yellow, firm. The taste and juiciness are slightly above average. It is freestone cultivar. The taste is acidic sweet, very good in full maturity. It should be harvested as late as possible, regardless of early coloration of fruit, to achieve the taste quality of fruits. Fruits hang well on the tree without premature drop. The cultivar ripens nine days after 'Domestic plum'. Trees come into bearing early, the second year after plantation. Yields are high and regular. The cultivar is suitable for direct consumption and for processing. The cultivar is recommended for intensive growing systems. Trees demand appropriate and regular pruning. The cultivar' is recommended for warmer and moderate growing zones. 'Tophit' extends the range of cultivars ripening after 'Domestic plum'. 'Tophit' is partially self-fruitful. Pollinators are recommended. The cultivar is tolerant of sharka disease [caused by Plum pox virus] (Paprstein, 2007).

8.39. Guangongli

Derived from the cross between Maihuangli and Sanyuetian cultivars, made in 1988. The clingstone fruit mature in mid May, are large, weighing 75 g on average, but reaching 125 g, having a solid bright blush on the whole skin. Flesh is yellow, fine, juicy with a soluble solids content of 12.5% and a pleasant acid sweet flavor, and are of very good eating quality. The trees are very productive. The production of 5-year-old trees could reach 45-60 t/ha. This cultivar has a late flowering habit, which reduces the possibility of damage by late frosts (Yuan *et al.,* 2005).

8.40. Hybrid 60/10

It was obtained as a result of hybridization between *Prunus spinosa x P. domestica.* Its fruits ripen during the second half of August and are dark blue, globular, 15.8 g in weight and contain 11.16% total sugar, 10.26% invert sugar, 0.86% sucrose, 1.84% acids and 0.337% tannins. At full maturity, fruits have a pleasant tart flavour and are suitable for consumption. Trees are characterized by moderate growth. When grown on wild plum rootstock, 13-year-old trees reach 3.75 m in height and 38 cm in trunk circumference. The crown is broad, 2.75 m high, 3.60 m wide and has a volume of 9.3 m3. The hybrid has field resistance to mottling disease (Minev, 2001).

8.41. Flavor Supreme

This new plumcot (*Prunus domestica x P. armeniaca*) variety originated from a cross between plum selection 5G1230 and plumcot line 4G1180. Borne regularly, fruits are medium in size (mean 57.2 mm diameter, 54 mm long), globose to slightly flattened on stem end, shallow suture, medium thick skin, moderate to heavy bloom and coloured violet brown (RHS 187A) to purplish-grey mottled with lighter dots. Flesh is firm but delicate intermediate between plum and apricot, flavour sweet, few fibres and colour deep red to orange. Fruits are clingstone in type with stone size 23.8 x 7.9 mm. Keeping quality and shipping quality is rated as good (Zaiger's, 2000).

8.42. Valjevka

This prune cultivar was bred in Yugoslavia from the cross Agen 707 x Stanley. Tree vigour is medium, but latterly (in full bearing stage) it is rather weak. The tree shape is broadly pyramidal and its canopy is medium dense. Fruit spurs are relatively short and abundantly ramified. Size of fruits is medium to large and their shape is elongated baggy. The colour of the skin is violet to blue with a heavy wax bloom. Flesh is firm, golden yellow and its taste is acid sweet and very good. Fruits are suitable both for fresh fruit and for processing including drying. They have good transport quality and are resistant to bruising. The time of ripening in average climatic conditions of the Czech Republic is in the second half of September, about one week after Domestic Prune. It is very precocious, first fruits appearing in the second leaf. The productivity is above average, not much less that of Cacanska najbolja or Stanley. Fruit bearing is mostly regular. This cultivar is self-compatible and it is a good pollenizer. Rather heavy and regular pruning is recommended with an aim to improve the size of fruits. The best performance of the cultivar is in warmer regions and on fertile soils for optimum fruit quality (Paprstein, 1999).

8.43. Mildora

It is a new plum cultivar derived from crossing Large Sugar Prune and Cacanska lepotica made in 1981, is characterized by moderate vigour, a pyramidal, moderately dense crown, and early flowering. This self-fertile cultivar has abundant pollen, with an average germination of 39.4%. It could be a regular or heavy cropper, with small fruits (each weighing 24.2 g, on average), that are slightly long and roundish. Its stalk is short, moderately thick, adhering tightly to the fruit and fruiting branch. Its stone is small, elliptical, compact and tan-coloured. It has a medium thick, firm, reddish-blue skin when fully ripe. In the initial phase of ripening, the fruit has poor appearance (whitish-green colour), but takes on a very attractive appearance when fully ripe. Its flesh is firm, yellow, fine-textured and has a very sweet flavour. It has 26% total soluble solids, on average, 12.65% total sugars and 0.57% total acids. It has a particularly high sorbitol content and resistance to the local strain of sharka virus (plum pox virus). Fruits of Mildora are very suitable for drying. When dried, Mildora is an amber-coloured fruit with a high quality (Ogasanovic *et al.,* 2005).

8.44. Victory

This variety (*Prunus domestica*) was developed from cv. Vision pollinated by cv. Valor. Victory produces large (62 mm long x 52 mm wide) fruits weighing on average 59 g. Fruits are heart shaped with violet blue skin colour and have juicy, firm, greenish-yellow flesh of fair to good flavour. The average maturity date in comparative fruit growing trials at Vineland was 22 September, the same as for Bluebell and 4-5 days later than Stanley. Yields were intermediate between those of the other reference varieties: cumulative yield in 1986-91 was 107 kg, compared with 133 kg for Stanley and 70 kg for Vision (Tehrani and Lay, 1992).

8.45. Fla. 1-2

This is a new self-incompatible cultivar producing medium-sized, bright crimson, flavoursome, firm fruits with yellow-orange flesh, which weigh more than those of the local standard, Kala Amritsari (28 vs. 14-16 g). Fla. 1-2 is compatible with Kala Amritsari and a mixed plantation of the 2 cultivars is recommended for successful pollination (Dhatt *et al.*,1992).

8.46. Jojo

Work at the University of Hohenheim, Germany, on breeding plum (*Prunus domestica*) for resistance to sharka, caused by plum pox potyvirus, resulted in the development of the new cultivar Jojo from the cross Ortenauer x Stanley made in 1981. Trees are moderately vigorous, high yielding, self fertile and completely resistant to sharka. Flowering is mid-early. Fruits are oval-oblong, 34-40 mm in diameter, dark blue-purple, mid-late to late maturing, weigh 40-50 g and have juicy, sweet flesh.

8.47. Fermareine

The cultivar, derived from the cross Reine Claude Verte (green gage) x Reine Claude de Bavay made in 1970, is early bearing, regularly produces a heavy crop and has large (60 g) fruits of good quality. The fruit skin is spotted pink over a yellow-green background, and the flesh is yellow, firm, juicy and of good flavour. It matures 20 days after Reine Claude Verte and 8 days before Reine Claude de Bavay. The fruits are moderately resistant to cracking and brown rot. Fermareine shows good graft compatibility with Marianna GF8-1, Myrobalan MyrocalReg. (*Prunus cerasifera*), IshtaraReg., and the pentaploid rootstocks P8-13 and P10-2. It yields more than Reine Claude Verte. Pollenizers are Reine Claude Verte, Stanley and other cultivars that flower at the same time (Renaud, 1994).

Grzyb and Rozpara (2004) observed three new plum cultivars: 'Promis', 'Tolar' and 'Nectavit' belonging to *Prunus domestica* L. and selected from 'German Prune' types, four rootstocks for plums and apricots: 'Alina', 'Agata', 'Amelia' and 'Anna', selected from *P. cerasifera* Ehrh. var. divaricata Led. These cultivars, rootstocks and interstock are virus free and are being introduced into orchards and nurseries in Poland.

9. Propagation

9.1. Seed propagation

Several factors are important concerning germination of plum seeds, and one of the most critical ones is the control of humidity. In the reported experiments, a certain loss of germination capacity was found after drying of seeds followed by water-saturation of embryos and stratification. An experiment with different stratification media illustrated that a mixture of sand and perlite (ratio 1:1) resulted in the highest germination percentage, while the organic medium "Moder jord" gave the poorest germination. A 4-week period of 20 degrees C preceding

3 months stratification recorded higher germination than no such warm period. Cracking the stones followed by a removal of the seed coat and sowing of embryos directly after harvest recorded the highest percentage of germination. This procedure, however, resulted in a high loss of seedlings due to dormancy and poor seedling growth. The highest germination of stones was observed when seeds were kept in plastic containers during stratification compared to plastic bags. The percentage of germination ranged from 0 to 85%, and clearly demonstrated the importance of a defined protocol (Hjeltnes and Nornes, 2007).

9.2. Vegetative propagation

Commercially, plums are propagated vegetatively by budding and grafting in autumn or in spring on seedling rootstocks or clonally propagated ones. The scion sticks are bundled and kept under moist condition partially inside soil under the tree. Grafting is done by tongue method with the advent of spring when sap flow of rootstocks starts, by using dormant scion stick. The seedling rootstock should be of pencil size diameter.

9.2.1. Propagation by cuttings

Rooting in cutting is a complex process that is determined by the relation of many factors. Rooting rate is highly influenced by the time of cutting collection, which includes the state of the mother plant and the environmental factors on that day. Szecsko *et al.* (2004) demonstrated that rooting ability of plum hardwood cuttings is associated with the activities of peroxidase and polyphenol oxidase [catechol oxidase] (PPO), phenolic content and bud dormancy of hardwood cuttings. Rooting percentages were higher in October and February, and lower in January. Phenolic content was low in September, reached a peak in November, decreased until the middle of December, peaked for the second time at the end of January and finally decreased in February. PPO activity also showed a seasonal pattern. The phenolic content of buds and the rooting of Marianna GF 8-1 had a linear correlation. These components were correlated with dormancy. Both the beginning and end of endodormancy were associated with increased PPO activity.

Gill and Chitkara (1990) standardized the propagation of plum by semi hardwood cuttings. The cuttings taken in November and December were treated with IBA (250-1000 p.p.m.) alone or in combination with Ethrel [ethephon] at 50 p.p.m. or with Ethrel only. All cuttings were planted on raised ridges 45 cm apart with 20 cm between cuttings. The basal cuttings (taken only from the basal part of 1-year-old wood) gave higher rooting percentages, especially when treated with IBA at 1000 p.p.m. + Ethrel. Cuttings taken in November gave better rooting than those taken in December. Mirabdulbaghi and Zarghami (2011) reported that hardwood cuttings treated with IBA at 3000 ppm IBA induced 47% rooting and found to be the best treatment among the other ones.

9.2.2. Propagation by budding

9.2.2.1. Steps in T or shield budding

Cut a slice of bud with bark from the parent tree. Cut a similar sliver of the rootstock, making a little lip at the base to slot the scion into. Join the two together and bind. In time, the scion bud will grow into a shoot, which will develop into the desired tree. For this seedling stocks are raised in warmer low hill areas under assured irrigated condition to enable the seedlings to assume rapid growth for purpose.

9.3. Micro-propagation

In recent years, micropropagation technique has been developed for producing plants in large numbers within shortest possible time to overcome the limitation of availability of planting materials and clonal rootstocks.

Plum (*Prunus salicina*) is one of the important fruit trees in China. However, plum trees possess not only longer juvenility but also higher pistil sterile and heterozygosity). In open field, Chinese plum trees are often infected by several viruses including plum pox virus and plum necrotic ringspot virus (Sun, 2004), which severely inhibit plum development in China. Tissue culture provides opportunity to produce the seedlings free from the viruses. The procedures for the production of plants from mature seeds of both *P. domestica* and *P. salicina* and from cotyledons of *P. persica*, P. domestica and *P. cerasus* have been described (Tian and Sibbald 2007). The application of in vitro methods in the production of own-rooted planting material of plum significantly increased the production capacities and contributed to the improvement of its quality and health status (Popov and Kornova, 2009). For micropropagation it is often preferable to use explants from mature plants where superior characteristics are evident (Jain and Babbar, 2003). However, it is difficult to culture in vitro mature trees including plum due to the higher incidence of microbial contamination, vitrification, difficulty in root induction and high levels of polyphenol exudation .Studies on regeneration in vitro from mature European plum *(P. domestica)* and ornamental Prunus species have been succeeded (Barbara *et al.,* 2004; Kalinina and Brown, 2007), but there are less reports on in vitro regeneration of Chinese plum species. Herein, an efficient protocol was established for micropropagation. of 'Gulf-ruby' Chinese plum (P. salicina Lindl. cv. 'Gulfruby') using mature stem segments as the explants.

Ruzic and Vujovic (2007) developed a protocol for the micropropagation of plum (*Prunus domestica* L.). Shoots were regenerated from aseptic cultures of plum cultivars Cacanska Rodna and Valjevka established in Murashige and Skoog's (MS) medium containing 2.0 mg BA [benzyladenine], 0.5 mg IBA, 0.1 mg GA3 [gibberellic acid], 7000 mg agar and 20 000 mg sucrose/litre.

Regenerated shoots were multiplied in MS medium supplemented with 1.0 mg BA, 0.1 mg IBA, 0.1 mg GA3, 7000 mg agar and 20 000 mg sucrose/litre. The cultivars, especially Valjevka, recorded high regenerative rates in the 3rd subculture, but exhibited reduced regenerative abilities in the 9th subculture. Thus, at this stage, plant growth regulators were applied. However, only Cacanska Rodna showed a positive response to the presence of growth regulators in the medium. The rooting medium consisted of half-strength MS medium supplemented with 1.0 mg NAA, 0.1 GA3, 7000 mg agar and 20 000 mg/litre. Rooting in both cultivars was up to 85.0%. Rooting parameters, including the number and length of roots, occurrence of secondary roots, amount of callus at the base of shoots, and height of rooted plantlets, were monitored. Under the mist system, 58 and 64% of Cacanska Rodna and Valjevka plants, respectively, were acclimatized.

Azar *et al.* (2010) carried out an investigation to evaluate factors affecting establishment and proliferation of two apricot x plum (*Prunus armenica x Prunus domestica*) interspecific hybrids HS405 and HS706. These hybrid rootstocks are resistant to diseases and pathogens and have been developed in East Azarbayjan Agricultural and Natural Resources Research Center, Tabriz, Iran. Propagation of these hybrids through cutting and grafting proved to be difficult due to some physiological disorders. Therefore, vegetative propagation using tissue culture can be a reliable method for mass production of healthy plant materials from these new rootstocks. Nodal explants were surface sterilized by immersion in 70% ethanol for 30 seconds, followed by 2.5% sodium hypochlorite for 7 and 10 min depending on the tissues, then rinsed three times with sterile distilled water. Cultures were established on MS and WPM media supplemented with BAP (0, 1 and 2 mg/l)+0.05 mg/l NAA and effects of 3 sampling dates (spring, summer and autumn) were evaluated. After four weeks, surface and internal infection, shoot necrosis, percentages of active buds were recorded and a scoring index system (1-4) was developed for determination of buds growth quality. WPM medium supplemented with BAP (0, 1, 2 and 4 mg/l)+0.5 mg/l GA3 was used in proliferation stage and after four weeks number of shoots, average shoot length, shoot quality index, leaf number, leaf length and leaf chlorosis index were recorded. All media were supplemented with 3% (w/v) sucrose and 0.7% agar and the pH was adjusted to 5.7 before agar addition and autoclaving. All the cultures were maintained in growth room at 24+or-1 degrees C and 16-h photoperiod under 40 micro M m-2 s-1 illumination created by white fluorescent lights. In establishment stage the highest percentage of active buds and bud quality index achieved from explants collected in spring. HS405 genotype had higher number of active buds compared to HS706. WPM medium was better than MS in establishment stage. The highest percentage of active buds was achieved in hormone-free medium that without significant difference with the medium containing 2 mg/l BAP+0.05 mg/l NAA. In proliferation stage, genotype HS706 had higher shoot length and quality index as compared to HS405. The best treatment for shoot proliferation was 4 mg/l BAP+0.5 mg/l GA3without significant difference with the medium containing 2 mg/l BAP.

The use of the in vitro method creates the possibility of obtaining virus-free plants in a short-term period and in large quantities. Kornova and Popov (2013) established that the successful disinfection of the source plant material could be achieved by treatment with 8-10% of potassium hypochlorite for 6-10 minutes depending on the studied cultivars. The optimal multiplication rate was observed when growing the microplants in a modified MS medium supplemented with BAP 0.5-0.7 mg/L. Increasing the cytokinin above 0.7 mg/L caused an increase of the proliferation effect but initiated vitrification. A high percentage of rooting was established when cultivating the microplants in a MS nutrient medium supplemented with a total or 1/4 content of macroelements and IBA 1.0-1.5 mg/L. The average height of the stem part, the number and length of the formed rootlets depended to a significant degree on the specific cultivar. The optimal percentage of adaptation to ex vitro conditions was obtained when the rooted plants were planted in a mixture of peat and perlite in a glass-and-steel greenhouse.

9.4. Rootstock

There are number of rootstocks having special characteristics (Table 2) which should be used as per local need.

Table 2: Rootstocks of Plum with different characteristics

Characteristics of rootstock	Name of rootstock
Dwarfing rootstock	St.Julien k, St. julien, GF 655.2, Pixy, Prune GF 43, Hybrid P 2038, citation, common Mussel, Broadleave Mussel, SK orospolka krasnaya, P 2088-IA(France)
Medium vigrous rootstock	Ishthars, Fereley, St.julien GF 655/1(France), St. julien w.61(Switzerland), Brampton, Ackermann, Wangenheim, Novinka and 140-1 (Russia)
Vigorous rootstock	Myrobalan B, St.julien C, Myro 2261,GF 1246, OPA-15-2 and OP-23-23 (Russia)
Drought tolerant rootstock	Myrobalan 27
Cold hardy rootstock	St.julie A, Mariana 2624,Mariana GF8/1
High soil moisture tolerant rootstock	Mariana GF 8/1, Mariana 2621, Damas GF 1869
High soil pH tolerant rootstock	Myroblan 29C, mariana GF 8/1, Damas GF 1869
Disease resistant rootstock	
a. Resistant to crown gall	Mariana GF 8/1, mariana 2624, Myroblan GF 31, Myrobalan 29c
b. Resistant to canker	Myrobalan B, Pixy
Nematode resistant rootstock	Mariana GF 8/1, Mariana 2624

9.4.1. Rival

It is a clonal plum rootstock, derived from *Prunus insititia,* which has medium vigour, good compatibility with various plum cultivars, good tolerance to major diseases and pests, and is thornless. Stem cutting propagation is good (over 79% rooted cuttings). This rootstock has a high cutting yield obtained in the mother orchard (149 000 pcs./ha). 'Rival' can be used for intensive plum culture (Botu *et al.*, 2007).

9.4.2. Ademir

Ademir is a myrobalan (*Prunus cerasifera*) rootstock developed for European plums and prunes (*P. domestica*). It was selected from open pollinated seedling populations of myrobalan on the basis of good rooting ability, low vigour, excellent plum graft compatibility and better graft compatibility with apricot cultivars than is evident with Myrobalan B and Marianna rootstocks. In orchard or nursery trials over 3-4 years, Ademir has shown good graft compatibility with European plums and prune cultivars, and also shows compatibility with several P. salicina cultivars. Ademir is suitable for plum cultivars to avoid excessive vigour or to increase planting density when tree size needs to be controlled. It adapts well to highly calcareous and compact soils, being resistant to iron chlorosis and root asphyxia while under waterlogged conditions no trees of Ademir died (Moreno *et al.* 1995).

9.4.3. Adesoto 101

Adesoto 101 (*Prunus insititia*) was derived from an open pollinated population of cv. Pollizo de Murcia. It suckers less and roots more easily from hardwood cuttings than most of the Pollizo population from which it was derived. In tests, good compatibility was shown with 30 peach and nectarine cultivars as well as with almond, apricot, European plum and prune (*Prunus domestica*) cultivars and cultivars of Japanese plum (*P. salicina*). It is well adapted to highly calcareous and compact soils, and is resistant to root asphyxia and Fe chlorosis. Adesoto 101 advances fruit harvest by 3-7 days, especially when compared with peach seedlings and peach x almond hybrids as rootstocks for peaches. Cropping efficiency and fruit size were also higher or similar to that on these rootstocks and other plum rootstocks. Adesoto 101 is resistant to *Meloidogyne arenaria, M. incognita and M. javanica* (Moreno *et al.*,1995).

10. Planting

Two year old feathered trees and one year old whips of plum propagated on seedling or clonal rootstocks are suitable for planting in winter when plants are dormant but before the plants sprout and before the onset of spring with a spacing of 6m in both way. Increase or decrease in spacing will be governed by vigour of rootstocks used and fertility of soil. A closer spacing of 5.4×5.4m may be adapted on sandy soil. The most common system of planting in the valley, undulated terrain and gentle slope of hill surface is square system. In the slopy hills, spacing is maintained only in one direction along the contour. The common pit size is 1×1×1m. While digging the pits, top soil up to the depth of 30 cm and the sub-soil below this should be placed separately. The pits are filled with top soil first. Pits should be filled up completely and tightly up to 10 cms higher than the ground level. Before the planting is taken up, the pits should receive a couple of rain or should be irrigated so that the soils in the pits settle down properly. Before planting the trees or whip, all broken roots are trimmed and the plants are placed in the centre of the pit with the use of planting board and trampled all around the plant properly so as to remove all air pockets and to keep them erect. While planting, grafted or budded points should be at least 15-20 cm above the ground level. A few inches of loose soil is left on top to act as a mulch and trees should be headed back at the time of planting to 60-70 cm height.

11. Training

Training in plum is an important operation in order to provide strong framework capable of supporting crop load and to facilitate easy management practices besides admittance of light and proper aeration inside the plant. Plum trees are generally trained to modified central leader or to open centre system depending on the growth habit of the variety. The varieties with spreading growth habit (Burbank and many Japanese varieties) are better trained to open centre system whereas those with upright growth (Santa Rosa, Stanley and Wickson) should be trained to the modified leader system. In Europe, open centre system is more common for plum, whereas in India, the modified central leader system is common practices of training.

At the time of planting, the un-branched 'whip' is beheaded at about 0.7- 1.0 m height to emergence of lateral branches while in branched plant, 3 to 5 appropriately placed wide angled branches are selected as scaffold each of which is cut back to one-half or one-third of its length. In open centre system of training, 4-5 scaffolds are developed along with few secondary branches. All the branches which are growing straight or downward to upward and outward growing branches and lateral shoots as well as water sprouts should be removed. European varieties mostly develop well shaped tree with little pruning while more pruning is required in Japanese plum to obtain the desired shape.

11.1. Modified leader training

It combines the best qualities of the central leader and open center systems. In the modified central leader system of training, equally spaced 3 to 5 scaffold branches arising with wide angled crotches from the trunk are selected in the first 2-3 years. The scaffold branches are preferably spaced vertically about 15 to 25 cm apart, spirally around the trunk for balanced distribution of crop load. The growth of the central leader in checked or modified at 5 m above the ground so that it can be no longer grow straight. For convenience of orchard operations, the lowest limb is usually kept 60 cm from the ground level. The other systems of training also adopted in European plum growing areas are Hedge rows with Central Leader Pyramid, Summer Hedged, Palmate etc. which show promise for mechanically handled trees.

11.2. Central leader spindle

A new training system - central leader spindle was applied by Mika and Buler (2011) A new training system was applied to obtain trees suitable for mechanical harvesting. The leader was not headed after planting and summer training treatments were performed in May/June. From the third year onward renewal pruning was done after fruit harvesting. The new training and pruning system resulted in very rapid tree growth, much young wood, fruit bud formation on young wood and early bearing. Trees appeared to be suitable for hand and mechanical harvesting within four years from planting.

12. Pruning

In general, plum trees require very less pruning. Plum fruit are generally borne laterally on one year old wood and on spurs of older wood. Plum trees should be pruned in such a way so as to make an annual extension of growth of about 25-50 cm which encourages more spur development. Older trees of European and Japanese plums require about 15 cm and 25 cm annual extension growth, respectively for proper fruiting. The extent of pruning in two types of plums differs largely. Japanese plum varieties tend to overbear and they need heavy pruning because of having too many fruit buds and spurs on the tree tending to produce too many flowers and fruits. The European plum trees are pruned lightly since they bear on spurs which required to be retained to maintain production. Light pruning is done in such case to renew the old spurs. In general, pruning in young and mature trees should be as little as possible to achieve the desired results.

12.1. Steps in Pruning

i. Leave only one trunk for the central leader.

ii. Remove branches with crotch angles less than 60 degrees.

iii. Remove all branches directly across from one another on the leader.

iv. Space lateral branches uniformly around the leader to prevent crowding as the limbs grow in diameter.

Pruning of bearing trees is done to secure a balance between vegetative growth and fruitfulness and to maintain the profitable yields and quality of fruit. It involve the removal of the upright branches and centre top to open out for better admittance of light, induction of vigorous spur and better colouration of fruits. When the plants bear many fruits, there is short terminal growth and spur growth becomes weak. In such trees, small thickly spurred branches should be pruned off. The lower tops of main limb should also be thinned out and headed back to vigorous laterals. All the cut ends should be covered with fungicidal paste to prevent entry of disease organisms.

12.2. Intensity of pruning

Cacanska Rodna is a highly cropping and good quality plum cultivar in Yugoslavia with medium-sized and firm fruits of pleasant sweet-acid taste. It produces fruits of uneven quality, *i.e.,* size, taste and degree of maturity with irregular and light pruning. The optimal pruning intensity for Cacanska Rodna was determined by Mitrovic *et al.* (2001) to obtain fruits of better and more uniform quality, both fresh and dried. The pruning severity treatments were: (1) severe pruning (1/3) - removal of 1/3 of skeletal and bearing branches, removal of 3-5 semi-skeletal branches and thinning; (2) moderately severe pruning (1/4) - removal of 1/4 skeletal and bearing branches, removal of 1-3 semi-skeletal branches and thinning; (3) light pruning (1/8) - removal of 1/8 skeletal and bearing branches and thinning; (4) thinning - without cutting back of skeletal and bearing branches, only superfluous branches and watersprouts were removed; (5) control - only water sprouts were removed. The differences of pruning severity in Cacanska Rodna significantly affected qualitative properties of fresh fruits. Variants with severe pruning (1/3 and 1/4) produced 77.55 and 75.56% of fruits suitable for drying with regards to size and the required degree of maturity. Variants with lighter pruning (1/8 and thinning) produce very low percentage of large-sized fruits (>25 g) which are suitable for pruning. The variant with no pruning produced fruits which did not fully meet the requirements of the quality needed for drying. Variants with severe pruning (1/3 and 1/4) are recommended for producing fruits which give high quality prunes.

Kumar and Thakur (2012) carried out a pruning trial with four pruning treatments viz., unheaded, 25 per cent head back, 50 per cent head back and 75 per cent head back were given to six year old plants of plum cv. Santa Rosa. The maximum yield (9.50 kg/tree, 10.00 kg/tree and 10.50 kg/tree, respectively) was obtained with unheaded plants during three years and the yield decreased with the per cent head back. The maximum fruit length (4.72 cm, 4.43 cm and 4.26 cm, respectively), breadth (4.86 cm, 4.48 cm and 4.17 cm, respectively) and weight (66.07 g, 55.60 g and 47.19 g, respectively) were obtained with 75 per cent head back and lowest with unheaded plants during three years. In unheaded plants yield were high but of poor quality. Whereas, in 75 per cent headed trees the quality was very good, but production was very low. Twenty five per cent head back resulted in good production and fruit quality. The leaf nutrient status was also affected by the different levels pruning intensities. Leaf nitrogen and potassium contents increased with the increase in severity of pruning, whereas leaf phosphorous content remained unaffected.

12.3. Time of pruning

Sosna (2010) investigated the effect of different pruning time on cropping and fruit quality of several early ripening plum cultivars in Poland and concluded that time of pruning had no influence on yielding and chemical composition of fruit. Soluble solids, vitamin C, magnesium, calcium, potassium and phosphorus contents in fruit were similar independently of pruning time. Mean fruit weight was significantly higher from trees with dormant pruning.

13. Fertilization

Fertilizer application varies from place to place and depends on many factors such as climate, plant species, plant age, rootstock, soil type, total nutrients in the soil, soil moisture, soil pH, and cation exchange capacity of the soil (Vitanova, 1990; Taylor, 2009). In a study, Vitanova (1985) indicated the amount of N, P, K, Ca, and Mg needed each year to replace the removal by the plum trees in the form of fruits, fallen leaves, and pruning wood (Table 3).

Table 3: Estimated annual removal of nutrients by mature plum trees.

Element	Amount (kg/ha)	Element	Amount (kg/ha
N	24.6 – 62.6	Ca	3.9 – 11.6
P	4.6 – 13.1	Mg	2.1 – 6.7
K	23.3 – 61.8		

Source: Bhutani and Joshi, 1995

Bajwa and Mishra (1970) recommended application of 120g of CAN, 95g of single superphosphate, and 60g of murate of potash for one year old tree and increased by same quantity every year. The fertilizer doses are stabilized at the age of 6 years. Teskey and Shoemaker (1978) recommended application of 224 – 336 kg/ha of complete fertilizer in plum orchards. Azad and Sharma (1987) recommended application of 60 kg FYM and 500, 250, and 600 g of N, P_2O_5, and K_2O to the fully grown plum tree. The entire dose of P and K along with farmyard manure is applied in January, whereas, half of the N is applied before flowering and the remaining half after fruit set. Under rainfed conditions, N fertilizers are applied in one lot, 15 days before bud break. Normally the fertilizer is spread evenly over the surface under the branches.

In non-irrigated areas, two foliar sprays of urea (1 kg/100 liters) should be applied in March (Thakur and Thakur, 2014). Under Meghalaya condition, for bearing trees of 10-13 years age, a dose of about 100 g N, 200-205 g P and 20-100 g K per tree per year is considered optimum for maximum yield with good physico-chemical parameters. In the plain areas of Punjab, application of 200 g N, 75 g P_2O_5 and 200 g K_2O per tree with a basal dose FYM at 35 kg/tree, applied in December together with the entire dose of P and K and half dose of N one month later (Shear and Faust, 1980). Response to fertilizer application by plum grown under clean cultivation system is better than permanent sod system. All trees receiving basal P_2O_5 at 30 or 60 kg/ha; those under permanent sod received N at 30-270 kg and K_2O at 30-250 kg/ha in various combination and those under clean cultivation received N at 120-270 kg and K_2O at 150 kg/ ha. The best tree growth and highest yields were obtained from trees under clean cultivation receiving N: P_2O_5: K_2O at 120: 30: 150 kg/ha. (Tariq *et al.*2008). A manurial schedule for different age group of plants has been presented in Table 4.

Micronutrient deficiency sometime occurs in plum in some areas. Zn deficiency may be corrected by dormant spray with zinc sulphate before the bud break. When Cu deficiency appears, young leaves develop intervienal chlorosis or complete yellowing and in older leaves necrotic spotting occurs between veins. It is corrected by soil application of 1-2 pounds copper sulphate per tree. Boron deficiency causes hard pockets in fruit flesh and corrected by half pound borax application once in every third year per tree.

Table 4: The manuring schedule (per plant)

Age of plant year	FYM(kg)	CAN(g)	SSP(g)	MOP(g)	Remarks
1st	10	280 (70)	220 (35)	170 (100)	Apply FYM during December-January along with full dose of P and K
2nd	15	560 (140)	440 (70)	340 (200)	
3rd	20	840 (210)	660 (105)	500 (300)	
4th	25	1120 (280)	880 (140)	680 (400)	
5th	30	14000 (350)	110 (175)	850 (500)	
6th	35	1680 (420)	1320(210)	1000 (600)	Apply ½ N in spring before flowering and ½ N a month later. The dose should be stabilized after 7 year.
7th and above	40	2000 (500)	1540(250)	1200 (700)	

Source: http://www.fruitipedia.com

For normal growth, optimum production and to minimize adverse environmental effects, fruit trees require 13 essential nutrients in varying amounts. Balancing of nutrients is very important; deficiency or excess of one element can negatively affect plant processes and thereby inhibit optimum uptake, utilization, or distribution of other elements. Therefore, efficient nutrient management calls for supplying the right nutrient in right quantity and at right time (Bhutani and Joshi, 1995). Most of the elements essential to trees need not be supplied in a regular fertilizer programme as those may be present in soils to provide an adequate supply or some nutrients in substantial proportion from the leaves are recycled back to the tree tissues before leaf fall. Nutrient problems (deficiencies and excess) can often be identified visually, but it is advisable to chemically analyze leaf samples to confirm the problem and quantify its severity. Even without obvious problem, leaf samples should be taken to check the effectiveness of the nutrients applied and to watch the gradual changes in nutrient levels (Table 5).

Table 5: Nutrient level in plum

Element	Optimum range in leaves	Range in mature fruit
N	2.3 – 2.8 %	1.0 – 1.5 %
P	0.1 – 0.3 %	0.1 – 0.3 %
K	1.6 – 3.0 %	1.5 – 2.5 %
Ca	1.0 – 1.5 %	0.05 – 0.15 %
Mg	0.3 – 0.8 %	0.05 – 0.15 %
Fe	100 – 250 ppm	20 – 80 ppm
Mn	40 - 160 ppm	5 – 10 ppm
Zn	20 - 50 ppm	10 – 20 ppm
B	25 – 60 ppm	20 – 50 ppm
Cu	6 - 16 ppm	5 – 10 ppm

Source: Olmstead *et al.*, 2016

13.1. Deficiency symptoms of macro and micronutrients

13.1.1. Nitrogen deficiency

The leaves are small, narrow, is light green in color. Older leaves may be yellowish orange or reddish purple and are shed early. Leaf stalks are branches with a view to form an acute angle, thin and short, and if the deficiency is seen very violent death stalks. The shoot growth is weak. The number of flower buds will be less, short period of pollination of flowers. Fruit color before ripening.

Nitrogen is the only major element that must be added regularly to typical stone fruit orchards. Deficiency symptoms are similar for peach and plum showing pale green leaves near the terminal of the shoot and yellow leaves at the base. Shoot growth and leaf size are reduced, but not nearly to the extent of zinc deficiency. Premature leaf drop often occurs. Fewer flower buds are produced and the resulting fruits are smaller and more highly colored (Taylor, 2009). Correction is easily achieved with any nitrogen containing fertilizer. Early spring application of fertilizers is effective. For sandy soils, split the application to minimize leaching losses. Little nitrogen is taken up when leaves are absent from the tree, so a dormant application is inefficient due to leaching, volatilization, and denitrification losses. A rate of 112 to 168 kg actual nitrogen per hectare is usually adequate for maintaining trees at optimum level (Johnson and Uriu, 1987). However, excessive nitrogen supply can cause negative impact on the fruit quality by reducing flesh firmness and sweetness (Rettke *et al.*, 2006), low color development and increasing susceptibility to postharvest diseases (Daane *et al.*, 1995). Cuquel *et al.* (2011) reported that for better postharvest quality and storability, N should be applied at the lesser rate of 40 kg/ha.

13.1.2. Phosphorus deficiency

It appears first in young leaves of plum, dark green, purple, bronze or become stained. Petiole they stand upright to make connections with a narrow moon is reddish branches. Rarely has a leaf system. The edges of older leaves dark brown necrosis occurs. Early defoliation occurs. Trees with phosphorus deficiency bear few flowers and fruits. Fruits remains small and flows from maturing. Fruits are lifeless color, hard and are often a tissue, they are tasteless. The excessive amount of phosphorus than nitrogen would still be appreciated in the flesh coarsely textured.

Phosphorus deficiency has seldom been observed in stone fruits, possibly because phosphorus is very efficiently recycled in senescing leaves. Also phosphorus levels in fruit tissues are not very high, thus, minimal amounts are removed in harvest. Leaves on deficient trees are dark green, eventually turning bronze and develop a leathery texture. A purple or red coloration appears on the leaves, petioles, and young shoots. Leaf size may be reduced and premature defoliation may occur, beginning with basal leaves. The fruits are more highly colored and ripen earlier but exhibit surface defects and poor eating quality. In fact, annual requirements of fruit trees in general range from 6 to 11 kg/ ha (Johnson and Uriu, 1987).

13.1.3. Potassium deficiency

Plum and dark brown will occur at the edges of the leaf chlorosis and potassium deficiency dry regions. In contrast, the leaves can stay in this state for too long on the tree. Fruits are small and pale, become thick-skinned. The amount of sugar would be small and sour taste. Lack of a plum tree in the pot begins to yellowing leaves firstly, after which the leaves are completely dry brown dead. Lack of potash in arid year would be more harmful. Potassium tree photosynthesis is an element which the influence of making the starch and sugar. Flowers prevent potassium periodicity due to play a role in the formation of the eyes.

Potassium deficiency is sometimes found in plums, although it is not nearly as much a problem as it is in prunes. Symptoms first develop in early summer on mid-shoot leaves, which exhibit a pale color similar to that of nitrogen deficiency. Leaf margins become chlorotic and then necrotic, leading to a marginal scorch that is particularly evident in plums (Johnson and Uriu, 1987). This necrosis eventually extends inwards, leading to cracks, tears, necrotic spots. Terminal leaves are least affected and defoliation of older leaves may or may not occur. Under potassium deficiency, stone fruit productivity, quality and storage potential are negatively affected (Chatzitheodorou *et al.,* 2004; Ruiz, 2006). Correction can be obtained with a soil application of potassium sulphate at the rate of 110 kg/ ha (Cuquel *et al.* 2011).

13.1.4. Magnesium deficiency

In the long shoots of magnesium deficiency in plum especially in older leaves, between veins, irregularly shaped, light green, yellow sometimes occurs grayish green spots. Vascular patches extending up from the leaf margin in some cases. Reddish brown spots quickly turn into necrosis. The leaves were then fades, curls, dry and fall early. Fruits and become unpleasant odor. Golden Delicious Plums are sensitive to magnesium deficiency more.

Magnesium deficiency is usually caused by high potassium in soil rather than by naturally low magnesium levels. Trees are usually vigorous, despite deficiency, and yield, fruit size and leaf size are not affected. The characteristic pattern on the leaf is a marginal chlorosis, leaving an inverted green 'V' around the midrib. These symptoms appear first on the basal leaves and develop sequentially up the shoot. As the season progresses, the chlorotic area increases and develops into necrosis (Weir and Cresswell, 1993). When correction desired, soil application of magnesium sulphate, magnesium oxide, and dolomite can prove effective.

13.1.5. Manganese deficiency

In case of mild to moderate deficiency in young leaves, opening the vessel is quite light in color tint them out of the opening, however, it is very visible when the leaves exposed to light. In case of more severe deficiency increases the opening of leaf colors and takes the network image. At a later stage, whitish-yellow spots on the leaf surface for containers.

Manganese deficiency symptoms in plum include an interveinal chlorosis extending from midrib to margin. Relatively wide bands around the major veins remain green, giving a 'herringbone' pattern in contrast to many other plants, stone fruits show symptoms first on older basal leaves, with terminal leaves often appearing more green. To correct deficiency, foliar sprays of about 0.4 to 0.6 kg/100 l of manganese sulphate may prove effective. Soil application of same thing is often ineffective, since manganese is quickly oxidized to less available forms.

13.1.6. Boron deficiency

Flowers wilt and suddenly they get damaged from the cold colors like black coffee. But while they stand at a branch in this state, not fall. Frost damage but create the same image, frost flowers are poured immediately affected. In cases where the iron deficiency chlorosis come forth from the green leaf veins, Vegetative growth point dies, becomes short shoots, leaves become small and misshaped. However, chlorosis of leaves is not visible. The fruits are smaller than normal and sometimes cracks happen. Recent research indicates that boron is very mobile in sorbitol-rich plants. An easily translocated compound is formed from boron and sorbitol that accounts for this mobility (Brown and Shelp, 1997).

13.1.7. Iron deficiency

Chlorosis symptoms arise if plum is called or jaundice. Symptom manifests itself more in terms of lime and nutrients in poor soils. In particular symptoms are exacerbated during the rainy season. In cases where the iron deficiency chlorosis come forth from the green leaf veins, whereas tissue between veins is yellow, symptoms begin in young leaves, move towards the old leaves. The edges of the leaves that occurs once or reddish brown and dry. Photosynthesis slows down development because of the disruption, the yield falls and eventually tree may die.

Iron deficiency symptom is the loss of chlorophyll, leading to chlorotic leaves. However, in early stages, only the interveinal areas of the leaf are affected, leaving the veins dark green. Young leaves are the first to show symptoms. Applying inorganic iron to the soil seldom corrects the deficiency. Synthetic iron chelates applied to soil or foliage have been effective, but may be costly, and need to be careful with soil application, especially in situations of lime induced chlorosis. Another promising approach is to inject or place materials into holes drilled in the trunk. Materials that have shown correction of symptoms include ferric phosphate, iron citrate, ferric ammonium oxalate, ferrous sulfate, and ferric ammonium citrate (Johnson and Uriu, 1989).

13.1.8. Zinc deficiency

Symptoms include short internodes, small narrow leaves, and interveinal chlorosis with shoot and branch dieback. In advanced stages, small, narrow terminal leaves are arranged in whorls or "rosettes." This results in the typical "rosette" and "little leaf" description for zinc deficiency.

Stone fruits are chiefly sensitive to zinc deficiency. There is rootstock and variety variability, especially in plum the disorder has often been called 'little leaf' because of the small pointed leaves produced. These leaves occur in rosettes on the tips of the shoots and young spurs because of shortened internodes. The affected leaves become chlorotic with an interveinal mottling (Horst, 2008). Leaf margins are often crinkled or wavy. These symptoms along with delayed foliation often occur early in spring. To prevent or correct the deficiency, treatments can be put on at several different times. Zinc sulphate (36% metallic zinc) applied at a rate of 11 to 17 kg/ha from mid-October or about 50% leaf fall through the dormant season is very effective. The dormant spray should not be applied with or within several weeks of dormant oil spray. Neutral of basic zinc (52% metallic zinc), also at a rate of 11 to 17 kg/ha, is effective when applied in spring or summer. However, it may leave spray deposits on fruit and cause leaf burn and defoliation if rain occurs shortly after application. Mixing an insoluble form of zinc (zinc oxide) into the soil of a planting hole prior to planting provides a substantial boost in leaf zinc concentrations that lasts for several years (Johnson and Day, 2013).

13.2. Time of application and source of nitrogen

From an investigation carried out at Solan (Himachal Pradesh, India), Saini *et al.* (2013) concluded that half dose of nitrogen should be applied during spring before flowering and remaining half dose to be applied one month after first application. Leaf nutrition status improved with urea+lime at 600 g N/tree. They also found that urea +lime at 600 g N/tree is the best as source of nitrogen as compared to other sources like ammonium sulphate, DAP+urea, calcium nitrate, 12:32:16+urea and 10:26:26+urea.

13.3. Effect of nitrogenous fertilizers on soil microbes

Pesakovic *et al.* (2012) showed that the use of high mineral fertiliser rates (800 kg ha^{-1} and 1000 kg ha^{-1}) induced a decrease in the number of *Azotobacter* and *oligonitrophyles*. On the other hand, the use of lower fertilizer rates (400 kg ha^{-1} and 600 kg ha^{-1}) enhanced the number of the studied group of microorganisms. Considering the soil biological properties, the N2 treatment (600 kg ha^{-1}) can be recommended as the most favourable one for 'Cacanska Rodna' growing in the stated agro ecological conditions at Serbia. Earlier Pesakovic *et al.* (2009) concluded that the high fertilizer rates used brought about a decrease in the total number of microorganisms as well as in the number of ammonifiers.

13.4. Integrated nutrient management

To know the response of plum to integrated nutrient management, a field experiment was conducted by Thakur and Thakur (2014) at, Solan (India) comprised of eight treatments having various combinations of inorganic fertilizers (urea, SSP and MOP), FYM, vermicompost, biofertilizers and green manures. Among all the treatments, 75% NPK + biofertilizers at 60 g each/tree basin +green manuring (Sunhemp @ 25 g seeds/tree basin) performed best where highest annual shoot growth, tree height, tree volume, fruit set, fruit yield (28.11 kg/tree), net income (Rs. 499.62) and benefit cost ratio (3.75) were observed.

13.5 Foliar feeding

Soil application of nutrients may not be likely under certain circumstances such as steep gradient of the soil, poor aeration, drought conditions, rootstocks, and nutrient interactions. Under such conditions foliar feeding has been reported to provide a better possibility to supplement the nutrient requirements. Foliar application of plant nutrients is helpful in satisfying plant requirements and can be highly efficient and attractive remedy especially in arid zones under low rainfall conditions where the lack of water in summer drastically restricts nutrient absorption by the tree (Inglese *et al.*, 2002). Potassium is adapted to foliar fertilization since when sprayed on leaves it is quickly translocated to other plant parts (Mengel, 2002). Mimoun *et al.*, (2009), obtained significant increase in fruit weight of 'Black Star' plum with the foliar application of 3% potassium sulphate sprays thrice during May and June. They also reported that foliar application of potassium is a significant method of fertilization especially during stage III of fruit growth.

13.6. Fertigation

Southwick *et al.* (1999) concluded from an experiment at Davis in California, USA that annual fertigation of 2- to 5-year-old plum cv. French trees with 0.11-0.23 kg N/tree is a sufficient amount of N to provide for good fruit yields, tree growth and orchard health, and minimize N leaching into groundwater. Nutrient application through fertigation has been found to increase the fruit yield and tree growth sharply in high density plum orchard (Rubauskis *et al*., 2003).

14. Irrigation

Plum is mostly grown in rainfed condition in India. The hilly regions of India where plum is grown receive more than 120-150 cm rainfall distributed during the growing season .The most critical periods, when dry spells prevail, are April to June. Watering to young plants in early periods of their establishment is highly essential to overcome water stress. Reports are available to suggest that the irrigation to bearing plum orchards could increase yield by more than 50 percent (Treder *et al*., 1999). It is recommended to apply irrigation at 50 percent field capacity. 2-3 irrigation applied during April and May were found to be highly useful. The irrigation can be done by furrow, portable pipe or basin flooding; drip irrigation is widely used because it requires less water per year than flood or furrow irrigation.

15. Fruit thinning

Demand for large fruit of uniform size is increasing in the market; thinning is a means to achieve consistently large fruit and to overcome possible alternate (biennial) bearing for the small-fruited European plum (*Prunus domestica* L.). However, chemical thinning agents for stone fruits are scarce and/or often ineffective. Seehuber *et al.* (2011) concluded that the most efficient method of flower removal (at full bloom stage) and fruit mass enlargement was mechanical blossom thinning with 400 rpm, which may provide a suitable replacement for chemical and/or manual thinning.

For highly productive varieties of plums, load regulation is generally necessary to meet the quality requirements for a table fruit. The manual thinning is very time consuming. The mechanical thinning with a string instrument 'Tree Darwin' can reduce the investment of time significantly, provided the trees have a narrow shape. The hand-held, battery-operated device Electro'flor can provide good thinning, fruit quality and a marked decrease workload compared to hand thinning. Electro'flor is compatible with different types of trees and its primary purpose is the application in small installations (Widmer *et al.,* 2013).

Hand thinning is simply pulling off flowers or fruits with fingers. Mechanical thinning is done with power shaker requiring considerable skill to prevent over thinning. Chemical thinning from full bloom to 4 weeks after petal fall is done. Various chemical are used for flower and fruit thinning in plum (Table 6). The advantage of chemical thinning over hand or mechanical thinning are reduced thinning costs, better fruit size and improved fruit quality Gonkiewicz and Nosal (2006).

Table 6: Flower and fruit thinning in plum by chemicals

Chemical	Concentration (ppm)	Time of application
NAA	10-15	Petal fall to pit hardening
NAAM	20-100	Petal fall to pit hardening
3-CPA	100-300	Full bloom to petal fall
CEPA	100-500	Petal fall to pit hardening
Paclobutrazol (PP333)	1000-2500	20-30 Days after full bloom

Khalil *et al.* (2005) evaluated the effect of chemical (ethephon at 50, 100 and 150 ppm) and hand thinning (10, 20 and 30%) on fruit retention, maturity and quality of plum cv. Santa Rosa in Shalimar, Jammu and Kashmir, India. Ethephon at 150 ppm and hand thinning at 30% improved the physico-chemical characteristics of plum fruits, and advanced fruit maturity. Hand thinning was labour intensive and only a small portion of the orchard may be thinned at an optimum time, while spraying with 150 ppm ethephon was efficient and cost effective.

Drkenda *et al.* (2006) observed that fruit thinning within a range of 0.5-1.0 kg/m branch had significant influence on the yield decrease in 2000 compared to the control variant. Manual thinning had a significant impact on the increase of fruit weight and sugar content; the drop in acid content was significant. No significant influence was noted on the fruit firmness. Considering the above interpreted research results, the optimal fruit load of 1.0-1.5 kg/m branch can be recommended for cultivar Elena in terms of yield and fruit quality.

16. Soil management and weed control

Soil management is primarily aimed at weed control but it is also concerned with two other objectives which are closely related to a) retaining good soil structure by preventing compaction and b) possible use of cover crops to improve fertility and to prevent erosion. Control of weeds in plum orchards is an important practice. Weed control may be accomplished by mechanical means (tillage, mowing etc.), use of herbicide, growing cover crops or combination of two or more methods. The conventional method of hand weeding is costly and at times injuries to the surface feeder roots of the fruit plants. Cover crops are not often used in non-irrigated planting because they compete with fruit plant for moisture. When irrigation is available, a permanent cover crop may be used to prevent soil erosion and provide base to run orchard equipment when the soil is wet. The cover crops namely sun hemp, cowpea, clover etc. can be used to enrich the soil with nitrogen and organic matter. With the development of selective and efficient herbicides, it enabled grasses and broad leaf weeds to be controlled. Glyphosate also effectively reduced weed density, as it acted both as contact as well as systemic herbicide. Diuron @ 2.4 kg/ha also significantly suppressed the weed count as compared to sod culture and manual weeding (Kaur and Kaundal, 2009).

The investigation on weed management and economics of various herbicidal treatments in plum cv. Satluj Purple was conducted by Kaur and Kaundal (2013). The results indicated that maximum fruit weight, pulp weight and fruit yield was obtained under followed by glyphosate @ 1.6 lha^{-1} as post-emergence and diuron @ 2.4 $kgha^{-1}$ as pre-emergence application. Based on the weed control efficiency, weed index values, black polythene mulch proved to be most effective. Considering net benefit from economic analysis, it appeared that the herbicide glyphosate @ 1.6 l/ha is the best possible option for effective and economic weed control in high density planted plum. Black polythene mulch is not at all cost-effective.

17. Intercropping

8 intercrop combinations, viz wheat (*Triticum aestivum* L. emend. Fiori & Paol.)-maize (*zea mays* L.) pea (*Pisum sativum* L. var arvense L.)-tomato (*Lycopersicon esculentum* Miller nom. cons), lentil (Lens culinaris Medikus)-blackgram, (*Phaseolus mungo* L.), lentil-maize, wheat-blackgrame clover (*Trifolium repens* L.) setaria (*Seteria anceps* Stapf ex Massey) and control, on growth, yield and fruit quality of 'Santa Rosa' plum (*Prunus salicina* Lindley) grafted on wild aprocot (*Prunus armeniaca* L.) rootstocks. the highest trunk girth (59.7 cm), shoot length (132.9 cm) and fruit yield (86.4 kg/tree) were observed with pea- tomato intercrop, following by lentil-blackgram. The lowest trunk girth (41.2 cm), shoot lenght (87.2cm) and fruit yield (41.2 kg/tree) were recorded with setaria grass. The treatment pea- tomatao also resulted in the highest fruit length (4.82 cm), breadth (4.93 cm) and fruit weight (67.7 g.). The lowest fruit length (4.37), breadth (4.22 cm) and fruit weight (49.3 g) were observed with setaria grass, However, the control resulted in the highest total soluble solids (13.8°B), reducing sugars (5.54%) and total sugars (6.75%)

18. Mulching

An experiment was conducted on eight year old trees of plum cv. Santa Rosa on seedling root stock of uniform size and vigour planted at a distance of 3x3 m receiving uniform cultural practices were selected for the experimentation. The plants were mulched in the month of March, 2012 with different types of mulching materials viz., transparent polyethylene, black polyethylene, pine needles, mowed grass, saw dust and chinar leaves besides control. The maximum annual shoot extension growth (62.50 cm), leaf area (44.30 cm2), fruit length (4.66 cm), fruit diameter (4.55cm), fruit firmness (2.48 kg cm-2), fruit weight (51.28 g), and yield (7.15 kg tree-1) were recorded in plants under black polyethylene mulch. Highest return bloom density (29.50%) was recorded from the trees under transparent polyethylene mulch treatments. Maximum fruit set and fruit retention were recorded from the trees under pine needles having 57.85 and 25.92 per cent respectively (Mehraj *et al*., 2014).

19. Use of plant growth regulators

Chilling injury (CI) is the primary factor limiting postharvest life during storage and transport of plum fruits at low temperatures. CI is usually manifested as dull coloration or breakdown and browning of internal tissues, loss of firmness and increased susceptibility to decay. The effect of preharvest 100 ppm GA3 sprays at three stages of fruit development combined with postharvest modified atmosphere packaging (MAP), intermittent warming (IW) on CI and fruit quality of plums were studied by Wang *et al.* (2013). CI index and browning index (BI) of plums were decreased ($p<0.05$) by preharvest GA3 treatments at three stages (a week before fruit expanding stage, fruit expanding stage, a week after fruit expanding stage). The optimum treatment was spraying 3 times at all three stages. CI index and BI in this treatment were 36 and 49%, respectively compared with the control 49 days after postharvest MAP with PE film (0.04 mm) and IW (15 days interval between 0 degrees C for 14 days and 20 degrees C for 1 day) storage. Moreover, the treatments also effectively maintained the firmness and soluble solids content (SSC), delayed the decrease of titratable acidity (TA) and ascorbic acid (ASA) content of the fruit.

The effects of foliar applications of growth regulators (NAA and triacontanol) and CaCl2 on the size and quality of Satluj Purple fruits were investigated by Mahajan and Sharma (1999). NAA at 25 and 50 ppm, triacontanol at 10 and 20 ppm, and CaCl2 at 0.5 and 1.0% were sprayed on 6-year-old trees at the pit-hardening stage of fruit development. Fruit size and weight increased significantly with both growth regulators. Triacontanol also increased TSS content. It was concluded that triacontanol at 10 ppm was the most effective treatment.

Chauhan *et al.* (2010) used Panchagavya for better plant growth, yield and fruit quality in plum cv. Santa Rosa. The highest shoot extension growth and leaf area was recorded with 4 per cent Panchagavya sprays. The fruit yield, fruit weight and size were significantly higher under Panchagavya spray treatment (4%). Fruit quality was measured in terms of acidity, TSS, Sugars, ascorbic acid and anthocyanin content was significantly improved by the 4 per cent Panchagavya spray followed by vermiwash spray (1:5 dilution). Lowest growth, yield and quality of fruit were observed in control plants.

Fruit drop is a major constraint in the cultivation of plum (*Prunus salicina* Lindl.) cv Satluj purple under the arid irrigated region of Punjab. Excessive fruit drop occurs during April. To check the fruit drop, the effect of growth regulators, viz., 2,4-dichlorophenoxycetic acid (10 and 20 ppm) and naphtyl acetic acid (20 and 30 ppm) were studied by Kaur *et al.* (2010) on eight year old plum trees. Fruit drop decreased significantly with the application of growth regulators over control and maximum fruit drop check (30.06%) was observed with 2,4-D (10 ppm). Fruit yield, fruit weight, fruit size and TSS/acid were recorded to be maximum with the foliar application of 2,4-D (10 ppm) during April.

The effects of prohexadione Ca (1.25 and 2.50 kg/ha) and paclobutrazol (0.2%) on the yield and quality of plum (cultivars Tuleu Gras and Stanley) were studied by Chitu *et al.* (2009) in Pitesti-Maracineni, Romania. The growth regulators were applied when the shoot length reached 8-10 cm and at 3 weeks thereafter. Foliar spraying with paclobutrazol reduced shoot length in both cultivars. Paclobutrazol resulted in the highest fruit yields and number of flower buds, and was the most effective in the enhancement of fruit quality.

Effect of pre-harvest application of putrescine on Japanese plum fruit ripening and quality was investigated by Khan and Singh (2010). The whole trees were sprayed with an aqueous solution containing putrescine (PUT) at 0, 0.1, 1.0 and 2.0 mM one week before anticipated commercial harvest for each cultivar separately. The PUT at 2.0 mM treatment reduced the respiration rate and ethylene production more than untreated fruit during fruit ripening at ambient temperature (20+or-1 degrees C). Activities of ACS and ACO enzymes, and ACC content were delayed and suppressed in all plum cultivars and these effects were more pronounced with higher PUT concentrations. PUT application reduced the fruit softening, soluble solids content, titratable acidity ratio, the levels of ascorbic acid, total carotenoids and total antioxidants in all plum cultivars. In conclusion, pre-harvest application of 2.0 mM PUT delayed the fruit ripening with reduced respiration rate, ethylene production, activities of ACS and ACO enzymes, ACC content, fruit softening, levels of ascorbic acid, total carotenoids and total antioxidants in Japanese plum.

Action of auxin application in accelerating fruit drop in plum was investigated by Gonkiewicz and Nosal (2006). The aim of auxin application in thinning of flowers and young fruits is to reduce the number of set fruits and to obtain a high quality fruit yield. Six-year-old plum trees (cultivars 'Stanley' and 'Iroquois') treated with a solution of potassium salt of alpha-naphthaleneacetic acid (NAA) after blooming were used in the experiment. The applied concentrations varied from 30 to 160 mg NAA l^{-1}. Microscopic preparations of flowers and young fruits were made using fluorescence and paraffin methods. Results showed a slight stimulating effect of NAA treatment on the fertilization process. It was also found that the applied auxin induced callose saturation and clogging up of conductive bundles and the chalazal part of the ovule. The inhibition of assimilate translocation to the ovule due to the clogging up of the chalazal part brought about an inhibition in endosperm development and resulted in embryo degeneration.

Thinning of plum trees promotes balance between vegetative growth and production, increases the fruit size and avoids alternation of production. Pavanello and Ayub (2014) evaluated the effect of different ethephon concentrations as chemical thinning of plum cv. 'Irati' and cv. 'Reubennel', in Arapoti-PR. Five treatments with ethephon were evaluated at dosages (0, 60, 80, 100, 120 micro L L^{-1}) in cv. 'Irati' and four treatments at dosages; 0; 60; 80; 120 micro L L^{-1} in cv. 'Reubennel', all applied when fruits diameter was 5 mm followed by measurements plant yield, weight, size, soluble solids, firmness, titratable acidity, pH and color of fruits. At a concentration of 120 micro L L^{-1} ethephon, cv. 'Irati' produced 20 kg $plant^{-1}$ and the cv. 'Reubennel' 56 kg $plant^{-1}$ presenting satisfactory quantity of fruit size III and IV. There was no significant difference between treatments for variable firmness, soluble solids, titratable acidity, pH and color of fruits evaluated in the post-harvest for the two plum cultivars.

ReTain plant growth regulator is a commercial formulation of aminoethoxyvinylglycine (AVG). AVG is known to competitively inhibit the activity of the enzyme ACC (1-aminocyclopropanecarboxylate) synthase which is the rate limiting enzyme in the ethylene biosynthesis pathway. By inhibiting the activity of ACC synthase, ethylene mediated ripening processes can be delayed. The postharvest benefits of inhibiting ethylene biosynthesis with AVG were studied by Jobling *et al.* (2003) for Tegan Blue plums, O'Henry and Desert Red peaches, Arctic Snow nectarines, Pink Lady apples and Cavendish bananas. In these studies AVG inhibited ethylene production and slowed ethylene-mediated ripening processes, particularly the loss of flesh firmness in Tegan Blue plums and Pink Lady apples.

'Black Splendor' (BS) and 'Royal Rosa' (RR) plums were treated preharvest with methyl jasmonate (MeJA) at three concentrations (0.5, 1.0 and 2.0 mM) along the on-tree fruit development: 63, 77 and 98 days after full blossom (DAFB). Both control and treated fruit were harvested at the commercial ripening stage and stored in two temperature conditions: 9 days at 20 degrees C or at 2 degrees C+1 day at 20 degrees C for 50 days. Preharvest MeJA at 2.0 mM significantly accelerated whereas 0.5 mM delayed the postharvest ripening process for both cultivars, since ethylene production, respiration rate and softening were reduced significantly at the two storage conditions for 0.5 mM. In these fruit, total phenolics, total antioxidant activity (hydrophilic fraction, HTAA) and the antioxidant enzymes peroxidase (POD), catalase (CAT) and ascorbate peroxidase (APX) were found at higher levels in treated than control plums during postharvest storage, which could account for the delay of the postharvest ripening process and the extension of shelf-life (Zapata *et al.*, 2014).

20. Diseases and their management

20.1. Bacterial Canker or Bacterial Gummosis (*Pseudomonas syringae or P.morsprunorum*)

The oozing of gum (Gummosis) on bark or outer sapwood and fruits is common, but is not always found with all cultivars and species. Disease development is rapid during the cool, wet period of early spring. Also, frosty weather during bloom and early growth favour the incidence of bacterial canker.

In the Netherlands, bacterial canker of plum trees (*Prunus domestica*) caused by *Pseudomonas syringae* or *Pathovars syringae* and *morsprunorum* is a recent and serious problem. The trunks of the affected plum trees are girdled by cankers resulting in relatively sudden death of the trees 1 to 4 years after planting (Wenneker *at al.,* 2012).

The disease may be controlled by cleaning the weeping wounds and smear with Mashobra paste during dormancy break. Repeat the same process on the new lesions in the following months or before the rainy season. Before rainy season, spray streptocyclin (3 g/ 10 litres water). After leaf fall, spray of copper fungicides such as copper oxychloride (0.3%) is effective.

20.2. Peach leaf curl (*Taphrina deformans*)

Affected young leaves develop curling, puckering and distortion turning to pink or red bronze colouration. It can be controlled by spraying of Ziram or Dithane Z-78 or Captan at 200 g per 100 litres water at bud swell stage.

20.3. Crown gall (*Agrobacterium tumefaciens*)

Crown gall organisms are widely distributed in the temperate fruit areas. The organisms live in the roots of many plants and in the soil. Infection takes place through wounds. Galls stimulated by the presence of bacteria result from hyperplasia and hypertrophy of cells. Control is achieved by using gall-free nursery stock since nothing can be done after the orchard is planted. Wherever possible, gall-resistant rootstocks should be used to avoid later infection.

20.4. Oak root fungus (*Armillarella mellea*)

The fungus infects the roots and lower parts of the trunk. Infected trees show progressive yellowing and wilting of the foliage and stunted growth because of trunk girdling and death of the root system. Control is achieved by soil fumigation in autumn after fruit harvest with carbon bi-sulfide. The fumigant should not be applied closer than 2-2.7 m from any living trees, as the material is toxic to living roots.

20.5. Brown rot (*Monilina fructicola*)

In brown rot, both blossoms and fruits are attacked. First symptoms are browning and death of blossoms, which persist on the trees. On fruit, it begins as a tiny brown spot and rapidly enlarges to a large soft watery, brown area. Within a few days the entire fruit is infected. The disease extends from them into shoots, causing wilting and cankering on the young branches. May-de-Mio *et al.* (2008) reported that mineral nutrition of host plants is known to influence the severity of numerous plant diseases. They noted that incidence of brown rot was higher when K was at 110 kg ha^{-1} $year^{-1}$ and when N was applied at 160 and 200 kg ha^{-1} $year^{-1}$ compared to the other levels tested.

The disease can be controlled by spraying trees with Difolatan (0.15%) or Captan (0.3%) about 3 weeks before harvest. Diseased and mummified fruits on trees as well as from the orchard floor should be collected and brunt.

20.6. White root rot *(Rosellinia necatrix)*

This generally causes weakening of leaves and twig. Infected roots become brown to very dark brown colour with the appearance of white fungus on affected roots during rainy season. The fibrous roots are completely devoured.

Drainage of orchard has to be improved and remove infected roots. Cut ends and healthy roots should be applied with Chaubattia paint (Red lead, copper carbonate and linseed oil in ratio 1:1:1/4) or Brassicol paint (10 g/ litre linseed oil) during November or December. Affected plants should be drenched 3 times at an interval of 15-20 days during monsoon with Bavistin (10 g/litre water) to be applied in holes 15-22 cm deep at a distance of 30 cm each throughout the drip of a tree.

20.7. Plum pox virus

Jakab *et al.* (2011) used Transmission electron microscopy (TEM) in order to detect the presence of Plum pox virus (PPV), Prunus necrotic ringspot virus (PNRSV) and Prune dwarf virus (PDF) particles in plum (*Prunus domestica* L.) shoots and leaves. The presence of virus particles was confirmed by DAS-ELISA serological technique. Filamentous particles approximately 750 nm long were found in ultrathin sections of leaves, shoots in plum. Spheric-concentrical particles were found in 'Renclode Althan' cultivars infected with Prunus necrotic ringspot virus. The TEM aided morphological analysis and measurements together with DAS ELISA serological technique confirmed the presence of Prunus necrotic ringspot virus and Plum dwarf virus. No virus particles were found in ultrathin sections of growth tips in the analyzed plum cultivars of 0.2-0,4 mm diameter. Meristematic tissues does not contained the three viruses, confirming the fact that due to the high division rates at the apical dome, the multiplication of the virus is suppressed and do not reaches the central zone of the meristem (CZ). Embryonic cells of growth tips of plum up to 0,1 mm in diameter are PPV, PNRSV, PDV free in the analyzed blocks. Therefore it highly recommended that the meristem diameter for virus elimination should be 0.1-0.2 mm in order to achieve potential virus free plants.

21. Pests and their management

21.1. Peach leaf curl (*Brachycaudus helichrysi or Anuraphis padi*)

Leaf develops characteristic curling due to sucking of sap from growing buds. This results weak flower buds, poor setting and premature fruit drop. Maximum injury is caused at pre bloom and blooming stage.

It can be controlled by spray with dimethaote (500 ml Roger 300 E.C.) or demeton (500 ml metacystox 25 E.C.) or fenitrothion (500 ml acethio/ folithio/ sumithion 50 E.C.) in 500 litres of water @ 4-6 litres per tree about 7-10 days before flowering at pink bud stage with fine nozzle.

21.2. San Jose scale (*Quadraspidiotus perniciosus*)

San Jose scale is a serious pest of fruits including plum and is distributed throughout the fruit-growing areas of the world. The adults, which feed on sap from limbs, twigs, and fruits, are small, circular, and gray. Dead leaves adhering to fruit spurs during the dormant season indicate the presence of San Jose scale.

Effective control is usually obtained by spraying 10 litres miscible oil (Hindustan Petroleum spray 2% oil E.) in 490 litres of water or diesel (light or high speed) oil emulsion 8 times with the water to be sprayed @ 4-6 litres per tree during dormant season at the end of February. Universal oil should be sprayed (IL 50 L water) during the spring. Alternatively spray with suitable insecticides such as chlorpyriphos (0.4%), fentitrothion (0.05%) dimethoate (0.03%), or phosphamidon (0.03%) to kill the crawlers during May before their hatching.

21.3. Stem borer (*Anarsia Lineatella*)

Stem borers are known to cause damage by boring into shoots or stem or roots and attack the fruits and twig terminals which weaken the plant and in severe cases death of plant may occur. To control this pest, the hole is cleared with flexible wire and then a cotton wick soaked in petrol, kerosine or methyl parathion (0.2%) is inserted and the hole is plugged with mud. In case of fruit infestation, 0.05% endosulfan or fenitrothion is sprayed.

21.4. Blossom thrips

Very small, slender insect of black, pale brown or yellow colour which suck the sap within flower bud causing weak and poor setting. Flower being heavily infested fail to open given dull brownish in appearance and fail to set fruit. They also suck the sap from tender leaves at growing point causing curling and distorted leaf growth. Pre-bloom spray of insecticides as in leaf curling aphid is effective for thrips also.

21.5. Plum scale (*Eulecanium* sp.)

The insect thrives mostly on new growth in the form of small spherical brown nodule like outgrowth. Adult females suck sap from tender shoots while the nymphs suck sap from the leaves. Heavy attack causes poor fruit set, small fruit size and reduced tree vigour. It has only one generation in a year.

Dormant oil spray as in case of San jose scale can also be used for controlling the pest. If dormant oil spray could not be done then spraying of monocrotophos (500 ml azodrin/ nuvacron 40 EC), quinalphos (1litre Ekaluxx 25 EC), fenitrithion (500 ml acethion/ folithion/ sumithio 500 EC) or methyl demeton (500 ml metacystox) in 500 litres of water as post bloom spray in May should be done. A waiting period of 3 weeks should be allowed before the fruits are plucked when trees are sprayed with monocrotophos and methyl demeton and for others two weeks period should be allowed.

21.6. Plum fruit moth (*Laspeyresia pamonella*, L. *funebrana*)

The plum fruit moth larvae enter both immature and mature fruit and tunnel deeply into the fruit tissue. The entrance is usually surrounded by grass and droplets of gum. For effective control, carbaryl (0.1%) is applied about a month before the anticipated date of harvest.

21.7. Plum sawflies (*Haplocampa minuta*, H. *flava*): Plum sawflies lay their eggs on flower sepals. The white grubs that emerge damage the young fruits by boring holes into them and feeding on the fruit tissue and kernel. For effective control, spraying with insecticides based on carbaryl, demeton, or dimethoate at just after petal fall is suggested.

22. Physiological disorder

22.1. Internal browning, chilling injury, dry fruit, mealiness, woolliness

This disorder is common in Plum and fresh prune fruits and is considered as number one problem in shipping stone fruit, and the most frequent complaint made by consumers and wholesalers.

22.1.1. Symptoms

Flesh browning, flesh mealiness, black pit cavity, flesh translucency, red pigment accumulation (bleeding), and low flavour.

22.1.2. Causes

These symptoms normally appear after placing fruit at room temperature, while some ripening is occurring, following cold storage. For this reason, this problem is usually experienced by the consumer, not the grower and/or packer.

Stone fruit cultivars vary greatly in susceptibility to internal breakdown injury. Some of them show no apparent susceptibility when grown under California climatic conditions. Among peaches and nectarines early season cultivars are least susceptible and late-season cultivars are most susceptible. Among plum cultivars there is no seasonal pattern of susceptibility.

Even under the best storage and handling conditions, stone fruit have a limited market life. Market life means the maximum number of weeks for each cultivar under continuous exposure to 32°F (0°C) and 90% RH (where the fruit can be stored until at least 85% remain marketable). These ideal conditions do not usually occur during the storage, transportation and handling at the retail end, the maximum potential postharvest life of stone fruit cultivars is never reached and therefore is normally shorter.

22.1.3. Control

The ideal way to eliminate this problem would be to produce cultivars resistant to it. In the meantime, temperature management is the best commercial tool available to delay the onset of internal breakdown. Storage below 0°C (32°F) but above the freezing point is beneficial to delay chilling injury symptoms and extend market life.

Use of controlled atmosphere (CA) conditions in combination with temperatures close to 32°F has been beneficial in extending the postharvest life of plum, nectarine and peach cultivars. CA treatment reliability can be attained by understanding the role of fruit size, canopy position, cultivar market life and shipping period.

1. Avoid cultivars susceptible to internal breakdown (I.B.)
2. Market susceptible cultivars according to their potential postharvest life.
3. Segregate fruit according to their postharvest potential life.
4. Pick fruit "well matured."
5. Enforce proper postharvest handling during transport and at the retailer.
6. Keep fruit near 32°F during storage and transportation.
7. Avoid 36-46°F temperatures during retail handling.
8. Educate warehouse and retail managers on I.B.

23. Harvesting

Plums generally comes to bearing 3-5 years after planting and remains in productive life upto 25-35 years depending on the care and management practices. It comes to full bearing stage around 7-9 years after planting. European plum comes to bearing about 2 years after Japanese plum. Optimum maturity standards are dependent on the cultivar and the intended use of the fruit chiefly whether fruits are to be consumed in the local or distant markets, canned or dried. Those to be consumed fresh locally can be harvested more mature than those to be sent to distant market. For long distant markets, fruits are harvested at less mature stage when 50% colouration of fruit appears but remain firm. For canning, the fruits should be ripe and firm with maximum sugar content.

Mika *et al.* (2012) reported that the self propelled straddle combine harvester was able to harvest 2-3 tons of plums per hour compared to 30 kg with hand picking where plum trees are and trained to central leader spindle tree. Harvesting effectiveness was 90-95%. The quality of mechanically harvested plums was a little worse than of those hand picked, but fruits were suitable for processing. The 'Common Prune' and the prune type small plums were the most suitable fruit for mechanical harvesting.

24. The maturity indices used for plums and prunes include

24.1. Colour development: Every variety of plum has its distinctive skin colour at maturity and when the full colour develops, it is considered ready for harvest.

24.2. Flesh colour: It is changes from yellow-green to amber and appears to be good visual measure of maturity. In oriental plums, colour varies from light green to yellow, red blushed or solid red. Beauty having 85% surface yellowish green or trace of red and Santa Rosa with 40% of surface red colour or full light greenish yellow are ready for harvest. Kelsey cultivar is harvested when surface develops a light green colour with the traces of red colour.

24.3. Flesh firmness: It is also used to determine harvest maturity. Depending on the variety, firmness of the fruit (per 0.8 cm diameter) at which they are harvested for distant market without pre-cooling ranges from 4.1-9.1 kg. Firmness in Beauty is 5.0±0.4 kg

24.4. Days after full bloom: For maturity of fruits of a variety, it is more or less fixed for a particular zone. It is also a reliable index for harvesting (Table 7).

24.5. Total soluble solids: It should reach the peak value at maturity. TSS while related to maturity but fluctuates with climate and crop load. Attainment of threshold value of 12.5 % TSS in European varieties has been reported to be a good standard for maturity and in Italian prune 17% is the minimum maturity level for acceptable quality.

24.6. Soluble solid/ acid ratio: It is also used as a standard for fixing maturity. The soluble solid/ acid ratio between 12-15 is considered satisfactory at maturity.

24.7. Maturity time: The optimum maturity varies from cultivar to cultivar. For Santa Rosa plums, it was calculated to be 94±3 days from full bloom .

Plums usually ripen unevenly over the tree so fruits should be harvested in two or three pickings. Plums which are picked for canning are usually harvested at one time. They are picked by hand into baskets with padded liners or into picking bags. Immediately after harvest fruits should be kept preferably under shade of the tree to prevent direct exposure of fruits to sun and care should done while handling the fruits. For distant market, fruits may be harvested early in the day with immediate transfer to the packing shed or picking can be done in the later afternoon and left in the orchards to cool overnight.

Table 7: Plum cultivars with bloom time and fruit maturity

Cultivar	Time of full bloom	Time of fruit maturity
Red Beaut	2nd-3rd wk of March	1st wk of June
Mathley	2nd-3rd wk of March	2nd wk of June
Burmosa	2nd-3rd wk of March	2nd wk of June
Santa Rosa	3rd wk of March	3rd wk of June
Frontier	2nd wk of March	2nd wk of July
Mariposa	2nd wk of March	2nd wk of July
Stanley (prune)	1st wk of April	Mid August

25. Yield

Yield will greatly depend on the age of tree as well as the management practices given. Under Indian condition, an average yield of 40-60 kg of fruits per tree is observed. The average yield of plum in India is quit less around 60-70 quintal per acre in comparison to that of Europe where the average yield ranges from 120-150 quintal per acre. High density orchards may produce 16 tonnes or more per acre.

26. Grading and packaging

Grading of fruits into different sizes is an important aspect and it forms an integral part of better marketing. Fruits picked at appropriate stage of maturity combined with size grading command better market price than ungraded fruits since it creates appeal to the consumer who is prepared to pay more for better sized fruits. Three standard or grades are recommended under 'Ag-mark' system of grading plum and box sizes used for marketing are given as under Table 8.

Table 8: Grading and packaging standards for plum

Grade	Fruit size dia (mm)	Box size inner (inches)	No. of layers	No. of fruits per layer
Special	42 and above	14.5 x6.5 x 6.5	3	28-32
Grade 1	36-42	14.5 x 6.5 x 6.5	4	38-42
Grade 11	Below 36	14.5 x 6.5 x 6.5	4	50-56

During handling the fruits for packing, care should be taken so that the fruits are subjected to less handling friction in order to conserve the bloom which creates appeal to consumers for better marketing. The interior of wooden boxes all round is given wrapping with news paper so as to prevent the fruits from coming in contact with the rough surface of the box which may cause injury to the fruits during transit. The bottom of the box is spread with a layer of pine needles or other available soft packing materials. The fruits are placed in layers. After each layer is completed, it is covered with a sheet of paper and a layer of pine needles which act as cushion for next fruit layer is used. After the fruits are packed in the box, the top layer is finally covered with a sheet of paper and pine needles and the lid of box is nailed. The use of fibre box is also increasing day by day for tight- fill and tray packs. Plastic film is most effective in board cartons parchment wraps and other specially treated packing materials to retard moisture loss significantly. Plums are packed in shallow crates about 4-5 inch deep packing not more than 3 layers of fruits.

After packing, the fruit should be cooled immediately to 0°C which would stop the ripening process in plum for a period of about 12 days and then for distant markets, transported in refrigerated vans or trucks or trains but for local markets, ordinary vans or trucks are used.

27. Storage

Plum is perishable fruit and its shelf life is short. Storability depends much on the stage of maturity of fruits. Picking at the correct stage of maturity is one of the best ways of ensuring optimum storage. To extend the storage life of plum fruits, its temperature must be lowered as soon after harvest as possible. Pre-cooling which removes the heat from the fruit should be done immediately after harvesting, packaging and palletizing. The ideal storage condition for plum is a temperature of -0.6 °C to 0°C and relative humidity of 85-90%. The storage period depends on the cultivar. Most cultivars cannot be stored for longer than 3-4 weeks but other can be stored for upto 3-4 months. Some varieties exhibit chilling injury if stored at 0°C for a long period. To prevent chilling injury, the temperature is raised from -0.6°C to about 7°C after 7-10 days and the temperature is maintained for the rest of the storage period.

The shelf life of the fruits can be further extended with the use of better storage technique namely controlled atmosphere (CA) storage. In CA storage, the storage is airtight and in the technique not only the fruits are cooled to reduce respiration, but the oxygen (O_2) and carbon dioxide (CO_2) gas mixtures in the store are carefully regulated. In CA storage, CO_2 at 2-8% and O_2 at 2-3% are maintained and fruits could be retained for duration of 2-3 months.

'Akihime' plums picked at mature stage (50-70% red coloration of the fruit skin) and CA (controlled atmosphere) storage regime of 3% O2+5% CO2 was found the most effective for maintaining overall fruit qualities and also delayed the development of chilling injury including lack of juiciness and flesh browning Lee *et al.* (2013).

Mahajan *et al.* (2008) reported that shelf life of fruits of Satluj Purple cultivar of plum can be extend up to 28 days by dipping the fruits in 2% $Ca(NO_3)$ for five minutes followed by air dried to remove the surface moisture and packed in corrugated fibre board carton and stored at 0-1^{0}C and 90-95% RH. They noted only 2% spoilage in the treated fruits against 15% in control after 28 days of storage.

Plum has a very short storage life. The role of pre- or post-harvest applications of putrescine (PUT) and low temperature storage on fruit ripening and quality was investigated by Khan *et al.* (2008) in plum fruit (*Prunus salicina* Lindl. cv. Angelino. Pre-harvest application of 2.0 mmol L^{-1} PUT 1 week before the anticipated commercial harvest was more effective in delaying plum fruit ripening and can be used to extend the storage (0+or-1 degrees C) life of plums for up to 6 weeks with minimum losses in fruit quality.

'Black Amber' plums were harvested at the commercial stage and treated at the same date of harvest with doses of 0.5 and 1 micro l L^{-1} 1-methylcyclopropene. Fruits maintained at the same conditions without any treatment were used as Control. Fruits were stored at 1+or-0.5 degrees C and 90% RH during 30 days of storage. Effects of 1-MCP on fruit firmness (FF: kg), weight losses (WL: %), titratable acidity (TA: %), and total soluble solids (TSS: %), were determined during 15 and 30 days of storage. The firmness maintained better than Control in both doses of 1-MCP applications, overall quality parameters were enhanced by 1-MCP treatments (Ozkaya *et al.*, 2010)

Sharma *et al.* (2009) suggested that the fruits of Santa Rosa plum should be harvested with pedicel and to be stored under zero energy cool chamber in the area where facility of cold storage is beyond the reach of marginal farmers. By this way fruits can be stored up to 9 days.

28. Value addition

28.1. Plum beverages

Novel yeast, *Candida colliculosa* was isolated from Iranian white cheese whey and its desired property of natural carbonation was exploited to produce a non-alcoholic naturally carbonated plum beverage. The fruit beverages were prepared using two plum varieties; yellow Golden Drop (*Prunus domestica* ssp. Syriaca/Mirabelle) and red Ghermez (*Prunus divaricata* var caspica). The prepared beverages had acceptable sensory quality, high carbonation (~1.6 bar) and physico-chemical properties at the end of 90th day of storage under refrigerated condition (Jooyandeh, 2013)

28.2. Anthocyanin pigment

Anthocyanins are responsible for the colour in plum fruits. The waste part of the plum after processing is thrown out by the processing industries which contain sufficient quantity of anthocyanin pigment which can be further processed for extraction of anthocyanins. Devi and Joshi (2012) standardized the extraction and concentration of anthocyanin from plum var. Santa Rosa. The treatment of plum extract with pectin esterase produced a clear extract without any gel formation. Four different extraction methods involving Ethanol (50% and 100%) and Citric acid (0.1% and 0.2%) with two levels of concentration of anthocyanin extracts (8:1 and 10:1) were tried. The best result was obtained with 50% Ethanol+0.2% Citric acid with 10:1 concentration ratio in terms of maximum anthocyanin content (325 mg/100 ml) and highest sensory evaluation (hedonic) score (8). Extraction of anthocyanin reflects the potential of plum waste to serve as a source biocolours.

28.3. Production of yeast

The production of organic acids from several yeasts isolated from mature plums on media containing glycerol as carbon source was analysed by HPLC-UV. The yeasts isolated were identified by sequencing the 5.8S internal transcribed spacer as *Pichia fermentans*, *Wickerhamomyces anomalus* and *Candida oleophila*. The organic acid profiles of these strains comprise acetic, citric, succinic and malic acids that qualitatively and quantitatively vary between different species as well as among strains from the same species. The production from glycerol of succinic, acetic, citric, malic and oxalic acids from *C. oleophila* and *W. anomalus*, and that of succinic, oxalic and acetic acids by *P. fermentans* is reported for the first time in this work, as is the production of oxalic acid from glycerol in yeasts (Garcia *et al.*, 2013). Results showed that mature plum fruits can be a good source of new yeasts able to metabolise glycerol, producing different organic acids with industrial and biotechnological interest.

29. Economic of cultivation

Plum is a labour-intensive crop. Labour costs are the major cost factor. It is thus important to analyze labour input and to determine which cultivars and production forms are the most efficient. Recording of labour input has shown 'Jubileum' to have the lowest labour requirement. Based on project data, as well as data from other sources, the cost of producing 1 kg of 'Jubileum', the cheapest to produce, was NOK 14.63, while 'Victoria', the most expensive to produce, was NOK 19.18 per kg. Labour costs significantly affected the gross margin and internal rate of return. At a yield of 7 t/ha and a plum price of NOK 15 per kg, the internal rate of return would be 14.4% at an hour wage of NOK 55, and decrease to 5.2% at an hourly wage of NOK 100. The gross margin per hour is somewhat lower in growing than for other fruit crops. The gross margin per hectare, however, is about the same for plums as for apples (Knutsen and Haukas, 2007).

30. Effect of climate change

In 1976-1999 an investigation was conducted by Grabowsky and Zielenkiewicz (2000) to determine the effect of 19 meteorological factors (the temperature values and sums, precipitation sums and sunshine duration) during the growing season on the yield of 4 plum cultivars growing near Olsztyn. The effect of late spring frosts on flowering and fruit bud formation as well as conditions of flower pollination by bees was studied. It was found that the weather pattern, mainly in April and May and during flowering influenced the yield of plums. Of the factors analysed at flowering late spring frosts, precipitation, rainy days and low noon temperatures had most negative effects on percentage fruit set and yield. The highest yields (more than 15 t/ha) were noted in 1976, 1977, 1980, 1983, 1985 and 1993 and the lowest (below 5 t/ha) in 1981, 1984, 1987 and 1990. Low yields were caused by unfavourable meteorological conditions during flowering. It was found that the greatest reduction in yields was due to late spring frosts. A single frost reduced the yield of plums by 2.46 t/ha.

The correlations between yield and climatic factors and plum fruit quality were studied by Vangdal *et al.* (2007) in 11 plum cultivars in Norway. A negative correlation was found between the yield and soluble solids content in 8 out of 11 studied cultivars. The titratable acidity contents were not influenced by yield. Up to 44% of the variation in soluble solids content was related to the variations in yield. Including climatic factors in the regression equations, 60-94% of the variation in soluble solids content was explained by variations in the yield, monthly mean temperature and precipitation. The soluble solids content was positively influenced by high temperatures in April and June and cool and wet weather in May. In years with high precipitation in August, the titratable acidity of the plums was high.

Cosmulescu *et al.* (2010) studied study the phenological changes in plum tree species (*Prunus domestica* L.) that recently appeared in Oltenia region (Romania). It was found that in the years with early spring season and high temperatures, vegetation phenological phases have developed much earlier than in normal years. Analysis of data obtained showed that “flowering time duration” is a feature which is influenced by climatic factor and genetic factor. The influence of meteorological factor is manifested in different years, thus causing, in the same cultivar, different time durations between the beginning and the end of flowering. In general, the earlier the flowering is developing, the shorter its time duration.

Dugalic *et al.* (2014) evaluated changes of sorbitol, glucose, fructose, sucrose, and total sugars of plum fruits in different years under different climatic conditions i.e. air temperature and precipitation. The cultivars investigated in the research were: Topstar, Toptaste, Jojo, Haganta, Tophit, and Top 2000. The fruits were harvested at optimal stage of maturity in the experimental orchard of the Agricultural Institute Osijek, Croatia. Content of glucose, fructose, sucrose, and sorbitol in the fruits were identified and quantified by high performance liquid chromatography (HPLC). Significant differences were found between sucrose, glucose, fructose, and sorbitol depending on climatic conditions and cultivar. Sucrose was the highest in almost all years. Glucose was the predominant sugar in cultivars Jojo and Top 2000. Fructose and sorbitol contents were lower than sucrose and glucose. Average sugar contents in 2008-2012 periods were compared with respect to climatic conditions. A significant effect of year was found for sucrose, sorbitol, total sugar, and dry matter, whereas no effect was found for glucose and fructose contents. Individual sugar contents correlated significantly with each other. The principal component analysis (PCA) showed that plum cultivars were clearly differentiated according to variability of sugar in fruit caused by climatic conditions. The obtained results highlighted that climatic conditions may have significant effect on plum fruit quality and, therefore, it is important to have adequate testing before recommending a cultivar for planting.

31. Technology generated for organic plum culture

31.1. Suitable cultivars

Due to environmental reasons, the importance of organic farming in Estonia is growing. Kahu and Klaas (2007) studied performance of 5-cultivars suitable for organic cultivation in Estonia. No pesticides and fertilizers were used in the experiment. The best average yield per tree was found in 'Perdrigon', The relatively resistant to the leaf spot were cultivars 'Ave', 'Liisu' and 'Queen Victoria'.

Glowacka and Rozpara (2014) recommended the cultivar 'Herman' for central Poland for organic production of plum. The cultivar was characterized by an early fruit ripening (came into bearing fruit in the third year after planting). The largest amounts of fruit were harvested from the trees of the cultivar 'Herman' as compared to other cultivars like 'Cacanska Rana', 'Valjevka' and 'Yellow Afaska', grafted on Myrobalan seedlings.

31.2. Nutrient application

The effect of green manure with winter forage peas and a peas-rye mixture in a plum orchard was studied by Vitanova *et al.* (2009) The green manure influenced positively the humus content, a very important element of the soil fertility. The quantity of the mobile forms of nitrogen, phosphorus and potassium in the soil of experimental plots with green manure was optimized in favor of the plum trees development. Their yield, without mineral fertilizers, was large - 15.6 t/ha for the variant with winter forage peas and 16.2 t/ha for the variant with the peas-rye mixture. The content of sugars, dry matter and organic acids in the fruits was optimal. The green manure with winter forage peas and the peas-rye mixture is very suitable for biological plum production.

31.3. Control of insect pest

Rizzo *et al.* (2012) conducted in an organic plum orchard in Sicily to evaluate the damage caused by the plum moth *Grapholita funebrana* Treitschke; to three plum cultivars (Angeleno, Friar, and Stanley) and to evaluate plum moth control provided by mineral oil applied every 20 days (Moil), spinosad applied every 10 days (S10), or spinosad and mineral oil applied alternately (S/Moil). Treatment application began when eggs were first observed on fruits. S10 and S/Moil effectively reduced plum moth damage on all cultivars but Moil did not. Plum moth infestation was significantly lower on Stanley than on Friar and Angeleno.

32. Future research thrust

Highly perishable nature, delicate skin, chilling sensitivity, and other postharvest physiological disorders, and the consistent supply of high quality fresh and safe fruit to the consumer are great challenges to postharvest physiologists. Development of non-destructive maturity assessment, quality evaluation methods for plum fruit warrants investigation. Molecular and conventional breeding approaches to regulating plum fruit ripening and improving their shelf-life have not resulted in substantial success. Determining whether ethylene is directly involved in promoting pigment accumulation or is triggered by another effector needs further investigation. Various factors that may affect the development of postharvest oxidative stress such as fruit ripening and storage disorders in plums (including genotype, harvest maturity, storage temperature, storage period and atmosphere composition) warrant further investigation (Singh and Khan, 2010).

References

Anon. 2016. Statistics and Market Information. National Horticulture Board, India.

Aulakh, P S. 1999. Evaluation of plum (*Prunus salicina* Lindl.) cultivars under arid irrigated region of Punjab. *Journal of applied horticulture,* 1(1):48-50.

Azad, K.C. and Sharma, R.P. 1987. Fertilizing Fruit crops in Himachal Pradesh. Dept. of Horticulture, Shimla, India, pp. 112.

Azar, E. V; Heravan, E. M; Azghandi, A. V; Dejampour, J. 2010. Establishment and proliferation of two apricot x plum hybrid rootstocks for in vitro culture. *Seed and Plant Production Journal,* **25** 2(4):465-470.

Bajwa, G S, Bindra, A. S., Bal, J. S. and Minhas, P. P. S. 1991. Problems of pollination and fertilization in plum. *Acta Horticulture*, 283:157-162.

Bajwa, M.S. and Mishra, K.A. 1970. Manuring the plum orchards. *Punjab Hort. J.* **10**: 19.

Balal, R. M , Gimeno, V,. Shahid M, Lidon A. V., Garcia, A. L., Abbas, T., Sanchez, F. G. and Ghazanfer, U. 2011. Effects of phosphorus fertilization on growth, leaf mineral concentration and xylem-phloem nutrient mobility in two rootstocks of prunus (*Prunus persica* × *Prunus amygdalus*) and (*Prunus insititia*) in the Mediterranean area. *AJCS*, **5**(12):1542-1549.

Barbara, N. M. Kazimierz and Ludwik H. 2004. Sugar uptake and utilisation during adventitious bud differentiation on in vitro leaf of 'Wegierka Zwykla' plum (*Prunus domestica*). *Plant Cell Tiss. Org. Cult.,* **76**:255-260.

Barroso J.1998. Evaluation of different rootstocks for 'Greengage' plum. *Acta Horticulturae*, **478:** 331-339.

Barua S C. 1998. Productivity of Santa Rosa plum trees in response to triacontanol. *Journal-of-Interacademicia*. 2(3): 124-129.

Bhutani, V.P. and Joshi, V.K. 1995. Plum. *In*: Handbook of fruit science and Technology, by Salunkhe, D.K. and Kadam, S.S. (eds.). CRC press, pp. 203-242.

Botu, I; Preda, S; Turcu, E; Achim, G. and Botu, M. 2007. Rival - a new rootstock for plum. *Acta Horticulturae,* **732**: 253-256.

Brown, P.H. and Shelp, B.J. 1997. Boron mobility in plants. *Plant and soil.* **193**: 85-101.

Chatzitheodorou, I.T.; Sotiropoulos, T.E. and Mouhtaridou, G.I. 2004. Effect of nitrogen, phosphorus, potassium fertilization and manure on fruit yield and quality of the peach cultivars 'Spring Time' and 'Red Haven'. *Agronomy Research.* **2**(2): 135-143.

Chauhan Archna ; Joolka, N. K. and Sharma, S. D. 2010. Effect of biostimulants sprays on growth, yield and physico-chemical characteristics of plum cv. Santa Rosa. *Horticultural Journal,* **23**(1): 6-11.

Chitu, V; Butac, M; Chitu, E. and Nicolae, S. 2009. Effect of growth retardants treatment on plums yield and fruit quality. *Lucrari Stiintifice Universitatea de Stiinte Agronomice si Medicina Veterinara Bucuresti Seria B, Horticultura,* **53** : 433-436.

Chung, K. H., Jihyal, J. and Sangjo, K. 1998 .Selection of suitable pollinizers for major Japanese plums. *Journal of the Korean Society for Horticultural Science,* **39**(5):560:563.

Cosmulescu, S; Baciu, A; Cichi, M. and Gruia, M. 2010. The effect of climate changes on phenological phases in plum tree (*Prunus domestica* L.) in south-western Romania. *South Western Journal of Horticulture Biology and Environment*, **1**(1): 9-20.

Costa, G. and Grandi, M. 1982.Contribution to the knowledge of pollination requirements of new plum cultivars. *Fruitticoltura,* 44 (12): 74-76.

Cuquel, F.L.; Motta, A.C.V.; Tutida, I. and DeMio, L.L.M. 2011. Nitrogen and potassium fertilization affecting the plum postharvest quality. *Rev. Bras. Frutic.* **33**(sp1): 328-336.

Daane, K.M.; Johnson, R.S.; Michailides, T.J.; Crisosto, C.H.; Dlott, J.W.; Ramirez, H.T.; Yokota, G.T. and Morgan, D.P. 1995. Excess nitrogen susceptibility to raises nectarine disease and insects. *California Agriculture*. **49**(4): 13-17.

Devi, M.P. and Joshi, V. K. 2012. Effect of different extraction methods and concentration of extracts on yield and quality of anthocyanin from plum var. Santa rosa. *Journal of Crop and Weed,* **8**(2): 7-11.

Dhatt A S, Chanana Y R, Minhas P PS and Bindra A S .1992.Table plum 'Fla. 1-2'. *Indian Horticulture*, **37**(2): 41.

Drkenda P, Brtschinger L, Stadler W and Schwitzer T. 2006. Manual fruit thinning at "Elena" plum cultivar for the purpose of fruit quality regulation. *Radovi Poljoprivrednog Fakulteta Univerziteta u Sarajevu Works of the Faculty of Agriculture University of Sarajevo,* **57**(2) : 45-52.

Dugalic, K; Sudar, R; Viljevac, M. and Cupic, T. 2014. Sorbitol and sugar composition in plum fruits influenced by climatic conditions. *Journal of Agricultural Science and Technology*, **16**(5): 1145-1155.

Ebert, G, and Lenz,F. 1991. Jahresverlauf der Wurzelatmung von Apfelbäumen und ihr Beitrag zur CO2-Bilanz. *Gartenbauwissenschaft,* **56**:130-133.

FAOSTAT. 2017. Food and Agriculture Organization Statistical Databases, United Nations. http://www.fao.org.

Garcia Fraile, P; Silva, L R; Sanchez Marquez, S; Velazquez, E. and Rivas, R. 2013. Plums (*Prunus domestica* L.) are a good source of yeasts producing organic acids of industrial interest from glycerol. *Food Chemistry*, **139**(1/4): 31-34.

Ghanshyam, S., Chanda, T. R. and Dhar, R. P. 1977. Studies on pollination requirements of 'Santa Rosa' plum. *Progressive Horticulture*, **9** (I): 77-80.

Gill, D. S. and Chitkara, S. D. 1990. Propagation of peach and plum by semi hardwood cuttings. *Research and Development Reporter*, **7**(1-2): 154-158.

Glowacka, A. and Rozpara, E. 2014. Growth, yielding and fruit quality of four plum (Prunus domestica L.) cultivars under organic orchard conditions. *Journal of Research and Applications in Agricultural Engineering*, **59**(3): 56-60.

Gonkiewicz A, Nosal K.2006. NAA-induced ethylene evolution by plum flowers and fruitlets *Folia Horticulturae*, **18**(1):63-72.

Gonkiewicz,A. and Nosal, K. 2006. Anatomical changes in plum fruit *Prunus domestica* L. after fruit thinning using potassium salt of alpha-naphtaleneacetic acid. *Electronic Journal of Polish Agricultural Universities*, **9**(4): art 50.

Grabowsky, J. and Zielenkiewicz, J. 2000. The effect meteorological factors during growing season on yield of plums. *Sodininkyste ir Darzininkyste*, **193**(2): 203-209.

Grzyb Z S and Rozpara E . 2004. New Polish cultivars, rootstocks and interstocks of fruit trees. *Acta Horticulturae,* **663**(2): 953-955.

Guerra, M. E., Wunsch, A., Carrales, M. L. and Rodrigo, J. 2010. Flower emasculation as the major cause for lack of fruit set in Japanese plum crosses. *Journal of the American Society for Horticultural Science*, **135**(6): 556-562.

Hartmann W. 1999. The first European plum totally resistant to sharka. *Rivista di Frutticoltura e di Ortofloricoltura*, **61**(9): 68-69.

Henkel G, Willmer C, Monien S, Mester E, Kaland B, Golecki B and Wunderlich M . 2010. Evaluation of the status of "Plum Pox Virus? in Schleswig-Holstein. *Julius Kuhn Archiv*., **428:** 408.

Hjeltnes, S. H. and Nornes,L. 2007. Germination of plum seeds. *Acta Horticulturae*, **734**: 83-87.

Horst, R.K. (ed.). 2008. Nonparasitic diseases. Westcott's Plant Disease Handbook. 7th ed. Springer, NY, pp. 413-438.

Hu H, PengXia Li and YuNing W. 2011.Effect of different O2 and CO2 composition on physiological and quality changes in plum fruit storaged under low temperature. *Acta Botanica Boreali Occidentalia-Sinica*, **31**(10): 2035-2044.

Ilyefalvi J, Pamfil Z D and Craciun C. 2011.Transmission electron microscopy of Plum pox virus, Prunus necrotic ringspot virus, Prune dwarf virus in plum (*Prunus domestica* L.). *Journal of Horticulture, Forestr –and Biotechnology*, **15**(1): 120-125.

Inglese, P.; Gullo, G. and Pace, L.S. 2002. Fruit growth and olive quality in relation to foliar nutrition and time of application. *Acta Hort.* **586**: 507-509.

Jain, N. and. Babbar, S. B. 2003. Regeneration of 'juvenile' plants of black plum, Syzygium cuminii Skeels, from nodal exp of mature trees. *Plant Cell Tiss. Org. Cult*., **73**:257-263.

Jobling, J; Morris, S.C. and Rath, A.C. 2003. The effects of ReTainReg. (AVG) on the postharvest storage life of plums, peaches, nectarines, apples and bananas. *Australian Postharvest Horticulture Conference, Brisbane, Australia, 1-3-October,* 2003. Pp. 132-133.

Johnson, R.S. and Day, K.R. 2013. Zinc deficiency and correction in California plum orchards. *ActaHortic*. **985**: 189-192.

Johnson, R.S. and Uriu, K. 1987. Mineral Nutrition. In: Peaches, Plums, and Nectarines: growing and handling for fresh market, by LaRue, J.H. (ed.). UCANR.Vol. **3331**:68 – 81.

Jooyandeh, H. 2013. Production of non-alcoholic fermented plum beverages. *Middle East Journal of Scientific Research,* **13** (6): 836-843.

Kahu, K. and Klaas, L. 2007. Organic farming of plums in Estonia. *Acta Horticulturae,* **734**: 453-456.

Kalinina, A. and Brown, D. C. W. 2007. Micropropagation of ornamental *Prunus* spp. and GF305 peach, a *Prunus* viral indicator. *Plant Cell Rep.*, **26**: 927-935.

Kan Juan; Che Jing; Liu Jun; Jin ChangHai. 2012. Study on the physiological characteristics and ethylene production during plum fruit ripening and softening. *Journal of Yangzhou University, Agricultural and Life Sciences Edition*, **33**(2): 67-72.

Kaur K and Kaundal G S. 2009. Efficacy of Herbicides, Mulching and Sod Cover on Control of Weeds in Plum Orchards. *Indian J. Weed Sci.,* **41** (1 & 2): 110-112.

Kaur Kirandeep and Kaundal, G.S. 2013. Economics and effect of various herbicidal treatments on fruit quality and yield of plum cv. Satluj Purple. *Asian Journal of Horticulture,* **8**(1): 101-105.

Kaur Nirmauit, Aulakh, P. S. and Monga, P. K. 2010. Control of fruit drop in plum cv Satluj purple. *Environment and Ecology*, **28**(4): 2347-2348.

Kemp, H. 1996 .Flowering phenology in european countries: apple, pear, plum and cherry. *Acta Horticulture,* **423**: 299-302.

Khalil A, Srivastava K K, Nazir N, Jabeen M and Hameed S.2005. Effect of chemical and hand thinning on fruit retention, maturity and quality of plum cv. Santa Rosa. *JNKVV-Research Journal*, **39**(1): 64-68.

Khan, A. S. and Singh Zora 2010. Pre-harvest application of putrescine influences Japanese plum fruit ripening and quality. *Food Science and Technology International,* **16**(1): 53-64.

Khan, A. S; Zora Singh; Abbasi, N. A. and Swinny, E. E. 2008. Pre- or post-harvest applications of putrescine and low temperature storage affect fruit ripening and quality of 'Angelino' plum. *Journal of the Science of Food and Agriculture*, **88**(10): 1686-1695.

Knutsen, H. and Haukas, T. 2007. Economics of plum production in Norway. *Acta Horticulturae,* **734** : 241-247.

Kornova, K. M. and Popov, S. K. 2013. Obtaining virus-free planting material of plum cultivars by the in vitro method. *Acta Horticulturae,* **981**: 507-512.

Kumar J, Rana S S, Verma H S and Parmar D K. 2001. Long-term effects of intercrops on growth, yield and fruit quality of plum (Prunus salicina). *The Indian Journal of Agricultural Sciences.***71**:38:45.

Kumar, J. and Thakur Disha. 2012. Effect of different pruning intensities on growth, yield, fruit quality and leaf macronutrient content of plum cv. Santa rosa. *Asian Journal of Horticulture,* **7**(2): 484-487.

Lee EunJin; Choi SunYoung; Cho MiAe; Hong YoonPyo; Choi JiWeon and Chung DaeSung. 2013. Quality maintenance and suppression of chilling injury of 'Akihime' plum fruits stored under controlled atmosphere. *Korean Journal of Horticultural Science and Technology*, **31**(6): 732-739.

Lin ShengHua; Fang ChengQuan; Pu FuShen; Zhang JiaYan; Gao XiuYun. 1994. Chromosome numbers and karyotypes of Prunus in China. *Advances in Horticulture,* 78-85.

Mahajan, B.V.C. and Sharma, R.C. 1999. Effect of foliar application of growth regulators and calcium chloride on the size and quality of plum (*Prunus domestica* L.) cv. Satluj Purple. *Horticultural Journal*, **12**(1): 73-76.

Mahajan, B.V.C., Randhawa, J.S. Kaur, Harminder and Dhatt, A.S. Effect of post harvest application of calcium nitrate and gibberellic acid on the storage life of plum. *Indian Journal of Horticulture,* **65**(1): 94-96.

May-de-Mio, L. L; Tutida, I; Motta, A. C. V; Dolinski, M. A; Serrat, B. M; Monteguti, D. 2008. Rate of application of nitrogen and potassium in relation to brown rot and scab in plum 'Reubennel' in the region of Araucaria, Parana. *Tropical Plant Pathology*, **33**(1): 35-40.

Mehraj S, Peer F A, Bilal A, Pandit I A, Bisati and Ganai I H.2014. Effect of organic and inorganic mulches on growth, yield and quality attributes of plum cv. Santa Rosa under temperate conditions. *Green Farming*, **5** (6): 1048-1051.

Mengel, K. 2002. Alternative or complementary role of foliar supply in mineral nutrition. *Acta Hort*. **594**: 33-47.

Mika A, Buler Z and Chlebowska D. 1999. Effects of planting density and training system on growth and fruiting of plum trees grafted on vigorous and semidwarf rootstocks. *Journal of Fruit and Ornamental Plant Research,* **7**(2): 61-69.

Mika, A. and Buler, Z. 2011. Intensive plum orchard with summer training and pruning. *Advances in Horticultural Science*, **25**(3): 193-198.

Mika, A; Wawrzynczak, P; Buler, Z; Konopacka, D; Konopacki, P; Krawiec, A; Bialkowski, P; Michalska, B; Plaskota, M; Gotowicki, B. 2012. Mechanical harvesting of plums for processing with a continuously moving combine harvester. *Journal of Fruit and Ornamental Plant Research*, **20**(1): 29-42.

Mimoun, M.B.; Ghrab, M.; Ghanem, M. and Elloumi, O. 2009. Effects of potassium foliar sprays on olive, peach and plum. *International Fertilizer Corrospondent*. **19**: 14-17.

Minev I . 2001. A promising early ripening interspecies hybrid .*Rasteniev"dni-Nauki*, **38**(3/4): 170-172.

Mirabdulbaghi M and Zarghami R. 2011. Effect of propagation methods on rooting ability of 'Tanasgol', a natural hybrid plum x apricot *(Prunus domestica x P. armeniaca). Acta Horticulturae*, **923**: 117-120

Mitrovic, O; Mitrovic, V; Stanojevic, V; Micic, N. and Kandic, M. 2001. The effect of pruning severity on the quality of dried plums cv. Cacanska Rodna. *Jugoslovensko Vocarstvo,* **35**(3/4): 97-104.

MiZhen Z, XiQiang Liu, Guo-Honga and Du-Ping. 2001. Experiment of pollination for two plum varieties. *China Fruits*, **2:** 15-18

Mladenovic, M; Grajic, N; Macukanovic Jocic, M; Vulic, T. and Rasic, S. 2013. Impact of honey bees on pollination of some plum varieties. *Fourth International Scientific Symposium "Agrosym-2013"*, Jahorina, Bosnia and Herzegovina, 3-6-October,-2013, *Book of Proceedings,* Pp. 1076-1081

Moreno, Miller Y M, Azarenko A N and Potts W .1992. Genotype, temperature, and fall-applied ethephon affect plum flower bud development and ovule longevity. *Journal-of-the-American-Society-for-Horticultural Science*. **117**(1): 14-21

Moreno, M. A; Tabuenca, M. C; Cambra, R. 1995. Adesoto 101, a plum rootstock for peaches and other stone fruit. *Hort. Science*, **30**(6): 1314-1315.

Moreno, M.A; Tabuenca, M. C; Cambra, R. 1995. Ademir, a myrobalan rootstock for plums. *Hort. Science,* **30**(7): 1475-1476.

Nicotra A and Moser L .Plum rootstocks. 1999. *Informatore Agrario Supplemento*, **55**(6): 46-50.

Nikolic, D. T., Rakonjac, V. S. and Aksic, M. M. Fotiric. 2012. The effect of pollenizer on the fruit set of plum cultivar Cacanska Najbolja. *Journal of Agricultural Sciences, Belgrade,* **57**(1): 9-18.

Nyeki, J. and Szabo, Z. 1997.Cross incompatibility in stone fruits. *Acta Horticulture,* **437**: 213-217.

Ogasanovic, D; Rankovic, M; Paunovic, S; Mitrovic, O; Stamenkovic, S. 2005. Mildora - a new plum cultivar for drying. *Vocarstvo,* **39**(3): 251-256.

Olmstead, M.; Miller, E.P.; Andersen, P.C. and Williamson, J.G. 2016. Growing plums in Florida. IFAS Extension, University of Florida. USDA. http://edis.ifas.ufl.edu/pdffiles/HS/HS25000.pdf

Ozkaya, O; Dundar, O. and Demircioglu, H. 2010. Effects of 1-methylcyclopropene on overall quality of 'Black Amber' plum during storage. *Acta Horticulturae,* **877**: 383-386.

Paprstein F . 1999. Plum cultivar 'Valjevka'. *Vedecke Prace Ovocnarske*, **16**: 121-126

Paprstein F, Kloutvor J and Plasilova J. 2007. Plum cultivar 'Tophit' . *Vedecke Prace Ovocnarske*, **20:** 137-139.

Pavanello, A.P. and Ayub, R. A. 2014. Chemical thinning of plum fruits with ethephon. *Ciencia Rural*, **44**(10): 1766-1769.

Pesakovic, M; Djukic, D; Mandic, L. and Rakicevic, M. 2009. The effect of mineral fertiliser use on the microbiological activity and productivity of alluvium under plum. *Acta Horticulturae*, **825**: 313-318.

Pesakovic, M; Miletic, R; Karaklajic Stajic, Z; Lukic, M; ukic, D. and Mandic, L. 2012. Effect of NPK fertilizer on the soil microorganisms growth in 'Cacanska Rodna' planting. *Acta Horticulturae*, **968**: 247-252.

Popov, S.K. and Kornova, K. M. 2009. Shoot growth structure of own-rooted micropropagated plum cultivars in the juvenile period. *Acta Hort.*, **825**:497-502.

Randhawa, G. S. and Nair, .K. R., 1960. Studies on floral biology of plum grown under sub-tropical conditions. *Indian Journal of Horticulture*, **17**(2): 83-95.

Ravelonandro, M; Scorza, R; Polak, J; Callahan, A; Krska, B; Kundu, J; Briard, P. 2013. Honey Sweet" plum - a valuable genetically engineered fruit-tree cultivar. *Food and Nutrition Sciences,* **4**(6A): 45-49.

Renaud R.1994. Fermareine: a new European plum variety. *Acta Horticulturae*, **359**: 93-96.

Rettke, M.A.; Pitt, T.R.; Maier, N.A. and Jones, J.A. 2006. Quality of fresh and dried fruit of apricot cv. Moorpark in response to soil applied nitrogen. *Australian Journal of Experimental Agriculture*. **46**(1): 123-129.

Rizzo, R; Verde, G. lo. and Lombardo, A. 2012. Effectiveness of spinosad and mineral oil for control of *Grapholita funebrana* Treitschke in organic plum orchards. *New Medit.*, **11**(4 Special issue): 70-72.

Rubauskis, E.; Skrivele, M. and Dimza, I. 2003. Growth and yielding of plum trees in experiment with fertigation and rootstocks. *AGRIS*. **5:** 59-65.

Ruiz, R. 2006. Effects of different potassium fertilizers on yield, fruit quality and nutritional status of 'Fairlane' nectarine trees and on soil fertility. *Acta Horticulturae*. **721**: 185-190.

Ruzic and Vujovic, T. 2007. A protocol for the micropropagation of plum (*Prunus domestica* L.). *Vocarstvo,* **41**(1): 79-85.

Saini Pooja, Pandey Avnish and Rehalia, A. S. 2013. Influence of various sources of nitrogen fertilizer on nutritional status of leaves in plum cv. Santa Rosa. *Annals of Biology,* **29**(2): 208-213.

Sapir, G,, Goldway, Shafir, M. and Stern, R. A. S. 2007. Multiple introduction of honeybee colonies increases cross-pollination, fruit-set and yield of 'Black Diamond' Japanese plum (Prunus salicina Lindl.). *Journal of Horticultural Science and Biotechnology,* **82**(4): 590-596.

Seehuber, C; Damerow, L. and Blanke, M. 2011. Regulation of source: sink relationship, fruit set, fruit growth and fruit quality in European plum (*Prunus domestica* L.) - using thinning for crop load management. *Plant Growth Regulation,* **65**(2): 335-341.

Sharma, S.K., Rao, V.K. and Dimiri, D.C. 2009. Effect of pedicel retention and temperature on storage behaviour of Santa Rosa plum. *Indian J. Hort.,* **66**(3): 358-362.

Shear CB and Faust M. 1980. Nutritional ranges in deciduous tree fruits and nuts. *Horticultural Reviews*, **2**: 142–163.

Singh Zora and Khan, A.S. 2010. Physiology of plum fruit ripening. *Stewart Postharvest Review*, **6**(2): article 3.

Sosna, I. 2010. Effect of pruning time on yielding and fruit quality of several early ripening plum cultivars. *Acta Scientiarum Polonorum Hortorum Cultus*, **9**(1): 37-44.

Southwick, S. M; Rupert, M. E; Yeager, J. T; Lampinen, B. D; Dejong, T. M; Weis, K. G. 1999. Effects of nitrogen fertigation on fruit yield and quality of young 'French' prune trees. *Journal of Horticultural Science and Biotechnology*, **74**(2): 187-195.

Sun Wei; Gao QingYu; Wu ZhenLin; Wu LiRen; Du HanJun; Liu WenDong and Gao-JiaYuan. 2010. A new cold-resistant and early-maturing plum cultivar 'Sui Li 5'. *Acta Horticulturae Sinica,* **37**(7): 1191-1192.

Sun, L. Y. 2004. Occurrence and protected research of main diseases of plum. *Chin. Agric. Bull.,* **20**:263-264.

Szecsko, V; Hrotko, K; Stefanovits Banyai, E. 2004. Phenolic compounds, bud dormancy, and rooting ability of plum hardwood cuttings. *Acta Horticulturae,* **658**(2): 679-687.

Tariq M, Shah Z and Ali A. 2008. Micronutrients status of plum orchards in Peshawar valley. *Soil and Environment*. **27**: 223–227.

Taylor, K.C. 2009. Cultural management of the bearing peach orchard. Athens, University of Georgia. Circular-879, pp 4.

Tehrani G and Lay W. 1992'.Victory' plum. *Hort Science*, **27**(11): 1232-1233.

Teskey, B.J.E. and Shoemaker, J. 1978. Tree Fruit Production, 3rd ed. AVI, Westport, CT, pp. 358.

Thakur Nidhika and Thakur, B. S. 2014. Studies on the effect of integrated nutrient management on growth and yield of plum cv. Santa Rosa. *Asian Journal of Horticulture*, **9**(1): 112-115.

Thakur, N and Thakur B S. 2014. Studies on the effect of integrated nutrient management on growth and yield of plum cv. Santa Rosa. *Asian J. Hort.,* **9**(1): 112-115.

Tian, L. and Sibbaldm S. I. 2007. Micropropagation of Prunus domestica and Prunus salicina using mature seeds, pp. 373- 379. In: Jain S. M. and Haggman H. (Eds.). Protocols for micropropagation of woody trees and fruits. Springer, Netherland.

Topp, B. L. and Russell, D. M.1990. 'Queensland Bellerosa' plum. *Hort. Science*, **25**(7): 814.

Treder W, Grzyb Z, Rozpara E. 1999.The influence of irrigation on growth and yield of plum trees cv. Valor grafted on Myrobalan and Wangenheim Prune. *Acta Agrobotanica*, **52(1-2)95-102**.

Vangdal, E; Doving, A. and Mage, F. 2007. The fruit quality of plums (*Prunus domestica* L.) as related to yield and climatic conditions. *Acta Horticulturae,* **734**: 425-429.

Vitanova, I. 1985. Effectiveness of mineral fertilization of plum trees. *Resteniev Nauki*. 22(8): 105.

Vitanova, I.M. 1990. Determination of needs for fertilizer of plum trees. *Acta Horticulturae*. **274**: 501-508.

Vitanova, I; Marinova, N. and Dimkova, S. 2009. The green manure - an important factor for biological plum production. *Acta Horticulturae*, **825**: 441-445.

Wang, Y; Li, Y. C; Bi, Y; Wu, L. F. and Ding, B. 2013. Preharvest GA3 sprays reduced chilling injury and maintained quality of plums during storage. *Acta Horticulturae,* **1012**: 271-276.

Weir, R.G. and Cresswell, G.C. 1993. Plant nutrient disorders. Temperae and subtropical fruit and nut crops. Inkarta press. Melbourne, Australia.

Wenneker, M; Janse, J. D; Bruine, A. de; Vink, P. and Pham, K. 2012. Bacterial canker of plum caused by *Pseudomonas syringae pathovars*, as a serious threat for plum production in the Netherlands. Journal of Plant Pathology, **94**(1, Supplement): S1.11-S1.13.

Widmer, A; Schwizer, T. and Stadler, P. 2013. Fruit thinning in plum cultivation. *Obst und Weinbau,* **149**(22): 10-13.

Yu X, Sun M and Liu W. 2008.Investigation on the fruit set rate of 174 isolated self pollinated plum germplasm. *South china Fruits,* 5: 58-59.

Yuan W , JianWen W, XinLie Z Z, JieQun and FaZhong C.2005. Guangongli", an extra early plum cultivar. *China Fruits*, **2**: 20-21.

Zaiger's. 2000. Variety: 'Flavor Supreme'. *Plant Varieties Journal*, **13**(3): 74-75.

Zapata, P.J; Martinez, Espla, A; Guillen, F; Diaz Mula, H.M; Martinez Romero, D; Serrano, M. and Valero, D. 2014. Preharvest application of methyl jasmonate (MeJA) in two plum cultivars. 2. Improvement of fruit quality and antioxidant systems during postharvest storage. *Postharvest Biology and Technology*, **98**: 115-122.

13

Quince (*Cydonia oblonga* Mill)

Amit Kumar and Shabbir Hussain

1. Introduction

Quince is a highly nutritious underutilized fruit of sub-Himalayan region which has great economics and medicinal potential and it belongs to the group of the oldest cultural plants. Quince is known with different names, Urdu name 'Bahee Dana', Farsi name 'Beh', Greek name 'Strythion', Kashmiri name 'Bamchunth' and Hindi name 'Bihi' (Torkelson, 1995, Khoubnasabjafari and Jouyban, 2011). Quince is very much like pear in shape and gets a golden yellowish outer layer when mature. Quince fruit is a valuable dietary product and even then it less suitable for the direct consumption (Hricovsky *et al.*, 2003) and for that reason they are mostly used either cooked (Alvarenga *et al.*, 2008) or preserved (Kyzlink, 1990). However, as compared with other species, this fruit tree species is less popular throughout the world. Its predisposition to turning brown and decay is a limiting factor in the long-term storage of this fruit (Gunes, 2008). The nutritional value of the fruit is high (Bucsek *et al.*, 1996; Kopec and Balik 2008). It is well known that quince fruit shows antioxidant effects (Silva *et al.*, 2002) which result from the presence of a number of polyphenolic substances (Fattouch *et al.*, 2007), e.g. flavonoids quercetin, rutin, kaemp ferol etc. (Silva *et al.*, 2005), and also from increased levels of vitamin C (Tetera, 2006). Quince has the important nutritional and flavor composition more than many fruits. Vitamin C content of quince is nearly two times higher than in apple (Biro and Lindner, 1999).

Quince plants are grown for their fruit, but they are also commonly grown as a dwarfing rootstock for grafting pears (Postman, 2009). It is an important source of sugars, crude fibers, organic acids, minerals like calcium, phosphorus and potassium and it has very low amounts of fat (Shinomiya *et al.*, 2009; Rodriguez-guisado *et al.*, 2009; Sharma *et al.*, 2011). The content of pectin in fruits ranged between 1.75 and 3.51 g/100 g of fruit weight and cultivar Hruskovita accumulated pectin in greatest amount. A medium sized fresh quince weighing 300 g contains nitrogen (62.57 g), phosphorus (12.95 g), potassium (165.87 g), calcium (9.31 g), magnesium (4.36 g), and sodium (1.16 g), per mg/100 g of fruit weight of Asenica variety (Rop *et al.*, 2011). Quinces are rich in phytic, malic and quinic acids and also in fructose, sorbitol and glucose, which gives them an interesting equilibrium among sourness, astringency and sweetness *(Przemys³aw et al.,* 2014; Hamauzu *et al.,* 2006). Quince fruit are reported comparatively higher amount of antioxidant activity and phenolic concentration (Hamid and Khadijeh, 2014). At present, the possibilities of a commercial use of quinces are intensively studied (Adler, 2001).

The nutritional values of the individual quince varieties may be really considerable (Gunes and Dumanoglu 2005), not only in fresh condition but also in relation to their processing (Alvarenga *et al.*, 2008). Nutritional value of quince has been presented in Table 2. Quince has a long history of cultivation in the Middle East where it is native and throughout Mediterranean region. Quince is wild in part of Asia including Armenia, Georgia, Turkistan, Iran and South Arabia. Greeks and Roman grew quince for its fragrant fruit and attractive pink flowers. It is still important fruit crop in its native region and in South America. Commercial production seems to be mostly in Easter Europe and Asian minor.

1.1. Uses

All parts of quince plant are utilized for different purposes. Wood is used for furniture, cabinet etc making (Bernardo, 2008). Some quince varieties are palatable uncooked, but most quince are quite firm when ripe and have an astringent flavour when not cooked. When cooked however, the fruit develops a pretty pink blush, becomes tender and develops a sweet and slightly tart flavour (McCandlish, 2009; Durand, 2014). Fruit is soft and juicy and is suitable for eating raw or when cooked (Launert, 1981). The cooked fruit adds a luscious taste to cooked apples (Simmons, 1972). The influential and characteristic odour of quince fruit has attracted great attention as a source of flavour in recent years, predominantly valued in jams, candies, sweets and brandy (Nisarg *et al.*, 2011; Ercisli, 2004; Usmanghani *et al*., 1997). Fruits are also used for preparation of marmalade, jam and sauce. Fruit may be used as an additive in tea (Ahmad *et al.*, 2014). In the Greece, tea prepared by boiling dry seeds of quince in water is given cholecystitis to humans (Ahmad *et al.*, 2004). Traditionally, quince after harvest is locally marketed and is useful for culinary purposes in the form of cooked delicacy dish in 'Kashmiri Wazwan'. It is also utilized for making preserve (muraba), which is known to possess beneficial medicinal properties in unani medicine. The kernel oil is used for massage. It was found that the triple combination of Semperfresh, ascorbic acid and cold storage provides

Table 1: Medicinal usages of different parts of quince

Part used	Medicinal uses	Reference	Part used	Medicinal uses	Reference
Leaves	Antidiabetic	Aslan *et al.*, (2010) and Palmese *et al.*, (2001)	Fruits	Antidiabetic	Tahraoui *et al.*, (2007)
Leaves	Antidioxidant	Aslan *et al.*, (2010)	Fruits	Cystitis	Sezik *et al.*, (2001)
Leaves	Antihemolytic and free radical scavenging	Costa *et al.*, (2009)	Fruits	Emollient for the skin	Pieroni *et al.*, (2004)
Leaves	Antihyperglycemic	Teresa *et al.*, (2001)	Fruits	Headche	Gorji (2003)
Leaves	Antihyperlipidemic	Khademi (2009)	Fruits	Laxative	Agelet and Valles, (2003)
Seeds	Conjunctivitis, Cough, Bronchitis	Siddiqui *et al.*, (2002)	Seeds	Migraine, Nausia, common cold and influenza	Hilgert (2001)
Seeds	Constipation	Ghanadi (2003)			

Table 2: Nutritive value of quince per 100 g of dry weight

Principle	Nutritive value	% of RDA	Principle	Nutritive value	% of RDA
Energy	176 Kcal		Protein	0.4 – 0.6	
Water	86.9 g		Fat	0.1 -	
Carbohydrates	15.3 g		Fibre	1.9 g	
Ash	0.6 g		Total lipid	0.10 g	
Vitamins					
Thiamine B1	0.02 mg	2 %	Vitamin B6	0.04 mg	3 %
Riboflavin B2	0.03 mg	3 %	Folate B9	3 µg	1 %
Niacin B3	0.2 mg	1 %	Vitamin C	15 µg	18 %
Pantothenic acid B5	0.081 mg	2 %			
Minerals					
Calcium	11 mg	1 %	Potassium	197 mg	4 %
Iron	0.7 mg	5 %	Sodium	4 mg	0 %
Magnesium	8 mg	2 %	Zinc	0.04 mg	0 %
Phosphorous	17 mg	2 %	Tannins	0.8%	
Total soluble solids	14.2 %		Total sugars	9.0 %	
Pectin	1.8 %		Crude fibre	1.6 %	

Source: Skurikhin and Tutelyan, 2007; Rodriguez-Guisado *et al.*, 2009; Sharma *et al.*, 2011; Lim, 2012

high microbial, chemical and sensorial qualities for the quinces, leading to an extension in shelf life when compared with untreated quinces (Yurdugul, 2005).

1.2. Medicinal uses

Quince fruit has several medicinal usages, such as carminative, expectorant, anticancer (Duke *et al.*, 2002), antibacterial (Fattouch *et al.*, 2007), antidiabetic (Tahraoui *et al.*, 2007) and laxative (Agelet and Valles, 2003), being also used for the treatment of skin lesions (Hemmati *et al.*, 2012), migraine, cold, influenza (Hilgert, 2001), inflammatory bowel disease (Rahimi *et al.*, 2010) and conjunctivitis (Siddiqui *et al.*, 2002), among other disorders. In the indigenous system of medicine, the quince is used in treating diarrhoea, dysentery and constipation (Velickovic *et al.*, 2001; Rodriguez-Guisado *et al.*, 2009). The fruits are known to be hypoglycaemic, anti-inflammatory and anticarcinogenic and act as a tonic for heart and brain (Sharma *et al.,* 2011).

1.2.1. Control of colon cancer

Plant proanthocyanidins, including procyanidins, display various biological activities. Gorlach *et al.* (2011) reported an inhibition of human colon cancer Caco-2 cell growth by the extract from Japanese quince fruit and the procyanidin-rich fractions of the extract. They concluded that the Japanese quince procyanidins exhibited proapoptotic activity in Caco-2 cells within a submilimolar concentration range.

2. Origin and distribution

The quince tree is native to a wide area that includes Caucasus, Transcaucasia and Central Asia (Georgia, Armenia, Azerbaijan, Uzbekistan, Turkmenistan, Tajikistan, Iran, Afghanistan and Pakistan) were gradually penetrated also to other parts of the world. During ancient times, quince spread from its wild centre of origin to the countries bordering the Himalaya Mountains to the east, and throughout Europe to the west. Now-a-days, there are still wild quince plants in Dagestan, Azerbaijan, Turkmenia and Iran and the centre of origin is considered to be the Trans Caucasus region now-a-days, there are still wild quince plants in Dagestan, Azerbaijan, Turkmenia and Iran (Postman, 2012). During early times, this plant extended from its wild centre of origin to the countries neighboring the Himalayan Mountains to the east and throughout Europe to the west, but it is probable that Quince reached the Mediterranean region only in classical times; it was used by the Romans (Vaughan and Geissler, 1997). The fruit was known to the Akkadians, who called it supurgillu; Arabic safarjal quinces collective plural. The cultivation of the quince likely preceded that of the apple, but it is difficult to ascertain. Early writings which referred to the 'apple' were likely referring to the quince instead. The state of Kydonia on the island of Crete was where the quince was improved to the level we know today. The scientific name of the quince is *Cydonia* and is an alternate spelling of Kydonia.

Cultivation of quince may have preceded apple culture and many references translated to apple such as the fruit in Song of Solomon, may have been to a quince. In India it is cultivated mainly for fruits and rootstocks for pear and plum. It is mostly grown in Jammu and Kashmir and some parts of Himachal Pradesh and Uttarakhand most frequently in backyards and fence corners. *Cydonia oblonga* is grown and cultivated in grounds or gardens under warm temperature and grows up to 8 m in height and 4 m width. The young branches are covered with pale greyish wool, leaves are elliptical, flowers are pink or white, fruits are bright yellowish and usually pear shaped (Khoubnasabjafari and Jouyban, 2011).

3. Acreage, production and export value

Worldwide, there are about 106, 000 acres of quince. The world production of quince has been increasing over the past decades and according to FAO (2012) current production is 596,532 metric tons. There are significant productions of this fruit in about fifty countries. The top two producers of quince are Turkey and China. Other major producers are Uzbekistan, Morocco, Iran, Argentina, Azerbaijan, Spain and Serbia (Table 3). Quince is grown in many countries for use as a dwarfing pear and loquat rootstock.

Table 3: Production of quince in world

S. No.	Country	Production (MT)	S. No.	Country	Production (MT)
1.	Turkey	135,406	6.	Argentina	27,500
2.	China	125,000	7.	Azerbaijan	27,140
3.	Uzbekistan	80,000	8.	Spain	14,000
4.	Morocco	46,000	9.	Serbia	10,795
5.	Iran	36,500	10.	Algeria	10,516

Source: FAO, 2012

4. Botany

Cydonia oblonga is a monotypic genus belonging to family Rosaceae, subfamily Spiraeoideae, tribe Pyreae, and subtribe Pyrinae (USDA 2009a). It is a thorn less bush-like or multi-stemmed shrub or small deciduous tree mainly grown for its edible fruits, has pubescent to tomentose buds, petioles, leaves, and fruit. The tree has crowded, gnarled branches and a crooked growth habit. The trees can reach heights of 5–8 m and can live for periods in excess of 50 years and have an economic lifespan of approximately 25 years. The leaves have a smooth upper surface and hairy lower surface and are ovate to oblong, about 5 cm across and 10 cm long.

The tree produces single white-light pink colored flowers on tiny shoots which are 1½ to 4 to 5 cm across, have 5 petals, 20 or more stamens, 5 styles, an inferior ovary with many ovules, and are borne on current season growth, after the leaves (Whealy and Demuth, 1994). Bloom time overlaps with that of apples, usually beginning mid April in the central latitudes of the northern hemisphere. The fruit is initially covered in dense gray-white hairs but these disappear as the fruit ripens. The ripe fruit is a golden yellow color and resembles a pear or apple depending on variety. It has a soft, yellow pulp with a grainy texture protected by a thin skin. The fruit is a fragrant, many-seeded pome about 8 cm in diameter. Fruits contain seeds which are poisonous. The fragrance of a few ripe quinces can fill a room with an intoxicatingly sweet aroma. Shape ranges from round to pear-like, flesh is yellow, and the Baileys refer to it as "hard and rather unpalatable" (Bailey and Bailey 1976; Rehder 1986). Fruit size and leaf size of cultivated varieties can be many times larger than the wild type described above.

All varieties are self-pollinating with a good fruit set in both cool and hot climates, pollination takes place by bees. Based on the shape of fruits it is possible to distinguish two basic form lines of quinces. The pomes of the first one are similar to apples (*Cydonia oblonga* subsp. *maliformis*) while those of the second one are pear-shaped (*Cydonia oblonga* subsp. *pyriformis*). The apple-shaped quinces have an aromatic, drier and harder flesh while that of the pear-shaped is fine and softer (Hricovsky *et al.*, 2003). The fruits which are pear or apple-shaped, are golden yellow in colour, juicy, fragrant, sub acid or sour in taste and contain a large number of plano-convex, mucilage-coated seeds, closely packed in two vertical rows (Michal, 2001; Ahmad *et al.*, 2004).

4.1. Flowering

The quince flower is similar to the apple, but in general it is coarser and more colourful. It develops on first year growth and therefore appears later in the season than the apple blossom. Quince trees may bloom from February to May depending upon the species and geographical area. Blooms on a tree may last 11 to 20 days with full bloom last 6 to 10 days.

The external morphology of the quince flower differs from that of the apple in that it has a bowl-shaped floral tube which is larger than the cone-shaped floral tube of the apple. In the quince there are a larger number of ovules per locule and the separation of the edges of the carpels making up the compound ovary is more distinct. In the vascular, structure of the quince flower the stamen traces are given off from the sepal and petal bundles at a somewhat lower level and are shown as distinct from the petal bundle at level C. Another minor difference is the branching of the dorsal carpellary bundle in the base of the style, to form three bundles. Considering the vascular structure as a whole, the quince would be rated phylogenetically more primitive than either the apple or the pear because of the separation of the stamen traces at a lower level. In general, the flowering period for most quince cultivars is concentrated between the end of August and early October, whereas the flowering periods are relatively large and slow, almost never occur significant losses due to frosts (Fioravanço and Simonetto, 2006). Quince has a bipolar (hermaphrodite) flower type. Stamens are composed of a large number of filaments (15– 20) and light yellow anthers, which are arranged in three circular rows. The pistil is made up of five fertile leaves (carpels) consisting of stigma, style and ovary. Quince has yellow pollen grains. Shape and size of pollen often changes with variable air temperature and humidity and other environmental factors (Nagy-Deri, 2011).

4.2. Pollination

Cross pollination is crucially important for quince. The fertility relations of this fruit tree remain a matter of controversy in the literature. Although three groups, self-fertile, partly self-fertile and self-sterile cultivars are usually distinguished, the fruit set of quince is very low with natural autogamy so it requires cross-pollination to achieve sufficient yield. On the other hand, for an adequate fruit quality more than 70 ovules need to be fertilized in a quince flower, so it needs more pollen for cross-pollination than other Maloideae fruit crop plants. Furthermore, flower density of quince is low as compared with other fruit species so it needs at least 20-25 per cent fruit set to reach an adequate yield. Flowers offer pollen and nectar as reward for honey bees and bumble bees, which usually serve as pollinators for quince. Accordingly, the study of nectar and pollen, which ensure the insect attraction of flowers, is very important (Nagy-Deri, 2013).

4.3. Pollen morphology and pollen studies

Morphological characterization of pollen grains is very important in the identification of individual species and cultivars of fruit trees, including quince. Analysis of the morphological characteristics of pollen by scanning electron microscopy (SEM) was performed for the identification and characterization of individual species and cultivars of quince (Evrenosoðlu and Misirli, 2009; Nagy-Deri, 2011). As material for these studies, the pollen of eight quince cultivars (Leskovacka, Vranjska, Morava, Pazardzijska, Hemus, Asenica, Portugal, and Triumph) was used (Radovic *et al.,* 2016). During the 3 year study period (2010–2012), using scanning electron microscopy (SEM), the following characteristics were investigated: the size and shape of pollen grains (length, width, and length to width ratio), colpus length, colpus width, mesocolpium width and characteristics of exine pattern. The tested quince cultivars differed significantly in terms of morphological characteristics of pollen. The exceptions were colpus and mesocolpium width, as well as the ridge and furrow width on the exine, which showed no statistically significant difference. Pollen grains of quince are characterized by great length (over 53 μm). The shape of pollen grains varied from prolate (Leskovacka, Vranjska, Morava, Hemus, Asenica, and Triumph) to perprolate (Pazardzijska and Portugal). The exine pattern of all cultivars was striate with longitudinal ridges, which were more parallel in cultivars Vranjska, Hemus, Portugal, and Triumph and less parallel in cultivars Leskovacka, Morava, Pazardzijska, and Asenica. Three groups of similar cultivars are separated on the basis of hierarchical cluster analysis results. Some pollen morphological characteristics, such as the size and shape of pollen grains and colpus length, can be used in the identification of quince cultivars.

Nectar production of quince was found to be small compared to other temperate zone fruit tree species, since quince flowers contained 1.07+or-0.06 mg of nectar on the average. Bee visitation was highly intense, ~5.5 bee visits were counted on the average at 50 open quince flowers in 10 minutes and ~9.7 flowers out of 50 were visited. This equals to about 7 bee visits per flower per day in good weather. Some cultivars tended to be more and less visited by honey bees. Most honey bees tended to collect pollen (51.6% on average for the 3 years) and usually fewer gathered only nectar (19.9%), while the rest had a mixed foraging behaviour gathering both (28.5%). Quince is highly sensitive to partial and/or complete limitation of the insect (bee) pollination period. Benedek *et al.* (2001) observed that complete limitation of insect pollination resulted in no yield and partial limitation of the insect pollination period depressed the final set and yield by at least 60-70% or more.

The fertilization system in Japanese quince (*Chaenomeles japonica*) was studied by Kaufmane and Rumpunen, (2002) using fluorescence microscopy. In a compatible combination, pollen grains germinated fast and pollen tubes grew rapidly through the stylar tissue. Most of the pollen tubes reached the base of the ovary within 2-6 days. In the ovary, pollen tube growth was considerably retarded, and fertilization of ovules started at the earliest 3 days after pollination. In an incompatible combination, only few pollen grains germinated, and pollen tubes grew slowly and were often completely arrested in the style. The pollen tubes reached the ovary in 6-7 days. Fertilization did not take place until 7-9 days after pollination, and the percentage of fertilized ovules was very low. The same results were obtained when selfing self-incompatible genotypes. However, a small percentage of pollen tubes reached the ovary within 3-4 days in some genotypes including self-pollination (as in a compatible cross-pollination), but fertilization was not observed or considerably delayed compared to cross-pollination. *C. japonica* thus shows the characteristics of a gametophytic self-incompatibility system. The viable period of the embryo sac was 6-7 days in most of the investigated genotypes, and the effective period of pollination was estimated to 5-7 days.

in-vitro pollen quantity, viability and germination tests in quince (*Cydonia oblonga* Mill.) cvs. Ekmek, Esme, Limon, Ege 2, Ege 22, Ege 25 and Quince-A rootstock pollens were collected in April from the unopened pink balloon-stage flowers on mature trees (Dalkilic and Mestav, 2011). The pollen amount was between 20063 ('Ege 25') and 11906 pollen/flower ('Limon') with hemacytometer. The lowest and highest pollen viability were determined from 79.8 (Quince A in day 14) to 97.8 % ('Ege 2' in day 0) by IKI (0.5 g iodine (I) + 1 g potassium iodide (KI) solution was prepared by dissolving I and KI in 100 ml distilled water). The germination rates in different sucrose and boric acid concentrations in 1 per cent agar changed from 0 ('Limon' in 0 % sucrose) to 91.7 % ('Ege 2' in 20 % sucrose) and from 32.7 (Quince A in 0 mg/L boric acid) to 94.9 per cent ('Ege 22' in 100 mg/L boric acid), respectively.

Cetin (2006) studied pollen germination rate (0 to 15 % sucrose in hanging drop) and pollen tube growth in pistil in 15 quince cultivars. Fruit set was compared among selfing and crossing combinations. Sharafi (2011) investigated pollen viability and pollen tube longevity in five different quince cultivars and genotypes in Iran. The highest pollen germination rate changed from 37.2 (CO-2) to 65.3 per cent (Gortoune-Sfahan) in quince. Yolacti (2006) observed that the highest pollen germination in quince was obtained from control treatment (22 ± 1°C) with 91.47 per cent.

5. Taxonomy

Cydonia is an important member of subtribe Pyrinae, tribe Pyreae, subfamiliy Spiraeoideae (earlier Maloideae), family Rosaceae, class Magnoliopsida, and division Magnoliophyta. Sax (1931) concluded that the genera of the Maloideae form a closely related group based on the chromosome number and the presence of intergeneric hybrids. The species was formerly named *Cydonia vulgaris* Pers. and *Pyrus cydonia* L. within the species, various botanical varieties, subspecies or form a have been described (Table 4). Many of these subspecies, botanical varieties and forma should probably not be considered as formal taxa but merely represent genetic variation within the species. Lobachev and Korovina (1981) divided the genus into two subspecies, *C. oblonga* subsp. *oblonga* and *C. oblonga* subsp. *integerrima*.

Quinces were originally thought of as a type of pear and were first called *Pyrus cydonia*, or the *Cydonian* pear. Later they were given a separate classification and are now properly called *Cydonia oblonga*. They are sometimes called tree quinces to distinguish them from *Chaenomeles* or the flowering quince, which is again a separate plant. In the genus *Cydonia*, species *oblonga* Pers; (*C. vulgaris* Pers.) is by far the most well defined and widely grown species and has number of botanical varieties (Sharma *et al.*, 2009).

i. *Cydonia oblonga* var. *pyriformis* Rehd (*C. vulgaris* var. *pyriformis* Kirchn). Fruits of this botanically are typically pear shaped.

ii. *Cydonia oblonga* var. *pyramidalis* Schneid. The plant shape is like a pyramid.

iii. *Cydonia oblonga* var. *lusitanica*, Schneid. (*C. lusitanica* Mill.) The fruit shape is like that of pear, fruits are markedly ribbed, leaves are larger and growth is more vigorous.

iv. *Cydonia oblonga* var. *maliforms* Schneid (*C. maliformis* Mill.) The fruit shape is like that of an apple.

v. *Cydonia oblonga* var. *marmorata* Schneid. The leaves of this botanically are variegated with white and yellow in colour.

vi. *Cydonia oblonga* var. *campanuloformis*. The fruits are of bell shaped.

5.1. Flowering quince (*Chaenomeles speciosa*)

It is a deciduous shrub valued for its masses of showy white, red or pink blooms that brighten the landscape in late winter and early spring. Most varieties grow to 6 to 10 feet tall, with a similar spread, although some top out at only 3 feet. It flowers best when the shrub is in full sun, but it will tolerate partial shade. It is a versatile shrub that grows in most soil types, including loam, clay-based soil and sandy soil. The plant prefers neutral to acidic soil with a pH of 3.7 to 7.0 and doesn't perform well in alkaline soil.

5.2. The Chinese Quince (*Pseudocydonia sinensis*)

This Chinese relative of *Cydonia* presently belongs to the genus *Pseudocydonia*, but has previously been assigned to both *Chaenomeles* (*Chaenomeles sinensis*) and *Cydonia* (*Cydonia sinensis*). Chinese quince has attractive single pink flowers that appear earlier than those of *Cydonia* but not as early as most *Chaenomeles*. The fruit is a large, oval, aromatic yellow pome that ripens in the fall. The shiny, leathery leaves develop nice red-orange fall color. But its most interesting characteristic is the exfoliating bark that reveals brown, green, orange, and gray patches. Chinese quince's attractive bark rivals that of many stewartias. The trunk often becomes fluted with age, adding even more textural appeal (Dirr, 1997).

5.3. Japanese quince (*Chaenomeles japonica*)

It is a small type of flowering quince, only growing to about 3 feet tall. Unlike the other types, which are native to China, this quince is native to Japan. Fresh Japanese quince accumulated high amounts of ascorbic acid, 55 mg 100 g-1 in hybrid clone C.47 and 106 mg 100 g-1 in 'Lichtar'. The fruit of Japanese quince 'Lichtar' had higher content of phenolic compounds (422.6 mg 100 g-1) and stronger antiradical activity (12.35 micro mol TE/g) (Rubinskiene *et al.*, 2014).

Table 4: Subspecies, botanical varieties and forms of *Cydonia oblonga* Mill.

Cydonia oblonga subsp. *oblonga*
Cydonia oblonga subsp. *integerrima* Lobachev
Cydonia oblonga subsp. *maliformis* (Mill.) Thell
Cydonia oblonga subsp. *pyriformis* Medik, ex Thell
Cydonia oblonga var. *onlonga*
Cydonia oblonga var. *biserrulata* Kakhadze
Cydonia oblonga var. *integerrima*
Cydonia oblonga var. *integerrimo-sepala* Lobachev
Cydonia oblonga var. *integerrimoespala* Kakhadze
Cydonia oblonga var. *lusitanica* (Mill.) C.K. Schneid
Cydonia oblonga var. *moliformis* (Mill.) Rehder
Cydonia oblonga var. *obpyricarpa* Lobachev
Cydonia oblonga var. *obpyriformis* Lobachev
Cydonia oblonga var. *orbiculato-complanata* Lobachev
Cydonia oblonga var. *ovalicarpa* Lobachev
Cydonia oblonga var. *ovalis* Lobachev
Cydonia oblonga var. *plano-cyclocarpa* Lobachev
Cydonia oblonga var. *pomiformis* Lobachev
Cydonia oblonga var. *rotundata* Kakhadze
Cydonia oblonga var. *serridata* C.K. Schneid
Cydonia oblonga var. *typica* Kakhadze
Cydonia oblonga var. *urceolata* Lobachev
Cydonia oblonga forma *lusitanica* (Mill.) Rehder
Cydonia oblonga forma *marmorata* (Dippel) C.K.Scheid
Cydonia oblonga forma *pyramidalis* (Dippel) Rehder
Cydonia oblonga forma *pyriformis* (Dierb.) Rehder

Source: Encyclopaedia of Life (2009)

6. Cytogenetics

Like other pome fruits, quince is a diploid with 2n = 34. The nuclear DNA content is 0.73 pg (Dickson *et al.*, 1992). The basic chromosome number (n = 17) of Cydonia is considered to be a secondary unbalanced number and the genus is probably a secondary polyploidy (Blando *et al.*, 1992).

7. Climate

The quince is very adaptable. It grows well in a range of conditions, from the cooler parts of subtropical regions to cool temperate regions and has a culture similar to that of apples and pears. Coastal, tablelands and inland districts of New South Wales can produce good quinces. However, a disease called 'fleck' becomes more of a problem under moist coastal conditions. The tree needs some winter chilling for good fruit production. Under dry, warm to hot inland conditions, irrigation will be required to produce good crops.

Trees will grow more vigorously and produce more fruit in full sun and will grow optimally at temperatures averaging 15°C. For the Japanese quince orchard, a sunny site on a slight slope should be chosen to reduce the risk of frost damage to both flowers and fruits. However, the orchard could if necessary be established on plain ground since the long and abundant flowering of Japanese quince plants reduces the risk of spring frost damage. Quinces are the last of the deciduous fruits to flower; therefore a late frost can still damage the flower. Quince trees are easily damaged by strong winds and should be planted in a sheltered location which will also protect them from rapid temperature fluctuations which is also detrimental to their growth.

8. Soil

Most soils are satisfactory except those that are light, shallow and likely to dry out. Heavy, moist soils are particularly good. Well-grown fruit is often seen on trees growing wild near creek banks where this type of soil is often found. Quinces produce best on deep, heavy loam, but they will grow on many soil types. Established trees can withstand periods of very wet conditions, but this does not mean that quinces like poorly drained situations. Quinces are particularly susceptible to iron deficiency under alkaline conditions; hence they prefer slightly acid soils. Japanese quince plants have a widespread and deep root system. The plants can therefore be grown in rather dry conditions, and Japanese quince has been recommended for cultivation on slopes to prevent soil erosion.

Soil salinity is a major constraint to the cultivation of horticultural crops. It is important to select a site with proper soil for optimum growth of quince plants. Well drained, acid mineral soils with some content of organic material and some clay should be selected. Quinces prefer slightly alkaline soil (pH 6.5 to 7.0 or higher). A soil pH of 7.0 or higher releases extra calcium, preventing bitter rot on ripening fruits. In dry alkaline soils, plant chlorosis is likely to be more pronounced than in alkaline soils with sufficient moisture. In the case of alkaline conditions, elemental sulphur or iron sulphate may be applied for soil acidification before planting. Iron was the first element to be considered a micronu-trient in plants; it is a component of clayish materials and, under some circumstances, is found to occur in levels insufficient to meet the needs of plants (Prado, 2008).

9. Rootstocks and Propagation

9.1. Rootstock

There is lack of information regarding the rootstock of quince, however till date most of the plantations are of seedling origin and still the seedlings are used as rootstocks for grafting/budding purpose. Quince seedlings are satisfactorily used as rootstocks with the advantage of that they do not produce suckers. For raising the seedling rootstocks, quince seed is extracted from mature fruit, cleaned, stored in a cool place or held moist in a plastic bag in the refrigerator until stratification which is done in late November or December for 80 to 100 days and the sprouted seeds are planted in early spring. If the seedlings are well grown, they will be large enough to bud with the desired varieties during the following late rainy season or grafts after one year. The Angers clonal quince selection, Quince A, can also be used as a rootstock for the various quince varieties. 'Japones' quince cultivar (*Chaenomeles sinensis* Koehne) use as rootstock for other quince cultivars and pear and loquat trees (Abrahao *et al.*, 1991, 1996). Pio *et al.* (2008) collected seeds of five cultivars ('Mendoza Inta-37', 'Provence', 'Portugal' and 'Japones') from ripe fruits and subjected to cold stratification. Raised seedlings were grafted by cleft grafting method using scion sticks of Portugal quince and it was observed that seedling rootstocks of 'Japones' showed the best performance.

However, clones of quince namely Quince A, B and C itself being commercially used as a rootstocks for most of the pear cultivars. The most widely used rootstocks for the nursery production of quince are taken from certain cultivars quince, *Cydonia oblonga* P. Mill. (Necas *et al.,* 2008). Mass (2008) observed that 'C-132' a quince rootstocks shows promise because of its control of tree growth in combination with good fruit size. Other quince rootstocks used are EMC, Adams and CPP.

Pio *et al.* (2008) evaluated the compatibility of pear (*P. calleryana*) rootstocks Taiwan Nashi-C and Taiwan Mamenashi with quince cultivars and found that Taiwan Mamenashi rootstock can be an excellent option for Portugal and Japones quinces.

9.2. Propagation

Quinces can be propagated by seeds, cuttings, budding and grafting. By grafting/budding the selected varieties onto this rootstock, the trees will bear a little earlier after planting out and when fully grown, will be smaller than those established on their own roots or seedling rootstocks. Trees on seedling rootstock should produce some fruit by about the fifth year. Those from cuttings will fruit sooner.

9.2.1. Seed propagation

Pio *et al.* (2010) evaluated the influence of gibberellic acid (GA3) and cold stratification of Japonese quince tree seeds (*Chaenomeles sinensis*), on the growth of rootstock seedlings. Seeds were collected from fruits, rinsed in water, and dried for 48 hours. Soon after, the seeds were submerged in gibberellic acid concentrations of 200 mg L^{-1}, for 24 hours, with an oxygenation system. The seeds were placed in Petri plates between moistened cotton and submitted to cold stratification in a B.O.D. chamber at 4 degrees C) for 60 days. At the end of the period, the seeds were distributed in polystyrene trays containing 72 cells, with vermiculite as substrate. Good germination percentage and seedling growth was observed after 60 days of sowing.

Seeds can be grown but will not breed true to type and should not be used for commercial purposes where fruit quality is very important. Germination behaviour quince seeds was studied by Nunes *et al.* (2014). The seeds are photoblastism neutral and do not germinate at 15 degrees C. The 25 degrees C showed a higher rate of speed and percentage of germination and shoot length of seedlings. The seeds are sensitive to salinity caused by NaCl, KCl and CaCl2, which negatively affect seed germination and seedling growth.

9.2.2. Propagation by cutting

Quinces can be propagated by hardwood cuttings, although cuttings from some varieties will not root easily (e.g. Champion). The cuttings should be about 25 cm long and taken during the winter period December-February (no later than the end of June). This is the most convenient method of propagation, but the disadvantage is that resultant trees tend to produce suckers, which need to be periodically removed. Two groups of rootstocks were used to evaluate different techniques for their vegetative propagation using hardwood cuttings. The first group comprised pear rootstocks: Pyrodwarf, Pyroplus (Rhenus 3), OH x F 333, Fox 11, the pear seedling (*Pyrus communis*) and 'Cydomalus' hybrids between *Malus communis* x *Cydonia oblonga*. The second group included quince rootstocks: 'BA 29', 'Sydo', 'MA' and 'S1'. 15-20 cm length of hardwood cuttings were taken in January-February and subjected to 1.0 per cent NAA and 0.5 per cent IBA treatments. Half the treated cuttings were kept at 21°C for a period of 25–35 days, until callus formed, and the other half were kept at 5°C in a cooling box. The rootstocks 'Cydomalus' and 'Pyrodwarf' performed best regarding callus formation (100 % of cuttings) followed by Pyroplus (75 %) and OH x F 333 (76 %). The highest overall success rates were seen in Pyrodwarf (94 %) and 'S1' (67 %) (Necas and Kosina, 2008).

Rooting of quince cv. Portugal cuttings was standardized by Pio *et al.* (2006). Half of the woody cuttings (20 cm in length, without leaves) from the middle portion of the branches were immerged in IBA solution at concentrations of 2000 or 3000 mg/l, for 5 seconds. After that, they were put in plastic sacks (0.65 dm3 capacity) filled with soil + sand (as substrate) at 2:1, v/v and placed in a greenhouse with 50% brightness for rooting.

9.2.3. Propagation by grafting

Pio *et al.* (2008) practiced cleft grafting for propagation of Quince cultivars onto Taiwan Nashi-C and Taiwan Mamenashi seedlings and maintained in plastic bags during the dormancy period. The scion sticks used had three buds each and were collected from the mother plants. Good combination was noted after 150 days of operation in both the rootstocks.

In an another experiment Pio *et al.* (2010) used 'Japones' quince (*Chaenomeles sinensis*) as rootstock, due to its good plant vigor for grafting, rusticity, and adequate performance in the field, mainly in the first years after planting. To determine the best grafting period and grafting method for this rootstock, three grafting methods and five quince scion cultivars on 'Japones' rootstock was tried. The quince cultivars 'Provence', 'Mendoza Inta-37', 'Portugal', 'Smyrna', and 'Japones' were grafted by winter cleft grafting, winter budding, and summer budding on 90 cm long plants of 'Japones' quince rootstock seedlings, cultivated in 3-liter plastic bags. Cleft grafting resulted in the highest graft survival rate for 'Smyrna', 'Mendoza Inta-37' and 'Japones' cultivars. Although the higher graft survival rate was recorded in the winter, the buds grafted in the summer had better development.

9.2.4. Micropropagation

Regeneration processes in differentiated or undifferentiated tissues have attracted the attention for their potential of application as tools for plant propagation and for genetic engineering. It is known that the regeneration of adventitious organs and/or somatic embryos differs according to species and to a number of endogenous and exogenous factors, among which hormonal balance has a primary role.

In particular, the auxin–cytokinin ratio appears to be the most important factor in channeling regeneration response towards a specific in vitro morphogenic process rather than another (Christianson and Warnick, 1983). Among the fruit tree species that have shown regenerating capacity, the quince (*Cydonia oblonga* Mill.) has produced somatic embryos by a combination of kinetin and naphthaleneacetic acid (NAA) (D'Onofrio *et al.*, 1998; D'Onofrio and Morini, 2002b) or adventitious shoots by treatments with thidiazuron and naphthaleneacetic acid (Baker and Bhatia, 1993) applied to *in vitro* cultured leave.

The frequency of adventitious shoots of Quince A regeneration was high (78 % of the cultured leaves with 3.2 shoots per regenerating leaf) at 32 μM TDZ (Thidiazuron) plus 0.3 μM NAA on young leaves obtained from micropropagated shoots. Shoots were rooted by culturing them first on medium containing 5 μM NAA for one week and then on auxin free medium for four weeks. The regeneration protocol may be useful for selection of somaclonal variants with increased tolerance to low Fe and for transformation mediated by Agrobacterium (Dolcet Sanjuan *et al.*, 1991). The best shoot proliferation and development were obtained from 30 g/l sucrose, 5 g/l agar and 5.5 pH on *in-vitro* shoots of Quince A (*Cydonia oblonga* Mill.) (Gulsen and Dumanoglu, 1991).

Successful rooting was obtained on half strength MS with Lee-Fossard vitamins IBA (1mg/L), GA_3(0.1mg/L), (NaFeEDTA (3, 2 mg/L), sucrose (20 g/L), with 8 days dark and photoperiod of 14 or 16 hours. NAA was found to be the most effective auxin with 1.0 mg/l giving the highest rooting rate (26.7%) on *in vitro* rooting of micropropagated shoots of East Malling Quince A (*Cydonia oblonga* Mill.) (Dumanoglu and Gulsen, 1994).

A procedure for adventitious shoot regeneration from leaf explants using thidiazuron (TDZ) was developed by Baker and Bhatia (1993). Excised leaves of cultures grown on MS medium containing 5 micro M benzyladenine (BA) and 0.9% Gibco Phytagar were used. Several experiments were conducted to determine optimum concentrations of TDZ, NAA and sucrose. When the medium contained 1.5 micro M TDZ and 2.5 micro M NAA, 85% of the discs regenerated shoots with an average of 8 shoots/leaf disc. An incubation period of 3 weeks in the dark was necessary for optimum shoot regeneration. Leaves excised from 4 to 6-week-old cultures gave a higher percent shoot regeneration than leaves from cultures older than 6 weeks. Regeneration percentages were significantly reduced when sucrose concentration in the medium was less than 3%. A significantly higher percentage of shoots regenerated when leaf discs were placed on the regeneration medium abaxial side down as compared to the adaxial side. Regenerated shoots were cultured on MS medium containing 5 micro M BA and rooted on half-strength MS medium containing 10 micro M NAA. Rooted plantlets were acclimatized to greenhouse conditions for evaluation of any somaclonal variation.

Micropropagation of Japanese quince was optimized by Panavas (1994). The culture was initiated from shoot tips of mature plants plated on MS medium supplemented with benzyladenine (BA). The medium containing 0.5 mg BA/ litre was the optimum for shoot elongation at this stage. *in vitro* grown shoots were multiplied in medium containing reduced strength MS basal salts with 0.1 to 2 mg BA/litre and 15-35 g sucrose/litre; the optimum for shoot elongation and proliferation was 2 mg BA and 35 g sucrose/litre with 3/4 strength basal salts. Rooting of regenerated shoots in medium with 1 mg IBA/litre was low in frequency (10%); the results were better (25%) when shoots were rinsed for 2 h in IBA solution at 250 mg/litre

There is less tendency to produce an increasing number of somatic embryos when the leaves were pre-treated for 3 and 6 days with Kin + NAA would seem to indicate that the embryogenic process had a cytokinin requirement gradually increasing with time, satisfied at first by Kin, which has a lower biological effectiveness than BA (Khanam *et al.*, 2000). The type of growth regulator was also responsible for the different shoot and root regeneration response. When the leaves were transferred to GR-medium, shoot regeneration was never observed, but on increasing the period of culture on Kin + NAA, a large number of morphologically undefined structures, reddish and nodular in shape, were produced by the leaves; they exhibited a histological configuration very similar to shoots precociously arrested in their development (Morini *et al.*, 2000).

The effects of NaCl and $CaCl_2$ on shoot regeneration from quince (*Cydonia oblonga* BA L29 clone) leaves were investigated by D'onofrio and Morini (2002a). Caulogenesis was induced on *in vitro* grown leaves treated for 2 days in liquid Murashige and Skoog (MS) medium with 11.3mM 2,4-dichlorophenoxyacetic acid and cultured on MS gelled medium supplemented with 4.5 mM thidiazuron and 0.5mM naphthaleneacetic acid. Three experiments were performed: in the first, it was compared the effects of NaCl at 0, 25, 50, 100, and 200mM in factorial combination with 3, 9, and 27mM $CaCl_2$. In the second, NaCl was tested at 0, 5, 10, 20, 40, and 80mM with $CaCl_2$ at 0.3, 1.0, and 3.0mM. The third experiment was carried out with the same experimental design as the second one but replacing NaCl with Na_2SO_4. Shoot regeneration was evaluated after 50 days of culturing: 25 in darkness and 25 in white light. In the first experiment, shoot regeneration was very poor and was observed only at the lower salt concentrations. In the second experiment, the percentages of caulogenic leaves were much higher, but decreased with increasing NaCl concentration. The more pronounced negative effect of the highest NaCl concentrations appeared to be partly mitigated by $CaCl_2$ at 1mM and 3mM. The presence of 3mM $CaCl_2$, in the experiment with Na_2SO_4, appeared to be even more effective in reducing the adverse effect of sodium stress on caulogenesis. This result was attributed to the lower chloride concentration in the growth medium, which resulted from replacing NaCl with Na_2SO_4. NaCl applied at low concentrations (5 and 10mM) in combination with 3mM $CaCl_2$ exerted a favorable effect on adventitious shoot regeneration. As regards the Na^+ and Ca^{2+} interaction, when the Na+/Ca2+ ratio was below roughly 35 and 20, with NaCl and Na_2SO_4, respectively, at least 60 per cent of leaves showed regenerating capacity, but optimal values of this ratio were not derived.

Standardization of explant orientation in the culture medium, sucrose and of agar concentration are important aspects for *in-vitro* multiplication of quince shoots. Erig and Schuch (2004) observed that the *in-vitro* multiplication of shoots of quince cv. MC was promoted by the use of culture medium solidified with 5 g agar/l, added with 25 or 30 g sucrose/l, and cultivation of the explants in the horizontal orientation. The culture medium used was the MS with 125% salt concentrations, added with myo-inositol (10 mg/l), BAP [benzyladenine] (13.3 micro M), IAA (0.6 micro N) and GA3 (0.3 micro M) and pH adjusted to 5.7.

D'onofrio and Morini (2005) studied the effects of three cytokinins (kinetin 4.5 *m*M (Kin), 6-benzylaminopurine 4.5 *m*M (BA) and N-phenyl-N'-1,2,3- thiadiazol-5-yl-urea 4.5 *m*M (TDZ)) and their duration on the regeneration of adventitious shoots from quince (*Cydonia oblonga* Mill.) leaves. Results demonstrate that reddish-colour structure nodules (RNS) were adventitious shoots blocked at an early developmental stage on account of insufficient cytokinin activity. BA was less effective than TDZ in inducing shoot regeneration. Finally, both Kin and BA applied after 2,4-D treatment promoted somatic embryo induction. Giorgota *et al.* (2009) establish the technology for *in vitro* propagation of 'BN 70' and 'A Type' quince rootstocks and of 'Aurii' quince cultivar. Worst results during the multiplication step were obtained with WPM (Woody Plant Medium) but the rest of the media produced very good multiplication rates, although plants showed hyperhydratation symptoms in Lepoivre media.

in vitro grown quince leaves to simultaneously regenerate somatic embryos, adventitious roots and shoots and to evaluate the variations induced on regeneration response by treatments of different length with growth regulators (D'onofrio and Morini, 2006), the type and the amount of regenerated morphogenic structures may be related to the biological effectiveness of the growth regulators utilized. The greater number of somatic embryos regenerated when the leaves were directly transferred to GR + medium (0 day of kinetin + naphthaleneacetic treatment), compared to the number of somatic embryos obtained on GR- transfer medium, demonstrates that somatic embryogenesis benefited from the combination N6 benzylaminopurine (BA) (2.66 mM) + GA_3 (0.58 mM) + IBA (0.3 mM), where BA probably had a predominant role.

Hatamian *et al.* (2014) investigated to optimize adventitious shoot regeneration in quince (cv. Isfahan, known as 'KVD2'). The results indicated the maximum frequency of regeneration (46%) was obtained from the whole leaf explants cultured on NN medium containing 40 micro M TDZ and 1 micro M NAA. The highest mean number of shoots (7.46) was obtained on MS medium containing 5 micro M TDZ and 1 micro M NAA.

A semisolid rooting MS culture medium supplemented with 0.1 mg/l NAA; 0.3 mg/l IBA and 3 per cent sucrose (pH 6.5), was used as a reference medium (Flores *et al*. 2015). Four different variations in the sucrose concentration (1 %, 2 %, 3 %, and 4 %) were performed in liquid medium and each trial was evaluated with *in-vitro* plants which had been previously exposed to the medium of the corresponding treatments for 15 day long period. The comparison of the root formation percentages evidenced the clear effect of sucrose concentration used, with the best results obtained when using the 2 per cent sucrose trial with no pre-treatment (73.3 %).

10. Varieties

Although the number of quince cultivars is quite lower than in other pome fruit crops, even then there is a great diversity was found in genotypes of quince (Ahmad *et al*., 2004; Ahmad *et al*., 2008). Some authors consider a larger number of subspecies, botanical varieties and forms, based on fruit shapes: pyriformis or typical (pear-shaped), maliformis (apple-shaped), lusitanica (the so-called Portuguese ribbed, pear-shaped fruit), marmorata (variegated) and pyramidalis (pyramidal fruit) (Mratinic, 2010, Bell and Leitao, 2011). The fruit pulp varies in colour, density, juiciness, flavour, presence of granulation (stony cells) and taste. Most varieties are too hard, astringent and sour to eat raw unless 'bletted' (softened by frost and subsequent decay). Some cultivars have little or no astringency and the fruit can be eaten fresh. Most cultivars are considered self-fertile but cross-pollination seems to increase the productivity of the orchards.

10.1. Champion: Green-yellow flesh with pear-shaped fruit. Harvest in early October. This nineteenth-century American cultivar is very productive, with large roundish fruits, which are mildly flavoured and ripen towards the end of the season. The blossom has pretty stripy buds which open into pale pink flowers. The tree will bear fruit when young, but is susceptible to blight

10.2. Orange: Orange-yellow flesh with golden skin having rich flavor. Low chill fruits. Harvest Early Sept. medium size fruit are round with a golden coloured skin and tender orange-yellow flesh. Excellent cooked in sauces, butters or preserves. Medium size fruit are round with a golden coloured skin and tender orange-yellow flesh. Excellent cooked in sauces, butters or preserves.

10.3. Van Deman: Colour of the fruits is pale yellow with coarse flesh. Orange skin that turns red when cooked, harvest in early Sept.

10.4. Aromatnaya: Round yellow quince that is deliciously sweet when eaten fresh as well as cooked. Extremely aromatic, with a fresh pineapple-like flavor, it will have a dense texture when first picked (October) but will soften up in a few days and is self-pollinating.

10.5. Arthur Coby: A very old cultivar named in honour of Dr Arthur AS Coby, Urbana Illinois, derived from *Chaenomeles lagenaria* and *C. japonica.* Small bush like, grows upto 5-6 feet, upright and spreading, flower bright red in colour two or more inch across, showy with ornamental value. Fruit large round flat greenish yellow, smooth, late in ripening, good for making jelly.

10.6. Pineapple: Self-fertile cultivar with large, smooth golden-yellow fruit. Low chill fruit, white flesh has a distinct pineapple flavor. Excellent for jellies, preserving and baking. Pineapple-like flavor in jelly. Fruit ripens in September to first week of October.

10.7. Smyrna: large, elongated fruit are lemony yellow in color with a tender, highly fragrant flesh. Excellent flavor- one of the best for jellies and preserves. large fruit with brown pubescence. Light, tender flesh. Yellow skin. Low chill fruit. Harvest Early Oct.

10.8. Cooke's Jumbo: It is a self fertile cultivar with very large, pear shaped yellowish green fruit with white flesh. Used for preserves, jelly and baking. Ripens in September.

10.9. Rich's Dwarf: Self fertile cultivar, having extremely large fruit with lemon yellow, fuzzy skin. Fragrant white flesh. Can be kept at 8-10' tall and wide.

10.10. De Vranja: This is one of the most commonly found cultivars and is deservedly popular. It originates from Vranje in south Serbia and was introduced into Britain in the 1920s. The buds are a stripey barber's pale pink and white and the blossom is sweetly scented. The fragrant pale green fruit ripens to a rich golden yellow. This will grow into a fairly upright tree, but can be fan trained if you start when the plant is young. It is the quince most resistant to blight.

10.11. Leskovac: This is one of the hardiest cultivars and crops well, with large apple shaped fruits. On dwarfing rootstock it grows into a neat shape and is perfect for small gardens and growing in containers.

10.12. Meech's Prolific: This cultivar was raised in 1850 in America by the Rev. W. W. Meech. It fruited better that the already popular 'Orange', giving rise to its name; prolific. It has big flowers and large pear-shaped fruits which can reach 150 mm long. The fruits tend to ripen early and have an excellent flavour. The tree will also bear fruit early, often after only three years.

10.13. Bereczki: This cultivar is Serbian in origin and is called after the

Hungarian pomologist Bereczki. It is a very good cropper, with large fruits and is one of the best varieties for fan training.

10.14. Iranian Quince: This cultivar, fairly obviously, comes from Iran. The fruit is sweeter and less gritty than many of the other varieties and it keeps its shape well when cooked.

10.15. Isfahan: This cultivar comes from Iran and is romantically named after the ancient city of Isfahan.

10.16. Ivan: This is a recent introduction from Russia which ripens early. It is said to be possible to eat the fruit straight from the tree after a warm summer.

10.17. Maliformis: This cultivar has apple shaped fruits with a fine flavour. It is very productive and the fruit ripens well in colder areas.

10.18. Shams; This fruit on this cultivar from Iran is particularly sweet and less gritty than many others.

10.19. Sobu: This is a large cultivar from Turkey which crops reliably; unfortunately the pear-shaped fruit can lack flavour.

10.20. Morava: It is a hybrid, evolved from a cross between Reans Mammoth (female) and Leskovacka (male). It was characterized by large fruit (mean weight 335 g) of good quality and high yield. The fruits can be readily peeled and have firm, compact, juicy flesh of good chemical composition, suitable for fresh consumption and processing. The trees flower early and are moderately vigorous. Pollen germination is good. Morava is precocious and gives regular high yields (15.08 kg/tree, or 7.54 t/ha, in the 7th year) (Stancevic, 1990).

10.21. Hemus: Hemus is found to have high levels of field resistance to fire blight (Bobev and Deckers, 1999).

10.22. Triumph: Triumph is found to have high levels of field resistance to fire blight (Bobev and Deckers, 1999).

10.23. Morova: It was evolved from a cross between Leskovacka X Reans Mammoth in Yugoslavia (Stancevic and Nikolic, 1992). Morova has an average fruit weight of 250 g; the fruits have good storage potential and almost surpass those of the best cultivar, Leskovacka, in quality.

10.24. Portugal: This variety is also known as Lusitanica. This cultivar was brought to England by John Tradescant in 1611, when it was described as the best quince for baking. This is a vigorous tree but can be slow to crop. It is not particularly hardy and needs to be grown in a sheltered spot. Given the right conditions it is an excellent choice, being very disease resistant, second only to 'Vranja'. The blossom is large and a beautiful pale rose colour. This is followed by largish (100 mm/4 inches), fragrant, pear-shaped fruits. These turn a deep orangey-yellow and are covered with a thick woolly down. The flesh is tender and juicy and is still regarded by many as the best for all types of cooking. Regular supply of water must be provided otherwise the fruit has a tendency to split.

Other's varieties are Agvambari, Fuller's, Apple Shaped, Powell's Prize, Missouri Mammoth, Rea's Mammoth, Serbian Gold, Pear Shaped, Ekmek, Early Prolific, Arthur Coby, Dwarf Orange, Gamboa, Le Bourgeaut, Ludovic, Morava, Perfume, Siebosa, Vrajna (Westwood, 1978; Baxeter, 1997).

10.25. Fire blight resistant cultivars

Papachatzis *et al.* (2011) investigated to identify quince cultivars that combine fruit quality characteristics and resistance to fire blight, considerable variation in fire blight resistance was observed that 'PI 26', 'PI 37', 'PI 41' and 'PI 49' were highly resistant to fire blight and 'PI 7', 'PI 22', 'PI 47' and 'PI 50' also exhibited notable disease resistance. In addition, the production efficiency and the fruit quality of the above genotypes were higher than the other genotypes evaluated.

10.26. Promising cultivars in different countries

10.26.1. Brazil

Bettiol-Neto *et al.* (2011) concluded that the quince 'Fuller' 'Smyrna', 'Portugal', 'Provence', and 'Mendoza Inta-37' showed higher productive performance. The quince 'Mendoza Inta-37' showed quality attributes most suitable for the fresh market, i.e., higher soluble solids content, lower acidity and firmness, and the highest ratio, followed by quince 'Fuller', 'Provence', and 'Portugal'.

10.26.2. Ukrainian

Ten varieties were selected for a combination of useful traits. Four of them were released for cultivation: Krymskaya Aromatnaya, Mir, Otlichnitsa and Krymskaya Rannyaya. The varieties Mechta, Novoselovskaya and Yaitsevidnaya Krymskaya were recommended for state varietal trials (Khrolikova and Baskakova,1989).

10.26.3. Moldova

The cultivar **Nakhodka** (obtained by open pollination of an unknown Caucasian cultivar) is early (ripens in Sep.-Oct.), dwarf (3 m), with a round crown, bearing large (320 g), bright green, apple-shaped fruits with an easily removed pubescence. The cultivar Volgogradskaya Myagkoplodnaya (obtained from crossing cv. Krasnoslobodskaya with cv. Myagkoplodnaya 1) produces medium-sized (260 g), round, pear-shaped fruits, bright yellow in colour, with an easily removed pubescence. The trees are dwarf (3-3.5 m) with a round crown. Fruits of both cultivars can be used fresh and for processing (Rotaru and Lobachev,1990).

10.26.4. Romania

The best for yield were Aromate, de Portugalia, Delicioase Beretzki and Local de Tirgu Jiu. The largest fruits with the highest dry matter content were found in de Husi. On the basis of these trials, de Husi, Aurii, Aromate and Local de Tirgu Jiu were released in the fruit-growing area of Bacau, Romania (Rati *et al.*, 1993).

10.26.5. Croatia

Nikolic (1996) recommended 'Morava', 'Trijumf' and 'Asenica' for propagation in Croatia. . The earliest cultivar to ripen was Morava and the latest was Trijumf. The largest fruits were recorded in Asenica (311 g). The highest quality fruits, in terms of the main chemical components, were Morava, Asenica, Trijumf. Trijumf and Asenica were noted for good productivity.

10.26.6. Russia

Avrora, Zolotoi Shar, Podarochnaya, Rumo and Novogodnyaya (Mozhar, 1999). Azamatov (2012) reported twelve varieties as most promising for the steppe zone of Kabardino-Balkaria in the northern Caucasus area of southern Russia. These are: Izobil'naya Krymskaya, Izobil'naya Krymskaya 16-45, Izobil'naya P13, Krymskaya Aromatnaya, Krymskaya Yaitsevidnaya, Limonno-Zheltaya, Sladkaya, Izobil'naya 11-13, Molodezhnaya, Muskatnaya, Yuzhanka, Nakhodka and Mir, the first 7 of which were considered the most suitable for use in the establishment of new orchards in the region.

10.26.7. China

The biological and fruit characteristics of **Xipi Shenghua, Douqing, Xipizi and Cupi Shenghua** of Chineses quince (*Chaenomeles sinensis*) were described by Cao *et al.* (2007). **Douqing** is a large fruit sized cultivar. Fruit mature in early-mid October, weighing 700-1500 g, but reaching 3500 g, having a waxy bumpy surface in bean-green colour. Its keeping quality is good. Fruit of **Cupi Shenghua** cultivar mature in late September to early October, are large, weighing 500 g on average, but reaching 2000 g, having a deep yellow skin with calyx. Flesh is very fine with strong aroma. It could be stored at room temperature to March-April. The trees are vigourous, highly adaptable to growing conditions. **Xipi Shenghua** is a small fruited cultivar, its fruit weighing

500 g on average, having a deep yellow skin with strong aroma, suitable for processing and ornamental. **Xipizi** is a high quality Chinese quince (*Chaenomeles sinensis) [Cydonia sinensis*] cultivar grown in Heze area, Shandong province. The trees are moderately vigorous and thornless. The branches are red-brown in colour and the flowers are light red. Fruits mature in late September to early October, and are large, weighing 250 g on average, but reaching 500 g, with a deep yellow colour and strong fragrance. The flesh is yellow-white, fine, juicy, suitable for processing or for keeping in a room to provide a fragrant aroma (Cao *et al.* 2007).

Bio-technology

11. Varietal identification

Randomly amplified polymorphic DNA (RAPD) technology was used by Bayaztto *et al.,* (2011) to study the genetic relationships among 13 quince accessions selected from different parts of Turkey. Thirty decamer primers were used and 14 of them did not produce any polymorphism. The remaining 16 primers ranged in their amplification fragments between one (P-402, P-437, OPA 10, OPA 16, OPA 18 and OPA-19) and five (OPA-06 and OPA-07). The size of fragments varied from 100 to 1500 bp. Similarity values among the studied genotypes ranged between 0.483 and 0.925. The resulting dendrogram clustered into two groups (0.69 similarity value) based on evaluation of genetic similarities and differences. The results suggest that RAPD analysis could be used to distinguish and determine genetic variation among quince accessions. Also, the obtained clustering based on RAPD markers agreed to some extent with the geographical origin of the studied set of quince accessions.

12. Crop improvement

The possibility of using *A. tumefaciens* to transform quince (*Cydonia oblonga*) was investigated using beta -glucuronidase (GUS) gene expression. Two strains of A. tumefaciens, LBA4404 and EHA101, were used, both with an intron containing the gus [uidA] gene and the neomycin phosphotransferase II (nptII) selectable marker gene. Leaf pieces were inoculated with *A. tumefaciens* for 30 min then transferred to a leaf initiation medium and incubated in the dark. At 24 intervals, 3-8 days after co-cultivation, leaf pieces were tested for GUS activity using a histochemical stain. The number of GUS expressing regions/ leaf increased over the 8-day period for both strains although the total number of GUS regions was generally greater for leaf pieces inoculated with EHA101. GUS activity was localised around the cut edges and where tissue damage had occurred during plant manipulation. No natural GUS activity was found in control leaves (Davidson *et al.,* 1998).

13. Cryopreservation

Quince shoot tips and nodal sections are successfully cryopreserved. Sustained re-growth of quince 'Angers A' was observed after encapsulation-osmoprotection/dehydration, encapsulation-dehydration and PVS2 vitrification. The highest regrowth rate (80%) was obtained from explants excised from cold hardened shoots and cryopreserved using encapsulation-osmoprotection/dehydration and vitrification protocols. The optimised vitrification protocol in combination with shoot cold hardening and a MS recovery medium without activated charcoal and auxin resulted in satisfactory re-growth of shoots from six quince cultivars. The morphology of acclimatised plants derived from cryopreserved shoots was comparable with non-cryopreserved plants (Lynch *et al.*, 2014).

14. Land preparation/pit digging and planting

Dig a planting hole approximately at the same depth as the root ball. Set that soil aside and mix it 50/50 with aged mushroom compost, aged manure, or rotted pine bark and aged manure/compost. Take the plant from the nursery gently loosen the root ball and place in the planting hole. To avoid burying too deep, make sure plant is positioned with the top most roots at the soil line. Fill the planting hole with the mix of soil and organic matter; gently tamp it in. Water should be applied lightly and thoroughly to settle the roots and eliminate air pockets. Only apply fertilizer if it is the correct time of year. If desired, construct a water basin around the base of the tree approximately 36 inches in diameter. Mulch of polythene or grass is to be applied in spring and summer with approximately 4-6 inches. Pull mulch a couple of inches away from the trunk for good air circulation.

15. Planting time and care

Planting is best done during the early spring or in autumn season when trees are completely dormant i.e. from mid-January to mid-February in Himachal Pradesh and Uttarakhand and in the month of March in Jammu and Kashmir. In areas with unstable snow cover and cold winters (temperatures below -20 °C) spring planting is recommended since young plants are vulnerable to winter frost damage. Spring planting may also be advantageous if the orchard is not fenced and protected against hares and rabbits. If spring planting is preferred, special attention should be paid to the storage conditions for the plants. Quince plants start to grow very early because of a short period of dormancy. Late spring planting may result in a reduced number of established plants, in particular if irrigation is not possible. It is also necessary to store plants properly (-1°C) since plants are very susceptible to grey mould (*Botrytis cinerea*). If planting is done during late autumn, the plants should be mulched using peat, wood chips or sawdust. The mulch layer must be at least 10 cm thick to prevent frost damage to the roots. A suggested planting distance is 5 x 2.5 or 5 x 3 metre, which will require some 667–800 trees per hectare. Where seedling rootstock has been used, especially where soils are deep and fertile, the trees can be expected to grow larger and a planting distance of 6 x 4.5 metres should be used, giving about 370 trees per hectare. Planting techniques and post-planting care are the same as for most other deciduous fruit trees.

16. Irrigation

After planting and upto the first year, there is a critical time for the establishment of a new quince plants. Water thoroughly twice a week on light soils and once a week on clay soils. Soak the entire root system deeply; this usually takes 40-50 minutes. Quinces should receive at least 1 inch of water each week for best growth and fruit production. Water regularly, especially during dry periods. Fruit may drop prematurely if insufficiently irrigated during dry spells.

17. Training and Pruning

Quince trees are usually developed as vase or open-centered trees or can be trained flat against a wall, but have to start at an early age because of their naturally twisty growth. The main aim is to create an even spread of main stems with short, productive branches growing out from them. It should be remember that the main stems will not remain productive indefinitely so we need to keep a supply of shorter stems ready to train out. The young shoots are quite pliable and will need to check every couple of months to ensure the plant spreads out properly. Fix the stems to a firm framework such as strong wires or trellis. Use soft garden twine as this will hold the branches in place but will not damage them as they grow. After the first three years the basic framework will be established and will need to follow the general pruning. Under open vase shape training system there is tendency of branches to weep and become leggy so therefore it is recommended to head back branches to support heavy fruit. Quinces tend to be bushy and ungrafted quince will form suckers if left unpruned. Quinces can be trained to a single trunk by frequent removal of suckers while grafted trees do not produce suckers.

Quince trees can also be trained on fan or espalier training system which have advantage that the tree will take up very little space and will flower and crop well against a sunny wall. For freestanding trees a goblet shape is best as this allows light and air through the tree and is reasonably easy to maintain. For this in the first 4–5 years, cut back the leaders of the main framework branches by half the previous summer's growth. Side shoots should be cut back to 2–3 buds.

After the initial formation, pruning is fairly simple and is best carried out on a 'little and often' basis. Light annual pruning of established trees encourages fruit production. Quince trees bear blossom (and fruit) on the spurs and tips of the previous summer's growth so remove only which is absolutely necessary. Pruning should be carried out in winter when the tree is dormant. Suckers round the base should be cut away, as should any shoots which appear on the clear part of the trunk. Remove any dead or weak stems and in particular cut away any stems which start to grow crosswise into the middle of the plant which keeps the tree open so it gets plenty of light and ventilation which will prevent disease and help the fruit ripen. Fruit does not need to be thinned unless there is a very heavy crop.

18. Weed control and Mulching

The ground should be kept free from weeds, at least in a strip one metre wide within the rows. Between the rows, a cover of grass may be established, which will provide a firm surface for field transport. To reduce competition for water and nutrients during plant establishment, it is recommended that the grass is not sown until the second year.

Mechanical weeding is the common method for controlling weeds in most orchards. For the dwarf and spreading Japanese quince plants, mechanical weeding is less efficient and must be complemented by hand weeding during the first three years. The equipment may also damage fruits during late stages of fruit development and mechanical weeding cannot be resumed until after harvest.

In an integrated growing system, herbicides may be used to control weeds efficiently until August. At the same time and labour costs are reduced compared to mechanical weeding practices. During fruit ripening, weeds must be controlled mechanically and herbicides should be avoided in young orchards since they reduce growth of plants.

For Japanese quince, any mulch tested was superior to herbicides and mechanical weeding in promoting a high yield. This may be partly explained by increased moisture and soil temperature, and partly by the fact that mechanical weeding and herbicides may negatively influence root growth.

In Japanese quince orchards, mulching with organic material alone does not control perennial weeds. Weeds probably start to grow already at the beginning of the second vegetation period. To prevent germination of annual weeds, 10 cm mulch should be provided in the first year and then 5 cm should be added each year. When wood chips or sawdust are used, nitrogen must be supplied to compensate for losses due to mineralization.

The best weed control is achieved by synthetic mulches (e.g. black plastic and woven plastic), which at the same time promote plant growth and yield. To improve durability, allow for traffic and mechanical harvesting, the synthetic mulch may be covered by gravel or organic mulch. Another advantage provided by the mulch is that the fruits are prevented from contamination by soil. The use of organic synthetic mulches may also reduce infection by fungal diseases.

19. Nutrient Management

Nutritional needs will vary according to soil fertility, tree age, density and productivity. However, results from fertilizer trials with varieties of quince plants are not yet available. The preliminary advice for nutrient management is therefore based on fruit contents of macronutrients, yield estimates and experiences from apple and pear. Leaf colour, growth and productivity are also best guides. The fertilizer recommendations should be adjusted according to nutrients content in the leaves (Table 5).

When the mineral composition of Japanese quince fruits was analysed for fruit flesh and seeds, it was observed that the composition of macronutrients was close to that of pear. Thus, the estimated nitrogen requirement seems to be 2–3 times higher than for apple, mainly because of the high content of seeds (with proteins) in the Japanese quince fruit. Rather high amounts of potassium, calcium and magnesium were also detected in the fruit flesh. To compensate for yield losses alone at a yield of 10 tons per ha, 80 kg N, 15 kg P, 120 kg K, 12 kg Ca and 6 kg Mg should be provided per ha. Thus, depending on the results of the soil analysis and expected yield, it may be recommended to supply 100–180 kg of nitrogen and 140–260 kg potassium annually per ha. In the case of slightly alkaline conditions, the use of ammonium sulphate as the nitrogen source will help to maintain a low soil pH. In sandy, acid soils (at a pH below 5.5), deficiency of Mg may appear and additional Mg should be provided as dolomitic limestone. Deficiency of Mg can also be induced by too high amounts of potassium fertilizer. In alkaline soils deficiency of iron may be compensated for by repeated foliar application of iron chelates, since soil application of iron chelates is not economically viable in practice.

As a rule, mixed fertilizers are the best. Other than these, chemical and organic fertilizers can also be applied. Selection of the mixed fertilizers should be based on the nitrogen content of the fertilizer, which should be higher than the phosphorus and potassium content. Apply a selected mixture to give about 350 g N annually to each tree of full bearing capacity. In case of applying single fertilizers, apply a heavy application of nitrogenous fertilizer only in the spring. Such single fertilizers should be supplemented by an application of 50–100 kg P/ha in the autumn. The fertilizer programme should also include the supply of micronutrients such as iron, zinc, manganese, magnesium, molybdenum, copper and boron through organic inputs. These minor elements are very important to plants and most soils are low in these elements.

Spread the fertilizer evenly under the entire canopy of the plant avoiding a 5 inch area around the trunk. Apply the recommended fertilizers three times a year i.e. in late February, late May and late July/early August. Never fertilize after August as this is the harvesting period and will also promote new growth late in the year which will be subject to freeze damage. Over-fertilization with nitrogen may contribute to fruit drop and promote fire blight. If fruit drop occurs, switch to a bloom special-type of low nitrogen fertilizer.

Table 5: Deficit and optimum values of different nutrients in quince leaves

Nutrient	Deficit value	Optimum value
Nitrogen (%)	1.81	4.88
Phosphorous (%)	0.04	0.19
Potassium (%)	2.11	4.45
Calcium (%)	0.91	2.36
Magnesium (%)	0.12	1.57
Zinc (ppm)	35.97	71.94
Iron (ppm)	18.15	57.48
Boron (ppm)	44.25	80.69

Mirabdulbaghi and Abdollahi, 2014

19.1. Deficiency symptoms of nutrients

19.1.1. Nitrogen

Nitrogen deficiency is common in fruit and nut trees. However, it is only element required for proper nutrition of the quince plants. Symptoms are usually fairly uniform over the entire plant. The symptoms may develop at any time during the growing season depending on weather conditions. In severely deficit plants root development was inhibited due to which plant growth and development suffers. Leaves may develop necrotic areas along their margins. Fruits are usually smaller, mature early and their quality deteriorates.

19.1.2. Phosphorous

Severe deficiency symptoms of phosphorous are rare in quince plants. Its deficiency occurs under high pH and is exhibited by leaf margin and tip burn, shoots terminal growth, die back of new growth and failure of fruits to mature normally. Lateral buds may remain dormant or die and as a result few lateral shoots appear. Blossoming and fruiting are reduced and bud opening in the spring may be delayed (Shear and Faust, 1980).

19.1.3. Potassium

Marginal leaf scorching appearing first on older leaves is a well-known potassium deficiency symptom in quince which was also noticed in other pome and stone fruits. Young leaves generally show less severe symptoms although they may be smaller than normal. A heavy fruit crop usually accentuates the symptoms (Ebert, 2009).

19.1.4. Calcium

Under calcium deficiency, leaves and fruits of quince has low Ca content, hence cork spot and reduction in firmness of fruit are the prominent symptoms. Cork spot may appear early in the development of the fruit. Calcium deficit fruits are also more susceptible to sunburn (Shear and Faust, 1980).

19.1.5. Magnesium

A mottled pale green or chlorotic pattern in the interveinal tissue of older leaves, often starting near the margin has been noticed under magnesium deficiency in quince. Its deficiency symptoms usually appear late in the season. However, it is difficult to distinguish between Mg and K deficiencies in quince (Shear and Faust, 1980).

19.1.6. Boron

Boron is required only in very small amounts in quince, but when it is deficient disorders can occur like bitter-pit. The affected fruit developed necrotic tissue which became obvious mainly after harvest, during storage (Holevas and Biris, 1980). Waterlogged and very dry soils can lead to boron deficiency, as roots are unable to take up nutrients. Symptoms may develop on fruit, twigs and leaves. Fruits with serious external symptoms of boron deficiency are affected quite young and are strictly an orchard problem because they would never reach the market.

19.1.7. Iron

Quinces are particularly susceptible to iron deficiency under alkaline conditions which results in reducing yields and fruit quality. Iron in interaction with other nutrients may become scarcely available to the plant and also lead to nutritional disorders (Prado, 2008). Chloroplast pigments decreased in quince plantlets under iron deficiency and also reduce fruit sucrose and total carbohydrates concentration (Tagliavini, 1995). Prado and Vara (2011) reported that quince plants were susceptible to iron deficiency and may have a low ability for nutrient absorption from the soil at low concentrations.

19.1.8. Zinc

Quince is sensitive to Zn deficiency and this is a major problem in areas where quince is an important fruit crop. Zinc deficiency is mainly a problem in alkaline soils that are low in organic matter. Its deficiency can also become a problem when soils are cool and wet during the vegetative phase. Its deficiency symptoms vary between species but several effects can be generalized. Young leaves of quince are most affected. Over time, the chlorotic spots can turn a bronze color and necrotic spots may start to develop from the margins.

20. Role of bio-regulators

Plant bio-regulators both natural and synthetic affect growth and development processes in plants. Plant growth regulators are the key factors in manipulation of any plant part and genetic manipulation through tissue culture. To achieve in-vitro rooting of micro-propagated shoots of East Malling Quince A (*Cydonia oblonga* Mill) with different auxin treatments (IBA, IAA and NAA at 0.0, 0.1, 0.5 and 1.0 mg/l) and found effective in mass multiplication (Dumanogl and Gulsen, 1994).

21. Use of growth promoting rhizobacteria (PGPR)

Floral and foliar applications of *Bacillus mycoides, Bacillus subtilis* OSU-142 single and in combination on quince significantly increased yield per tree, number of fruit, fruit weight, fruit width and fruit height respectively, compared with the control. It was determined that the bacteria applications did not changed importantly rate of soluble dry matter, titratable acidity and pH while increased width of fruits and height of fruits compared with the control. Arkan *et al.* (2013) suggested that use of *Bacillus mycoides, Bacillus subtilis* OSU-142 single and in combination have a great potential to promote growth and development of quince cultivation.

22. Fruit bagging

Mediterranean fruit fly damage to quince fruit is wide spread in central Greece. Bagging of each fruit with a paper bag resulted in more yellow fruit and less starch that could be perceived as more advanced maturity, but with lower percent dry matter and organoleptic quality. Nevertheless, bagged fruit were free of fruit fly damage (Pastopoulos *el al.,* 2012).

23. Insect-pests and their management

23.1. Apple aphid

Nymphs and adults of apple aphid (*Aphis pomi*), apple grain aphid (*Rophalosiphum fitchii*) and woolly apple aphid (*Eriosoma lanigerum*) can cause injury to quince trees. Aphids damage leaves on water sprouts and rapidly growing terminal shoots. Young quinces may be deformed when infestations are severe. In the fall, females lay elongate shiny, black eggs on the bark. The nymphs hatch from eggs in the spring. Both the nymphs and adults of this small green aphid pierce quince tissue to suck the sap in stems, foliage and fruit. During the growing season, apple aphids have many generations, during which young are produced asexually. In the fall, sexual forms appear and after mating, the females lay the eggs that will overwinter. Heavy fertilization with nitrates may increase aphid infestations. To control the aphids spray of horticultural oil, malathion, diazinon is recommended. Dormant sprays of horticultural oil have been marginally effective because of later migration. Sprays of malathion and diazinon control aphids but are detrimental to natural enemies. Natural enemies, such as syrphid flies, ladybird beetles, and lacewings, sometimes adequately control apple aphids.

23.2. Codling moth (*Cydia pomonella*)

Caterpillars of the codling moth damage the quince. This moth is one of the three most destructive apple insects that feed directly on quince. Pinkish caterpillars that have overwintered under bark or debris form pupae in spring. Females emerge around bloom to lay eggs on fruit. The grayish adults are slightly more than 1/4" long with wavy lines and a dark band at the tips of the wings. Caterpillars of the first of the two generations enter the fruit shortly after the blossoms fall. Those of the second generation cause injury to fruits in August and September. The amount of damage varies greatly from year to year. Caterpillars of the codling moth are distinguished from those of other boring caterpillars by having a dark brown anal comb at the rear of the body and by tunneling directly to the core of the fruit. Pheromone traps may be used to monitor adult flights. Applications of carbaryl, phosmet or ryania are among the insecticides use against this pest which are applied about 21 days after full bloom.

23.3. Oriental fruit moth (*Grapholita molesta*)

Caterpillars of this moth are among the most destructive pests of quince. The principal damage is 'wormy' fruit, but shoots may also be killed. Mature caterpillars overwinter in impervious silk cases on the bark of the trunk or upon debris in orchards. These caterpillars are pink and about 1/2" long. In spring, they form pupae in the cases and the mostly gray adults emerge in May, starting when peaches bloom. The moths have wingspan of just less than 1/2". The females lay eggs upon the undersides of the leaves and in some cases, on the twigs. The caterpillars of the first generation tunnel into the tender shoots early in the season and later they leave the twigs to enter the fruit. As twig borers, the caterpillars kill the tips of shoots and cause the shoots to branch. The caterpillars of later generations go directly into the fruit after hatching. Insecticidal applications of phosmet or diazinon are recommended to control this pest which is applied from petal fall until four weeks before harvest. Use of pheromone traps will assist in timing the sprays.

23.4. Quince curculio (*Conotrachelus crataegi*)

This beetle is one of the most destructive pests of quince fruit. After spending the winter in the soil, the larvae form pupae in spring. The adults appear in midsummer, usually about the middle of July. These brownish gray beetles are about 1/4" in long. Females lay eggs singly in fruit. After the eggs hatch 7-10 days later, the larvae enter into the flesh of the fruits, rarely reaching the core. The flesh-colored larvae are grub-like and without distinct legs. In about one month, usually in August, the mature larvae leave the fruit to make cells 2-3" deep in the soil where they remain until the next year. Adult curculios can be controlled by applying phosmet at petal fall and again about 10 days later.

23.5. Resplendent shield bearer (*Coptodisca splendoriferella*)

The caterpillars of this moth cause minor foliar damage to quince and apple. In May, the tiny adults emerge from pupae that have overwintered in cases on the tree. The forewings of these moths are lead gray on the basal half and golden with streaks of dark brown and silver at the tip. The females lay eggs on the leaves. After hatching, the caterpillars form tiny blotch mines as they feed on leaves. When the caterpillars are mature, they cut out shield-shaped portions of the leaf for cases and fasten them to the bark. They have at least two generations, with pupae of the last remaining the leaf cases until May of the following spring. Sprays applied for other foliar pests have controlled this pest.

23.6. San Jose scale (*Quadraspidiotus perniciosus*)

San Jose scales commonly infest quince. Large infestations of the San Jose scale can cause branch and even shrub death. Partially-grown scales overwinter under their circular gray covering or scale on the twigs and the branches of trees. They begin to feed as the sap starts to flow. When apple trees bloom, the males emerge from under their scales to mate with the immobile females. Females are circular and cone-shaped and their circular scales are about 1/16" in diameter, with a raised center or nipple. The males are smaller and elongate, with the nipple not centered on the scale. Females give live birth to tiny bright yellow crawlers in June. The young crawlers quickly settle, insert their long mouthparts into the twigs, and then suck sap from branches. As they grow, the crawlers secrete a waxy filament that becomes their scale or covering. Overlapping generations are present from June through September. Scales may be controlled by applying horticultural oil at a dormant timing or in June when crawlers are active. Diazinon can also be applied against the crawlers in June. To detect the yellow crawlers, wrap black tape coated with vaseline around small branches. Adult flights may be detected with pheromone traps.

23.7. Apple maggot flies

Apple maggot flies also attack the quince fruit, laying their eggs beneath the skin of the fruit. The maggots will hatch and burrow through into the soil to continue their life cycle. Gathering fallen fruits at least twice a week can cut down on the number of flies present around the tree. Common insecticides containing organophosphate will control the apple maggot. Avoid over spraying within two weeks of harvesting the fruit.

24. Diseases and their management

24.1. Black spot

This fungal disease is caused by *Fabraea maculate* which can cause severe injury if not controlled. The disease shows small circular spots on leaves, reddish-brown spots on the periphery, with black centers. In the centers, numerous minute black spore-bearing pustules can be seen. A heavy infection will cause extensive defoliation. The fruit infections appear as irregular sunken black spots scattered at random over the surface. The spots may become confluent and cause cracking of the fruit. The causal fungus overwinters on the old leaves, producing two types of spores in early summer. These start new infections on the foliage and young fruit. Spray of mancozeb fungicide will control the disease.

24.2. Quince rust

It is a fungal infection caused by *Gymnosporangium clavipes*, also seen on hawthorn, juniper and crab apple trees. This rust occurs on the leaves, stems and fruit. On the leaves it produces small orange-colored spots, on the stems it causes swollen gall-like growths and on the fruit it appears as circular orange-colored areas which may comprise half the surface area. On all infected parts of the plant the brilliant orange spores are borne in very small, white, toothed cups. Control quince rust by pruning out infected branches and spraying with a fungicide.

24.3. Fire blight

Fire blight is more common and serious disease of quince. It is a dangerous bacterial disease caused by *Erwinia amylovora* which will crack and distort all parts of the tree. It develops in early spring, during rainy warm weather. The blossoms of the tree begin to shrivel and turn black. The tips of twigs will start to curl and become distorted and dead leaves will cling to the tree as the infection worsens. As the bacteria spread, they form black cankers along the edges of the bark, giving the appearance of scorched wood. The bacteria are easily transferred between trees through insects and rain. Fire blight can be controlled with terramycin and Bordeaux mixture (copper spray). Balajoo *et al.* (2012) observed good control of fire blight with the application of Bion (active ingredient acibenzolar-s-methyl) at 200 mg/l in both glasshouse and screen house conditions

Organic growers use pruning as their control. All pruning tools need to be disinfected after each cut. All infected debris should be burned. Copper sprays should be applied at a time when they can dry quickly as they may cause scarring on the quince rust.

24.4. Powdery mildew

Powdery mildew is a fungal growth that can cause the fruits to become discolored and growth to be stunted. Symptoms usually begin with small patches appearing on the leaves that spread into black spots that rapidly reproduce on the buds and branches of the tree. Infected blossoms become unable to produce fruit and fall off and leaves often become distorted. Application of copper and sulphur based fungicide can control the occurrence.

24.5. Scabs

Quince trees along with many other fruit bearing trees often suffer from scabs such as the apple scab. Scabs begin to appear as small dark spots on the leaves that eventually begin to spread. As it develops the scab will appear on the fruits as raised black spots. This infection will begin to kill the tree's leaves as it worsens. Concentrated dormant oils may be used, which will smoother and kill the infection fungi.

24.6. Blue mold

It is caused by *Penicillium expansum*, is the most common post harvest disease of quinces. Control strategies include careful handling to minimize wounding, prompt cooling to 0°C and maintenance of optimum temperature and relative humidity during storage.

24.7. Fruit rot (C.O. *Botryosphaeria obtuse*)

The affected fruits showed a soft rot that evolved rapidly covering the entire fruit surface and causing mummification. A fungal stroma developed on the fruit surface but no spores, conidiomata or ascomata were observed. Some mummified fruits dropped prematurely, although most of them remained attached. There were no symptoms on leaves or branches of the affected trees (Moral *et al.,* 2007).

In Italy Marinelli et al. (2012) reported another strain that causing fruit rot. Affected fruits showed rotting that evolved rapidly, covering their entire surface and causing mummification. Some mummified fruits dropped prematurely, although most of them persisted on the trees. No symptoms were observed on leaves or branches of the affected plants. The causal pathogen was identified as *Fusicoccum aesculi*, the anamorph of *Botryosphaeria dothidea,* based on morphological, genetic (ITS) and pathogenicity analyses.

24.8. Anthracnose (C.O. *Colletotrichum clavatum*)

The symptoms were observed on fruits approximately 2 months after harvest. Affected fruits showed sunken, dark brown to black lesions with orange conidial masses produced in black acervuli. Based on morphological characteristics, molecular investigations (sequence analysis of the rDNA internal transcribed spacer region and beta -tubuline 2 gene) and pathogenicity tests, the causal agent was identified as *Colletotrichum clavatum* (Zivkovic *et al.,* 2014).

24.9. Diseases caused by viruses

Paunovic and Rankovic (1997) reported the occurrence and symptoms of virus diseases in all the major quince growing areas of Yugoslavia. The majority of trees of cv. Leskovacka exhibited fruit deformation accompanied by sooty ringspot of leaves, bark splitting, stunting and progressive tree decline. Comparative analyses of dsRNAs isolated from infected quince trees and plants infected with PVYV ((pear vein yellows virus) confirmed the causal agent of quince fruit deformation and sooty ringspot of leaves to be a strain of PVYV. Symptom expression on fruits and leaves or only on fruits in different cultivars is obviously a cultivar-specific reaction.

Virus diseases *viz*., Apple stem pitting virus (ASPV), Apple stem grooving virus (ASGV) Apple mosaic virus (ApMV) and Apple chlorotic leaf spot virus (ACLSV) are known as viral pathogens that can affect quality and quantity of quince production. Birisik and Baloglu (2010) observed leaf mosaics, leaf deformation, fruit malformation, gummy fruit, dwarfing of the tree, bud-union abnormalities and trunk deformations in commercial quince orchards in Turkey.

25. Physiological disorders

25.1. Brown core and Stem cavity browning

Brown core is characterized by a browning of the flesh around the seed cavities. The browning may involve all or part of the core area and may extend as discolored wedges beyond the core line into the surrounding flesh and sometimes in the underlying flesh, properly called stem cavity browning may be associated with severe brown core. While these two symptoms may not always occur together, it is believed that they are expressions of the same disease. Brown core may be confused with boron deficiency cork. Cork, however may be distinguished by corky spots or streaks scattered through the flesh in addition to the corky discolored core tissues. The discolored core tissues of brown core are softer and wetter than those of boron deficiency cork.

Brown core is a low temperature storage disorder of pome fruits that usually does not develop until January. It becomes most serious in fruits stored for long periods at (31^0 to 32^0F) and becomes more pronounced after the fruits have been removed from the cold. An extended period of cloudy rainy weather when apples are maturing appears to predispose the fruits to brown core. Delay in harvesting beyond prime maturity and an excess of nitrogen fertilization may also increase susceptibility.

Brown core may be greatly reduced by storing susceptibility varieties at (38^0 or 40^0F), but the storage life of the fruits is seriously shortened. Store the fruits at 38^0 in controlled atmosphere to prevent brown core and also to extend the storage life of the quince fruit. The controlled atmosphere contains 2 to 3 per cent CO_2 for the first month, then 5 per cent for the rest of the storage period. A content of 3 per cent oxygen is maintained throughout the storage period.

25.2. Fruit deformation

Deformed quince fruits were observed on the cultivar Leskovacka in several locations in Yugoslavia. Symptoms were also observed on leaves and bark of affected trees. Tree growth was stunted and branch dieback and progressive tree decline were observed. Grafting of infected material reproduced symptoms on leaves and fruit of healthy Leskovacka quince plants (Paunovic and Rankovic, 1992).

25.3. Internal browning

Fruits of quince cv. Gigante di Vrania grown on 2 farms in the Pianura Padana (Italy) and picked on 3 dates at weekly intervals were placed in cold storage for 6 months. At the end of storage fruits picked early showed less internal browning than those picked later. The activities of superoxide dismutase, catalase and glutathione peroxidase were higher in fruits showing internal browning than in healthy fruits, suggesting that lipid peroxides were involved in the development of internal browning in quinces during storage (Testoni *et al.*, 1996).

26. Harvesting

Most of the quince varieties when approaches to maturity, the deep green of immaturity gradually lightens and passes through various colour changes until the yellow gold of full maturity is reached. Maturity can also be commenced when the seed coat turns dark brown, the skin gets sticky and the fruit becomes fragrant. In spite of the very firm fruit flesh, the fruit skin is very sensitive to mechanical damage. Despite its apparent firmness, the fruit will bruise and the skin will easily become marked by handling. Ideally, fruit for the fresh market should be harvested and handled when some colour approaching maturity is apparent, but before it is fully mature. Because of the limited fresh fruit market for quinces, some of the larger growers begin marketing as soon as the fruit is fully developed but still fairly green. This extends the marketing period. By growing a range of varieties with differing maturity dates, the harvesting period may extend over a period of two months. Under north Indian conditions of the country the quince fruits are ready to harvest in October or November, however in warm to hot districts, harvesting may commence as early as mid-February extending to April. In Turkey, The harvest time lasts from mid-September to the middle of October based on the different cultivars. Quince fruits show a climacteric respiratory pattern (Gunes, 2008). Fruit should be cut from the tree using a sharp knife as pulling them can cause damage to the skin. While picking, long pieces of stem should be removed so that they will not cause skin damage to other fruits. Pick only undamaged fruits.

27. Yield

Young and mature plants with about 250 trees per hectare will yield about 25-35 tones fruit per hectare. In high density orcharding, where a greater number of trees are now planted per hectare than in the past, yields per tree may be less but the total yield per hectare will be greater. However, from a single tree we can get 80 to 100 kg fruit per tree under heavy cropping varieties with good management practices.

28. Storage

Although no storage trials have been conducted for quince fruits as quince is considered as a underutilized/minor fruit crop and mostly this fruit is consumed freshly even then it can be successfully stored at cool and dry place for a period of 2 months at a temperature of 0°–1°C with a relative humidity of 85 per cent (the temperature used for cool storing apples). Gunes (2208) reported that the fruit can be stored at 0 or 2 degrees C for 2-6 months and storage can be prolonged up to 7 months by controlled atmosphere conditions of 2% O2 + 3% CO2 at 2 degrees C. The formation of flesh browning is the most limiting factor in the storage of quince fruit. Some varieties can develop superficial scald and flesh browning when in cool storage. Quinces have a distinctive fruity odour, especially when confined in a storage chamber. Storing with other fruits may lead to problems of flavour tainting enclosing the stored quinces in suitable polyethylene can usually overcome this problem. Obviously a cautious approach to storage is required at this stage.

Gunes (2010) studied the effect of 1-MCP on quality of 'Esme' quince (*Cydonia oblonga* Mill.) fruit during storage. Immediately after harvest, fruits were treated with 625 ppb 1-MCP at 12 degrees C under gas tight conditions for 24 hours and subsequently they were stored at 2+or-1 degrees C and 85-90% RH for six moths. As a result, 1-MCP mostly affected respiration rate, flesh browning and decay in fruits from both regions. Moreover, they showed higher soluble solid content and sensory evaluation value. Based on sensory characteristics and inhibition on flesh firmness, flesh browning data, pre-storage 1-MCP treatment may be a promising postharvest treatment to keep postharvest quality of 'Esme' quince fruit.

Akbari and Ebrahimpour (2014) investigated the effects of potassium permanganate treatments and packing types on fruit quality and storage life during storage. Results showed that fruits treated with potassium permanganate at 5 g/kg and stored in sealed polyethylene bags resulted in the lowest decrease in flesh firmness, skin color change and astringent content.

29. Grading and packing

There is no fix standard grades, sizes or packaging for quince. However, quality standards for quince fruit should be marketed at optimum maturity and which are free from insect damage, mechanical damage and decay (Kader, 1992). Since fruits are rather tender and bruise easily, the shipping container should be lined with soft pads on both ends. The blossom end is turned upwards (Auchter and Knapp, 1929). Fruit should be handled carefully during harvesting, grading and packaging to avoid mechanical injury, if they are not immediately processed or frozen. For fresh fruit markets, the smaller sizes of quinces can be packed in the fibreboard carton, but larger sized fruit will not fit in a tray pack; these can be fitted in a single-layer tray. Care must be taken when loose filling.

30. Utilization/processing/value addition

Sood and Bhardwaj (2015) evaluated the quince fruit for its quality parameters and develop products namely pickle, preserve, jam, *chutney*, squash and candy and reported that these products are well suited for table use as they are rich in various constituent especially fiber and low in fat content i.e. almost equivalent to zero.

Processed product like jam and jelly were prepared and assed for their nutritive quality and acceptability. The total soluble solid of jam and jelly were recorded to be 70° and 66°Brix. It was found that quince jelly contains more amount of ascorbic acid (12.60 mg/100 g) and pectin (1.32 %) compared to jam (Sharma *et al.,* 2011). Quince fruits were harvested in the course of October and thereafter analysed at the stage of consumption ripeness after storage at the temperature of +2°C and a relative air humidity of 85 per cent. The contents of pectins in fruits were high – the cultivar Hruskovita contained 3.51 ± 0.19 g/ 100 g of fresh weight. The contents of vitamin C were also high in cultivar Muskatova containing as much as 79.31 ± 2.01 g/100 g of fresh weight. The affinity of chemical properties of the individual cultivars was expressed by means of cluster analysis and it was found out that there were no marked differences between pear-shaped (*Cydonia oblonga* subsp. *pyriformis*) and apple-shaped (*Cydonia oblonga* subsp. *maliformis*) forms of fruit.

30.1. Dried quince: Dried quince has been used as a tea for centuries. Gheisari and Abhari (2014) observed that oven drying of fruits can better preserve the antioxidative activity of the quince than sun drying.

31. Marketing and Trade

As quince is considered as underutilized fruit crop of Northwestern Himalayas, so no proper orchards are there and only scattered plantations are found, mostly in the home gardens, bunds of apple, pear orchards. Fruits of quince are sold in the local markets @ 40 - 60 rupees for fresh consumption as vegetables or for some processed products are mentioned above. In California, quince fruit are harvested from mid August to early November and can be available for purchase up through January (Karp, 2009). Quince is mainly sold fresh at local farmers markets, but it can also be found at specialty grocery stores. Processed quince products can be found through all marketing channels viz. farmers markets, specialty grocery stores and even online.

32. Future thrust areas

Quince possesses tremendous potential as a crop of next generation. However, so far the research work done in this crop with respect to production technology is extremely limited and is also unorganized. The following areas can be considered as the thrust areas for further studies.

1. Survey and identification of high yielding genotypes with good quality fruits.
2. Development of technology for proper growth of the seedling or grafted plants.
3. Standardization of protocol for micropropagation.
4. Canopy management for high yield and better quality fruits.
5. Standardization of GAP.
6. Standardization of the maturity indices for proper harvesting.
7. Innovative value addition and processing.

References

Abrahao, E., Alvarenga, A.A. and Souza, M. 1991. Marmeleiro (*Chaenomeles sinensis*) cv. Japones – Porta-enxerto para marmeleiros, pereiras e nespereiras. *Revista Brasileira de Fruticultura,* **13**(2): 181-182.

Abrahao, E., Souza, M. and Alvarenga, A.A. 1996. A cultura do marmeleiro em Minas Gerais. Belo Horizonte: EPAMIG, 23 (Boletim Tecnico, 47).

Adler, M. 2001. Quince (*Cydonia oblonga* Mill.) and its growing and economic descriptions. In: Proceedings 9th International Conference of Horticulture, 3–6 Sept., Lednice, Czech Republic: 3–7.

Agelet, A. and Valles, J. 2003. Studies on pharmaceutical ethnobotany in the region of Pallars. Part II. New or very rare uses of previously known medicinal plants. *Journal Ethnopharmacology*, **84**: 211-227.

Ahmad, M.F., Sharma, A.K. and Jabeen, A. 2004. Genetic diversity in quince (*Cydonia oblonga* Mill). *Indian Horticulture,* **48:** Cover II.

Ahmad, M.F., Sharma, A.K. and Jabeen, A. 2004. Genetic diversity: Quince (*Cydonia oblonga* L.). *Indian Journal of Horticulture,* **48** (4): 21-22.

Ahmad, M.F., Wani, N., Hassan, G. I., Zafar, G. and Khan, I.A. 2008. Variability in quince (*Cydonia oblonga* Mill.) population from Baramulla district of Kashmir valley. *Indian J. Genet.*, **68**(2): 215-218.

Ahmad, W.S., Ejaz, K.A. and Ashraf, M. 2014. Exploration of the *in vitro* cytotoxic and antiviral activities of different medicinal plants against infectious bursal disease (IBD) virus. *Central European Journal of Biology,* **9**: 531-542.

Akbari, H. and Ebrahimpour, H. 2014. Potassium permanganate and packing types impacts on postharvest quality and storage period of quince fruit (*Cydonia oblonga* Mill.). *International Journal of Advanced Life Sciences IJALS*, **7**(2): 267-275.

Alvarenga, A.A., Abrahao, E., Pio, R., Assis, F.A. and De-Oliveira, N.C. 2008. Comparation among marmalades produced from different fruit quince species (*Cydonia oblonga* Mill. and *Chaenomeles sinensis* Koehne) and cultivars. *Ciencia e Agrotecnologia,* **32**: 302–307.

Arkan, S; Ipek, M; Prlak, L. 2013. Effects of plant growth promoting rhizobacteria (PGPR) on yield and fruit quality of quince. International Proceedings of Chemical, Biological and Environmental Engineering-IPCBEE, **60**: 97-100.

Aslan, M., Orhan, N. Orhan, D.D. and Ergun, F. 2010. Hypoglycemic activity and antioxidant potential of some medicinal plants traditionally used in Turkey for diabetes. *J Ethnopharmacol,* **128**(2): 384-389.

Auchter, E.C. and Knapp, H.B. 1929. Harvesting, Storing and Marketing and Establishing the orchard. *In*: Getman, A.K. and Ladd, C.E. (eds). Orchard and Small Fruit Culture. John Wiley & Sons, NY. Pp. 1-132 and 138-176.

Azamatov, M.A. 2012. Results of quince variety trials in the steppe zone of Kabardino Balkaria. *Vestnik-OrelGAU*, **3**: 73-75.

Bailey, L.H. and Bailey, E.Z. 1976. Hortus Third: A Concise Dictionary of Plants Cultivated in the United States and Canada. John Wiley and Sons, Inc.

Baker, B.S. and Bhatia, S. 1993. Factors effecting adventitious shoot regeneration from leaf explants of quince (*Cydonia oblonga*). *Plant Cell Tiss. Org. Cult.,* **35**: 273–277.

Baker, B.S. and Bhatia, S.K. 1993. Factors effecting adventitious shoot regeneration from leaf explants of quince (*Cydonia oblonga*). *Plant Cell Tissue and Organ Culture,* **35**(3): 273-277.

Balajoo, O.M; Kesahavarzi, M; Zahabi, A; Danesh, Y.R. and Haghjuyan, R. 2012. Protective effect of acibenzolar-s-methyl on fireblight severity in quince and characterization of the Erwinia amylovora strains involved. *Journal of Plant Pathology,* **94**(1): 211-214.

Baxeter, P. 1997. Growing Fruit in Australia. 5th edition, Thomas Nelson Australia, Melbourne, Pp: 134-139.

Bayazt, S; Imrak, B; Kuden, A. and Gungor, M.K. 2011. RAPD analysis of genetic relatedness among selected quince (Cydonia oblonga Mill.) accessions from different parts of Turkey. *Horticultural Science*, **38**(4): 134-141.

Bell, R.L. and Leitao, J. 2011. Cydonia *In*: Kole, C., (ed.) Wild Crop Relatives: Genomic and Breeding Resources. Berlin, Germany: Springer-Verlag. p. 1-16.

Benedek, P., Szabo, T. and Nyeki, J. 2001. New results on the bee pollination of quince (*Cydonia oblonga* Mill.). *Acta Horticulturae,* **561**: 243-248.

Bernardo, G. 2008. Trees in Patagonia. 1st edition. ISBN10: 3764388374. pp: 13-14.

Bettiol Neto, J.E; Pio, R; Sanches, J; Chagas, E.A; Cia, P; Chagas, P.C; Antoniali, S. 2011. Production and quality attributes of quince tree cultivars in the eastern of the State of Sao Paulo. *Revista Brasileira de Fruticultura,* **33**(3): 1035-1042.

Birisik, N. and Baloglu, S. 2010. Evaluation of the presence and symptomology of viruses in commercial quince orchards in Turkey. Julius Kuhn Archiv, **427** : 257-262.

Biro, G. and Lindner, K. 1999. Tápanyagtáblázat. Medicina, Budapest.

Blando, F., Giorgetti, L., Tonelli, M.G. and Nuti-Ronchi, V. 1992. Cytological characterization of cell suspension cultures of fruit tree. *Acta Horticulturae,* **300**: 377-380.

Bobev, S. and Deckers, T. 1999. Field susceptibility to fire blight of pome fruits in Bulgaria. *Acta Horticulturae,* **489**: 221-224.

Bucsek, M.J., Nyeki, J., Szabo, Z. and Kadar A. 1996. Quantitation of mineral elements of different fruit pollen grains. *Mikrochimica Acta,* **13**: 333–338.

Cao BaoQin; Zhang MingChun; Du GuiXi; Rong WenCun, and Wang ZhenXi. 2007. Description of the 4 Chinese quince cultivars. *China Fruits*, **3**: 57-58.

Cao BaoQin; Zhang MingChun; Zhao CuiHua; Du GuiXi, and Lu jingPing. 2007. "Xipizi", a promising Chinese quince cultivar. *China Fruits,* **6** : 67.

Cetin, M. 2006. Standart ayva çesitlerinin döllenme biyolojisi üzerinde arastýrmalar (with English abstract). *Bahce,* **35**: 83-95.

Christianson, M.L. and Warnick, D.A. 1983. Competence and determination in the process of *in-vitro* shoot organogenesis. *Dev. Biol.,* **95**: 288–293.

Costa, R.M., Magalhães, A.S., Pereira, J.A., Andrade, P.B., Valentão, P., Carvalho, M. and Silva, B.M. 2009. Evaluation of free radicalscavenging and antihemolytic activities of quince (*Cydonia oblonga*) leaf: A comparative study with green tea (*Camellia sinensis*). *Food Chem. Toxicol.,* **47**: 860-865.

D'onofrio C. and Morini, S. 2006. Somatic embryo, adventitious root and shoot regeneration in *in-vitro* grown quince leaves as influenced by treatments of different length with growth regulators. *Scientia Horti.,* **107**: 194–199.

D'onofrio, C. and Morini, S. 2002a. Increasing NaCl and $CaCl_2$ concentrations in the growth medium of quince leaves: I. Effects on Shoot Regeneration. *In-vitro Cell. Dev. Biol.-Plant,* **38**: 373–377.

D'onofrio, C. and Morini, S. 2002b. Increasing NaCl and $CaCl_2$ concentrations in the growth medium of quince leaves: I. Effects on somatic embryo and root regeneration. *In-vitro Cell. Dev. Biol.-Plant,* **38:** 366–372.

D'onofrio, C. and Morini, S. 2005. Development of adventitious shoots from *in vitro* grown *Cydonia oblonga* leaves as influenced by different cytokinins and treatment duration *Biologia Plantarum,* **49** (1): 17-21.

D'onofrio, C., Morini, S. and Bellocchi, G. 1998. Effect of light quality on somatic embryogenesis of quince leaves. *Plant Cell Tissue Org. Cult.,* **53**: 91–98.

Dalkilic, Z. and Mestav, H.O. 2011. *In-vitro* pollen quantity, viability and germination tests in quince (*Cydonia oblonga* Mill.) *African Journal of Biotechnology,* **10**(73): 16516-520.

Davidson, M.D; Chartier Hollis, J. M. and Lynch, P.T. 1998. The expression of GUS in quince (*Cydonia oblonga* Mill.) leaves after co-cultivation with *Agrobacterium tumefaciens. Tree biotechnology: towards the millennium,* pp.321-323.

Dickson, E.E., Arumuganathan, K., Kresovich, S. and Doyle, J.J. 1992. Nuclear DNA content variation within the Rosaceae. *American Journal of Botany,* **79**: 1081-1086.

Dirr, M.A. 1997. Dirr's Hardy Trees and Shrubs, An Illustrated Encyclopedia. Timber Press, Portland, Oregon.

DolcetSanjuan, R., Mok, D.W.S. and Mok, M.C. 1991. Plantlet regeneration from cultured leaves of *Cydonia oblonga* L. *Plant Cell Reports,* **10**(5): 240-242.

Duke, J. A., Bogenschutz, M. J., du Cellier, J. and Duke, P.A.K. 2002. Handbook of medicinal herbs. 2nd edition CRC. p 650.

Dumanoglu, H. and Gulsen, Y. 1994. *In-vitro* rooting of quince (*Cydonia oblonga* Mill.) *Acta Horti.,* **367**: 360-363.

Durand, F. 2014. Quince: The Tough fall fruit with a Secret Reward, thekitchn.com.

Ebert, G. 2009. Fertilizing for High Yield and Quality Pome and Stone Fruits of the Temperate Zone. *IPI Bulletin* 19: 75p.

Encyclopedia of Life, 2009. *Cydonia oblonga* Mill. Quince. http://www.eol.org/pages/637321?category. Accessed 3 May, 2009

Ercisli, S. 2004. A short review of the fruit germplasm resources of Turkey. *Genetic Resources and Crop Evolution,* **51**: 419-435.

Erig, A.C. and Schuch, M.W. 2004. Effect of the explant orientation in the culture medium, sucrose and of agar concentration in the *in vitro* multiplication of quince shoots (*Cydonia oblonga* Mill.) cv. MC. *Revista Cientifica Rural,* **9**(2): 80-88.

Evrenosoglu, Y. and Misirli, A. 2009. Investigations on the pollen morphology of some fruit species. *Turk J Agric For.,* **33:** 181-190.

FAO, 2012. www.fao.org.

Fattouch, S., Caboni, P., Coroneo, V., Tuberoso, C.I., Angioni, A., Dessi, S., Marzouk, N. and Cabras, P. 2007. Antimicrobial activity of Tunisian quince (*Cydonia oblonga* Mill.) pulp and peel polyphenolic extracts. *Jr. Agric. Food Chem.,* **55:** 963-969.

Fioravanco, J. and Simonetto, P. 2006. Phenological and productive behavior of quinces in Veranopolis-RS. *Science and Agrotecnologia,* **30**(1): 15-20.

Flores, M.D., Chacon, C.R., Alvarado, M.L., Schmidt, D.A. and Alvarado, U.C. 2015. Quince (*Cydonia oblanga*) *in-vitro* plant root formation through an automated temporary immersion system and its acclimation. *Brazilian Magazine of Fruit Culture,* **37**(3): 739-747.

Ghanadi, A.R. 2003. In: Iranian Herbal Pharmacopoeia (in Farsi). Tehran: Ministry of Health and Medical Education Press, pp: 176-182.

Gheisari, H. R and Abhari, K. H. 2014. Drying method effects on the antioxidant activity of quince (Cydonia oblonga Miller) tea. *Acta-Scientiarum-Polonorum-Technologia-Alimentaria,* **13**(2): 129-134

Giorgota, A., Preda, S., Isac, M. and Tulvinschi, M. 2009. Development of micro propagation protocol for the Romanian quince (*Cydonia oblonga*) cultivar 'Aurii' and Rootstocks 'BN70' and 'A Type'. *Acta Horticulturae,* **839**: 105-110.

Gorji, A. 2003. Pharmacological treatment of headache using traditional Persian medicine. *Trends Pharmacol. Sci.,* **24**: 331-334.

Gorlach, S; Wagner, W; Podsedek, A; Szewczyk, K; Koziolkiewicz, M. and Dastych, J. 2011. Procyanidins from Japanese quince (*Chaenomeles japonica*) fruit induce apoptosis in human colon cancer Caco-2 cells in a degree of polymerization-dependent manner. *Nutrition and Cancer,* **63**(8): 1348-1360.

Gulsen, Y. and Dumanoglu, H. 1991. The effect of sucrose, agar and pH on shoot multiplication and quality in quince micro-propagation. *Acta Horticulturae,* **289**: 115-116.

Gunes, N. T. 2010. Effect of 1-MCP and different ecological conditions on postharvest quality of 'Esme' quince fruit during long term storage. *Acta Horticulturae,* **877**: 387-394.

Gunes, N.T. 2008. Ripening regulation during storage in quince (*Cydonia oblonga* Mill.) fruit. *Acta Horticulturae,* **796:** 191–196.

Gunes, N.T. and Dumanoglu, H. 2005. Some fruit attributes of quince (*Cydonia oblonga*) based on genotypes during the pre-harvest period. *New Zealand Journal of Crop and Horticultural Science,* **33**: 211–217.

Hamauzu, Y., Inno, T., Kume, C., Irie, M. and Hiramatsu, K. 2006. Antioxidant and antiulcerative properties of phenolics from Chinese quince, quince and apple fruits. *J. Agric. Food Chem.,* **54**: 765-772.

Hamid, R.G. and Khadijeh, H.A. 2014. Drying method effect on the antioxidant of quince (*Cydonia oblonga* Mill.) Tea. *Acta Sci. Pol. Technol.,* **13**(2): 129-134.

Hatamian, H; Kermani, M.J; Zanjani, B.M. and Habashi, A.A.2014. Adventitious shoot regeneration and flow cytometry of *Cydonia oblonga* Mill. (cv. Isfahan). *Agricultural Communications,* **2**(4): 41-48.

Hemmati, A.A., Kalantari, H., Jalali, A., Rezai, S. and Zadeh, H.H. 2012. Healing effect of quince seed mucilage on T-2 toxin-induced dermal toxicity in rabbit. *Experimental and Toxicologic Pathology,* **64:** 181-186.

Hilgert, N.I. 2001. Plants used in home medicine in the Zenta River basin, Northwest Argentina. *Journal of Ethnopharmacology,* **76:** 11-34.

Holevas, C.D. and Biris, D.A. 1980. Bitter pit-like symptoms in quinces; effect of calcium nd boron sprays on the control of the disorder. *Acta Horti.* **92:** 319-326.

Hricovsky, I., Reznícek, V. and Sus, J. 2003. Jablone a hrusne, kdoulone, mispule. Príroda. *Bratislava*: 53–54.

Kader, A.A. 1992. Postharvest technology of Horticultural crops, Pub. No. 3311, Univ. Calif., DANR, Communication Services-Publications, Oakland, CA.

Karp, D. 2009. There's a new taste for quince, Los Angeles Times.

Kaufmane, E. and Rumpunen, K. 2002. Pollination, pollen tube growth and fertilization in Chaenomeles japonica (*Japanese quince). Scientia Horticulturae*, **94**(3/4): 257-271.

Khademi, F. 2009. The efficay of quince leave extract on atherosclerotic plaques induced by atherogenic diet in coronary and aorta, hyperlipidemia and liver in rabbit. MSc dissertation, Tabriz University of Medical Sciences, Tabriz, Iran.

Khanam, N., Khoo, C. and Khan, A.G. 2000. Effects of cytokinin/auxin combination on organogenesis shoot regeneration and tropane alkaloid production in Duboisia myoporoides. *Plant Cell Tiss. Org. Cult.,* **62**: 125–133.

Khoubnasabjafari, M. and Jouyban, A. 2011. A review of phytochemistry and bioactivity of quince (*Cydonia oblonga* Mill.). *Journal of Medicinal Plants Research,* **5**(16): 3577-94.

Khrolikova, A. Kh and Baskakova, V.L. 1989. Breeding quince in the Crimean steppe. *Sbornik Nauchnykh Trudov Gosudarstvennyi Nikitskii Botanicheskii Sad,* **107**: 133-139.

Kopec, K. and Balík, J. 2008. Kvalitologie zahradnických produktu. MZLU, Brno: 135–136.

Kyzlink, V. 1990. Principles of Food Preservation. Elsevier, Amsterdam

Launert, E. 1981. Edible and Medicinal Plants. Hamlyn. ISBN 0-600-37216-2

Lim, T.K., 2012. Edible Medicinal and Non-Medicinal Plants, **4**: 373.

Lynch, P.T; Siddika, A; Mehra, A; Benelli, C. and Lambardi, M. 2014. Cryopreservation of quince (Cydonia oblonga Mill.). *Cryo Letters*, **35**(3): 188-196.

Marinelli, E., Orzali, L; Scalercio, S. and Riccioni, L. 2012. First report of *Botryosphaeria dothidea* causing fruit rot of quince in Italy. Journal of Plant Pathology, **94**(4, Supplement): S4.92.

Mass, F. 2008. Evaluation of pyrus and quince rootstocks for high density pear orchards. *Acta Horticulturae,* **800**:

McCandlish, L. 2009. Demystifying the quince. National Public Radio.

Michal, A. 2001. Quince (*Cydonia oblonga* Mill.) and its growing and economics descriptions. *Proc. of Ninth International Conference of Horticulture,* **(1)**: 3-7.

Mirabdulbaghi, M and Abdollahi, H. 2014. Foliar nutrient response in some Iranian quince genotypes. *J. Horti. Fores.* **6(10):** 92-98.

Monka, A., Brindza, J., Olearova, H., Chlebo, P. and Eftimova, J. 2013. Comparison of quince (*Cydonia oblonga* Mill.) and Chinese quince (*Pseudocydonia sinensis schneid.*) in morphological and antioxidant characteristics. *Institut ochrany biodiversity a biologickej bezpeènosti*, FAPZ SPU v Nitre, Tr. A. Hlinku

Moral, J; Lovera, M; Benitez, M. J; Arquero, O; Trapero, A. 2007. First report of *Botryosphaeria obtusa* causing fruit rot of quince (*Cydonia oblonga*) in Spain. *Plant Pathology*, **56**(2): 351.

Morini, S., D'Onofrio, C., Bellocchi, G. and Fisichella, M. 2000. Effect of 2,4-D and light quality on callus production and differentiation from in vitro cultured quince leaves. *Plant Cell Tiss. Org. Cult.,* **63**: 47–55.

Mozhar, N.V. 1999. Promising quince cultivars. Sadovodstvo-i-Vinogradarstvo, **5/6** 12-13.

Mratinic, E. 2010. Dunja. Beograd, Serbia: Partenon (in Serbian).

Nagy-Deri, H. 2011. Morphological investigations on anthers and pollen grains of some quince cultivars. *Acta Biol. Szeged,* **55**(2): 231-235.

Nagy-Deri, H. 2013. Floral attractivity of *Cydonia oblonga* Mill. cultivars. Ph.D Thesis submitted to University of PECS.

Necas, T. and Kosina, J. 2008. Vegetative propagation of Pear and Quince Rootstocks Using Hardwood Cuttings. *Acta Horticulturae,* **800**: 701-706.

Nikolic, M. 1996. Major biological-pomological properties of some more recently developed quince cultivars and hybrids. *Jugoslovensko-Vocarstvo,* **30**(3/4): 237-244.

Nisarg, C.P., G.R. Bhoomika, V.N. Shah and A.N. Mahajan, 2011.*Cydonia vulgaris* Pers.: A review on diversity, cultivation, chemistry and utilization. *Der Pharmacia Lettre,* **3**(3): 51-61.

Nunes, D.P; Scalon, S. de P.Q; Bonamigo, T. and Mussury, R M. 2014. Germination of quince:

temperature, salinity and light. *Bioscience Journal,* **30**(6): 1737-1745.

Palmese, M.T., Uncini Manganelli, R.E. and Tomei, P.E. 2001. An ethno-pharmaco-botanical survey in the Sarrabus district (south-east Sardinia). *Fitoterapia,* **72:** 619-643.

Papachatzis, A., Kalorizou, H., Vagelas, I., Sotiropoulos, T. and Tsipouridis, K. 2011. Screening quince cultivars and hybrids for resistance to fire blight (*Erwinia amylovora*). *Acta Horticulturae,* **918**: 933-936.

Pastopoulos, S; Pliakoni, E. D. and Nanos, G.D. 2012. Kaolin sprays and individual fruit bagging effects on quince fruit quality. *Acta Horticulturae,* **940** : 381-385.

Paunovic, S and Rankovic,M. 1992. Infectious fruit deformation in quince cv. Leskovacka. *Zastita Bilja,* **43**(3): 227-233.

Paunovic, S; Rankovic, M. 1997. Virus diseases of quince in Yugoslavia. *Jugoslovensko Vocarstvo,* **31**(3/4): 231-243.

Pieroni, A., Quave, C.L., Villanelli, M.L., Mangino, P., Sabbatini, G., Santini, L., Boccetti, T., Profili, M., Cicciol, T., Rampa, L.G., Antonini, G., Girolamini, C., Cecchi, M. and Tomasi, M. 2004. Ethnopharmacognostic survey on the natural ingredients used in folk cosmetics, cosmeceuticals and remedies for healing skin diseasesin the inland Marches, Central-Eastern Italy. *J. Ethnopharmacol.,* **91**: 331-344.

Pio, R., Chagas, E.A., Barbosa, W., Alverenga, A.B., Abrahao, E., Feldberg, N.P. and Tombolato, A.F.C. 2008. Grafting of quince 'Portugal' on Cydonia and Chaenomeles rootstocks. *Rev. Bras. Frutic,* **30**(3): 850-852.

Pio, R., Chagas, E.A., Barbosa, W., Tucci, M.L.S., Filho, F.A.M. and Campagnolo. 2010. Production of quince nursery trees by different grafting methods. *Ciencia Rural, Santa Maria,* **40(5):** 1049-1052

Pio, R; Chagas, E.A; Barbosa, W; Alvarenga, A.A; Entelmann, F.A. 2008. 'Taiwan Nashi-C' and 'Taiwan Mamenashi' (*Pyrus calleryana*), new rootstocks for quince trees. *Acta Horticulturae,* **800 (Vol 2)**: 659-664.

Pio, R; Chagas, E.A; Barbosa, W; Tucci, M.L.S; Mourao Filho, F de A A. and Campagnolo, M. A. 2010. Production of quince nursery trees by different grafting methods. *Ciencia Rural,* **40**(5): 1049-1052.

Pio, R; Dalastra, I.M; Abucarma, V.M; Campagnolo, M.A; Dranski, J.A.L. and Pinto Junior, A.S. 2010. Gibberellic acid concentrations and the stratification on emergence of quince tree rootstock (*Chaenomeles sinensis*). *Revista Brasileira de Sementes,* **32**(4): 125-131.

Pio, R; Ramos, J.D; Chagas, E.A; Gontijo, T.C.A; Carrijo, E.P; Mendonca, V; Alvarenga, A.A. and Abrahao, E. 2006. Rooting of cutting of cv. Portugal quince submitted to application of indolebutyric and acid naftaleno acetic acid. *Revista Cientifica Rural,* **11**(2): 115-119.

Postman, J. D. 2012. Quince (*Cydonia oblonga* Mill.) centre of origin provides sources of disease resistance. *Acta Horticulturae,* **948**: 229-234.

Postman, J.D. 2009. *Cydonia oblonga*: The unappreciated quince. *Arnoldia,* **67**: 2-9.

Prado, R.M. 2008. Nutricao de plantas. Sao Paulo: Unesp, 407 p.

Prado, R.M. and Vara, E.A. 2011. Tolerance to iron chlorosis in non-grafted quince seedlings and in pear grafted onto quince plants. *J. Soil Sci. Plant Nutri.,* **11(4):** 119-128.

Przemys³aw, J.S., Picazo, S.M., Szumny, A. Carbonell-Barrachina, A.A. and Hernández, F. 2014. Quality parameters, bio-compounds, antioxidant activity and sensory attributes of Spanish quinces (*Cydonia oblonga* Mill.). *J. Scientia Horticulturae,* **165**: 163–170.

Radovic, A., Nikolic, D., Milatovic, D., Durovic, D. and Trajkovic, J. 2016. Investigation of pollen morphological characteristics in some quince (*Cydonia oblonga* Mill.) cultivars. *Turkish Journal of Agriculture and Forestry,* **40:** 441-449.

Rahimi, R., Shams-Ardakani, M.R. and Abdollahi, M. 2010. A review of the efficacy of traditional Iranian medicine for inflammatory bowel disease *World Journal of Gastroenterology,* **16:**

4504-514.

Rati, V; Acatrinei, F. and Brinza, D. 1993. Study of some quince varieties with a view to completing and diversifying the varietal assortment for Bacau district. *Buletinul Universitatii-de-Stiinte Cluj Napoca Seria Agricultura-si-Horticultura,* **47**(1): 291-295.

Rehder, A. 1986. Manual of Cultivated Trees and Shrubs Hardy in North America, 2nd edition. Dioscorides Press, Portland, OR.

Robinson, T.L. 2008. Performance of pear and quince rootstocks with three cultivars in four high density training systems in the North-Eastern United States. *Acta Horticulturae,* **800**: 793-802.

Rodriguez-Guisado, I., Hernandez, F., Melgarejo, P., Legua, P., Martinez, R. and Martinez, J.J. 2009. Chemical, morphological and organoleptic characterization of five Spanish quince tree clones (*Cydonia oblonga* Miller). *Scient Horti.,* **122**: 491-496.

Ronzio, R.A. 2003. The Encyclopedia of Nutrition and Good Health. 2nd Edition, Facts on File, Inc. p 547.

Rop, O., Balik, J., Reznicek, V., Jurikova, T., Skardova, P., Salas, P., Sochor, J., Mlcek, J., and Kramarova, D. 2011. Chemical characteristics of fruits of some selected quince (*Cydonia oblonga* Mill.) cultivars. *Czech J. Food Sci.*, **29:** 65–73.

Rotaru, G.I. and Lobachev, A Ya. 1990. Comparative anatomical characteristics of fruits of new quince cultivars Nakhodka and Volgogradskaya Myagkoplodnaya. *Izvestiya Akademii Nauk Moldavskoi SSR, Biologicheskikh-i-Khimicheskikh Nauk.* **1**: 16-21.

Rubinskiene, M; Viskelis, P; Viskelis, J; Bobinaite, R; Shalkevich, M; Pigul, M. and Urbonaviciene, D. 2014. Biochemical composition and antioxidant activity of Japanese quince (Chaenomeles japonica) fruit, their syrup and candied fruit slices. *Sodininkyste-ir-Darzininkyste,* **33**(1/2): 45-52.

Sax, K. 1931. The origin and relationships of the Pomoideae. *Jr. Arn. Arbor.* ,**12:** 3-22.

Sezik, E., Silada, E.Y., Honda, G., Takaishi, Y., Takeda, Y. and Tanaka, T. 2001. Traditional medicine in Turkey X. Folk medicine in Central Anatolia. *J. Ethnopharmacol,* **75**: 95-115.

Sharafi, Y. 2011. Investigation on pollen viability and longevity in *Malus pumila* L., *Pyrus communis* L. and *Cydonia oblonga* L. *in-vitro. African Jr. Med. Plants Res.,* **5:** 2232-236.

Sharma Rakesh; Joshi, V.K and Rana, J.C. 2011. Nutritional composition and processed products of Quince (*Cydonia oblonga* Mill.). *Indian Journal of Natural Products and Resources,* **2**(3): 354-357.

Sharma, G., Sharma, O.C. and Thakur, B.S. 2009. Systematic of fruit crops. New India Publishing Agency, New Delhi, Pp: 429-431.

Sharma, R., Joshi, V.K. and Rana, J.C. 2011. Nutritional composition and processed products of Quince (*Cydonia oblonga* Mill.). *Indian Jr. Natu. Prod. Resou.,* **2**(3): 354-357.

Shear, C.B. and Faust, M. 1980. Nutritional ranges in deciduous tree fruits and nuts. *Horti. Rev.,* **2:** 142-163.

Shinomiya, F., Hamauzu, Y. and Kawahara, T. 2009. Anti-allergic effect of a hot-water extract of quince (*Cydonia oblonga*). *Biosci Biotechnol Biochem.,* **73**: 1773-8.

Siddiqui, T.A., Zafar, S. and Iqbal, N. 2002. Comparative double-blind randomized placebocontrolled clinical trial of a herbal eye drop formulation (Qatoor Ramad) of Unani medicine in conjunctivitis. *Journal of Ethnopharmacology,* **83**: 13-17.

Silva, B.M., Andrade, P.B., Ferreres, F., Seabra, R.M., Oliveira, M.B. and Ferreira, M.A. 2005. Composition of quince (*Cydonia oblonga* Mill.) seeds: phenolics, organic acids and free amino acids. *Nat. Prod Res.,* **19**(3): 275-81.

Silva, B.M., Andrade, P.B., Mendes, G.C., Seabra, R.M. and Ferreira, M.A. 2002. Study of the organic acids composition of quince (*Cydonia oblonga* Mill.) fruit and jam. *Jr. Agric. Food Chem.,* **50**(8): 2313-2317.

Simmons, A.E. 1972. Growing Unusual Fruit. David and Charles. ISBN 0-7153-5531-7.

Skurikhin, I.M. and Tutelyan, V.A. 2007. Tables of chemical composition and caloric content of

Russia's food. DiLiPrint, Moscow. 276.

Sood, S. and Bhardwaj, M. 2015. Nutritional evaluation of quince fruit and its products. *Jr. Krishi Vigyan*, **3**: 67-69.

Stancevic, A. 1990. Morava - a new quince cultivar. *Jugoslovensko Vocarstvo*, **24**(3): 11-16.

Stancevic, A. and Nikolic,M. 1992. Quince breeding in Yugoslavia. *Acta Horticulturae,* **317**: 107-109.

Tagliavini, M., Rombola, A.D., Marangoni, B. 1995. Response to iron-deficiency stress of pear and quince genotypes. *Journal Plant Nutrition*, **18:** 2465-2482

Tahraoui, A., El-Hilaly, J., Israili, Z. H. and Lyoussi, B. 2007. Ethnopharmacological survey of plants used in the traditional treatment of hypertension and diabetes in southeastern Morocco (Errachidia province). *Journal of Ethnopharmacology,* **110**: 105-117.

Teresa, P.M., Uncini, R.E.M., Paolo, E. and Tomei, P.E. 2001. An ethno-pharmaco-botanical survey in the Sarrabus district south-east Sardinia. *Fitoterapia,* **72**: 619-643.

Testoni, A; Vanoli, M; Lovati, F; Zini, M; Visai, C. and Budini, R. 1996. Storability and physiological aspects of internal browning in quinces. *Atti-dell' Istituto Sperimentale per la Valorizzazione Tecnologica dei Prodotti Agricoli, Milano*. 1993 publ. **16**: 79-90.

Tetera, V. 2006. Ovoce Bílých Karpat. CSOP. *Veselí nad Moravou,* 278–279.

Torkelson, A.R. 1995. The Cross Name Index to Medicinal Plants.CRC Press, pp: 1707

USDA. 2009a. Germplasm Resources Information Network - (GRIN) Online Database. National Germplasm Resources Laboratory, Beltsville, Maryland.

Usmanghani, K., A. Saeed and M.T. Alam, 1997. Indusyunic Medicine, University of Karachi press, Karachi. pp: 368.

Vaughan, J.G. and Geissler, C. 1997. The New Oxford Book of Food Plants. pp 202.

Velickovic, M., Jelicic, S. and Radivojevic, D. 2001. Pomological, techonological and medicinal properties of quinces, Vranjska and Leskovic. *Jugoslevensko Vocarstvo,* **34** (3/4): 125-129.

Westwood, M.N. 1978. Fruit and nut species. *In*: M.N. Westwood. eds., Temperate Zone Pomology, pp. 41-76. W.H. Freeman, San Francisco, CA.

Whealy, K. and Demuth, S. 1994. Fruit, berry and Nut inventory. Seed Saver Publications.

Yolacti, S. 2006. Düsük ve yüksek sýcaklýklarýn bazý meyve aðaçlarýnda in vitro polen gelisimi üzerine etkileri (with English abstract). MS. thesis. Fýrat Üniv. Fen Bil. Ens. Elazýð. p. 35. (unpublished).

Yurdugul, S. 2005. Preservation of quinces by the combination of an edible coating material, Semperfresh, ascorbic acid and cold storage. *European Jr. of Food Res. Technol.,* **220**: 579–586

Zivkovic, S; Gavrilovic, V; Popovic, T; Dolovac, N. and Trkulja, N. 2014. First report of *Colletotrichum clavatum* causing quince anthracnose in Serbia. Plant Disease, **98**(9): 1272.

14

Strawberry (*Fragaria x ananassa Duch.)*

R.R. Sharma

1. Introduction

The modern cultivated strawberry (*Fragaria x ananassa* Duch.) is one of the most delicious, refreshing and soft fruits of the world. Worldwide it is also the most widely distributed fruit-crop due to its genotypic diversity, highly heterozygous nature and broad range of environmental adaptation. The plant is cherished in gardens and in commercial fields for its beautiful red fruit that has a tentalizing aroma (Sharma, 2002; Sharma and Sharma, 2004). Being a rich source of vitamins and minerals coupled with delicate flavour, strawberry has now become an important table-fruit of millions of people around the globe. It is amongst the few fruit crops, which gives quicker and very high returns per unit area on capital investment, as the crop is ready for harvesting within 6 months of planting. However, due to several technological advances in its cultivation like, introduction of day-neutral cultivars, and protected cultivation, strawberries remain available as fresh fruit throughout the year.

At present, strawberry is grown in wide climatic zones, extending to temperate, Mediterranean, sub-tropical and taiga zones (Sharma, 2002; Sharma and Sharma, 2004). Its cultivation is influenced by the specific regional adaptations due to critical photoperiods and temperature requirements and thus its cultural systems are highly variable. Due to the constant efforts of strawberry breeders, the world-wide interest for strawberry cultivation has boosted its production tremendously, which has resulted in widespread popularity of strawberry in the last 50 years.

1.1. Composition

The fresh-ripe fruits of strawberry are a rich source of vitamins and minerals. Among vitamins, it is a fairly good source of vitamin A (60 IU/100 g of edible portion) and vitamin C (30-120 mg/100 g of edible portion) (Table 1).

Table 1: Nutritive value of strawberry fruit

Nutrient	Content	Nutrient	Content	Nutrient	Content
Water (%)	87-90	**Organic acids** (mg /100 g of edible portion)		**Minerals** (mg/100 g edible portion)	
Protein (g)	0.25-0.7	Citric	420-1240 mg	Potassium	164 mg
Carbohydrates (g)	8.5-9.2	Malic	90-675 mg	Calcium	21 mg
Fats (g)	0.2-0.5	Succinic	100 mg	Phosphorus	21 mg
Fibre (%)	1.1	**Vitamins**		**Sugars**	
Total phenolics	60-120 mg	Vit-A	60 IU	Fructose	1.7-3.5 %
Anthocyanins	50-100 mg	Ascorbic acid	30-120 mg	Glucose	1.4-3.0 %
Energy	44 calories			Sucrose	0.5-2.5 %

Source: Hancock (1999); Sharma and Sharma (2004)

Strawberry is also a rich source of pectin (0.55%), available in the form of calcium pectate, which serves as an excellent ingredient for jelly-making. Besides, it is also a rich source of minerals like, potassium, calcium and phosphorus (Sharma, 2002; Sharma and Sharma, 2004). The total soluble solid (T.S.S.) content of fruit comprises sugars, acids and other substances dissolved in cell-sap. Water constitutes about 90% of fruit. The mature soft fruit contains about 5.0% total sugars and 0.90% to 1.85% acids. Fructose and glucose are the major sugars found in strawberry, with a small proportion of sucrose. Citric acid is the most abundantly found organic acid, followed by malic, succinic and oxalic acids. These acids determine pH, colour stability and inhibit enzyme activity of strawberry-fruit. Strawberry-fruit contains polyphenols like, chlorogenic acid, catechin and coumaric acid. The contents of the phenolic compounds, however decrease as fruit ripens due to the synthesis of anthocyanins.

Strawberries are known for their characteristic aroma, which is attributed to the presence of ethyl hexanoate, methyl hexanoate, ethyl heptanoate, ethyl propionate, ethyl butanoate, methyl butanoate, furanone and linalool (Hancock, 1999; Sharma and Sharma, 2004). The red colour of the fruit is mainly due to the presence of an anthocyanin, pelarogonidin 3-onoglucoside, and traces of cyaniding.

1.2. Uses

Fruits are mostly eaten fresh, and are consumed not for the food value but for the flavour. Besides dessert purpose, strawberries are processed into various value-added products, canned strawberry, jam, jelly, ice-cream, conserves, freeze strawberry, wine and other soft drinks. The strawberry jam is famous the world over.

2. Origin, history and distribution

Fragaria is a genus of perennial creeping herbs, found growing in wild in different climatic zones of the world. The cultivated strawberry (*Fragaria x ananassa* Duch.) is a monoecious octoploid hybrid of two largely dioecious octoploid species, *Fragaria chiloensis* L.) Duch. and *Fragaria virginiana* Duch (Darrow, 1966; Hancock, 1999; Sharma and Sharma, 2004). It is assumed that hybridization between *F. chiloensis* and *F. virginiana* took place spontaneously in Europe in the seventeen century when female plants of *Fragaria chiloensis* of Chilean origin were grown in proximity to male *Fragaria virginiana* plants of North American origin. The history of the strawberry goes back as far as the Romans and perhaps even the Greeks, but because the fruit had never been a staple of agriculture, it is difficult to find its ancient references.

Chandler strawberries

However, it is evident from literature that strawberry was in cultivation in Europe by the thirteen century, and then French began to transplant wood strawberry (*F. vesca*) from wilderness to garden. But the plant was considered more for ornamental flowers than for its fruits, although it was grown to some extent for table-purpose (Darrow, 1966; Hancock, 1999; Sharma and Sharma, 2004).

The first reference of strawberry cultivation dates back to 1368, when a gardener, Jean Dudoy of King Charles V had planted 1,200 plants of strawberry in royal gardens of the Louvre in Paris. In 1375, Chateau de Couvres had planted 4 blocks of strawberry in his garden in Duchess of Burgundy. Doubtless, its culture was crude, but it was highly appreciated by Duchess of Burgundy. England too was an early admirer of strawberry. The successive modifications of the name from "Streowberige:, "Streowberge", "Streberewyse" and "Strawberry" show a long familiarity with it. There are many theories in England about the derivation of the name strawberry. Anglo-Saxons called strawberry as "hayberry" because it ripened at the time, the hay was mown. Another guess is that the name has been derived from the way children strung the berries on straws or grass or hay to sell, a cushion, still practised at present also in different parts of Ireland. A more likely explanation is that the Anglo Saxons used the name strawberry to describe the way the runners strew or strang away from mother plant to find place in which to grow. Commercial strawberry cultivation in the USA had started at the beginning of 18th century, and within 25 years it had gained significant momentum. Most of the European countries had started growing strawberries commercially since 18th century, and of the total production (4,594,539 tones), today nearly 60% of the world's production is only from Europe (Table 2). In Canada, the main growing regions are Ontario, British

Columbia and Quebec. Strawberry is also grown in Israel, Japan, Turkey, Australia and Newzealand. Poland produces strawberries only for processing. In India, strawberries were first introduced by the IARI regional Horticultural Research Station, Shimla (Himachal Pradesh) in early sixties. But the early efforts to popularize its cultivation in Himachal Pradesh and Uttar Pradesh had received set back on account of the poor adaptability of cultivars, low returns per unit area and lack of technical know-how. However, after the introduction of cultivars like, Tioga, Torrey, Elista, Chandler, Shasta, Douglas, Fairfax, Senga Sengana, and standardization of different agro-techniques, its cultivation has got a boost during the last decade, and now its area and productivity is increasing slowly.

3. Area and production

At present, it is being grown in Shimla, Solan, Bilaspur, Kangra, Kullu, Palampur (Himachal Pradesh); Dehradun, Saharanpur, Muzaffarnagar, Ghaziabad (Uttar Pradesh); Hoshiarpur, Ludhiana; Jalandhar, Patiala (Punjab); Gurgaon, Hisar, Karnal (Haryana); Bangalore, Coorg (Karnataka); Kodaikanal, Palani hills and Servoy hills (Tamil Nadu); Pune, Mahabalweshwar (Maharashtra) on a small scale.

Table 2: Top-ten strawberry producing countries of the world

Country	Production (tones)	Country	Production (tones)
USA	1,312,960	Russia	184,000
Turkey	302,416	Japan	177,300
Spain	262,730	South Korea	171,519
Egypt	240,284	Poland	166,159
Mexico	228,900	Germany	154,418

Source : FAO Production yearbook-2012

4. Taxonomical and botanical description

4.1. Taxonomy and cytogenetics

Strawberry belongs to the genus *Fragaria*, tribe Potentilleae of the Rosaceae family. The cultivated strawberry *Fragaria x annanasa* Duch is considered to have arisen in France in the 18th century from a chance cross involving two native American species, *Fragaria chiloensis* (L) Duch with *Fragaria virginiana. Duchesnea* Smith and *Potentilla* L., are the related genera of *Fragaria* all belonging to the tribe Potentilleae.

Fragaria species constitute a polyploidy series ranging from diploid to octaploid, the basic chromosome set having 7 chromosome (X = 7). Many of these species have been cultivated at one time or the other and some are still being grown though on limited scale (Hancock, 1999; Sharma and Sharma, 2004). On the basis of chromosome number, the different species have been classified into four different groups viz., diploids, tetraploids, octoploids and a hexaploid. The cultivated strawberry, *Fragaria x ananassa,* is an octoploid, having chromosome number (2n) of 56. In addition to *F. x ananassa*, the genus *Fragaria* includes at least 11 other species. Out of 11, 4 species *viz. F. chiloensis, F. daltoniana, F. nilgerrensis* and *F. vesca*, have been reported to be grown in different parts of the world. The other species of *Fragaria* like *F. moschata* (the musky-flavoured strawberry), *F. virdis* (the green strawberry), *F. alba* and *F. sylvestris* have also been described.

4.2. Botany

The strawberry plant is a low-creeping perennial herb in which stem is compressed into a resettled crown with 2 mm long inter-nodes. The axillary buds in the leaf nodes of the crown either remain dormant, or develop into branch crowns, or stolons (runners), depending on the prevailing environment. Inflorescence is terminal, and vegetative growth continues by the uppermost axillary bud of the crown, resulting in a sympodial growth habit (Darrow, 1966; Hancock, 1999; Sharma and Sharma, 2004). Although, it is oftenly referred to as a herbaceous perennial, aging results in lignification of the crown, producing a hard woody tissue. With time, extensive branched crown development produces a structure resembling a highly compacted crown scaffold. However, vegetative and reproductive growths of strawberry are more sensitive to photoperiod and temperature than most of the other fruit-crops.

Strawberry has two types of roots, the large primary ones, which originate in crown and small secondary lateral roots that make up the mass of the root system, which arise from the primary roots. There are usually 20 to 35 roots but these can be up to 100 or more primary roots and thousands of small rootlets in a good root system. In general, strawberry roots grow in a year and die the next season. It happens only when fruits are taken during the season. However, if all flower clusters are removed from the plant, then most of its roots do not die in that season (Darrow, 1966; Hancock, 1999; Sharma and Sharma, 2004). The leaves are arranged in a 2/5 spiral, and each 6th leaf is just above the first, that is for maximum light exposure. In reality, the strawberry crown is very shortened stem of the plant. Sometimes, it may become as much as 2 feets in length and nodes several inches apart (as in sand-dunes of Pacific Coast). During frost injury, the crown is the first to get damaged.

Many investigators have found a certain type of fungus (Endogne) in strawberry roots, and have suggested that it is mycorrhizal in nature *i.e.* the fungus may be furnishing nutrients to the host plant The commonly associated mycorrhizal fungi are *Rhizophagus* and/or *Phycomycetous*.

4.3. Flowering, pollination and fruit-set

Flowering, pollination and fruit set in strawberry is influenced by photoperiod, temperature and photoperiod x temperature interactions.

4.3.1. Flowering

Light is the major environmental factor regulating growth and development of *Fragaria* species. Increased irradiant level may affect root and crown development, fruit-set, fruit size and stolon formation (Chandler *et al.,* 1995; Sharma and Sharma, 2004). Similarly, constant shading at 40% of ambient sunlight also affects fruit yield and leaf and crown weight. Variation in light quality can affect strawberry physiology. In floral inductive conditions (8-hr days), flower initiation in a strawberry cultivar was delayed or inhibited and leaf petiole length increased by extending day length with far red light (FR, 700-800 nm). Stolon formation, leaf lamina and petiole length all increased with exposure to far-red light during first half of the daily dark period and with the exposure to red light during second half of the dark period.

4.3.2. Photoperiod

Strawberry cultivars are often categorized on the basis of photoperiodic responses, particularly with regard to floral induction and floral bud initiation. Leaf size, petiole length and stolon development are also highly sensitive to photoperiod.

Strawberry cultivars have been categorized as short day (SD), long day (LD) and day neutrals (DN) on the basis of photoperiodic requirements for floral initiation.

The short day (SD) cultivars: Most strawberry cultivars are classified as short-day (SD), also known as single cropping, June-bearing or non-everbearing types, in which floral induction occurs with photoperiods of less than 14 hours (Larson, 1994). However, most short-day cultivar exhibit a facultative short-day response, in which floral induction occurs more or less continuously regardless of day-length, provided that temperature is less than 16°C (Darrow, 1966). Although, short-day cultivars may have different critical photoperiods, most of them bloom independently of day-length with moderately cool temperature.

The day-neutrals (DN) cultivars: Second photoperiodic group comprises everbearing (EB) cultivars, also referred to as day-neutrals (DN); they generally flower continuously regardless of day-length. Although temperature also modifies photoperiodic response of day-neutral cultivars but they are less sensitive than short-day cultivars.

The long-day (LD) cultivars: Some authors have recognized a third group as long-day (LD) types. This group is in between ever-bearing and day-neutral types (Sharma and Sharma, 2004). However, oweverHoweverHhthere is other group of researchers who consider this division as ill-defined ((Sharma and Sharma, 2004).

In general, flowering occurs under short light period (10 hr) with long dark period (14 hr), short light and dark periods (10 hr), and long light and dark periods (14 hr) (Hartman, 1947). Thus, duration of dark period rather than light period is the factor, which controls floral initiation in strawberry

Present-day cultivated strawberry, *Fragaria x ananassa*, is grown extensively in temperate and to some extent in sub-tropical culture i.e. short days in autumn and hard winter, or to meridional culture i.e. long days in autumn and moderate winter. Gorella, Redgauntlet, Senga Sengana strawberries grow under septentrional conditions (short days in autumn and hard winter), while Tioga and Torry are important for meridional climate. However, with the selection of several day-neutral cultivars (e.g. Selva, Fern, Hecker, Aptos, Tribute, Tristar, Fern, etc.), it is possible to grow strawberry throughout the year (Sharma and Singh, 1999; Sharma, 2002).

4.3.3. Temperature

Strawberry is highly sensitive to variations in air and soil temperatures. Simple effects of temperature as well interactions often are cultivar and species dependent. An optimum growing season temperature of 15°C has been reported for most of the strawberry cultivars and species (Larson, 1994), though the Primella cultivar grows best even at 25°C. Temperature effects on fruit-set may be cultivar dependent For example, fruit-set of Deutsch Evern cultivar was good with temperatures between 10 and 26°C, and it did not occur below 17°C in cv. Jucunda, possibly due to stamen abortion induced by low temperature (Kronenberg *et al.,* 1959). The soil temperature optima for vegetative growth of strawberry appear to vary with cultivar. For example, a soil temperature of 12.8 and 7.2°C has been found best for better growth in Shasta and Lassen respectively, but fruiting was greatest at 7.2°C and stolon development occur only at or above soil temperature of 18.3 °C, and was greatest at 23.9 °C. However, the optimum temperature for the period between onset of flowering and onset of ripening is 13.8°C for early, 14.4°C for mid-season and 15.1°C for late cultivars (Larson, 1994). Temperature below 15.6°C may inhibit pollen germination and pollen-tube growth, resulting in misshapen fruits. Temperature can also affect fruit-set and fruit quality due to indirect effect on bee-activity. Although, cultivated strawberry has perfect flowers and is self-fruitful, pollination is enhanced by bees (Bagnara and Vincent, 1988). Thus, cool temperature, rain, or strong winds that inhibit bee flights, results in decreased pollination and misshapen fruits.

Strawberries vary in their ability to withstand low temperature. Although certain cultivars reportedly withstand winter temperature of -50 °C (Darrow, 1966) yet strawberries, in general, are not as cold hardy as most temperate fruit crops. In autumn, exposure to cool temperature and short photoperiod results in floral induction, branched crown development, reduced leaf size, and with prolonged exposure, and there is a onset of dormancy in strawberry. However, unlike most temperate-zone fruits, dormancy in strawberry is solely controlled by external conditions /factors (ectodormancy), such as temperature and photoperiod. Dormant (resting) plants retain green leaves and can resume growth when environment becomes favourable. Thus, there is no true dormancy in strawberry (Saure, 1985). However, in the absence of sufficient chilling, dormant strawberry plants have low compact growth and yield poorly.

Optimum temperature between onset of flowering and ripening is 13.8°C for early cultivars, 14.4°C for mid season and 15.1°C for late cultivars. Temperature below 15.6°C may inhibit pollen germination and pollen tube growth, resulting in misshapen fruits (Sharma *et al.*, 2015).

4.3.4. Photoperiod and temperature interaction

This interaction plays a vital role in regulating vegetative and reproductive phases of strawberry. Short-day conditions usually favour floral initiation and inhibit stolon formation, regardless of temperature (Darrow, 1966). The optimum temperature for stolon formation and floral initiation is photoperiod dependent. In short-day conditions, floral initiation is greatest at 21°C, intermediate at 15.6°C and lowest at 12.8°C. With a 16-hr photoperiod, optimum temperature for floral initation was 15.6°C; floral initiation was greatest at 12.8 °C and lowest at 21°C. And if some varieties are subjected to 8-hr photoperiod and 20°C, then they will flower one month early than those subjected to continuous darkness and a temperature of 12.5°C. Cultivars also differ widely to floral initiation in response to photoperiod and temperature interaction (Sharma and Sharma, 2004). For example, in long-day conditions (14-16 hr) and at 21°C, Klondike produced only stolons, but Burrill produced equal number of stolons and flower clusters, and at 15.6°C, Klondike produced both flowers and stolons and Burrrill was almost entirely sexually reproductive.

4.3.5. Floral biology

Strawberry bears many flowers in clusters and a definite order of flowering. Flowers are white in colour. In octoploid strawberries, 3 types of flowers are found, pistillate, staminate and hermaphrodite (Sharma and Sharma, 2004). Rarely, plants may be polyamodioecious; with first few flowers on predominately pistillate plants producing fruits. However, most of the present-day cultivars bear hermaphrodite flowers and therefore require emasculation for cross-pollination.

The flower has many pistils, each having its own style and stigma attached to receptacle, and on fertilization, develop into a fleshy fruit. The true fruits of strawberry are nutlets (achenes) containing one seed each. Each flower may have 50-400 pistils that develop seeds. Usually primary flowers have the largest number of pistils, followed by secondary and tertiary flowers (Hancock, *et at.*, 1996)

The inflorescence is receme and has a dichotomic branching. A typical inflorescence bears primary, secondary, tertiary and quaternary flowers. The number of pistils and female tendency is strongest in the primary flowers and decreases successively in the secondary, tertiary and quaternary flowers. A definite order is observed in blossoming of strawberry flower cluster. It has two main branches at the crotch of which one flower is produced. It is known as primary flower, which bloom first. Each branch is similarly made up of two other branches, at the crotches of which secondary flowers are borne. Each of these four branches divides and at the crotches tertiary flowers occur, later flowers may occur on large cluster. Older flowers open first and produce large berries as compared to later produced flowers.

4.3.6. Pollination

Most strawberry cultivars produce hermaphrodite flowers and are self-fertile. However, some also produce male or staminate, imperfect, and female or pistillate flowers. Hermaphrodite flowers are self-fertile and pistillate flowers require cross pollination for fruitful production. In some varieties, partial sterility is reported (Darrow, 1966), which is quite persistent in some European varieties like Korona, Gorella and Confitura (Bagnara and Vincent, 1988). The first flush of flowers in some perfect flowered varieties open and set well, but the later flowers set only partially or do not set at all. In these varieties, the natural sterility is the primary cause. Sometimes, first flush of flowers develop into nubbins and later flushes produce good berries; the poor development is probably due to partial pollination (Darrow, 1966; Childers *et al.*, 1995). Petals and sepals may sometimes cover pistils and prevent pollination. Frost and insect injuries may also produce nubbins. Poor fruit development in some varieties may be due to the fact that they produce either pistillate flowers only, flowers with few stamens and stamens that fail to produce sufficient pollen for fertilization (Sharma and Sharma, 2004). Even plants of same variety vary in their capacity of pollen production in any one area as well as in different areas. The first to open flowers in the spring of many varieties may develop good anthers with abundant pollen but almost none in the following seasons. Thus, a variety may be self-fertile in a particular locality but may be self-sterile in the other (Darrow, 1966)

Honey-bees (*Apis ceranaindica* and *Apis mellifera*) are chief insect pollinators of the strawberries in different countries. Other pollinators are blow-flies, bumble-bees and *Lasioglossum* sp. Appropriate management and native conservation of pollinator should be practised in the vicinity of the strawberry plantations , and four strong honey-bee colonies should be placed in strawberry field /ha for proper pollination. This improves fruit-set, yield and quality and reduces percentage of malformed fruits (Sharma and Sharma, 2004). In general, the bee activity is hindered at temperature above 40°C and below 4°C; the optimum temperature being 10-15°C. Similarly, bee activity retards in rainy weather and in strong winds. Thus, weather also hinders pollination and fruit-set invariably by interfering with bee-activity (Childers *et al.,* 1995). In the absence of insect-pollinators, wind also plays vital role in pollination of strawberries.

4.3.7. Fruit-set

It is a characteristic feature of a cultivars, and is influenced by management practices, carbohydrate metabolism and climatic factors and activity of insect-pollinators. Strawberry inflorescence has one primary, two secondary, four tertiary and eight quaternary flowers (Darrow, 1966). But, maximum fruits are produced from tertiary flowers (Mitra, 1991). Even under proper management, most of the varieties do not give 100 % fruit-set. Increased fruit-set can be obtained with nitrogenous fertilizers, auxins and with insect-pollinators, especially, honey-bees, in the vicinity of the strawberry plantations (Sharma and Sharma, 2004).

5. Plant propagation

Strawberry can be propagated through sexual (seed) and asexual (vegetative) means. However, propagation by seed is not considered as a viable method of propagation for cultivated strawberry as they do not come true-to-type. Thus, strawberry is usually propagated through runners. Nowadays, a large-scale commercial propagation by tissue culture has also been used widely in strawberry industry.

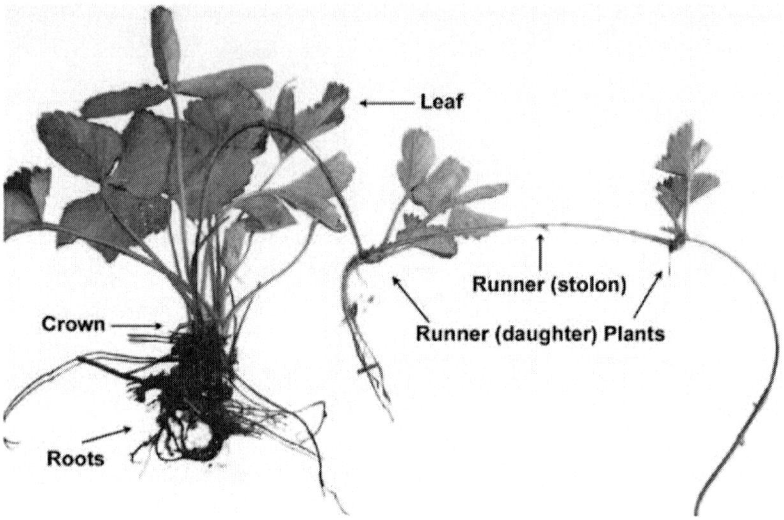

Runner production in strawberry

5.1. Vegetative propagation

The stolon, a creeping stalk, is produced in the leaf axil of the plant and grows-out from the parent plant during summer (Hancock, 1999; Lanz, 2012; Riotte, 1998; Sharma and Sharma, 2004). At the second node, a runner plant is formed and a new stolon arises on the runner plant to continue runner train that often branches. Initially, runner plant produces fewer roots but thereafter makes excessive fibrous roots. Thus, the plant naturally propagates itself by vegetative method of runner production. After the runner plant has acquired sufficient growth and roots, it is separated from mother-plant, and be planted elsewhere.

5.2. Micropropagation

It refers to the propagation from the smallest plant part under controlled nutrient, environmental and aseptic cultural conditions. Strawberry is perhaps the first fruit-crop in which micropropagation technique has been standardized (Sharma and Singh, 1999), and now its large-scale commercial propagation is being done through tissue culture. This fruit-plant requires very high number of plants/ha (50-60 thousand) and such a high demand can be easily met through micropropagation. Different explants like meristem-tip, anthers, immature embryos and first axillary buds of stolon have been used to produce millions of plants in a year (Jones *et al.*, 1988). Many culture media have been standardized, and most widely used media are Knoop's solution, Murashige and Skoog's medium (MS) and Linemaier-Skoog (LS) media with agar, sucrose and hormones at desired levels (Jones *et al.,* 1988). The addition of growth regulators into culture media has showed very stimulative effect on regeneration, growth and yield of tissue-cultured plants. The best shoot formation in Pajaro, Tioga, Tufts cultivars was obtained on solid MS medium, supplemented with 1 mg IBA + 1 mg GA_3 + 0.2 mg BA/litre (Sharma and Yamdagni, 1999) in tissue-culture raised plants

Many investigators have proved efficacy of tissue culture technique in ever-bearing strawberries, which are otherwise slow and expensive to propagate. Mahajan (1987) has deveolped *in-vitro* protocol for Chandler and Fern strawberries. Micropropagation is an efficient technique to produce quality plants from wilt-resistant varieties like Zenit and Kulon. Tissue-cultured plants have been reported to give very high yields as compared to conventionally propagated (runner) June-bearing cultivars. Badawi *et al.* (1990) had tried different concentrations of sodium hypochlorite and mercuric chloride for disinfection of meristems of Pajaro, Tioga and Tufts cultivars, and found that dipping meristems in 0.2% mercuric chloride for 20 minutes could result in the lowest contamination (5-15%) in all cultivars and addition of sodium chloride to the medium adversely affected shoot proliferation and percentage survival of explant (meristem). Their report emphasized that the best shoot formation in these cultivars could take place in MS medium, supplemented with 1 mg IAA + 1 mg GA_3 + 0.2 mg BA/l and the plantlet survival was highest in 2:1 peat:sand mixture in green-house.

Usually, growth of tissue-cultured plants is very good, producing lust green foliage. These plants produce high yield of good quality. In India, the regeneration of strawberry plants through micropropagation has been demonstrated by some scientists in various ICAR research institutes and Agricultural universities. However, Tata Energy Research Institute (TERI), New Delhi, has started strawberry multiplication through *in-vitro* techniques on the commercial scale. Plants raised through tissue culture techniques are very healthy, free from most of the diseases and perform exceptionally better in field conditions. The fruit yield of such plants is quite high as compared to those raised through conventional means

Advantages of micropropagation in strawberry

1. In micropropagation, only very minute plant parts are used.
2. With this technique, a large-scale multiplication in limited time is possible in *in-vitro* conditions.
3. Very limited space is required.
4. It is a labour-intensive technique.
5. It is possible to produce plants throughout the year.
6. The plants produced are healthy and usually free from diseases.

7. Ever-bearing strawberries perform better even under *in-vitro* conditions as compared to field propagation.
8. Tissue-cultured plants perform better in fields as compared to those propagated by conventional means.
9. Tissue-cultured plants have higher runnering capacity.
10. Screening against insect-pests and diseases is easy in tissue-culture systems as compared to field conditions.

6. Important cultivars

Strawberries are adapted to different climates, moderate, mediteranean, subtropical and can even be grown at high altitudes under tropical climate. Purposeful breeding has played leading role in the development of modern cultivars. Most important varieties of strawberry are: Chandler, Camarosa, Sweet Charlie, Tioga, Torrey, Etna, Fern, Catskill, Cambridge Vigour, Belrubi, Canoga. Gorella, Jewel, Mars, Pajaro, Ofra, Primetime, Premier, Selva, Sparkle, Senga Sengana etc (Hancock, 1999; Lanz, 2012; Riotte, 1998; Sharma and Sharma, 2004). Due to development of area-specific varieties, there has been tremendous increase in strawberry production. Several cultivars have been developed in different countries and were evaluated under local conditions (Awasthi and Badiyala, 1983; Beniwal *et al.*, 1989; Daubeny, 1980; Sharma and Badiyala, 1980). Nowadays, day-neutral strawberries are becoming favourite in the world because these are not at all affected by the day length, and thus can be grown throughout the year.

Chief characteristics of some of the commercially grown varieties

6.1. Adina

This is a new short-day variety. Its fruits are early ripening, very large, bright-red, medium glossy, short conic with pointed tip. Fruit pulp is medium-red and very firm.

6. 2. Allstar

It is a vigorous and productive, and June-bearing cultivar. Its plants are medium to large, vigorous, form runners freely and form a dense matted row. It produces medium-sized, firm fruits, which develop full red colour on ripening, and are tough and glossy. It is good for table and freezing purposes. It is highly resistant to red-stele root-rot.

6.3. Belrubi

Belrubi strawberries are large and conical in shape. Each fruit weighs around 15 gm. Fruits contain 11.8 % TSS; 0.98 % acidity and sugar content 6 %.

6.4. Blomidon

Its plants are vigorous, fruits are large to very large, and firm, glossy, with medium to deep-red exterior and medium red interior. Pulp is with white fibro-vascular bundles. Calyx is moderately difficult to remove. Fruits have usually tart flavour and ripen mid to late in season. It is moderately resistant to powdery mildew, *verticillium* wilt, and is resistant to *Botrytis* fruit-rot and red stele - root-rot.

6.5. Camarosa

Plant growth is similar to 'Chandler'. Compared to Chandler, leaf colour is distinctly lighter on the underside. Individual leaflets are larger, somewhat longer and narrower than Chandler. Leaves (including petioles) are similar in length to Chandler, but are much broader. It produces fruit early and has good storage properties.

6.6. Cambridge Favourite

It is a seedling selection from the cross between Avant Tout and Blackmore. It is an early to mid-season variety. Plants are fairly large, moderately open and vigorous. Fruits are large, rounded and plump; pinkish-scarlet, turning light-red on maturity. Pulp is firm, moderately juicy, pale pink to pale red with a moderate flavour. Fruits are suitable for processing. It is resistant to powdery mildew, but susceptible to *Botrytis* fruit rot but susceptible to many strains of red core. It produces runners profusely.

6.7. Canoga

Its plants are vigorous and upright with dark-bluish green foliage. The fruits are exceptionally large, oblate, and are characterized by an intense, brick-red colour. The pulp is dark red throughout and is very firm. Canoga is susceptible to red stele and *verticillum* wilt.

6.8. Catskill

Catskill was developed at Geneva from a cross between Marshall and Premier. The plants are vigorous having bright green foliage. The large, attractive fruit is good for the local market. Fruit is very soft and has poor shipping quality. The berries get darkened on holding and are rated unsuitable for processing. The plants are fairly resistant to *verticillium* wilt, powdery mildew, leaf -scorch and leaf-spot diseases.

6.9. Chandler

This is a strawberry with bright red color and strong flavor. Chandler strawberries are resistant to damages caused by rain and plants are highly resistant to viral diseases. These strawberry fruits are equally good for raw consumption and processing. Chandler strawberry is very large and each fruit weighs approx. 15 to 20 gm. Nutritional value of chandler strawberry is very high with 12% TSS, 0.85% of acidity, 55.5 mg/100g of Vitamin C and 6.1% of sugar.

6. 10. Dover

It is a high-yielding variety, which bears conic and slightly flattened fruits. Fruits are generally smooth and of uniform shape with slightly recessed seeds. Ripe fruits develop attractive deep red colour on ripening. The fruits have a strong epidermis and a firm pulp, which gives them the excellent resistance to bruising during harvesting and shipping. Dover is highly resistant to anthracnose and is good for fresh markets (Howard and Albregts, 1980).

6.11. Earlidawn

It is a cross between Midland and Tennesse Shipper, and produces fruits of good size and bright-red colour. Fruits are firmer than those of Catskill or Premier. Berries when frozen retain good colour, texture and flavour. This cultivar is very productive and is a very early variety. It is susceptible to powdery mildew and highly susceptible to *verticillium* wilt.

6.12. Etna

Etna, a mid-season variety, is a cross between Marlate and Belrubi. The plants are vigorous and spreading and fruits are large, which develop light-red colour on maturity. It can be grown in glasshouse conditions and in open as well. It is highly productive, medium-to-late ripening variety, which is resistant to *verticillium* wilt.

6.13. Fern

It is an early, day neutral cultivar developed in California (USA). The plants are vigorous but erect growing. It bears large, conical-shaped fruits. The pulp is firm with excellent flavour. It is suitable for fresh market and processing.

6.14. Festival

It is a short day strawberry variety. Average height and width of mature plants is 23 and 30 cm, respectively. Average petiole length and diameter is 120 mm and 3.5 mm, respectively. Petioles have medium pubescence. Average length and breadth of terminal leaflets is 78 and 73 mm, respectively. Leaflet margins are crenate and average 21 serrations per terminal leaflet, and 26 serrations per secondary leaflet. The upper leaf surface is dark grey green colour, and the lower leaf surface is a light grey green in colour. Petiole colour is medium yellow green.

6.15. Gilbert

Its plants are vigorous and produce abundant runners that root readily. This cultivar bears very large, firm, red fruits. Its plants are winter-hardy and resistant to leaf-scorch. The fruits are suitable for freezing, besides for fresh market.

6.16. Gorella

This cultivar was selected in Netherlands from a cross between Jespa and US 3763. Plants are moderate in growth, sparse with coarse thick foliage. The fruits are very large (15-20 g), conical, crimson-red, are sometimes with a green tip, and are well exposed beneath the upright foliage. Pulp is bright crimson, juicy with average flavour.

6.17. Jewel

Berries are large (11- 15 g), bright red, moderately firm, easily picked, are of high quality with very high and attractive flavour and suitable for fresh and frozen uses. This cultivar is resistant to post-harvest fruit-rot.

6.18. Joliette

Plants are vigorous and can tolerate temperature even below –25°C. Leaves are flat, medium, dark-green, obovate and with 21 sharp serrations. Fruits are large with a small white raised neck. Peel is glossy red at maturity and pulp is light-red throughout. Fruits are moderately firm and can be decapped easily. It is a short-day, June-bearing variety, resistant to leaf-spot and red stele root-rot and has been recommended for fresh and pick-your-own (PYO) markets.

6.19. Latestar

Plants are vigorous having very good runnering capacity. Fruits are attractive large, glossy-red, with firm pulp flesh and tough peel. Pulp is light-red, pleasant and mild, slightly acidic with fresh flavour. It is resistant to most of common leaf diseases and fruit-rots but is susceptible to leaf blight, incited by *Phomopsis obscurans*.

6.20. Lester

The plants are medium to small, runnering moderately to form a plant bed of medium density. Leaflets are medium dark-green on the upper surface and light waxy green on the lower side. Peel of fruits is glossy deep-red but pulp colour varies from medium-red in outer surface to pink or whitish at the core.

Both fruit peel and pulp are firm. Fresh fruit flavour is wild, sweet and pleasant. Its fruits are highly tolerant to fruit-rot.

6.21. Ofra

This is an early-ripening strawberry cultivar derived from cross between Parker and breeding line 111. Fruits are large, longer than broad, red, and strongly glossy. Fruit pulp is firm to very firm, orange red, and of medium sweetness and acidity.

6.22. Pajaro

Pajaro strawberries are susceptible to rain damage, resistant to plant viruses and good to grow under summer system. Each fruit weighs approx. 7.5g, with TSS content of 12.2%, acidity 0.97% and sugar content of 5.5%.

6.23. Pelican

Its plants are large, vigorous and erect. Leaves are large, cupped upward at the margins, are medium-green on the top and light-green beneath. Fruits are large, medium firm, very long, wedge-shaped with a glossy-red exterior. Interior pulp is pink and melting (uniform and smooth), juicy, with balanced sweet and acid flavour. It responds well to raise bed plasti-culture system. It has out-yielded Chandler and Sweet Charlie. The cultivar is highly resistant to anthracnose and red stele root-rot.

6.24. Primetime

Its plants are vigorous and produce runners well. It bears attractive large, symmetrical and good quality fruits. Fruits are medium in peel firmness and toughness, are bright scarlet, with pink pulp having a pleasant, mild and light aromatic flavour. It is a June-bearing, mid-season variety. It is resistant to red-stele root-rot, and to several leaf diseases and fruit-rots. It has been recommended for local markets but not for processing.

6.25. Puget Reliance

Its plants are vigorous with an erect growth habit. Leaves are medium in size and petiolar hairs are irregularly perpendicular to petiole axis. Fruits are usually smooth, glossy and symmetrically conic with inserted seeds. It is susceptible to aphids but highly tolerant to viral diseases, leaf-scorch and anthracnose fruit-rot and tolerant to powdery mildew and red stele root-rot.

6.26. Selva

It is a day-neutral variety. Its fruits can be produced in open or with low plastic cloches placed over them in early September or in peat bags in greenhouses.

This variety is considered as ray of hope in sub-tropical zones of India because it has performed very well under north Indian plains in recent years.

6.27. Sparkle

It is also called as Paymaster, which is a cross between Fairfax and Aberdeen. Plants are vigorous and spreading. Medium-sized, firm, rather dark and attractive berries are good for dessert and are acceptable for freezing. It is a fairly high-yielding variety in many areas. Plants are moderately susceptible to *verticillium* wilt, leaf-scorch and leaf-spot, but are resistant to powdery mildew.

6.28. Tioga

Tioga is an early variety of strawberry. Tioga strawberries are large and juicy with firm skin. Hence these are suitable for dessert and processing purposes. As far as nutritional values are concerned, tioga strawberries contain 12.2% TSS, 0.98 - 1% acidity and 6.2% sugar. Each tioga strawberry weighs about 9-10 gm.

6.29. Torrey

Torrey strawberry plants are very tough and highly resistant to plant viruses. They produce large good quality fruits which are suitable for both dessert and processing purposes. Each fruit weighs approx. 7 gm. Torrey strawberries contain 12 % TSS, 0.98 - 1% acidity and 6.2% sugar content.

6.30. Tribute

This was the first day-neutral (ever-bearing) cultivar developed in the USA. Plants are medium but vigorous. Fruits are large to medium, having irregular to symmetrical shape, with short conical wedge and pronounced shoulders. Fruit is glossy bright-red, not quite as deep as of Tristar. Peel and pulp texture is quite firm. Tribute gives high yield in matted-row system of planting. Roots are highly resistant to red stele root-rot and tolerant to *verticillium* wilt, and leaves are resistant to powdery mildew and blight, and tolerant to leaf-scorch.

6.31. Tristar

It is a sister seedling of Tribute. It is also a day-neutral cultivar. Plants are medium to small with medium vigour. Its plants bear heavy, with very early spring crop of small to medium sized, symmetrical, short conical fruits having marked shoulders and reflexed cap at maturity. Fruit peel and pulp are firm. Peel colour is glossy red with “Star” like reflecting pigment around achenes. Pulp colour is deep-red and flavour is excellent when eaten fresh. It is resistant to red stele root-rot with high degree of tolerance to *verticillium* wilt.

6.32. Selva

Selva is also a day-neutral variety of strawberry. Selva strawberry plants produce berries even during off-seasons. It produces large fruits with firm skin and good dessert quality which can endure long distance transportation. Hence these strawberries are best suited for exports. Each individual selva strawberry weighs around 15 to 20 gm and contains TSS 11 %, acidity 1 % and sugar content 5.5%.

6.33. Sweet Charlie

Plants of Sweet Charlie are smaller and more compact when compared to Camarosa. Leaves are generally slightly cupped, medium to dark green and semi glossy.

6.34. Winter Down

Highly productive. Fruits are medium to large in size and moderately resistant to Botrytis and anthracnose.

In India, strawberry is primarily grown in a few places only. The major growing areas with important varieties are as under:

Growing areas in India	Important cultivars
Palampur, Solan, Kangra, Kullu and Shimla (H.P.)	Tioga, Torrey, Chandler, Sweet Charlie
Dehradun, Saharanpur, Muzaffarnagar, Ghaziabad (Uttar Pradesh)	Camarosa, Fern, Etna, Chandler
Pune, Mahabalweshwar (Maharashtra)	Chandler, Sweet Charlie
Hisar, Gurgaon, Karnal (Haryana)	Camarosa, Chandler
Delhi	Winter Dawn, Sweet Charlie, Chandler, Camarosa, Cadila
Hoshiarpur, Ludhiana; Jalandhar, Patiala (Punjab)	Camarosa, Chandler
Bangalore, Coorg (Karnataka)	Chandler
Udhampur, Jammu, Srinagar	Chandler, Sweet Charlie
North east states	Camarosa, Chandler, Sweet Charlie

7. Soil and climatic requirements

7.1. Location and site

Selecting a good site with a suitable soil is the first step in growing strawberries successfully. Factors of prime importance in selecting a site for commercial strawberry plantations are (a) accessibility to markets, (b) transportation facilities, (c) adequate labour availability (d) community interest, and (e) appropriate climate. If strawberries are to be grown for general market, it is usually best to select a site, which is nearer to a town or city.

7.2. Soil

Strawberry can be grown on a variety of soils, ranging from heavy clay to gravelly soils (Hancock, 1999; Sharma, 2002; Sharma and Sharma, 2004). Strawberry has a fibrous root system and most of its roots are confined to the top 15-20 cm layer of soil, and thus it grows best in light porous soils that are rich in humus. However, in light soils, frequent irrigation is required for proper establishment of runners and to maintain berry-size and quality. In heavy soils, development and penetration of roots of runners is inhibited adversely. Though, strawberry is not sensitive to soil reaction (pH), it however, prefers slight acidic soils with pH of 5.8-6.5. Strawberry suffers badly in poorly drained soils. Water logging, even for a limited period, may kill plants. Poorly drained soils also encourage fruit-rot diseases, affect plant growth, fruit yield, fruit quality adversely.

7.3. Climate

Cultivated strawberry can be grown profitably under extremely different climates. Among the different climatic factors, temperature, and day-length affect considerably the growth and yield of strawberry (Hancock, 1999; Lanz, 2012; Riotte, 1998; Sharma and Sharma, 2004). However, a precise control on light and temperature is must.

On the basis of photoperiodic requirement for floral induction, strawberry cultivars have been categorized as short-day (SD), long-day (LD) or day-neutral (DN). In short-day cultivars, floral induction occurs with photoperiods of less than 14 hours, though these cultivars flower continuously regardless of day-length but environmental temperature should be less than 16°C. The day-neutral cultivars, usually called as ever-bearers, flower continuously, regardless of day-length, although temperature also modifies photoperiodic response but day-neutral cultivars are sensitive to lesser extent to high temperature than short-day cultivars. The third group is between short day and day-neutrals in which flowering takes place under long-day conditions.

8. Cultivation

8.1. Planting

Planting can be done either on furrows or on raised beds in a particular system of planting. In India, raised beds are usually preferred (Sharma and Singh, 1999; Sharma *et al.,* 2001; Singh and Sharma, 2005; Singh *et al.,* 2009). The planting may be done by hand or by machine but care should be taken to prevent damage and drying of roots of runners. After removing runners from the soil, they may be planted directly in the field or placed in a some shadow before transplanting. It is better if runners are kept in moist sphagnum moss grass. It will prevent drying of roots and can keep runners in a fresh condition for longer period. Planting should be done in early or late evening hours. It should, however, be avoided in sunny day and under wet or frosty conditions. Immediately after planting, a light irrigation is essential.

8.1. Systems of planting

Three planting systems are in general use in strawberry; (a) the hill, (b) the spaced row, and (c) the matted row system. Further, single hedge, double-hedge row, single row or double row hill are also followed and in backyard plantings, barrel or pyramids can be used (Hancock, 1999; Lanz, 2012; Riotte, 1998; Sharma and Sharma, 2004).

8.1.1. Hill system

This system of planting is commonly followed in cultivars, which produce few runners. In this system, plants are grown either in single or double rows on raised beds. The beds are usually made 15-20 cm high and runners are set 20-25 cm apart in twin rows, 30-35 cm apart and distance of 90-120 cm is kept between twin rows. With this system, a small garden tractor or a field tractor can be used for tillage; greatly reducing the expense of manual labour. In home garden, where hand labour is used, rows can be spaced closer, or about 45-50 cm apart. Sometimes, twin rows are set 40-60 cm apart, then a wide space is left and another two rows are set. In some cases, triple rows are set. However, these are only modifications of hill system with the plants set at the same distance apart in the row in each case. This system demands much higher inputs and large number of runners are set per unit area. As a result of crowding, removing of runners becomes difficult in this system.

8.1.2. Spaced row system

This system of planting is used for varieties, which are moderate to weak in producing runners and thereby daughter plants. In this, runners are usually set at 30-50 cm apart in rows with spacing of 90-100 cm between rows. The rows may be wider than 90 cm in rich soils. It helps in easy-picking of fruits by moving between rows easily.

A view of a strawberry field in Meghalaya

8.1.3. Matted row system

This system is adopted in areas where crown injury occurs in winter either due to freezing or temperature fluctuations. Runners are planted along the row at about 90 cm apart and at a space of 45 cm between plants. The width of mats is maintained between 40-45 cm. In this way, more number of plants can be accommodated per unit area, which may give higher yield under suitable conditions.

Although planting can be done at any time between July to April but early planting ensures good crop. In temperate humid climate, planting is usually done in spring. If weather is undesirable, runners can be stored in poly-bags at 0°C till conditions become favourable. Planting during July-August, however needs adequate care, particularly from strong sunshine or dry soil conditions. During this period, irrigation should also be given frequently otherwise leaves may desiccate and defoliate, which may result in delayed fruiting. However, planting in October and November usually does not require much special care for establishment. The planting season vary widely with elevation of planting site and nature of cultivars. For example, Chandler is sown during March under hilly tracts (Shimla) of India and the best time for it under subtropical zones is the last week of October or 1st week of November (Sharma and Sharma, 2004; Sharma *et al.,* 2001; Singh and Sharma, 2005).

8.2. Water management

Due to an evergreen, semi-herbaceous growth habit, rapid growth rates, high fruit productivity and relatively shallow root system, growth and development of strawberry plants are sensitive to variation in soil moisture. Strawberry responds to irrigation in a more profitable manner than any other fruit crops. Irrigation is useful in utilizing nutrients in the soil, preventing blossoms from frost damage, hastening growth of runners and increasing bearing size and yield (Sharma and Sharma, 2004; Sharma *et al.,* 2001; Singh and Sharma, 2005).

Strawberry is a shallow-rooted plant and roots are confined to the upper 25-30 cm of the soil surface. Due to shallow root system, irrigation is required more frequently than any other fruit crops. Shortage or excess of water should be avoided to harvest a good crop of quality fruits. Irrigation at the time of active vegetative growth and during flowering is necessary. It should, however, be avoided during ripening, as it may result in softening of fruits. The most commonly used irrigation systems are the surface and drip or trickle system. However, nowadays drip or trickle irrigation systems are gaining popularity. This method is more convenient to regulate water as well as to minimize amount of water. On an average, water used is 30-40% less for trickle irrigation as compared to furrow method.

8.3. Nutritional management

Practically, all strawberry plantings benefit from manures and fertilizers application, but fertillizers are not, as assumed, a cure-all, because several other factors like, fertility and physical conditions of soil, careful planting, thrifty disease tested plants, suitable cultivars, adequate moisture, the quantity, quality and time of applying mulch, and weed and insect-pest control, may be as important as the application of fertilizers. Certain cultivars are inherently more productive than others in a given region. Fertilizers do not normally contribute to cause yield of a low-yielding or a disease-susceptible variety to equal to a high-yielding and disease-resistant cultivar. Similarly, fertilizer will not make a low chilling cultivar to succeed in warm areas and vice versa.

Depending upon different factors, different doses of fertilizers have been recommended for profitable production of strawberry. An autumn application of 20 tonnes FYM, 40 kg P_2O_5 and 20 kg K_2O per hectare, followed by spring application of 20 kg N after the onset of growth but before flowering is a common practice followed in India (Sharma and Singh, 1999; Sharma, 2002). The rate of potash can be increased to 40 kg/ha in K deficient soils, though Indian soils are not usually K-deficient. In the above recommendation, FYM should be added as pre-planting dose during land preparation. Use of 75-100 kg

Black polythene mulch in strawberry

N, 80-120 kg P_2O_5 and 50-75 kg K_2O/ha has also been recommended for best growth, highest fruit yield and for the production of quality fruits (Sharma and Singh, 1999; Sharma and Sharma, 2004; Sharma *et al.,* 2001). Substitution of N,P,K with vermicompost @ 8 tones per hectare is highly useful in producing organic strawberries ((Singh *et al*., 2008; Singh *et al*., 2010).

8.3.1. Foliar feeding

Foliar application of nutrients is very effective tool for fertilizer supplementation as plant leaves readily absorbs the nutrients. Generally, urea and some micronutrients can be applied through foliar means. Sometimes, K and P can also be applied through foliar means. Foliar application of nutrients may result in early floral initiation, higher fruit-set and more yield per plant. Four foliar sprays of N (0.5%), P_2O_5 (0.2%) and K_2O (0.5) between August-February for supplementing N, P and K have been helpful for good growth, fruit yield, fruit firmness and storability of fruits (Sharma and Sharma, 2004; Sharma *et al.,* 2001). Pre-harvest foliar application of Ca and B helps in significant reduction in fruits disorders of strawberry (Singh et al., 2007). Sequential application of vermicompost leachates helped in quality production of fruits and reduction in fruit disorders (Singh *et al*., 2008; Singh *et al*., 2010).

8.3.2. Organic farming

It is perhaps the most wanted aspect in strawberry growing. As the roots of strawberry are confined to the top 25-30 cm layer of soil, the soil should be made friable by repeated ploughing and by supplementing with sufficient organic manures. Sufficient amount of organic manures (70-80 tonnes/ha) is recommended at the time of land preparation, as organic manures take more time for decomposition and to make nutrients available to plants. The adverse effects of chemical fertilizers on physical and chemical properties of the soil can be counteracted if these are applied with green manures, like cowpeas, sunhemp, dhaincha, vetch or cloves. Soils that contain more than 2.5% organic matter usually supply sufficient N for a good berry-crop, but optimum percentage varies on different sites (Sharma and Singh, 2003). Usually well-rotten farmyard manure (FYM) is considered the best for strawberry. Hughes *et al.* (1969) has reported the application of 70-80 tonnes/ha of FYM prior to planting of runners as the best dose. In addition, deep litter of poultry house, poultry manure and spent mushroom compost can also be used. These manures, however, contain more N, P and K and hence their doses should be reduced. Thus, 5-7 tonnes of poultry manure/ha is sufficient, equaling to 70-80 tonnes of FYM. Pandey and Mishra (1983) had recommended 200-300 quintals FYM prior to planting in the absence of green manuring. Thus, a pre-planting dressing with farmyard manure (FYM) or other manure often provides all the needs of the crop over several years on the fertile soil. The availability of nutrients though depends upon the number and spread of roots, which in turn, is decided by the soil depth and availability of soil water. Organic fertilization alongwith *Glomus* sp. has been reported to enhance yield and quality in strawberry (Gengott *et al.*, 2001).

Different forms and doses of manures have kbeen recommended for strawberry under protected cultivation. In greenhouses, strawberries can be grown profitably on pumice mixed with sphagnum grass, mixture of peat and styromull (75:25), and broiler litter and layer litter. Similarly, rockwool, peat bales, mushroom compost waste are suitable media for growing strawberries under controlled (protected) conditions (Sharma and Sharma, 2004).

8.4. Mulching and winter protection

Mulching is recognized as one of the most beneficial intercultural practices in strawberry as it helps in better conservation of soil moisture soil temperature, minimizes winter and frost injury, suppresses weed growth, reduces soil erosion, killing of harmful microbes and avoids contact of berries with soil thereby reducing number of dirty and rot infected berries (Badiyala and Aggrawal, 1981; Sharma and Singh, 1999; Sharma *et al.*, 2008). Mulching can be done by using black or white plastic sheet, or paddy straw, pine needles and/or with locally available material. Plastic mulch reduces the number of harmful arthropods in strawberry fields. Reflective mulch has been reported to suppress the density of tarnished plant bugs, *Lygus lineolaris.* Similary, the aroma of fruits under mulch is always better. As a winter protection, mulching checks damage to roots caused by freezing and thawing. It also prevents plants from drying out or freezing excessively. Sometimes, it makes picking pleasant, may improve fruit colour and may increase yield and quality (Sharma and Singh, 1999; Sharma *et al.*, 2008). However, there are certain objections on mulching in strawberry like the cost of mulching material and its application, difficulty in getting suitable material in some cases, possibility of weed seeds, injury to blossom clusters in areas having high incidence of strawberry weevil (clipper), and the picking may be somewhat delayed, but its advantages usually outweigh objections.

8.5. Flowering, pollination and fruit-set

Light is the major environmental factor regulating growth and development of *Fragaria* species. The optimal day-length for floral induction in short-day cultivars appears to be between 8 and 11 hours (Sharma and Sharma, 2004). Most strawberry cultivars produce hermaphrodite flowers and are self-fertile. However, some also produce male or staminate, imperfect, and female or pistillate flowers. Hermaphrodite flowers are self-fertile and pistillate flowers require cross pollination for fruitful production. The presence of the pollinators and pollinizer in the vicinity of the strawberry planting is important for cultivars whether self-fertile or self-sterile. In self-sterile cultivars, fruit-set improves considerably as cross-pollination is required for them. In the absence of a suitable pollinator and/ or pollinizer, or insufficient pollination, percentage of malformed or misshapen fruits increases. Honey-bees are chief insect pollinators of the strawberries in different countries. Other pollinators are blow-flies, and bumble. Four strong honey-bee colonies should be placed in strawberry field /ha for proper pollination (Sharma and Sharma, 2004).

Even under proper management, most of the varieties do not give 100 % fruit-set. Increased fruit-set can be obtained with nitrogenous fertilizers, auxins and with insect-pollinators, especially, honey-bees, in the vicinity of the strawberry plantations.

8.6. Major insect-pests and their control

About 100 insect-pests have been reported in strawberry but only about a dozen pests cause damage (Hancock *et al.,* 1990; Sharma and Sharma, 2004; Childers, 2003; Lanz, 2012). A brief description of some very important pests has been given hereunder.

8.6.1. Red spider mite (*Tetranychus urticae*)

It is the most serious pest of strawberry. Damage is caused both by nymphs and adults by sucking sap from leaves, causing a rusty brown colour. Plants are stunted and show reduced yields. Organophosphatic insecticides, Chloropyriphos and Carbamates, Aldicarb are effective for controlling red spidermite. Aldicarb (10G) @ 5 kg/ha is quite effective in controlling red spider mite.

8.6.2. Blossom weevil (*Anthonomus rabi*)

Blossom weevil is also called as elephant weevil. These cause damage by making holes in flowering-stalks and buds, and causes whole truss to wilt and die. Systemic pyrethroids like Deltamethrin (0.05%) and Sumicidin (0.025%) sprays just before flowering saves plants from blossom weevil.

8.6.3. Root weevil (*Otiorhynchus rugostriatus*)

This is a wingless weevil and its young ones (grubs) are more harmful than adults. Small white legless grubs feed on small rootlets in winter and spring and make deep tunnels in the crown base and finally plants collapse. Adults feed on foliage at night but damage is less severe than grubs. Carbofuran (6-10 kg/ha) and Parathion (0.017%) applied around plants, reduce pest population and damage.

8.6.4. White grubs

In some places white-grubs cause severe damage. Its larvae are dirty white with brown heads, usually coiled and with distinct legs. They eat off roots and kill plants, usually during planting time and runner formation. Drenching of soil with Chlorpyriphos (0.02%) reduces damage of white grubs. Chickens or hogs may also destroy grubs.

8.6.5. Thrips *(Thrips atratus)*

Nymphs and adults cause damage by sucking sap from growing points, leaves and fruits. The damage is more severe in late-season strawberries. Severe infestations may result in fruit distortion, leading to down-grading and finally financial losses to farmers. Frequent sprays of synthetic pyrethroids like Deltamethrin (2.25 g a.i. / ha) and Cypermethrin (16.8 g a.i./ha), reduce thrips incidence and number of distorted fruits.

8.6.6. Cut-worms and army-worms (*Agrotis* sp.)

Several species of cut-worms and army-worms infest strawberries. Large, stout, stripped mottled or grey caterpillars eat on fruits and make large holes. They usually feed at night and hide during day. They cut off young plants at the ground level, and may eat leaves and berries of established plants. Drenching soil with Chlorpyriphos (0.01%) during March is effective for cut-worms. Similarly 4-5 deep ploughings during land preparation also controls cut-worms to some extent.

8.6.7. Tarnished plant bug (*Lygus lineolaris*)

Adult bug is coppery-brown with piercing and sucking mouth parts. It causes damage by puncturing berries, causing uneven ripening or "catfacing". It also feeds on individual achenes (seeds) and destroys their contents. Spraying with Endosulfan (0.02%) controls it effectively.

8.6.8. Nematodes

Sting nematode (*Belonolamus sp.*) is more destructive to strawberries than any other nematode, and it occurs mainly in light soils. It feeds on the surface of roots and is usually dislodged when plants are dug. Root-knot nematode (*Meloidogyne* sp.) penetrates and feeds on small roots and causes knot-like enlargements. Affected plants weaken and produce few runners and fruits. Nematodes that enter roots can stay alive even when plants are dug, stored and shipped. It has been reported to infest all strawberry cultivars in north India. Lesion nematode (*Pratylenchus* sp.) also attacks roots and initially causes small spots that are amber to dark-brown. Roots may decay and may be much like caused by root-rot diseases. Length of leaf stalks of the affected plants shortens, and leaves become yellow, and plant growth is reduced. This nematode may increase chances of *verticillium* wilt. Bud nematode (*Aphalenchoides* sp.) lives in most succulent tissues of buds and very small leaves. Flower buds are often killed by feeding, resulting in flowerless plants. It causes bud disease 'spring crimp' or 'red plant'. Stem nematode (*Ditylenchus dipsaci*) lives in plant stems and causes damage by sucking sap. It results in shortening and thickening of stem and leaf stalks.

All soil-inhabitation nematodes can be checked by fumigating soil with nematicides, methyl bromide or by micro granular nematicides, Thimet or Temik. Root-knot nematode can be controlled by Basamid (dazomet) and solar-heat treatment. Nematodes in roots, leaves and buds can be killed by hot-water dip of dormant plants at 52.8°C for 2-3 minutes.

8.6.9. Birds

Sparrows, parrots and peacocks cause heavy losses to strawberry at the time of berry-ripening. The damage by birds is more alarming than the one caused by insect-pests or diseases. It can be reduced by using bird scare, reflecting ribbons and by beating drums. Covering field with nylon-netting is very useful. The birds are very clever and get acclimatized to the scaring measure used, and hence no scaring device should be continuously used for a longer time.

8.7. Major diseases and their control

Several diseases cause heavy losses to strawberry in different parts of the world. However, in India, only 5-6 diseases have been reported (Sharma, 2002; Sharma and Sharma, 2004; Singh *et al*., 2005). A brief description about these diseases is as under.

8.7. 1. Powdery mildew

Mildew may affect strawberries in nearly all regions during dry periods. It is a fungal disease caused by *Sphaerotheca macularis*. Its characteristic appearance is an upward curling of leaves and a cobweb-like mold on the lower surface as the fungus destroys surface layers of leaves. Catskill, Dunlop, Klondike, Marshall and Sparkle are notable resistant varieties. Baytelon or Karathane (0.05%) at 10-15 days interval between May to September is quite effective for controlling powdery mildew.

8.7. 2. Red stele

It is the most destructive disease of the plants in cool and moist soils and poorly drained soils in late winter and early spring. It is also called as 'red core' or 'brown core'. It is caused by a fungus, *Phytophthora fragariae*. The fungus enters central parts of roots, the stele, which turns reddish while its cortex appears normal. If the season is dry, infected plants may die before blossoming, and if the season is wet, plants may blossom, but they die before ripening of fruits. At first affected plants show dull, bluish green colour, which soon wilt and may die in a few days. In well-drained sandy soils, it is rarely serious. The disease is spread by drainage water, machines, tools and by infected plants. Improvement in drainage system, ridge planting, crop rotation, covering plants with cloches and use of resistant varieties are important cultural techniques to reduce red stele incidence. Soil drenching with disinfectant Ridomil and treatment of planting material before planting with bordeaux mixture (1%) or copper oxychloride (0.1%) for 15-20 minutes reduce the occurrence of red stele in field.

8.7. 3. Verticillium wilt

It is caused by a fungus, *Verticillium albo-atrum,* which is active in cool weather. Fungus invades roots and interferes with movement of water to leaves. Disease symptoms include internal browning of vascular tissues at the crown base. Outer older tissues wilt and dry at margins and become reddish to dark brown. In affected plants, only a few leaves develop, which tend to be stunted, and may curl up along the mid vein. The central young leaves are small, new roots from the crown are short with blackish tips, and plants look dry and may suddenly collapse. Soil fumigation is the best way to keep disease under control.

8.7. 4. Grey mould or botrytis rot

It is a serious disease of strawberry in field and storage. It is caused by a fungus, *Botrytis cinerea*. The disease appears as light-brown soft spots on the green and ripening fruits. The berries dry out, become tough and get covered by a dusty fungal growth. Infection is common on raw and ripe fruits resulting in complete rotting of fruits. Fruit loss of 50% in strawberry may only be due to botrytis rot. A single infected fruit may spread rot to all fruits during transit or storage. Following control measure should be followed for reducing its incidence (i) Don't allow fruits to touch the soil; (ii) provide suitable mulch; (iii) avoid excessive application of N fertilizers; (iv) avoid excessive irrigation; (v) ensure proper drainage; (vi) follow proper sanitary measures; (vii) spray carbendazim (0.05%) or captan (0.05%0 or thiram (0.05just after the opening of flowers%) (viii) remove all infected fruits before storage or transportation.

8.7. 5. Viral diseases

Several viral diseases damage strawberry plants. They may produce mottling, mild yellow edge, crinkle, vein chlorosis, leaf stunting or dwarfing. Viruses are transmitted by leaf hoppers and nematodes. Control of viral diseases is possible only if insect vectors are kept under control, following clean cultivation and by crop rotation. Initially, virus-free planting material should be used and infected plants may be uprooted and burnt to avoid further spread

8.8. Physiological disorders and their management

8.8.1. Fruit malformation

Malformed fruits can be commonly seen even in recognized varieties of strawberry. In this, common shape of berry is changed. Usually, the primary and secondary flowers produce more malformed fruits due to undeveloped achenes at the distal end of the receptacle than the tertiary and quaternary ones (Sharma and Singh, 2008; Sharma, 2011). Many factors have been found associated with malformation of fruits. However, planting of more vigorous runners and high N levels and insufficient pollination and lack of growth promoting substances are the ones chiefly associated with the production of malformed fruits. Provision of adequate pollinizer cultivars in commercial plantings of the main cultivar and honey-bee colonies reduce fruit malformation (Sharma and Singh, 2008; Sharma, 2011). Planting of young and less vigorous plants and avoidance of excessive N fertilizers also reduce it to a greater extent. Use of gibberrelic acid (75 ppm) during February helps to reduce fruit malformation significantly (Sharma and Singh, 2009).

8.8.2. Albinism

It is the most serious disorder of strawberry fruit, occurring primarily at the time of ripening. It has attained an alarming situation in the USA, Belgium and the Netherlands. Fruits suffering from albinism appear bloated, and these develop white or pink areas on their surface; the pulp remains pale. These fruits have poor flavour and tend to be acidic. Affected fruits develop normally but do not ripen uniformly and show waxy appearance and are liable to severe damage during harvesting and are highly susceptible to fruit rot during storage (Sharma and Sharma, 2003; Sharma *et al.*, 2006[a,b]; Sharma, 2011). This disorder tends to develop in densely growing crops, dense plantings or due to excessive fertilizer application. It generally develops in crops grown in sandy, low pH soils and soils with high N, K, Ca contents and sometimes P content also (Sharma *et al.*, 2006[a,b]; Sharma, 2011). Albinism is more severe in closed tunnels (with little ventilation). Similarly, its incidence is more in crops mulched with black film than those with white films, probably due to high N mineralization under black film, which generates higher temperature. Selection of a suitable variety, avoiding dense planting and excessive application of fertilizers particularly N and K etc are some measures to reduce its menace (Sharma and Sharma, 2004; Sharma, 2011).

9. Maturity

The crop is ready for harvesting in March-April, but in cool climates, it matures by mid-June. The fruits are usually harvested when 2/3rd to ¾ portion of the fruit peel has developed colour. For distant markets, berries are sometimes harvested while they are still green, or white and are hard. However, a little delay in picking may increase percentage of over-rotted berries. Berries should be harvested in shallow containers as they are highly perishable and damage early if deep and bulky containers are used.

10. Harvesting

Usually, berries should be picked up daily in warm weather and 2-3 times a week in cool. As far as possible, berries should be picked in early and late hours of the day. Keep picked berries in a shade and a dust-free place. A shelter bed, even a structure with simply a roof and a wall on one side of the prevailing wind is advantageous to reduce farm heat of berries and other adverse effects. Pick-your-own (PYO) method of harvesting strawberries is the most popular method of direct marketing in several advanced countries of the world (Sharma and Sharma, 2004). In this, strawberry farms are established near road-sides and consumers come and pick the wanted fruits in small baskets, get it weighed and pays money to the growers. This concept of PYO farms began when

A view of strawberry harvesting in India

harvesting labour became scarce and increasingly expensive. The advantages of PYO farms include no direct harvest labour and transportation costs, improved quality, increased product availability for consumers, and reduced packing costs.

11. Yield

Yield is influenced by soil and climate, variety, planting system, planting density, number of crowns/plant, number of fruits/plant, fruit size and various management practices. In general, primary and secondary flowers bear large-sized fruits, as compared to tertiary and quaternary flowers, thus affecting yield. Average strawberry yield is about 5-7 tones/ha.

12. Post harvest handling

12.1. Pre-cooling

Strawberries reach markets in a better condition if pre-cooled at 4.4° C within 2 hours and kept at this temperature. This is especially true during warm weather when fruit is picked during the heat of the day and fruit temperature is 26.4°C or more. At higher temperature, rate of respiration goes high, which result in further reduction in post-harvest life of berries (Hancock, 1999; Sharma and Sharma, 2004).

12.2. Grading and Packing

After harvesting, fruits should be graded according to size, shape and colour. Fruits of uniform size and colour should be selected for A grade packs and slightly damaged ones may be packed and sold out as B grade. Before grading, remove all cull, diseased or rotten fruits. In India, sorting is usually done manually. In some advance countries, pan-method of sorting is followed for grading. The grading pan is 25 cm long, 2.5 cm wide and 1.5 cm deep. Commercial growers may run fruits over a belt, grade it and then jumble back into cardboard trays of different sizes, which can conveniently be stacked for pre-cooling and marketing. Standardized grades promote honest and fair-dealing, and discourage careless and unscrupulous packing (Hancock, 1999; Sharma and Sharma, 2004). Strawberries may be packed in small baskets, tins, polyethylene over-wrapped baskets and crates. Nowadays, special packing cases of plastics, commonly called as 'plastic punnets', have been developed for packing strawberries. About 200-250 g fruits can be safely packed in one punnet. These packs provide proper ventilation for strawberries because they have many perforations (10-12 holes/punnet) (Sharma and Singh, 1999). After packing fruits in punnets, these are placed in corrugated fibre-trays. Punnet-filled trays should be kept under shade or shelter to reduce heat and water-loss from berries. In recent development, packing of strawberries in cellophane (cellulose film and chitosan films has been recommended for better post-harvest life of strawberry (Hancock, 1999; Sharma and Sharma, 2004). Properly filled baskets should neither be slack nor so full that berries are crushed by covers or dividers. Packs should be full enough to look attractive and maintain a well-filled appearance for the consumer. Similarly, the berries of one cultivar should be packed in one container and may not be mixed with the other. The pack should be properly labeled with name of variety, garden and date of packing.

12.3. Storage

The strawberry fruit is highly perishable. It should be marketed soon after harvesting. Due to perishable nature, refrigerated transportation and short-term storage of strawberries can give good money to grower. Further, strawberries are available only for a short time (15-20 days) and thus their proper storage can extend their availability (Hancock, 1999).

Many techniques like pre-harvest application of fungicides and other chemicals have been tried to prolong post-harvest life of strawberries, but with limited success. The better way of storage of strawberries is the controlled atmospheric (CA) storage. The requirement for temperature and concentration of CO_2 and O_2 in controlled atmosphere varies with cultivar but, in general, they can be stored for 10-12 days at 2% O_2 and 5% CO_2 at 3°C (Lanz, 2012). Some investigators have suggested that long-term storage of strawberries at 1% or lower concentration of O_2 may lead to off-flavour, and thus higher concentration of CO_2 (10% or more) should be restricted in storage for up to a week only, where adequate refrigeration is not available. The post-harvest life of berries in storage can further be increased if plants are sprayed with Bavistin (0.1%) during flowering and then storing of harvested berries in 20% CO_2 and 0-3% O_2.

13. Value addition

In addition to table and dessert purposes, various value added products can be prepared from strawberry. Berries can be used for canning, candy, jam and jelly-making and for flavouring ice-creams. Juice is also made from strawberries, which is considered as a refreshing and soothing drink in summer. Some developing countries also make wine and freeze products from strawberry. In our country, strawberry is consumed as a fresh fruit. However, it is also used for flavouring ice-cream. Several specific value added products have been developed through processing techniques like freezing, drying, canning in strawberry (Linda and Mitcham, 2007).

13.1. Frozen strawberries

Freeze-drying is used for the preservation of highly perishable fruits such as strawberry. Frozen strawberries can be used as such or as raw material for the development of different value added products. For example, individual quick frozen strawberries are used for freeze drying. The frozen strawberries can also be sliced or diced for use in ice cream, jelly and jam. It is important to select uniform sized, red coloured strawberries for freezing purpose. The process consists of following steps: Fruit reception '! precooling '! removal of stem and caps '! air circulation for removal of extraneous material '! inspection and grading '! washing with chlorinated water '!quick freezing of fruit (at -40^0 C) in a blast air freeze tunnel '! packing and storage under freezing temperature (at-18). Quick freezing reduces formation of large ice crystals, which causes drip loss upon thawing. Packing in heavy weight, moisture resistant and air tight packing material is helpful in minimizing the problems during storage (Linda and Mitcham, 2007).

13.2. Strawberry puree

Strawberries are processed to make puree or pulp and utilized in various products. It is critical that rotten fruits are not used for processing into puree. The fruit before processing are sorted, washed and separated the bruised one, however it is not necessary to remove the berry cap. The pulper meshes (mesh size 0.5-10 mm) the fruit into a smooth pulp or puree and also removes skin and seeds. Often the puree is pasteurized through a vacuum system to maintain colour and flavour (88^0 C for 2 minutes) and cooled to about 15^0 C. The final product is filled and stored at refrigeration temperature (Hui *et al.,* 2006).

13.3. Dried strawberry

Slices of strawberry fruits are processed in two stages. In the first phase most of the water is removed using sugar syrup as an osmotic agent. In the second phase air drying is done where the moisture content is further reduced to about 15%. The process is simple and involves operations like selection of fruits, cleaning, washing, peeling, curing and slicing and dicing. To remove water by osmosis the prepared fruit slices are steeped in sugar solution. The slices are then drained, dried in a hot air drier and dehydrated slices are packed in plastic polyethylene pouches. The osmo-air dehydrated product is near to the fresh fruit in terms of colour and flavour. The osmo-air dehydrated product can be used in ready-to-eat foods, ice cream, fruit salad, kheer, cakes and bakery products.

13.4. Canned strawberry

Canning is a method of food preservation in which the food material is subjected to elevated temperature and pressure, packed in hermetically sealed container which provides unfavourable conditions for microbial growth. Strawberries produce and ripen quickly over a period of just a couple of weeks, so canning is one of the best methods to preserve them. Strawberries are canned using the boiling water-bath method. Canned strawberry is suitable for cooking, used for preparation of pudding, cakes, can be mixed with ice cream and used with a lot of more products (Linda and Mitcham, 2007). The basic steps involved in canning process of strawberry fruits are sorting, grading, de-capping, syruping, filling in jars or cans, exhausting, sealing and storage.

14. Breeding and crop improvement

The increase in cultivable area under strawberry may be attributed to purposeful breeding and development of specific cultivars. During the last 75 years, strawberry breeding has been conducted by several research institutions of the world, and many superior cultivars possessing desired features have been developed.

14.1. Modern Breeding objectives

Breeding programme objectives are similar but their relative importance differs among strawberry breeders depending upon the conditions (Sharma and Sharma, 2004). For example, a disease or insect in a particular locality may be a serious problem, and which may not be in the other area. So a breeder in that particular locality will give more emphasis to develop resistance for that disease or insect in the variety. However, the commonly desired plant characters in breeding strawberry are yield, vigour, fruiting habit (long-day, short-day, or day-neutral), ripening time, winter hardiness, runnering capacity, high temperature tolerance, concentrated ripening, disease resistance (root diseases, foliage diseases, virus tolerance) and pest resistant (aphids, weevils and mites). Out of these, yield, vigour, fruiting habits are of primary importance, while others are considered as per local importance (Hancock *et al.*, 1996). The characteristics of fruits include size, symmetry, shape, colour, firmness, flavour, ease of capping, vitamin C content, TSS, acidity and resistance to rot. Out of these; size, firmness, and flavour are always important, while others are of variable concern, depending upon the locality.

14.2. Breeding (crossing) technique

Remove all flower parts by emasculation so that pistils are fully exposed for pollination. Emasculation should be done within 1-3 days before anthesis to prevent selfing. Emasculated flowers must be protected from foreign pollen by bagging or covering with a layer of cotton. Pollens of desired cultivar are collected 1-2 days prior to anthesis and kept in vials to dehisce. Pollens can be stored at 4°C under low humidity, and they are transferred to stigmas with a small camel's hair brush. After pollination, fruit ripens in 25-30 days under normal environmental conditions.

14.2.1. Seed collection

Seeds may be collected from the pulp, which can be stored for longer time at 1-4°C or can be sown as per requirement. However, to hasten germination, exposure of seeds to light or their scarification for 10-15 minutes with concentrated sulphuric acid is useful (Hancock *et al.*, 1996).

14.2.2. Seedling culture

Seeds are usually sown in sand, sterilized soil or shredded sphagnum moss in seed trays. Young seedlings are susceptible to damping-off and should be handled carefully during their early growth. These are transplanted into soil in flat pans or pots after 6-8 weeks in hill or matted-row system of planting. For screening seedlings against diseases, artificial inoculation with disease-causing pathogens, have been perfected. Similarly, for measuring resistance of seedlings to insect-pest in fields, greenhouse screening is done. Further, the tests for virus tolerance usually require planting or maintenance of seedlings in the field for at least 2 or more years in areas where virus and vectors are present in abundance.

The selected seedlings are then evaluated subjectively or objectively in single or replicated plots for yield, fruit size, fruit weight and fruit firmness.

14.3. Biotechnology

Elimination of viruses through thermotherapy has been achieved. After thermotherapy, shoot-tips of desired material are removed and then proliferated on tissue culture media (Sharma and Sharma, 2004). Strawberry is the first horticultural crop to be micropropagated commercially through meristems and meristematic callus. In general, tissue-cultured plants produce more flowers and have higher potential yield than conventionally produced plants, and now millions of tissue-cultured plants are available in nurseries. Even some success in single-cell regeneration through protoplast culture and recombinant DNA technology has also been achieved, which is considered as a pre-requisite for producing of transgenics.

14.4. Breeding systems

14.4.1. Out-breeding and inbreeding

The outcross method has been used mostly for improving strawberries (Darrow, 1966). In this, most characters segregate in each generation and selection pressure can be exerted simultaneously on each character. The seedlings with desirable combination of characters are selected. In general, wide parentage base is necessary in outbreeding to avoid chances of inbreeding, which often results in loss of vigour and reduction in yield. However, inbreeding has also been used occasionally as a method for evolving new cultivars in strawberry. Some investigators have tried to evolve inbred lines in strawberry but due to lanky and weak growth, lack of vigour and susceptibility to viral infection have made it difficult to maintain them (Hancock *et al.*, 1996).

14.4.2. Interspecific hybridization

All octoploid *Fragaria* species are inter-fertile. *F. chiloensis* and *F. virginiana* have been used extensively in cultivated strawberry improvement (Sharma and Sharma, 2004). Winter hardiness of *F. virginiana* ssp. *glauca*, resistance of *F. chiloensis* to red stele, strawberry aphid, two-spotted mite, resistance of *F. chiloensis* and *F. virginiana* to verticillium wilt; virus tolerance of *F. chiloensis*, the day-neutral character of *F. virginiana* ssp. *glauca*, large fruit size and firmness of *F. chiloensis* and high photosynthetic rate of *F. chiloensis* have been studies thoroughly (Hancock *et al.*, 1989). However, their incorporation into cultivated strawberry is not an easy task as some undesirable characters, small fruit size, partial fertility, soft flesh, black seeds, susceptibility to leaf diseases and dull fruit colour are also inherited to the progeny (Sharma and Sharma, 2004). Thus, a native octoploid species should be used for breeding programme of strawberries for raising at least 3-4 generations by outbreeding or back crosses to obtain a progeny of desired characteristics. It requires a large population to select a seedling with desired characters from segregating population (Hancock *et al.*, 1996)

14.4.3. Intergeneric hybridization

Hybridization between *Fragaria* and its related genus *Potentilla* has been tried but most hybrids die at an early age or were sterile. Several attempts to cross *F. vesca* with different species of *Potentilla* have been tried but all hybrids died in early or late stages. However, two fertile colchicine-induced amphidiploids have been produced, which could be used to transfer genes between two genera.

14.4.4. Breeding for specific plant character

Most of the plant and fruit characters of cultivated strawberries are inherited quantitatively (Galletta and Mass, 1990). Additive, dominance and epistatic components may act in expression of a particular character. It usually creates a complex breeding problem when maximum expression of a combination of characters is desired.

i. **Yield**. Yield is a component of number and size of fruit, plant vigour, hardiness, disease resistance of plant and crown number per row. Fruit size differs within a fruit cluster; primary berry being the largest, and others relatively smaller. Fruit size is inherited quantitatively and is controlled by 6-8 allelic pairs. Crown number per row is often a factor most strongly associated with fruit yield (Hancock *et al.*, 1983). The crown production can be increased by stolon formation, branch-crown formation or shifting to ever-bearing habit, and *F. virginiana* ssp. g*lauca* is a rich source of both types of genes (Bringhurst and Voth, 1981). For stolon production, Guardian and Honeoye are best cultivars (Hancock *et al.*, 1983). Lal and Seth (1982) have found that heritability estimates were high for a number of yield components, like berry weight, berries/flowers, yield/flower-stem, flower-stem number and suggested that when several characters are involved in a breeding programme, crosses should be designed to exploit all genetic variances, whether additive, dominant or epistatic.

ii. **Disease resistance**. Resistance to disease has been found quantitative and additive with a high heritability estimate, and is partially dominated over susceptibility, governed by several genes (Sharma and Sharma, 2004). Leaf spot is a problem in Europe, New Zealand, Korea and there appears to be considerable variability for resistance to it. It is not known whether this is due to environment or pathogen attack.

    Breeding for resistance to Ramularia leaf spot is generally restricted to races that occur locally, and breeding for resistance has, therefore, been on a regional basis, and cultivars that are resistant in one locality may be susceptible in another (Sharma and Sharma, 2004). No cultivar has been found immune, although Dabreak is most resistant. Other cultivars showing regional resistance to leaf-spot are Albritton, Elista, Hood, Howard 17, Lassen, Lateglow, Surecrop and Tangi (Hancock *et al*., 1996).

Powdery mildew is widely prevalent on strawberry leaves, and may infect flowers and fruits. The inheritance of powdery mildew resistance has not been fully understood, however, it may be partly due to thickness of cuticle (Sharma and Sharma, 2004). *Verticillium* wilt is a widespread disease of strawberry plants grown in soil previously occupied by Solanaceous crops. Fumigation controls it but resistant cultivars are a long-term practical solution. Resistance to *verticillium* wilt is partially dominant and is inherited quantitatively. Widely grown cultivars such as Allstar, Catskill, Delite, Earliglow, Gem, Guardian, Hecker, Hood, Tribute, Tristar and Totem are resistant to *verticillium* wilt (Hancock *et al* 1990; Maas *et al*., 1989).

Red stele is a destructive disease of strawberry in cool moist areas. Resistance to red stele is partially dominant and quantitatively inherited (Sharma and Sharma, 2004). Wild selections of *F. chiloensis* and *F. virginiana* and *F. ananassa* and cultivars Aberdeen and Frith have been used to incorporate red stele resistance into cultivars, and seedlings with varying degrees of resistance have been obtained. Gorella, Juspa, Lavo, Tamella, Templarl, Guardian, Hood, Midway, Redchief, Sparkle, Surecrop, Sunrise have been reported to be resistant to one or the other races of red stele.

Anthracnose attacks crowns, petioles and runners in warm humid areas. The fungus has many races and thus cultivars resistant in one location may be susceptible at another. Resistance to anthracnose has been incorporated into cultivars by using Sequoia and locally adapted material of cultivated strawberry (Maas and Galletta, 1989).

Botrytis-rot is the most destructive disease of strawberries in the field as well in storage. Two separate modes of inheritance appear to confer resistance to pre- and post-harvest rot although resistance to both these stages has been combined in Earliglow (Maas and Smith, 1978) and Tyee (Daubeny, 1980). However, the exact nature of the resistance is not clear but it appears to be quantitative and additive. In general, fruit firmness may play a vital role against botrytis-fruit rot resistance.

iii. **Virus tolerance:** Viral diseases cause heavy losses to strawberry where large aphids or nematode populations are found. For breeding for virus tolerance, plants are screened either in fields (Daubeny *et al.*, 1972) or by transmission tests in greenhouses. A high degree of tolerance in several clones of *F. chiloensis* has been reported and there exists wide differences in viral tolerance among cultivars. However, Cheam, Northwest, Chuksam and Totem are tolerant (Daubeny *et al.*, 1972). Virus tolerance among different cultivars differs in different areas, which may be due to the presence of different races of vectors especially aphids (Hancock *et al.*, 1996).

iv. **Pest resistance:** Resistance to two-spotted mites is highly heritable and quantitative. Sources of resistance to two-spotted mite are Lassen, Cardinal, Comet, Blackmore and Muir (Hancock *et al.*,1996). Inheritance of resistance to strawberry aphid is still emerging but it appears to be controlled by more than a single locus, with additive gene action. De Norte and Yaquina clones of *F. chiloensis* are found resistant. Similarly, Benton has also been found resistant to aphid colonization than Totem, Hood and Tyee (Crock *et al.*, 1982). Totem and Cheam are found most tolerant and Northwest most susceptible cultivars to strawberry aphid.

v. **Fruiting habit:** Three types of fruiting habits have been recognized in strawberry: single crop or June bearer, which is characteristic of most of the cultivars; ever-bearer, fruiting 2-3 times per season; day –neutrals (Sharma and Sharma, 2004).

   Many investigators have studied breeding behaviour of ever-bearing strawberries and it has been postulated that complimentary dominant genes govern inheritance and they segregate in an octoploid manner. Ever-bearing fruiting habit in diploids (wild type *F. vesca*) has been reported to be governed by a single recessive gene, which acts independently for a single recessive gene for non-runnering. Day-neutrality is regulated by a single dominant allele.

vi. **Winter hardiness:** Cultivars differ invariably in their winter hardiness. Inheritance of winter hardiness involving crosses with *F. virginiana* ssp *glauca* has showed that winter hardiness is partially dominant over non-hardiness (Sharma and Sharma, 2004). Usually, the clones of *F. virginiana* ssp *glauca* have very high degree of winter hardiness, and cultivars of European origin are more winter hardy; primarily due to large ancestral contributions from *F. virginiana*. Spring frosts are hazardous in most strawberry-growing areas and frost tolerance character is highly heritable (Darrow and Scott, 1947). Earlidawn and Howard 17 are frost-tolerant cultivars (Scott and Lawrence, 1975).

vii. **High temperature tolerance:** *F. virginiana* clones show a high degree of resistance to high temperature (30°C) and humid heat than *F. chiloensis* (Darrow, 1966; Caldwell *et al.,* 1990); and Blackmore and Missionary are as resistant as *F. virginiana.* Pusa Early Dwarf and Chandler, Fern, Douglas and Sweet Charlie can also tolerate high temperature of north Indian plains (Sharma, 2002).

viii. **Time of flowering and ripening:** Time of flowering and ripening are closely related in strawberry. Flowering time, ripening and rate of fruit development are inherited quantitatively with heterosis, and earliness has been reported to be partially dominant.

ix. **Adaptability to mechanical harvesting**: In the present-day breeding programmes, emphasis has been given on the development of cultivars suited for mechanical harvesting. Cultivars for mechanical harvesting should have high yields, uniform ripening, easy calyx removal, fruits with long petiole and necks (Sharma and Sharma, 2004). Ease of calyx removal has showed additive gene action and in some parents like Gorella, low capping force is dominant. Some cultivars like Pugget Beauty, Olympus, Totem, Red Chief are easy to cap. Saladin and Silver Jubilee have long (>40 mm) pedicels and Veeglow and Pugget Beauty have long necked fruits (Dale *et al.,* 1987), which make them ideal for mechanical decapping (Gooding *et al.*, 1983).

x. **Stolon production:** Stolon production rate is highly variable among different species and cultivars, and the breeding behaviour of the runnering character is very complex (Sharma and Sharma, 2004). Non-runnering charcater has been highly heritable in *Fragaria x ananassa* and is found closely correlated with ever-bearing character. With few exceptions, non-runnering plants are ever-bearers. When non-runnering plants are selfed, their seedlings have produced runnering and non-runnering plants, and their inbreeding to S_3 generation has resulted in high percentage of non-runnering seedlings, but no homozygous non-runnering genotypes could be obtained.

xi. **Edaphic adaptations**. *Fragaria* species differ widely in soil requirements. For example, *F. chiloensis* grows on much sandy and well-drained soils than *F. virginiana* or *F. vesca* (Hancock and Bringhurst, 1979). *F. chiloensis* is more drought tolearnt than *F. virginiana* (Darrow, 1966). Similarly, *F. orientalis*, *F. vesca* ssp *americana* and *F. virginiana* ssp *glauca* have also high tolerance to drought. *F. chiloensis* can tolerate very high salt levels (> 1,200 ppm) as compared to *F. vesca* (Hancock and Bringhurst, 1979).

Cultivated strawberries appear to differ in their relative sensitivity to soil reaction. British Sovereign can grow in delta soils with 4.5- 5.0 pH while the Marshall could not. In general, wild strawberries have been growing vigorously on acid, neutral and alkaline soils. Delite and Shuksan were found resistant to iron chlorosis when grown in soils with 7.9 pH (Scott and Lawrence, 1975).

14.4.5. Breeding for specific fruit characters

i. **Fruit size:** Large fruit size is preferred in strawberry. Fruit size usually differs within a fruit cluster, primary berry being largest and berries at other positions being progressively smaller. Decline in fruit size from primary to other positions is similar in all cultivars, but Moore *et al.* (1970) reported it to be a varietal character. It is inherited quantitatively, with 6-8 allelic pairs controlling fruit expression although heterosis has been observed for it in some progenies. Differences in combining ability for yield, fruit size and number of fruits/plant have been reported. Some large-fruited cultivars are Catskill, Douglas, Gorella, Pajaro, Totem and Vesper. Even in India, Belrubi, Chandler, Fern, Senga Sengana, Selva, Tioga have been reported large-fruited cultivars, and can be used in future breeding programmme (Sharma and Badiyala, 1980; Awasthi and Badiyala, 1983; Beniwal *et al.*, 1989).

ii. **Pulp firmness and peel toughness:** Berries with tough peel and pulp are considered best for post-harvest handling and value-addition. Both these characters are greatly influenced by temperature and relative humidity during fruit development and ripening (Sharma and Sharma, 2004). In general, high temperature and relative humidity result in softer fruits than cool temperature and low relative humidity. Cultural practices also influence these characters invariably. Fruit firmness has been found highly heritable, but was negatively correlated with verticillium wilt resistance. Genotypic correlations between firmness and yield and between firmness and size have been zero, indicating no genetic relationship between firmness and fruit size or yield. Soft-fruited parents could transmit soft-fruit character to a high percentage in the progeny, indicating that softness in fruits is controlled by many genes. Positive correlation between peel toughness and pulp firmness has been found. Blackmore was the first firm-fruited cultivar developed by the scientists. At present, Apollo, Holiday, Korona, Linn and Hapil have notable firm pulp (Hancock *et al.*, 1996), and Blakemore, Pajaro, Selva and Tioga have have good flavour besides firm pulp (Sharma, 2002).

iii. **Processing qualities:** A large amount of strawberries in Poland and the United States are processed into various value-added products. Fruits for freezing purpose should have uniform pulp colour, firmness, high flavour with tartness, and ease of capping or plugging. Moreover, frozen berries should retain its flavour after processing (Sharma and Sharma, 2004). Flavour is influenced by environmental and cultural practices, but it is a characteristics of cultivars also. Fairfax is considered as the best parent for incorporating flavour in hybrid seedlings. However, Albritton, Earliglow and Korona cultivars have also fresh market flavour. Some cultivars have attractive red pulp and flavour, and they retain it even after freezing. The notable examples are Cardinal, Cornwallis, Totem, Tyee, Vibrant and Veegem. The red colour of the peel and the flavour are quantitatively inherited. Similarly, capping character is also inherited quantitatively.

iv. **Quality attributes:** In addition to fruit size, shape, colour and flavour, total soluble solids (TSS), acidity and vitamin C contents are considered important quality attributes in strawberries. All these are influenced by climate, cultural practices and more on cultivar itself. Both TSS and acidity are controlled genetically with varying levels of additive and dominance control (Shaw *et al.*, 1988). High content of total soluble solids is dominant and the low titratable acidity is partially dominant (Lal and Seth, 1982; Shaw *et al.*, 1988). Vitamin C content is quantitatively controlled and high vitamin C content of the fruit is partially dominant.

15. Economics

During recent years, strawberry cultivation has emerged as one of the most remunerative farming enterprise. It is very difficult to work out the cost of cultivation for a particular locality because of changing rates and availability of labour, cost and planting of runners, cost of fertilizers, insecticides and fungicides etc. Further, the cost of cultivation is not a constant factor and it may change from place to place. In general, the cost of strawberry cultivation is higher in hills as compared to plains. By following scientific management practices, a farmer can earn as high as 2.50 lakhs from one hectare strawberry field. The rough estimate of cost of cultivation of strawberry is given in Table 3.

Table 3: Rough estimates of cost of cultivation of strawberry in India on per hectare basis

Sl.No.	Expenditure	Amount (Rs.)
1.	Cost of plants (50,000/ha) at Rs 2 per plant	100,000.00
2.	Land preparation	
	Ploughing (One tractor at 500/hr for 3 hours)	1,500.00
	Cost of FYM and insecticide (50 tones FYM at Rs 200/tonn)	10,000.00
	Furrow / bed preparation (50 labourers at Rs 70/day)	3,500.00
3.	Transplanting	
	a. Fungicide treatment (cost of fungicide + 2 labourers at Rs 70/day)	300.00
	b. Transplanting cost (50 labourers at Rs 70 per day)	3,500.00
	c. Watering with hand fountain (5 labourer at Rs 70 per day)	350.00
4.	Installation of micro irrigation unit	40,000.00
5.	Intercultural operations	
	a. Fertilizer application (Cost of fertilizers + application charges)	8,000.00
	b. Mulching (cost of mulching material + labour cost)	5,000.00
	c. Weeding (cost of weedicide + labour cost)	1,000.00
	d. Irrigations , 12-15(labour cost Two labourers at Rs 70)	1400.00
	e. Insect/pest and diseases control (cost of insecticides, fungicides + labour cost)	5,000.00
	f. Bird scaring and watch and ward (two labourers at Rs 70 for one months)	4,200.00
6.	Fruit harvesting (five labourers at Rs 70 per day for 30 days)	10,500.00
7.	Overhead expenses	5,000.00
8.	Bank interest on loan or otherwise	10,000.00
	Total expenditure	208,650.00
	Income	
	a. Sale of fruits (8 tonnes at Rs 50/kg)	400,000.00
	b. Sale of runners (20,000 plants at Rs 2/ plant)	40,000.00
	Total income	440,000.00
	Net profit	231,350.00

16. Future research thrusts

Though there has been phenomenal increase in the area and production of strawberry in India but still there is an urgent need to give more emphasis on the following points for commercial cultivation of this valuable crop.

16.1. Introduction of Floridian cultivars-Experiences of different research investigations have indicated that the varieties introduced from Florida (USA) like, Chandler, Douglas, Fern, and Sweet Charlie have performed well under tropical and sub-tropical climates of India. We need to introduce, evaluate and recommend introduced cultivars, if suitable, for cultivation in warmer areas, which constitutes the major cultivable area in India.

16.2. Intensification of research-At present, only one university (University of Horticulture and Forestry, Solan (Himachal Pradesh) is involved in breeding work of strawberry. We must intensify the work on breeding new cultivars which are best suited to our conditions. In our country, diseases like red stele, botrytis rot, verticillium wilt have been observed, but no systematic research work is being conducted on their causes and control. Thus, there is an urgent need to initiate work for breeding resistant varieties for various biotic and abiotic stresses.

16.3. Supply of genuine planting material-In India, availability of the genuine planting materials is always a problem because we do not follow standard procedures for raising new plants. In many advanced countries, only tissue-cultured plants are given to farmers, which perform better in fields, and give higher yields of better quality. In India, only 3-4 private institutions produce its plants through micropropagation. Thus, use of micropropagation technique should be encouraged for the production and supply of quality planting material.

16.4. Development and use of modern packaging: Strawberry is a delicate fruit, which can perish easily, and it is very difficult to send it to distant markets. Thus we need to intensify research for the development of suitable packaging system. Now specially designed "packs", commonly known as punnets are available and should be encouraged among farmers. Attractive, coloured and cushioning packaging materials should be used for attracting consumers.

16.5. Co-operative and contractual farming: Keeping the problems of marketing of this delicate fruit in mind, efforts should be made to encourage farmers for co-operative and contractual farming. This effort has been very successful for grape, mango and pomegranate fruit crops in Maharashtra, as result of which, there has been tremendous increase in production and productivity, as a result of which, Maharashtra has emerged as horticulture state of India.

16.6. Standardization of runner-production technique in plains: Farmers of plains usually purchase runners from hills. The cost of planting material for establishing one hectare orchard comes around 1-1.25 lakh. Moreover, new plants are required year-after-year. It is almost impossible to propagate strawberry in plains because the runners usually die in plains due to high temperature in summer. Thus, there is an urgent need to develop a technology for its propagation in plains too.

16.7. Data base: Many farmers are now growing strawberry in different parts of India but no data is available on its acreage and production. Data is an integral part of a sound research and extension programme. Thus, some efforts should be done in this direction.

References

Awasthi, R.P. and Badiyala, S.D. 1983. Spacing and performance studies in some strawberry cultivars. *Indian J. Hort.,* **40**(1 & 2): 29-34.

Badawi, M.A., Alphonse, M., Bondok, A.Z. and Hosni, Y.A. 1990. Propagation of some strawberry cultivars by means of tissue-culture technique. *Egyptian J. Hortic.,* **17**(1): 9-16.

Badiyala, S.D. and Aggarwal, G.C. 1981. Note on the effect of mulches on strawberry production. *Indian J. Agric. Res.,* **51**(11): 832-834.

Bagnara, D. and Vincent, C. 1988. The role of insect pollination and plant genotype in strawberry fruit set and maturity. *J. Hortic. Sci.,* **63**(1):69-75.

Beniwal, L.S., Daulta, B.S. and Bisla, S.S. 1989. Evaluation of different strawberry (*Fragaria x ananassa*) cultivars under Hisar conditions II. Yield and fruit quality charcaters. *Haryana J. Hortic. Sci.,* **18**(1-2): 56-60.

Bringhurst, R.S. and Voth, V. 1981. Breeding for high productivity, large fruit size. *In*: *The Strawberry Cultivars for Marketing,* Childers, N.F. (Ed). Horticulture Publications, Gainsville, Florida.

Caldwell, J. Hancock, J.F. and Flore, J. A. 1990. Strawberry leaf photosynthesis acclimation to temperature. *Hort. Science,* **25**: 166-170.

Chandler, C.K., Miller, D.D. and Ferree, D.C. 1995. Shade during July and August reduces growth but not fruiting of strawberry plants. *Hort. Science,* **27**(9):1044.

Childer, N.F. 2003. The strawberry: For growers and other. NF Childers Publications, USA.

Childers, N.F., Morris, J.R. and Sibbett, G.S. 1995. *Modern Fruit Science.* Horticulture Publication, Gainesville, Florida, USA.

Crock, J.E. Shanks, C. H. and and Baritt, B. H . 1982. Resistance in *Fragaria chiloensis* and *F. ananassa* to aphid *Chaetosiphon fragaefolii* and *C. thomasi*. *Hort. Science,* **17**: 959-960.

Dale, A., Gray, V.P. and Miles, N.W. 1987. Effects of cultural systems and harvesting techniques on the production of strawberries for processing. *Canadian J. Plant Sci.,* **67**: 853-862.

Darrow, G.M. 1966. *The Strawberry*: *History, Breeding and Physiology*. Holt, Rinehart and Winston, New York, Chapter, 20.

Darrow, G.M. and Scott, D.H. 1947. Breeding for cold hardiness of strawberry flowers. *Proc. Amer. Soc. Hortic. Sci.,* **50**:239-242.

Daubeny, H .A., Norton, R.A., and Barritt, B.H. 1972. Relative differences in virus tolerance among strawberry cultivars and selections in the Pacific Northwest. *Plant Disease Repr.,* **56**:792-795.

Daubeny, H. A. 1980. Tyee strawberry**.** *Canadian Journal of Plant Science,* **60**: 743-746.

Galletta, G.J and Maas, J.L. 1990. Strawberry genetics. *Hort. Science,* **25**: 871-79.

Gengotti, S., Tisselli, V., Lucchi, C. and Nasolini, T. 2001. Growing organic strawberries. *Inform. Agrario* **57:** 27, 45-48.

Gooding, H. J., McNicol, R.J. and Reid, J. H. 1983. Studies of strawberry fruit characteristics necessary for machine harvesting. *Crop Res.,* **23**: 3-16.

Hancock, J.F.1999. Strawberries. CABI, New York, USA.

Hancock, J.F. and Bringhurst, R.S. 1979. Ecological differentiation in perennial octoploid species of *Fragaria. Amer. J. Bot.,* **66**: 367-375.

Hancock, J.F., Flore, J.F. and Galletta, G.J. 1989. Gas exchange properties of strawberry species and their hybrids. *Sci. Hortic.,* **40**: 139-44.

Hancock, J.F., Maas, J.L., Shanks, C.H., Breen, P.J. and Luby, J.J. 1990. Strawberries (*Fragaria sp.*) *In*: *Genetic Resources in Temperate and Nut crops*. Moore, J.N. and Ballington, J.R. (Eds). International Society of Horticultural Science, Wageningen, the Netherlands.

Hancock, J.F., Scott, D.H . and Lawrence, F.J. 1996. Strawberries. *In*: *Fruit Breeding, Vol. III* : *Vine and Small Fruit crops*., Janick, J. and Moore, J.N.(Eds), John Wiley and Sons, Inc, Chapter 6, pp. 449-70.

Hancock, J.F., Siefker, J.H. and Schulte, N.L. 1983. Cultivar variation in yield components of strawberries. *HortScience* **18**: 312-313.

Hartman, H. T. 1947. Some effects of temperature and photoperiod on flower formation and runner production in the strawberry. *Plant Physiolology,* **22**: 407-10.

Howard, C.M. and Albregts, E.E. 1980. Dover strawberries. *HortScience* **15**(4): 540.

Hughes, H. M., Duggan, J.B. and Banwell, M.G. 1969. *Strawberry Bulletin No 95*, HMSO 10, 6 edition, Ministry of Agriculture, Fishery and Food, UK.

Hui, Y., Barta, J., Cano, M., Gusek, T., Sidhu, J. and Sinha, N. 2006. Handbook of Fruits and Fruit Processing., 1st ed. Blackwell Publishing.

Jones, O.P., Waller, B.J. and Beech, M.G. 1988. The production of strawberry plants from callus cultures. *Plant Cell Tissue Org. Cult.,* **12**: 235-41.

Kronenberg, H. G., Braak, J.P. and Zeilinga, A.E.1959. Poor fruit setting in strawberries II. malformed fruit in Jucunda. *Euphytica,* **8**: 245-48.

Lal, S.D. and Seth, J.N. 1982. Studies on combining ability in strawberry (*Fragaria x ananassa*) II. Fruit length, fruit diameter, fruit weight, ascorbic acid, total soluble solids and fruit yield. *Canadian J. Gene. Cytol.,* **24**: 479-483.

Lanz, H. 2012. Grow your own strawberries. Sea-to-Sea Publications, USA.

Larson, K.D. 1994. Strawberry. *In*: *Handbook of Environmental Physiology of Fruit Crops*. Volume 1: *Temperate crops*. Schaffer, B. and Andersen, P.C.(Eds). CRC, Press, Inc., pp. 271-297.

Linda, J. H. and E. Mitcham. 2007. Strawberries: Safe methods of store, preserve and enjoy. ANR Publication 8256. Division of Agriculture and Resources. University of California, USA.

Maas, J.L. and Galletta, G.J. 1989. Germplasm evaluation for resistance to fungus incited diseases. *Acta Hort.,* **265**: 461-472.

Maas, J.L. and Smith, W.I. 1978. Earliglow : a possible source of resistance to *Botrytis* fruit rot in strawberry. *Hort. Science* **13**: 275-276.

Maas, J.L., Galletta, G.J. and Draper, A.D. 1989. Resistance in strawberry to the races of *Phytophthora fragariae* and to isolates of *Verticillium* from North America. *Acta Hort.,* **265**: 521-526.

Mahajan, R. 1997. 'Studies on micropropagation of strawberry cultivars Chandler and Fern through vegetative buds'. M.Sc thesis, Dr Y.S. Parmar University of Hort. and Forestry, Solan, India.

Mitra, S.K. 1991. Strawberries. *In*: Temperate Fruit. Bose T.K., Mitra, S.K. and Rrathore, D.S. (Eds), Horticulture and Allied Publishers, Calcutta, pp.549-96.

Moore, J.N., Brown, G.R. and Brown, E.D. 1970. Comparison of factors influencing fruit size in large fruited and small fruited clones of strawberry. *J. Amer. Soc. Hortic. Sci.,* **95**:827-831.

Pandey, R.M. and Mishra, K.K.1983. Strawberry. *In. Phal Utpadan*, GBPUAT, Pantnagar, Uttar Pradesh, India, pp. 342-48.

Riotte, L. 1998. Grow the best strawberries. Storey Books, USA.

Saure, M.C. 1985. Dormancy release in deciduous fruit trees. *Horticultural Reviews* **7,** AVI Publishing Co. Chapter, 6.

Scott, D. H . and Lawrence, F.J. 1975. Strawberries. *In*: *Advances in Fruit Breeding*. Janick, J and Moore, J.N. (Eds). Purdue Univiversity Press, West Lafayette, Indiana, pp. 71-97.

Sharma Nirmal, Kumar Amit, Sharma, R.M. and Chauhan Neena. 2015. Strawberry. In: *Temperate Fruit Crop Breeding: Domestication to Cultivar Development (Part-I and Part-II),* Editors: S.N. Ghosh, M.K. Verma and A. Thakur, Jaya Publishing House, 27-B, Pocket-B, Dilshad Garden, Delhi-110095. Pp: 761-802.

Sharma, R.L. and Badiyala, S.D. 1980. A study on the performance of some strawberry (*Fragaria x ananasa*) cultivars in the mid-hills of north India. *Prog. Hort.,* **12**(2):17-24.

Sharma, R.M. and Yamdgni, R. 1999. *Modern Strawberry Cultivation*. Kalyani Publishers, New Delhi.

Sharma, R.R. 2002. Growing strawberries. International Book Distributing Co., Lucknow, India.

Sharma, R.R. 2011. Managing disorders in strawberry. *Indian Hort.,* **56**(4):12-13.

Sharma, R.R. and Sharma, V.P. 2003. Plant growth and albinism incidence in different strawberry cultivars under Delhi conditions. *Indian J. Hort.,* **61**(1): 92-93.

Sharma, R.R. and Singh, R. 2008. Fruit nutrient content and LOX activity in relation to the production of malformed and button berries in strawberry (*Fragaria* x *ananassa* Duch.). *Sci. Hortic.*, **119**: 28-31.

Sharma, R.R. and Singh, R. 2009. Gibberellic acid influences the production of malformed and button berries, and fruit yield and quality in strawberry (*Fragaria* x *ananassa* Duch.). *Sci. Hortic.,* **119:** 430-433.

Sharma, R.R. and Singh, S.K. 1999. Strawberry cultivation- a highly remunerative farming enterprise. *Agro India,* **3**(2): 29-31.

Sharma, R.R. and Singh, S.K. 2003. *Strawberry ki Kheti*: *Khoob Labh Deti*. *Asean Kisan* **1**(1): 42-45.

Sharma, R.R., Krishna, H., Patel V.B., Dahuja, A. and Singh, R. 2006. Fruit calcium content and lipoxygenase activity in relation to albinism disorder in strawberry. *Sci. Hortic.,* **107**: 150-154.

Sharma, R.R., Patel, V.B. and Krishna, H. 2006. Relationship between light, fruit and leaf mineral content with albinism disorder in strawberry *Sci. Hortic.,* **109** (1): 66-70.

Sharma, R.R., Sharma, V.P. and Meena, Y.R. 2001. Strawberry cultivation, a profitable business. *Intensive Agri.,* **38** (11-12) **:** 3-7.

Sharma, R.R., Singh, R. and Gupta, R.K. 2008. Row covers influence leaf physiology and physiological parameters in Sweet Charlie strawberry. *Fruits, **63:*** 143-150.

Sharma, V,P. and Sharma, R.R. 2004. The strawberry. DMKA, ICAR, New Delhi, India.

Shaw, D., Bringhurst, R.S. and Voth, V. 1988. Quantative genetic variation for resistance to lef spot (*Ramularia tulasnei*) in California strawberries. *J. Amer. Soc. Hortic. Sci.,* **113**: 451-456.

Singh, R. and Sharma, R.R. 2005. Strawberry cultivation with Hi-tech plasticultural techniques. *Agro India* (May issue), 18-20.

Singh, R., Gupta, R.K., Patil, R.T., Sharma, R.R., Asrey, R., Kumar, A and Jangra, K.K. 2010. Sequential foliar application of vermicompost leachates influence marketable yield and quality of strawberry (*Fragaria x ananassa* Duch.). *Sci. Hort.,* **124**:34-39.

Singh, R., Sharma, R.R. and Tyagi, S.K. 2007. Pre-harvest foliar application of calcium and boron influences physiological disorders, fruit yield and quality of strawberry (*Fragaria* × *ananassa* Duch.). *Sci. Hortic.,* **112:** 215-220.

Singh, R., Sharma, R.R., Kumar, A. and Singh, D.B. 2009. Package of practices for strawberry cultivation with modern techniques under northern Indian plains. *Acta Hort.*, **842**: 607-610.

Singh, R., Sharma, R.R., Kumar, S., Gupta, R.K. and Patil, R.T. 2008. Vermicompost substitution influences growth, physiological disorders, fruit yield and quality of strawberry of strawberry (*Fragaria x ananassa* Duch.). *Biores. Technol.*, **99**:8507-8511.

15

Walnut (*Juglans regia* L.)

Amit Kumar

1. Introduction

Persian walnut is of great importance across worldwide where nuts are harvested from wild stands, backyard gardens and rarely from commercial orchards that vary considerably in their size due to degree of management. The walnut is an old tree cultivated before the Greek age and appreciated for centuries for its fruits, wood and other parts utilized in several ways. Before the discovery of the phenomena of allelopathy, someone considered the walnut an un-healthful tree, since peasants used to wake up with migraine or fever after they had slept under a walnut tree. Some others considered the walnut a cursed tree, and in fact in Italy an old walnut tree growing in Benevento, has long been considered the place where the witches used to dance during the night of Saint John and, therefore, that tree was named "Tree of the Witches" (Avanzato, 2010).

Nuts are collected for home consumption, sale at local roadside stands and markets and for shipping to cities. Walnut trees are further utilized for their high quality timber to make a wide array of products. The leaves, bark and other plant parts are used for medicinal remedies and trees are grown and maintained for soil conservation purposes (Mamadjanov, 2006b). A diverse mix of seedlings, local selections, and advanced cultivars are grown for nut production, many with unknown origins, including some of which are the direct result of selection from the wild.

The origin of the word 'walnut' has been traced to 'Gaul nut' (O'Rourke, 1969). In the past, France was known as Gaul and the walnut as 'Gaul nut' which later on got converted into 'walnut' in English. Another view is that the name walnut has been derived from 'walnut', an old English word for 'Welsh-nut' (Beede, 1985). The nut is also called as 'Persian walnut' English walnut' or 'Carpathian walnut', but the use of term 'Persian walnut' has been considered more appropriate for the nut, because it indicates about the place of its origin (Ramos *et al.*, 1984; Serr 1969; Woodroof, 1979), 'English walnut' is a misleading term because the nut is in no way connected with English ships is probably the source of the origin of this name (O'Rourke, 1969). It was also easier to distinguish it from the American black walnut (Forde, 1975). The walnut grown in the North America and France were extremely susceptible to severe withers restricting the extension of their cultivation further into the north but the introduction of cold-hardy race of walnut from the Carpathian mountains of Poland greatly solved this problem and the term 'Carpathian walnut' thus, came into existence. According to O'Rourke which survives and produces a good crop even when the winter temperature falls to -35°C and it includes all the cold-hardy walnuts from the USSR, Czechoslovakia and Germany.

1.1. Nutritive value

Nutritive value of walnut has been presented in Table 1.

Table 1: Nutritive value per serving of 100g

Energy	2.738kj (654 kcal)	Alcohol	0
Carbohydrates	13.71	Caffeine	0
Starch	0.06	Vitamin-A	1µg (0%)
Sugars	2.61	Vitamin-A	20 IU
Lactose	0	Beta-carotene	12µg (0%)
Dietary fiber	6.7	Lutein and Zeaxanthin	9µg
Fat	65.21	Thiamine (Vit-B_1)	0.341 mg (30%)
Saturated	6.126	Riboflavin (Vit-B_2)	0.15 mg (13%)
Monounsaturated	8.933	Niacin (Vit-B_3)	1.125 mg (8%)
Polyunsaturated	17.174	Pantothenic acid (B_5)	0.570 mg (11%)
Protein	15.23	Vitamin-B_6	0.537 mg (41%)
Water	4.07	Folate (Vit-B_9)	98µg (25%)
Iron	2.91 kg (22 %)	Vitamin-B_{12}	0 ig (0%)
Magnesium	158 mg (45 %)	Vitamin C	1.3 mg (2%)
Manganese	3.414 (163 %)	Vitamin D	0 IU (0%)
Phosphorus	346 mg (49 %)	Vitamin E	0.7 mg (5%)
Potassium	441 mg (9 %)	Vitamin K	2.7µg (3%)
Sodium	2 mg (0 %)	Calcium	98 mg (10%)
Zinc	3.09 mg (33%)		

1.2. Uses

The Persian walnut is one of the few plants in Central Asia that is almost fully utilized, from the flowers to the roots. Walnut wood is prized for its beauty and is highly valued by woodworkers. The wood is strong, durable, resistant to pests, rarely cracks, and does not change its volume when heated (Bemmann, 1998). It was often used to manufacture furniture in Europe in the 17th and 18th centuries. The demand for walnut veneer remains great and is still used for making high-quality decorative furniture, gun boxes, and various craft projects. The trees sometimes form the so-called Kapa-galls, or burls, which contain a very dense, ornamental wood with curly grain and many eyes of heartwood due to the breaking of dormant buds under the bark. These burls can weigh from 40 to 500 kg (Yaskina, 1980; Bemmann, 1998) and are utilized by master woodworkers to make souvenir boxes, lockets, brooches, and snuff boxes.

In traditional folk medicine, young, green fruits were used as a health supplement. A decoction of nuts was imbibed to treat high arterial pressure, cardiac diseases, and to rinse the mouth to treat gum disease. Juice from the fruit husk was applied as an ointment to treat mild skin conditions, such as eczema. A tea made from the leaves has been used to treat diabetes (Burmistrov, 1996). The leaves have also been used as a vermifuge (a medicine used to expel intestinal worms) and to treat venereal diseases, tuberculosis, scrofula, and rickets, and the bark from the roots has been used to make a mild laxative (Akopov, 1981). The many folk medicinal uses of walnut are supported by the wide array of pharmacologically active health compounds more recently studied from the nuts, nut husks, leaves, and bark (Alasalvar and Shahidi, 2009).

Juglans regia Linn is a medicinal plant that has been widely used in traditional medicine for a wide array of ailments that include helminthiasis, diarrhea, sinusitis, stomachache, arthritis, asthma, eczema, scrofula, skin disorders, and various endocrine diseases such as diabetes mellitus, anorexia, thyroid dysfunctions, cancer and infectious diseases. The present review, attempts to provide comprehensive information on the ethnobotanical use, pharmacology, nutritional value, preclinical and clinical studies, toxicity, other uses and current research prospects of the Juglans regia L. Currently, there is a renewed interest in walnut, and several investigations aimed at scientific validation of its traditional uses and a humble scientific investigation aimed at isolation and identification of active constituents of crude extracts (Taha and Al-Wadaan, 2011)

1.3. Importance

Walnut has been regarded as a health food that is delicious and nutritious. Both preventive and therapeutic effects of walnut are well documented. Walnuts are rich in omega-3 fatty acids that are reported to have beneficial effects on brain function. The present work was designed to evaluate the effects of walnuts on learning and memory in male rats. The effect of oral intake of walnut was also monitored on food intake. Walnut was given orally to rats for a period of 28 days. Memory function in rats was assessed by elevated plus maze (EPM) and radial arm maze (RAM). A significant improvement in learning and memory of walnut treated rats compared to controls was observed. Walnut treated rats also exhibited a significant decrease in food intake while the change in growth rate (in terms of percentage) remained comparable between the two groups. Analysis of brain monoamines exhibited enhanced serotonergic levels in rat brain following oral intake of walnuts. The findings suggest that walnut may exert its hypophagic and nootropic actions via an enhancement of brain 5-HT metabolism (Saida, *et al.,* 2011).

2. Origin and distribution

The wild seedling of walnut (*Juglans regia* L.) have been found growing in vast region right from the Carpathian mountains in eastern Europe, across Turkey, Iraq, Iran, southern USSR and Afghanistan to the northwestern Himalayas (Woodroof, 1979, Forde, 1975; Serr, 1969). It is therefore, believed to have originated in Iran and the areas surrounding it. Therefore the returning armies of Alexander the Great brought it to Europe.

There are about eighteen more species of genus *Juglans* which are native to the North and South America, central China, Manchuria and Japan. Romans liked it so much that they named it in honour of the king of all their Gods, Jupitor's Acorn, or in Latin, jovis-glans which later on became *Juglans*. From Italy the walnut moved to Spain, France, Portugal, Germany and to the southern USSR and went to China from Tibet. According to Forde (1975), the walnut trees were introduced in England by 1562, where it was not successful because of very cold summer. The American walnut industry was, however, founded on the introduction of French strains to Chile and California but it flourished only when the 'Carpathian' and 'Kaghizi' (thin-shelled) walnuts were also brought into cultivation. In India, walnut remained confined to Kashmir and it appears that it has spread to Himachal Pradesh and Uttar Pradesh hills during the past century only because there has been lack of direct contact among these states. It is still to be tried in the northeastern Himalayan region and the hills of South India.

3. Area and Production

Persian walnuts are grown in almost all the temperate areas viz. North and South America, Europe, Asia and the former Soviet Republics and to a limited extent in Oceania and North Africa. The important walnut growing countries of the world as per area and production are given in Table 2.

Table 2: Area and production (top ten) of walnut growing countries (2013)

S.No.	Country	Area (ha)	Production (MT)
1.	China	425000	1700000
2.	Iran (Islamic)	57386	453988
3.	USA	113120	420000
4.	Turkey	108767	212140
5.	Ukraine	14100	115800
6.	Mexico	72563	106945
7.	Chile	18995	42668
8.	India	31000	36000
9.	France	19563	33716
10.	Romania	1478	31764

Source: FAO, 2015

Walnuts is the only temperate nut crops of India being exported to more than 40 countries with earnings of more than Rs. 32453.5 lakh of foreign exchange annually (NHB, 2014). It is mainly grown in Jammu and Kashmir, Himachal Pradesh, Uttarakhand and Arunachal Pradesh. The total area under the walnut cultivation in India is 121.87 thousand hectares having production of 240.63 thousand MT with a productivity of 1.97 (NHB, 2014). In India, Jammu and Kashmir is the major producing state contributing more than 85 per cent of total production of the country. It is grown in almost all the districts of Kashmir and Ladakh division; however under Jammu division major producing districts are Doda, Poonch and Udhampur with minor quantities in Rajouri and Kathua. Major walnut growing districts in Himachal Pradesh are Kullu, Mandi, Shimla, Kimmaur, Sirmour and Chamba and in Uttarakhand major producing districts are Dehradun, Pauri, Tehri, Nanital, Chamoali, Almora and Pithoragarh.

Table 3: Area, production and productivity in India

States	Area (ha)	Production (MT)	Productivity
Jammu and Kashmir	82045	146781	1.79
Uttarakhand	19260	8730	0.45
Himachal Pradesh	6540	1240	0.19
Arunachal Pradesh	2285	51	0.02
Total	110125	156801	1.42

Source: National Horticulture Board, 2014

Jammu and Kashmir is the leading state in area as well as in production comprising of 82.04 thousand ha and 146.78 thousand MT production with 1.79 productivity. Uttarakhand has an area of 19.26 thousand ha under walnut with production of 8.73 thousand MT and productivity of 0.45. Himachal Pradesh has an area of 6.54 thousand hectares under the nut with a production of 1.24

thousand MT and productivity of 0.19 and Arunachal Pradesh has 2285 hectares under this nut with production of 51.0 MT and productivity of 0.02 (Table 3).

4. Taxonomy and Cytogenetics

The family Juglandaceae consists of seven genera and about 60 species of deciduous, monoecious trees with alternate, pinnately compound leaves. In addition to the genus *Juglans* (walnuts), the family includes *Carya* (pecans and hickories), *Pterocarya* (wingnuts), *Platycarya*, *Engelhardia*, *Alfaroa*, and *Oreomunnea*. Members of the genus *Juglans* are trees or large shrubs possessing twigs with chambered piths, large aromatic compound leaves, generally solitary staminate catkins on one year old wood, and female flowers on current season's wood. The husked fruit is a false drupe containing a large, woody-shelled nut. All *Juglans* produce edible nuts, although size are extractability differ considerably. Most species are highly regarded for their timber. The genus *Juglans* consists of approximately 21 species native to parts of North America, the Andean region of South America, and the mountain ranges traversing Central Asia. These species have been grouped taxonomically into four sections: *Juglans, Trachycaryon, Cardiocaryon, and Rhysocaryon.*

All *Juglans* species examined have 32 (2n) chromosomes and are monoecious, i.e., the male and female flowers are borne separately on the same tree.

5. Botanical description

Walnut is a deciduous tree. Leaves alternate, odd pinnate, large, aromatic, estipulate, leaflets opposite, serrate or entire.

5.1. Flower

Flowers are monoecious. Flowers are typically produced on the tips of terminal shoots shortly after leaves emerge. The staminate flowers on the last year branchlets in lateral peduncles catkins, each consisting of a bract bearing 8-40 stamens, 2 bractlets and 1-4 calyx lobes. The male flowers are densely packed on catkins that hang from the tree in the spring. Each catkin has up to 40 sessile petal less florets each with numerous stamens. The immature naked catkin buds first appear in leaf axils in late summer and persist over winter maturing in the spring in the axils of leaf scars on wood from the previous season.

Pistillate flowers in few to many flowered terminal racemes with 4 calyx lobes and a 3-lobbed involucres consisting of a bract and 2 bractlets. Flowers appear before or after the leaves. Female flowers are borne on current season's growth in spikes of two (to five) flowers in *J. regia* and more in other species. In some cultivars, female flowers are also produced on the tips of lateral shoots. This type of flowering is termed 'lateral bud fruitfulness' and is associated with high yields when trees are young. Lateral buds are rare on mature trees. The female flower consists of a hairy involucre fused to four sepals enclosing the pistil, which has a swollen base, the ovary, and a short style with a forked stigma with two feathery stigmatic lobes. The ovary is surrounded by the ovary wall, and inside the single locule is divided into four parts by the major and minor septa. Attached to the ovary is a single ovule enclosed by a single integuement.

5.2. Flowering

Walnuts are heterodichogamous with male and female flowering occurring at different times. Some cultivars are protandrous, with the male flowering first, while others are protogynous, with the female flowering first. There is usually some overlap between male and female bloom, and walnuts are self-fertile, but the dichogamy promotes outcrossing. Flowering occurs in the spring from mid-April for early cultivars to mid-May for late-blooming cultivars.

5.3. Pollination

Pollen is wind-borne and can be carried great distances, but is relatively short-lived. The female is receptive when the two stigmatic surfaces separate to form a V-shape and secrete an exudates that makes them appear moist. Once the lobes have achieved a 45-degree angle, the surface begins to dry, and the female flower is no longer receptive. If it has not been fertilized, the flower will expand for the next 3 weeks to the size of a marble and abscise. The stigma surfaces stay moist if the flower has not been fertilized. Another cause of pistillate flower abscission (PFA) is caused by excess pollen. In this case, the flowers shrivel and abscise shortly after they lose receptivity. PFA occurs to some extent in all cultivars but is most serious in the cultivar 'Serr'. Gonzalez *et al.* (2008) carried out a study on symptoms of flower abortion in walnut cv. Serr, induced by excess or lack of pollen. Excess pollen abortion produced flowers with rapid damage. The growth rate was very low, involucre brightness was lost and the stigma turned necrotic, in contrast to the flower aborted from lack of pollen. Normal flowers continued growing to fruit harvest, without losing brightness. Excess pollen abortion induced flower drop, with or without attached peduncle. Two separation zones could be present: the distal and proximal areas of the peduncle. Peduncle scars allowed for differentiating abortion induced from excess or lack of pollen, or other reasons.

Liu *et al.* (2011) studied pollination activity of pistillate and staminate flowers in walnut excellent varieties. Stigma receptivity and pollen viability during walnut flowering can determine the fruit set percentage and the yield of current year directly. The results showed that the stigma receptivity period was ranging from the stigma slightly open to it with brown stripes. The best period for pollination was 3-5 days after the split stigma opened. The pollen can be stored for 96 hours under natural conditions and the viability decreased quickly to 33% after 4 hours. The pollen viability was the highest when the walnut anthers shedding, reaching 95%. Pollen viability could be reinforced at 4 degrees C and the average viability was 63.3% after 120 hours.

Walnuts are self and cross fruitful. Pollen of a *Juglans regia* variety is capable of fertilizing the eggs of same variety and of any other variety as well. Furthermore, the pollen of the other walnut species is capable of fertilizing eggs of *J. regia* varieties. Walnut flowers are anemophilous representing all characteristics of wind pollination i.e. non sticky; the pollen is light and produced in abundance; stigma is protruded, feathery and remains receptive for a few days. The stigma is most receptive for first 5 days after anthesis. The problem of dichogamy is further aggravated by short periods of both pollen viability and stigma receptivity. Walnut pollen had highest germination capacity when it was first shed, under simulated natural conditions lost its viability 2 to 3 days after collection and on exposure to light at 25-30°C, its viability was lost in 1 to 3 hours. However, its pollen germinated on the stigma within 1 to 3 days after pollination. The viability of walnut pollen lasted for 100 hours at 14 to 15°C and 55 hours at 22 to 25°C and zero degree celcius while temperatures above 25°C and humidity below 33 per cent during pollen maturation resulted in pollen sterility.

5.4. Fruit

Fruit a large indehiscent drupe. Nut thick walled, incompletely, 2-4 celled, indehiscent or finally separating into 2 valves. Seed 2-4 lobed, remaining within the shell in germination.

The nut is harvestable approximately 19-22 weeks after bloom. Timing of leaf emergence and harvest has always been an important consideration in selecting walnut cultivars. Early-leafing cultivars are more susceptible to frost and, in California, to walnut blight and insect problems. But early-leafing and flowering cultivars usually harvest early (in September) and have the advantage of early entry into the holiday nut trade. Late-harvesting (mid-October) cultivars can be subject to difficulties such as rain, which can interfere with harvesting operations. Vigorous late-harvesting cultivars that continue growing late into the autumn may also be more prone to injury by early frosts and winter injury.

6. Physiology of flowering, fruit set and development, abiotic factors limiting production

6.1. Physiology of flowering

Walnut is a deciduous and monoecious tree i.e. the pollen bearing male (staminate) flowers and the nut producing female (pistillate) flowers are borne separately on the same tree. The staminate flowers are borne on catkins that develop from lateral buds on the previous season's growth. Persian walnut generally has one or two buds in the axil of the leaf while other species commonly have three or more buds. Catkin buds differentiate early in growing season with the first flush of growth. In the following spring, each of these catkins produces a large amount of pollens. Each flower of catkin has 13 to 18 anthers and each anther has 900 pollen grains. It is estimated that about 1.8 million pollen grains are produced per catkin.

The pistillate flowers are borne at the end of spring shoots and usually more than one flower is borne on a terminal. Though the fruit bud formation in a pistillate flower is initiated during summer, yet evidence of flower part differentiation can be detected three to five weeks before they open in spring. The pistillate flowers are borne on shoots upto 60 cm length. Walnut shoots do not start new growth flushes at the same time. In moderately cool summers or in poor soils, the last flush is short and in the following spring its buds are appropriate to produce shoots that are terminated by flower clusters. While trees growing in good soil with late summer temperature favorable for strong growth, the shoots in the last growth flush of the season tend to be long and rather succulent and have buds that open into shoots in next spring.

6.2. Abiotic factors limiting production

Adverse weather conditions like snowfall, hail, frost or rain during flowering in spring result in crop failure in walnut. Pollen of late varieties is less susceptible to spring frost damage than those of early varieties. If the fruit set is inadequate even when the plant health is normal, inadequate pollination may be the cause and artificial pollination can be the answer for this problem. With this, the inadequate pollination due to dichogamy can also be rectified. The walnut catkins can be collected and stored for 2 to 3 weeks in a desicator at 0°C or in a refrigerator maintained at low humidity. Pollen could be stored even for 3 to 8 months at – 20°C. While comparing early, mid and late season cultivars it was observed and found that the optimum temperature for pollen germination was positively related to mean staminate bloom date. However, other found that *in vitro* pollen germinability was completely lost after 1 month of storage at 4° or – 20°C but it remains viable *in vivo* germination test on stigmas. But few reported that storage at -1°C to -2°C reduced viability and impaired fruit set.

6.3. Fruit set and development

Walnuts are naturally outcrossing species and varying degrees of dichogamy prevents self-pollination. Fruit set is generally high but flower and fruit abscission reduces the consequent yield considerably. The pollen can be used for pollinating the same cultivar or other cultivars at the same location or different locations. However, cross pollination results in higher fruit set which was clearer from the earlier studies. Kumar *et al.* (2005) reported that mean fruit retention in different cultivars/selections of walnut was 52.93 per cent under open pollination, 68.02 per cent under self pollination and 64.44 per cent under cross-pollination. The maximum fruit set (60.57 %) and fruit retention (50.82 %) was recorded in Payne and the minimum fruit set (19.74 %) and fruit retention (25.29 %) in Local Selection 2 (Pandey and Tomar, 2012). Even mixture of pollen from 2-3 cultivars or different species – *Juglans mandshurica* and *Juglans cinerea* increased fruit set. Atefi (1990) reported higher fruit set from artificial pollination than open pollination. Parihar and Bajpai (1985) reported 18.73, 34.50, 36.99, and 42.13 per cent fruit set under self, open, natural (cross) and hand (cross) pollination, respectively. Nedev (1975) and Rovira and Aleta (1997) reported up to 70-80 per cent fruit set under different modes of pollination in walnut.

Sartorius and Stosser (1992) studied the non-apomictic cultivar Weinsberg I and the apomictic cultivar Esterhazy II and showed the same fruit set whether or not open pollination was prevented by bagging. The apomictic cultivar No. 26, however, only set fruit when open-pollinated. A marked fruit drop occurred 4-6 weeks after anthesis, depending on the weather, which coincided with the beginning of intensive fruit growth. The growth curves for fruit and ovule length were sigmoid. No differences were found in fruit development between bagged and open-pollinated blooms.

Li *et al.* (2012) studied the fruit developmental process in 9-year-old precocious walnut cultivar 'Xiangling' and the dynamics of fat, protein and sugar content in the fruit, and the activities of enzymes related with sugar metabolism. The result shows that the developmental process of walnut fruit could be divided into four stages: Slow growth {within 30 days after florescence, (DAF)}, fast growth (30 to 60 DAF), fat accumulation (60 to 100 DAF) and fruit maturity (100 to 140 DAF). Fat content in walnut fruit increased continuously and the maximum increment occurred within 60 to 90 DAF. It amounted to 50.11% at 100 DAF and maximized with 63.13% at 140 DAF when the fruit ripens. Sucrose and glucose content showed similar trend. They were both low at earlier stage, then gradually increased and reached the peak at 90 DAF, followed by gradual descending. Fructose content topped at earlier stage and then gradually decreased. Acid invertase (AI, EC 3.2.1.26), neutral invertase (NI, EC 3.2.1.26), sucrose phosphate synthase (SPS, EC 2.4.1.14) and sucrose synthase (SS, EC

2.4.1.13) activity were higher during the earlier stage and then decreased until the minimum point at 80 DAF, thereafter these enzyme activities turned to ascend the crest value at 90 to 100 DAF, and then kept decreasing until the late maturity stage.

7. Climate

Spring frost, extreme summer temperature and insufficient winter chilling adversely affect walnut cultivation. The plants can tolerate as low as -11°C during deep dormancy but after the commencement of growth, the temperature of 2 to 3°C below freezing point kills leaves, shoots and flowers. Late spring frost results in secondary flowering and dichogamy is more marked on secondary than on primary inflorescence. High temperature of about 38°C causes sunburn of hulls and shriveling of kernels leading sometimes to the production of blank nuts. Short growing season is also a limiting factor in walnut production. In the absence of adequate chilling, bud opening and blossoming are irregular and delayed, resulting into poor crop and death of shoot tips. The chilling requirement varies with cultivars from 700 to 1,500 hours. In the drier regions, under severe drought conditions the plant may make fair growth but nuts do not fill well. The walnuts produced at an altitude of above 1500 m are considered of superior quality with a light coloured kernel and a thin shell. At lower elevations, the kernel usually turns brown because of high temperature at the time of ripening.

8. Soil

The soil most suitable for walnut is a well drained, deep silt loam containing an abundance of organic matter. Soil pH influences the availability of mineral nutrients and maturity of which are available to the plant in the pH range of 6-7. Soil should not have a fluctuating water level, hardpan and sandy sub-soil. A soil of 2-3 metres deep gives the best results because walnut roots penetrate upto a depth of 3 metres. The sub soil should be free from solid rock, impervious clay or gravel layers which restrict root growth. The trees growing on shallow soil with less moisture content are liable to suffer more from sunburn than the trees growing on a deep silt soil. Alkaline soils and sandy sub-soils should be avoided for walnut growing.

Black walnut (*Juglans nigra* L.) is sensitive to soil conditions and develops best on deep, well-drained, nearly neutral soils that are generally moist and fertile (Brinkman, 1965). Walnut grows best on sandy loam, loam, or silt loam textured soils but also grows well on silty clay loam soils (Losche, 1973). Soils with these textures hold a large amount of water that is available to the tree during dry periods of the growing season. Internal drainage and depth to gravel are especially important site characteristics for black walnut. Twenty five year old trees on deep soils having > 40 inch depth to gravel layer, were 51.4 inch in height and 6.4 inch in diameter at dbh (diameter at breast height) than trees on shallow soils less than 40 inch depth to gravel layer (Losche, 1973).

9. Propagation

Walnut can be propagated through sexual (seed) and asexual methods viz. grafting, budding and rarely through cutting and layering have also been reported. Walnut trees can be also raised by clonal means. The seedlings are highly variable producing the nuts of inferior quality with an average shelling percentage of 30-35 per cent as compared to 50-60 per cent of good commercial cultivars which are raised through asexual means.

9.1. Propagation by seed

The majority of trees in India are of seedling origin and such they bear nuts of variable quality. The seedlings are usually raised for rootstock purpose but in this case hard shelled nuts of inferior quality are sown. Although walnut germinates without stratification under some conditions, yet proper stratification is desirable for good germination and growth. The walnuts should be collected from vigorous and high yielding trees. The desirable characters to look for in the selection of nuts for rootstocks are big size, bright brown colour and good cracking quality of the shell, bright yellow colour, good taste and flavour of the kernel. Walnut plants raised from seedlings come to fruit at the age of 10-15 years but the productive yield comes at the age of 20-25 years, however, these trees may continue to provide yield upto the age of 100 years. Seed propagation resulted in heterozygosity in plant material coupled with long juvenile period and greater degree of variability in relation to nut shape, size, colour, quality and maturity. Besides, plants grow unmanageable size which necessitates spars plantations leading to low density and ultimately low yield.

9.2. Vegetative propagation

The commercial walnut can be propagated on a number of rootstocks such as seedlings of *Juglans regia, J. hindsii, J. nigra, J. microcarpa, J. californica, J. sieboldiana* and Paradox (*J. regia* x *J. nigra*). *Juglans regia* is the most easily available rootstock in India and is the only one used whenever walnut trees are raised by budding or grafting. This rootstock has better compatibility and is particular suitable for light soils. It is known for its resistance to crown rot disease. However, it is unsuitable for waterlogged, saline and nematode infected soils. Walnut can be propagated vegetatively by cutting, grafting, budding and stooling. The per cent grafting and budding in walnut is lower than in pome and stone fruits.

9.2.1. Rootstocks

The open pollinated seed of *J. regia* is used for raising seedling rootstock for walnut all over the world except the USA where *J. hindsii* Paradox *(J. hindsii* x *J. regia*) and Royal (J. *hindsii* x *nigra*) seeds also serve the same purpose. In addition, the seedling rootstock of *J. nigra, J. mandshurica, J. cathayensis* and *Pterocarya stenoptera* are employed to a small extent in other countries (McGranahan and Catlin, 1987; Ji, 1980). Most of the *Juglans* species, viz., *J. cinerea, J. major, J. hindsii, J. microcarpa, J. nigra, J. ailantifolia, J. australis* and several hybrids have been reported to be more or less compatible with Persian walnut, exhibiting some signs of incompatibility, such as, overgrowth of scion or stock, swelling at the union, lack of scion vigour, low percentage of union, etc. A correct assessment of compatibility is not possible because of the interference of blackline disease.

9.2.1.1. Characteristics of different rootstocks

9.2.1.1.1. *Juglans regia*

Persian walnut is the most common rootstock everywhere because it is readily available besides being fully compatible, resistant to crown rot and tolerant to blackline disease (McGranahan and Catlin, 1987). It is however, unsuitable for waterlogged, saline and oak root fungus and nematode infected soils. This rootstock is also slow to grow in the nursery. Inspite of a great diversity found in *J. regia* no particular source of seeds has been identified for rootsocks which are raised from inferior quality hard shell nuts in India and from Manregian, Eureka, Serr and Waterloo seeds in the USA.

9.2.1.1.2. *J. hindsii*

It is a common widely used rootstock of Persian walnut in California making a strong graft union. It is resistant to oak-root fungus and root knot nematode but susceptible to *Phytophthora*, root lesion nematode and blackline disease. Low availability of phosphorous and zinc in soil or waterlogging hampers the growth of plants on this stock.

9.2.1.1.3. *J. nigra*

It is a satisfactory rootstock for its own cultivars and has also been recommended for *J. regia* in some countries. In comparison to *J. regia* rootstocks this rootstock is more tolerant to crown gall, *Phytophthora* and oak root fungus and induces precocity (Delbos *et al*., 1983). J. nigra reduces the tree size both as rootstock and interstock. Lower yield efficiency and susceptibility to blackline disease are the main impediments in its popularity.

9.2.1.1.4. *J. microcarpa* (*J. rupestris*)

This rootstock is particularly suitable for high boron, chloride and soil pH conditions where it has given better performance than *J. regia* and *J. nigra* rootstocks (McGranahan and Catlin, 1987).

9.2.1.1.5. *J. sieboldiana* (*J. ailantifolia*)

It has been reported as a highly suitable rootstock from the USSR with a wide range of adaptability to varying soil and climatic conditions. It is cold hardy and resistant to *Phytophthora* (Matheron and Mircetich, 1985, Glonti *et al.*, 1987).

9.2.1.1.6. Paradox (*J. hindsii* x *J. regia*)

Paradox has performed better in California than its parents in respect of scion growth and tolerance to both nematodes and *Phytophthora* species. It also makes more growth in replanting of old orchards than *J. hindsii* (McGranahan and Catlin, 1987). It is however, susceptible to blackline disease and its seeds are not available in sufficient quantity.

9.2.1.1.7. *J. californica*

It was earlier used in California has been discarded due to its susceptibility to crown rot, root rot and wet soils.

9.2.1.1.8. *J. major*

It may find favour in good and high pH soils but its superiority over *J. hindsii* is still to be investigated.

9.2.1.1.9. *J. mandshurica*

It is used in China in cold but was unsatisfactory in USSR (Nedev, 1973).

9.2.1.1.10. Royal (*J. hindsii* x *J. nigra*)

It has been tried in California with promising results.

9.2.1.1.11. *Pterocarpa stenopetra*

It is commonly known as wingnut and a member of Juglandaceae, native to China. It is tolerant to *Phytophthora*, waterlogging and root lesion nematode but is susceptible to *Armillaria* and blackline disease. Since it is incompatible with many walnut cultivars, it should be used in rootstock breeding programmes (McGranahan and Catlin, 1987).

Among the various walnut rootstocks, *J. nigra* has only reduced tree size by 10-20 per cent but this effect is still to be confirmed (Germain *et al.*, 1975). *J. nigra* however, definitely dwarfens the tree when used as interstock (Serr,

1968). The dwarf trees are more yield efficient but dwarfing rootstock is not recommended for commercial purpose (McGranahan and Catlin, 1987).

9.2.1.2. Raising of rootstock

Generally hard shelled, medium size, unmarketable nuts are used for raising the seedling rootstock. Well sealed nuts from current season crop should be used and are collected directly from the plants in the month of the September-October. Hull the seeds and air dry in the open shed. However, collecting nuts from the ground to be used in sexual propagation has been shown to be linked to increased incidences of infection by *Agrobacterium* and crown gall in seedlings (Yakabe *et al.*, 2010). Though walnut germinates without stratification under some conditions, however, for good germination and better growth, stratification is desirable. Stratification of walnuts can be done by the end of December. For stratification nuts are placed in alternate layers with suitable moist medium such as sand, vermiculite or peat moss in the container which is then subjected to cold treatment (3-5°C) for a period of two-three months. During this period the nuts should be prevented from drying and should get proper aeration. Walnut seeds soaked for 24 hours in 125 to 250 ppm GA_3 or 30 minutes in 10 per cent H_2SO_4 germinate better. At higher elevation, there is no need to stratify as they are naturally stratified by exposure to outside winter temperatures in a cold room or shady open areas. In well prepared beds the hard shell nuts are directly sown at a distance of 30 x 30 cm and at a depth of 8-10 cm. The planting of large nuts at 7 cm depth in soil in vertical position gives better germination and seedling growth (Lal *et al.*, 1984). Dip the seeds for one minute in hot water before direct sowing or stratification. A large number of factors such as nut type, size, position of seed, planting depth, soil moisture and temperature and seed treatment influence the germination of seeds. The seedlings don't attain good height in the first year but their roots system is well developed. In the second year, the seedlings develop quickly and are ready for transplanting by the time the growth stops in the autumn. Higher germination of seeds sown in a vertical position was noticed than those placed flat in the soil.

After stratification, the stratified seeds or sprouted seedlings are planted in the well prepared nurseries. It takes more than two months for the seed to germinate in the nursery without stratification and while after cold stratification, the seeds germinate within 1-2 weeks after planting. Make sure that the nursery is free of weeds. Irrigate frequently and fertilize the seedlings to correct soil deficiencies. Where there is scarcity of rains or water mulching is advised during summer months to conserve moisture which also control weeds, enhance germination and boost the growth of the plants. If nursery is properly managed, seedlings attend grafting size within a year time. In order to ensure better establishment of grafts after the transplanting, the tap root of seedlings should be cut off 20-

25 cm below ground level in dormant season, a year before the actual transplanting. 2-3 year old seedlings under open conditions and 1-2 year old under controlled conditions give higher graft success.

Chen *et al.* (2004) advocated a method for quicker availability of rootstock for grafting. The method consisted of collecting walnuts when the fruit rind is changing from green to yellow. The seeds should be soaked in water for one week before sowing in deeply ploughed soil which has been sterilized with carbendazim or M-topsin. The sowing bed should have a width of 1 m, and a film shed should be placed over the soil before sowing. The temperature in the shed should be maintained around 25 degrees C. Sprouting occurs about 50 days after sowing, and 3 months later, when the seedlings are 6 cm in height, they are ready for tongue-grafting with a scion of the same thickness. Grafted seedlings are then transplanted in the field and should be ready for planting in spring.

Fan and Xi (2004) also suggested another way for quicker availability of rootstock for grafting. For raising walnut planting material, the rootstock seeds are usually collected in late September. Before sowing, it is important to soak the seeds in a solution of 300-400 mg GA3/litre for 6-8 days. The sowing bed should be well irrigated and then covered with film to maintain humidity and increase the soil temperature. In late October, holes are made to sow the seeds, which are then covered with fine soil. After 50% of the seeds have germinated, 0.5% KH_2PO_4 + 0.5% urea solution is applied once every 6 days. When 60-70% of the seeds have germinated (usually in late December), the seedlings are ready for cleft grafting. The thickness of the scion should be 0.6-1.2 cm. Using this method, the survival rate for 547 000 rootstock seedlings was 91.8%.

9.2.2. Propagation by cuttings

Although hardwood and semi-hardwood cuttings can also be used for rootstock propagation, they are not commonly used because success rates are often low. Such cuttings have rooting percentages from 30 to 40 per cent and often have poor initial survival. Use cuttings from only vigorously growing shoots for propagation and root them in individual liners to promote uniform and deeper root branching. The bases of cuttings are treated with KIBA (IBA supplemented with potassium salt) at 8-12 ppm soaked for 24 hours before sticking (Anonymous, 2013). Semi-hardwood cuttings are rooted in greenhouses on bottom heated mist benches in mid to late summer and hardwood cuttings on bottom heated mist benches in late fall and winter outdoors. The use of broadcast flats yields lower rooting percentages, shallower roots, and poor root development. Plant liner-sized rooted cuttings in the field in late February or early March. Direct field rooting is not commercially successful.

9.2.3. Propagation by Grafting

9.2.3.1. Different methods

Grafting in walnut is less successful compared to most other temperate fruits, especially in natural climatic conditions thus requiring more precision in grafting operation and handling of stocks and scions (Weber and MacDaniels, 1969). Therefore different methods are used for the production of transplants in controlled temperature, relative humidity, time of grafting and bleeding etc. The choice and combination of formerly enumerated factors affect the percentage of surviving plants in the different propagation methods in controlled environment. Favourable temperature and humidity conditions are essential for callus development in walnut which are usually not available outdoor, resulting in failure of graft unions. The ideal rootstock thickness get maximum graft success is 1.5 – 2 cm, at 15 – 20 cm above the ground level.

Better success in grafting by cleft, tongue and splice techniques was obtained when grafting was either done in a greenhouse having 28°C temperature and 80-100 per cent relative humidity or the grafts after preparation were placed in the incubator at 25 ± 1°C temperature with 75 ± 2 per cent humidity (Ji, 1980; Park, 1984; Sharma *et al.*, 1982).

9.2.3.1.1. Bark grafting

This method should be done only in the spring season when bark of rootstock slips readily whereas the scion used is in dormant condition. The scion should be collected either in late winter or in early spring and kept in dried placed until used for grafting. The bark of the rootstock is split downward up to 5 cm in length. The length of the scion should be 15-18 cm containing at least three dormant buds having pencil thickness. To cut scion one side should be cut about 1 to $1^{1/4}$ inches long and the opposite side about ½ to ¾ inch long. Place a scion against the stock and outline each edge of the scion with a knife. Pull the top of the bark strip away from the stock, insert the end of the scion (short cut side outward) and push downward until the scion is seated. Tied tightly both scion and stock together with polythene strip. After 10-12 weeks the grafts begin to grow. In old rootstock 2-3 scion can be inserted as it depends on the size of the rootstock.

9.2.3.1. 2. Modified bark grafting

This is an efficient and simple method of grafting. As compared to bark grafting, date of grafting was delayed for two weeks in modiûed bark grafting after rootstock heading back and graft unions were covered with moist sawdust for about three weeks prior to waxing the graft. Saw dust (mainly of spruce and ûr) was washed gently several times in a wooden container and ûnally rinsed with fungicide (1.5% benomyl) and drained for approximately 48 hrs. A small amount of sawdust, on one side of a plastic bag, was used to cover the graft union and gently wrapped and fastened around the union. After about three weeks, the sawdust cover was removed and the graft union was covered with grafting wax. The covering of the graft union with moist sawdust improves callus formation at the graft union. For grafting purpose scions (dormant one year old shoots) were cut to 15-20 cm in length with at least two buds. Scions were then wrapped in damp paper, placed in a plastic bag and stored at 2 ± 2°C until they were used for bark grafting. Modiûed bark grafting performed in mid-April gives an average grafting success of about 85 per cent.

9.2.3.1.3. Rapid grafting technique

A rapid grafting technique for walnut propagation was standardized by Xin *et al.*(2014). In order to produce seedling rootstocks of walnut, late, thick-shell, fresh walnuts were harvested, and were seeded directly in mid-September. The development of root system fully formed before the soil was frozen in the winter, the seedlings started to grow early in spring of next year and looked neat and sturdy. The growing season was extended about two months and the seedlings were grafted about one month in advance. The optimum period for grafting was from mid-June to early July. By the technology of single-edged blade patch bud grafting of walnut, the grafting survival rate could rise to 90% and one person could graft about 500 plants per day. The young trees were taken out from the nursery in the same year. The technique could shorten the walnut propagation period to 1 year, and reduce costs and improve the economic benefits of propagation of selected cultivars of walnut.

9.2.3.1.4. Single bud grafting

Single bud grafting (Chip budding) and *in-situ* grafting in walnut were reported by Srivastava *et al.* (2008) for regeneration of grafted sapling under open condition. Maximum bud take and sprout of 24.0 and 23.3 per cent, respectively, recorded when single bud grafting was performed on 15th September. Maximum linear growth of the scion (131.00 cm) was recorded on 30th September grafting. In case of *in-situ* grafting, maximum bud take and sprout was 47.14 and 44.58 per cent, respectively, on mid March grafting. On the basis of saleable plants, 70.00 and 95.75 per cent of plants become ready for sale in single bud grafting and in situ grafting during 15th September and 1st week of February, respectively. The single bud grafting in mid September and in situ grafting in mid March were successful in walnut regeneration.

9.2.3.1.5. Bench grafting

Erdogan (2006) reported a technique in areas where winter and late spring frosts limit the walnut propagation. Walnut seedlings were bench grafted with scion of some walnut cultivars and selections, and planted in plastic bags. Grafted plants were put in plastic boxes and placed in greenhouse. Hot callusing cables were lined next to the graft unions between bags and covered with growth medium of 5-7 cm thickness leaving the bud at the top of scion uncovered to keep graft unions moist and the temperature constant. Graft unions were heated at 25 degrees C for 33 days. The greenhouse was heated when air temperature decreased below 10 degrees C. Growth medium was removed after power cut. The average graft take was 82%. Hot callusing cable is now routinely used to enhance callus formation.

9.2.3.1.6. Epicotyl grafting

Epicotyl grafting in walnut (*Juglans regia* L.) was reported by Qureshi *et al.* (2014) from Jammu and Kashmir. The research findings indicated that walnut epicotyl grafting could be economically viable vegetative propagation method in Kashmir valley. Grafting at the point of attachment of cotyledons on 15-day-old rootstock is feasible method for mass propagation of walnut.

9.2.3.2. Factors responsible for grafting success

9.2.3.2.1. Effect of season on grafting

For higher success, bench grafting should be done 15 cm above ground level during mid-February to mid March. The most suitable time to graft walnut under ambient condition is towards the end of dormancy or early spring season (mid-March) especially in areas of 700-2100 m amsl (Gyambo, 2000).

Chebotar (1974) reported that the optimum period for bench grafting was between December and February while Zachej (1974) found most suitable time for grafting is March followed by November in Czechoslovakia.

Sharma *et al.* (1982) obtained higher survival of grafts when grafting operation was carried out in the middle of January rather than in the last week of December or in the middle of February. They attribute higher success in the middle of January to the favourable temperature and humidity conditions prevailing in the first week of February when the incubated grafting was planted in the field.

Avanzato and Tamponi (1988) reported 73 per cent grafting success by heating the unions at 27°C. They cleft grafted at the end of January and February and wrapped the graft union area with aluminum foil to maintain uniform temperature and humidity and heated the unions for 24 days.

The time of grafting also influenced the success of propagation. Epicotyl grafting performed at the end of March for two consecutive years i.e. 2007 and 2008 led to obtain higher survival percentage (75 and 71.2 %, respectively) of the grafted plants as compared to end of April under Plovdiv conditions of Bulgaria (Gandev and Arnaudov, 2011).

9.2.3.2.2. Effect of methods

In Bulgaria, when grafting was done by whip-tongue technique and success percentage was 77.9 per cent in 1973 and 88.6 per cent in 1974 (Nedev *et al.*, 1976).

In a comparison of seven grafting methods during two years, veneer grafting carried out in July gave more than eighty per cent success followed by side tongue grafting done in the late January with fifty percent (Chauhan and Sharma, 1980).

Unal (1995) claim that bench grafting has a number of advantages compared to bud grafting. The most popular bench grafting techniques are whip, cleft and omega (Germain *et al.*, 1999). Solar *et al.* (2001) reported that omega bench grafting technique has resulted in 83.0 per cent plants with callus formation on average of three years, of them, 64.0 per cent survival at the end of the second year with considerable variation of successful grafting during the separate years.

Ozkan and Gumus (2001) compared two grafting techniques cleft and whip-tongue during the months of January, February and March and reported the highest percentage of successful grafting in February for both grafting techniques. In Tokat variety 60.0 per cent in cleft grafting and 66.0 per cent in whip-tongue grafts and in variety Yalova, 63.0 per cent in cleft grafting and 70.0 per cent in whip-tongue grafting.

During recent years, hypocotyls grafting technique has been introduced (Atefi, 1997; Vahdati and Zareie, 2006; Gandev and Dzhuvinov, 2006). In this method, grafting takes place during vegetation with growing tip of soft wood cutting on cleft. Grafted plants should be kept in about 80 to 90 per cent relative humidity and temperature of 26-27°C. In Bulgaria, Gandev and Dzhuvinov (2006) obtained 83 per cent graft survival, the grafted plants being kept in a greenhouse under controlled temperature.

Different grafting and budding methods including modified bark grafting in mid April, cleft (modified and normal) and whip grafting in late March, shield/patch budding at three different periods (mid April, early July and late September) and chip budding in early October, were compared in terms of their success at KARS located in northwestern Iran on 2 to 3 year old walnut seedlings. Modified bark grafting in mid April was the most convenient and reliable method with 80 to 93 per cent grafting success (Rezaee *et al.*, 2008).

9.2.3.2.3. Effect of temperature

According to Wilbur *et al*., (1998), the optimal temperature for walnut callus formation was 26-27°C. The most popular and widely practiced method of callus formation in the walnut is to place the plants in woody boxes in a room with controlled temperature. Grafting takes place during winter dormancy with one year rootstock and grafts. This method is known as bench grafting. The success of grating depends not only on the grafting technique choice but the month as well.

In the recent years, local heating the graft union with a hot callusing pipe has been successfully applied (Achim and Botu, 2001; Avanzato *et al*., 2006 and Gandev, 2007). Methods for local heating of the graft union are different in the different experiments but the purpose is the same, to achieve a temperature of 26-27°C. A possible method of heating is with an electric cable, placed in a plastic U-pipe full of peat. The method is known by the name of hot callus. In this technology, the grafted plants are placed horizontally, the graft union being positioned on the heating cable. Achim and Botu (2001) obtained best results in walnut grafting at Valcea with two methods of callus stimulation after grafting viz. hot callusing at grafting point (26 ± 1°C) and classical method i.e. room temperature (26-28°C) under three grafting methods viz whip (86.0 and 63.0 %), cleft (89.5 and 67.2 %) and machine or nut and feder (68.3 and 48.6 %) methods, respectively.

9.2.3.2.4. Effect of condition

Erdogan (2006) reported that the method hot callusing cable is used for the annual production of about 20,000 plants in Turkey with an average graft survival of 82 per cent. Two grafting techniques i.e. hot callusing in winter and top grafting in spring-summer were adopted and 7 to 53 per cent and 75 per cent graft success was obtained, respectively (Avanzato *et al*., 2006). Under three environmental conditions (polyhouse, hot callusing cable and open field) and three rootstocks thickness (thin, medium and thick), maximum graft success was recorded under polyhouse (68.88 %) conditions and with thick (62.77 %) rootstock (Sharma and Dar, 2006). Gandev (2007) reported 74.2 per cent successful propagation in cultivar Izvor 10 by the methods of cleft grafting and hot callus.

One year old local walnut seedlings were grafted with scions of five walnut cultivars and selections for two consecutive years using whip and tongue techniques. The average per cent graft take was 70.0 per cent in 2003 and 83.4 per cent in 2004. The highest average graft success of 86.0 per cent in both years in all the tested cultivars and selections was obtained with the treatment involving dipping scions and graft un ions into paraffin, stratification in sawdust up to the union and covering with polyethylene foil (Paunovic *et al*., 2010).

9.2.3.2.5. Effect of time, method and condition

Mir and Kumar (2011) studied three methods (tongue, wedge and venner) with three timings (2nd, 3rd and 4th week of February) and two environmental conditions (open and polyhouse) and reported that under polyhouse the maximum grafting success (76.66 %) was obtained through wedge grafting when done in 4th week of February.

With three environmental (open field, poly-trench and polyhouse), two grafting methods (wedge and tongue grafting) and four grafting times (15 February, 1 March, 15 March and 1 April) was practiced by Ahmad *et al.* (2012) at CITH, Srinagar. They concluded that under polyhouse condition, highest percentage of sprouting (65.51 %), graft success (54.46 %) and plant growth (168.75 cm) were recorded. Among grafting methods and times, wedge grafting and 15 March recorded maximum graft success (52.45 % and 58.0 %, respectively).

9.2.4. Propagation by Budding

Once the rootstock is established, growing well in the nursery and has a diameter more than scion wood, the scion varieties can be budded between July-August. (patch or chip budding). A finished tree will be grown by the next fall if the buds heal over, remain dormant until the spring, and grow out in the following summer. The rootstocks that were initially too small or had failed in the grafting in the spring are budded with patch or chip budding (Gautam, 1990). The percentage of successful budding was influenced both by rootstock age and the time of pruning after budding.

Budding in the walnut is done mainly by the technique of patch budding. This is one of the oldest and popular techniques for propagation in a nursery in the open (Kuniyuki and Forde, 1985). Collect bud sticks when bark is slipping on both the rootstock and scion bud stick. To ensure the best take, the bud wood can be prepared by removing the leaves while still on the tree a few weeks before use. Patch or chip budding is done using a double bladed knife to cut a square piece of bark from the rootstock which is replaced with the same size patch or chip from the bud stick containing a well-developed bud, the only difference between both the methods is that in later method along with bud wood is also taken.

Scientific literature reports different data on the efficiency of this inoculation method in different countries (Nedev *et al.*, 1976; Ozkan *et al.*, 2001; Solar *et al.*, 2001). Nedev *et al.* (1976) reported that one month after budding inoculation success was about 70 to 80 per cent. While Solar *et al.* (2001) reported that the efficiency of patch budding technique in walnut in Slovenia is only 16 per cent. The percentage of successful inoculation in Turkey is higher and is characterized with decrease in time. It is 88.3 per cent after budding, 72.5 per cent in the spring of next year and 41.25 per cent before taking out the young trees from the nursery (Ozkan *et al.*, 2001).

Gandev and Dzhuvinov (2004) reported big difference between day and night temperatures, which is not favorable for successful transplants in the open conditions.

9.2.4.1. Different method of budding

9.2.4.1.1. Annular Budding

This method is also known as flute budding. In this method, instead of taking out complete circular ring of bark containing bud, the bud wood is taken out from scion shoot by giving a vertical slit in bud wood. On suitable portion of rootstock similar size bark is taken out from the rootstock. The scion is fitted on rootstock and wrapped using polythene tape of 300 guage. When bud sprouts the wrapped polythene is cut and removed. The portion of rootstock above bud is removed. In this method of budding, sometimes due to leaching of rain water through rootstock via slit of scion bud, rotting is noticed. To avoid such type of problem the vertical slit of the bud is sealed using paraffin wax.

9.2.4.1.2. Effect of method

Chip budding is another way for walnut propagation in the open. Walnut can also be propagated through annular budding and chip budding. Annular budding gives a very high percentage of bud take in walnut when performed during the month of July. Chip budding during May give good performance in walnut. According to Qureshi and Dadal (1985) except for patch budding with 30-35 per cent bud take, no other method, such as, budding, veneer grafting or bench grafting gave desirable success.

Chip budding is another way for walnut propagation in the open. Walnut can also be propagated through annular budding and chip budding. Annular budding gives a very high percentage of bud take in walnut when performed during the month of July. Chip budding during May give good performance in walnut. According to Qureshi and Dadal (1985) except for patch budding with 30-35 per cent bud take, no other method, such as, budding, veneer grafting or bench grafting gave desirable success.

9.2.4.1.3. Effect of time

Pathak and Srivastava (1975) found mid-June as an optimum time of budding. Removal of leaf blades fifteen days before budding operation enhanced bud-take. Chandel *et al.* (2006) reported that the optimum period for chip budding for the climatic conditions of North western Himalayas mid hills was from mid may to the first week of June and the right period for the patch budding was mid June to end of the June. In the experiment budding was done on one year rootstocks (*J. regia* L.) with buds taken in the current season. Successful inoculation percentage in chip budding was 89 per cent and in patch about 50 per cent during the mentioned periods of time.

Ozkan and Gumus (2001) applied chip budding to one year old rootstock in January, February and March. The highest success percentage was reported for cultivar Tokat in March 53.0 per cent and the surviving plants were 50.0 per cent in September. (Porebski *et al.*, 2002) obtained 26.9 per cent of the survival under chip budding done in winter.

9.2.4.1.4. Effect of condition

Achim and Botu (2001) reported that chip budding in the open was possible for the climatic conditions of Romania. Time from May 15 to June 15 was the best period for budding using buds collected during the winter dormant period of the trees and kept in a refrigerator at a temperature of 1 to 4°C. The percentage success during this period was 69 to 83.2 per cent.

Budding can be practiced not only during rainy season (i.e. July-August) but also during the dormant winter period of the trees and the budded plants are put in controlled temperature conditions. In this case, patch budding is not recommended due to the difficult separation of buds from the scion (Bayazit *et al.*, 2005). Such problem doesn't exist in chip budding.

Ahmad *et al.* (2007) reported 22.99 per cent budding success under controlled polyhouse condition compared to open field conditions recorded only 20.33 per cent success while 30 July was found to be the best both for bud take and budding success.

9.2.5. Propagation by Stooling

Stooling is another method of vegetative propagation which can be performed in walnut in the month of March-April. To obtain good rooting and better establishment of stool layers, the growth regulators like IBA, IAA and NAA can be used. The mother stool of the commercial cultivars will have to be raised by cuttings or layers in the first instance (Rashid, 1978). Vahdati and Khalighi (2001) reported that concentrations of 5000 and 7500 ppm of equally mixed IBA+IAA+NAA plus mounding with the mixture of perlite and sand (2:1) and wiring the base of the layers led to the most number and longest roots. Vahdati *et al.* (2008) studied the rooting ability of low-vigour and precocious seedlings of three year old Persian walnut were compared with semi-vigorous and high-vigour seedlings using the stool layering method. The results indicated that low-vigour seedlings rooted more (40 %) than semi-vigorous (31.42 %) and high-vigour (17.14 %) seedlings.

On the whole, cleft grafting performed in the month of January and thereafter incubating at 26°C with 90 per cent relative humidity for three weeks before planting in the nursery and patch budding in mid June using double blade knife hold promise as vegetative methods for propagating walnut on commercial scale.

9.2.6. Zero energy polyhouse system

Zero-energy polyhouse can be made either a wooden or GI pipe structure covered with UV protected white polythene. In order to achieve high graft success, the relative humidity (> 80 %) and temperature (25 ± 3°C) should be maintained inside the polyhouse. This can be possible by spraying water through sprinkler one or two times in a day or by automatic humidifier. However to increase the temperature no additional aid is fitted. In case the temperature inside the polyhouse exceeds 28°C, the both sides of the polyhouse should be open to maintain requisite temperature so the sprouting bud will remain safe. Approximately 450 number of seedlings at a distance of 6 inch from plant to plant and 1.5 ft from row to row can be accommodated in one polyhouse having a dimension of 14' x 50'.

9.3. Top working

Since a large number of young seedling trees producing inferior quality nuts are abundant in the hills of India and a high rate of mortality found in the transplanting of grafted or budded plants, the top-working obviously becomes very useful choice for walnut production. As the rootstock is already well established, the scion makes rapid growth and commences bearing earlier than the transplanted trees. Top-working is usually carried out by modified cleft grafting or bark grafting late in the spring season or when new growth has taken place. The dormant scion wood should be cut from parent tree quite in advance and be stored in refrigerator after proper packing. Bleedling is a problem in walnut top-grafting especially when it is done in early spring which can be avoided by heading back the stock before two weeks of actual operation or make few slanting cuts on the stock through the bark into the wood just below the grafting point so that if any bleeding occur latter, it occurs through these cuts. Select disease free young trees for top working. Collect the dormant scion wood from the parent tree; do bark grafting for top working in late spring (mid March to mid April) or when new growth is taken place. Wrap the graft union with polythene strip and paint the stem portion with white wash.

In order to find a suitable method and time for improving the quality and productivity of walnut trees through top working, a series of field trials were conducted by Rezaee *et al.* (2014). The results showed that, the modified bark grafting (BG) performed at mid-April showed the highest grafting success and survival percentage. Other grafting or budding methods gave poor results and/

or no survival. The top worked trees successfully bore fruits after 3 to 4 years. Dehghan *et al.* (2009) demonstrated that temporary covering of grafting point with moist sawdust resulted in improving callus formation and grafting success in top worked Persian walnut

Sometimes for orchard establishment plant the walnut seedlings directly in the production field without nursery grafting and top worked two years later. This is best method of orchard development in walnut. In top-working grown up trees, the frame working or grafting of branches is desirable for quicker healing of wounds and earlier larger yields. After grafting, the open wounds must be covered by grafting wax and, it required, re-waxing may be done.

9. 4. Micro-propagation

Micro-propagation has increased in popularity over the past two-three decades because walnut is difficult to propagate clonally from cuttings. Earlier discussed methods of vegetative propagation are labour intensive, time consuming, costly and having less success in grafting or budding or cuttings (McGranahan *et al.*, 1988, Land and Cunningham, 1994).

Rodriguez and Sanchez-Tames (1981) and Rodriguez (1982 a, b) were among the first to report establishment of walnut cultures *in vitro* and to describe the development of shoots or roots from cultured walnut embryos. Later a large amount of work was conducted on different walnut species using different types of explants, media, culture conditions and rooting techniques, with encouraging results (Jay-Allemand and Cornu, 1986; Gruselle *et al.*, 1987; Cornu and Jay-Allemand, 1989) and presented in Table 4 and Table 5.

Most of the works were based on a medium developed by Driver and Kuniyuki (1984) (DKW medium) for the culture of apical and lateral meristems of Paradox walnut. They also reported *in vivo* rooting of tissue culture derived shoots. Heile-Sudholt *et al.* (1986) reported the quicker growth of axes on solidified woody plant medium than on Driver Kuniyuki's Walnut medium (DKWM), but auxiliary shoot proliferation and elongation was better on liquid DKWM. They also reported better rooting of axillary shoots in sterile vermiculite following a 15 second dip 10m M IBA and transferring of micropropagated plants to the greenhouse.

Vahdati *et al.* (2009) observed better growth of shoots having green-colored leaves on x1.5 DKW medium. Multiplication rates were different among the cultivars and 'Vina' had the maximum stem length and number of auxiliary buds on x1.5 DKW. The number of auxiliary buds per shoot and stem length were enhanced with higher concentrations of Cu and myo-inositol.

McGranahan *et al.* (1987) reported the successful micropropagation of Parado x *J. hindsii and Pterocarya stenoptera* and asserted that clonal rootstock for *J. regia* would be commercially available in coming five years.

To micropropagate walnut, surface-sterilize individual stem segments containing one bud (no leaves) to remove surface bacteria and fungus. Next, place the basal portion of the stem segments into an agar medium containing a mixture of plant hormones, nutrients, and sugars to promote bud growth. Many clonal microshoots can be produced from a single bud. These microshoots can be used for rooting (clonal rootstock or own-root trees). For rooting, the bases of microshoots are treated with potassium indolebutyric acid (KIBA) *in vitro* for 5 to 7 days at the concentration of 10 ppm and then stuck in a peat:perlite medium on a fog bench in a greenhouse (Vahdati *et al.*, 2004).

Navacchi *et al.* (2009) developed a protocol of the micropropagation of walnut cv. Chandler, which is currently one of the most requested cultivars in the Italian market, was developed. Following the introduction and establishment of shoot tips in vitro, high level of proliferation was obtained in the DKW medium containing 1.5 mg BA [bezyladenine] and 0.01 mg IBA/litre. After culture in a medium containing 0.5 mg IBA and 0.01 mg IBA/litre, elongated shoots (4-5 cm) were rooted in DKW (with macroelements reduced to 1/4 of the original concentrations) with 5 mg BA and 40 g sucrose/litre in the dark for 7 days. After root induction, the culture vessels were exposed to light for 7 days. The shoots were then transferred in the greenhouse on a peat and perlite (10%) medium. Plant survival during acclimatization was 60%, on average.

Quantitative and qualitative improvements in root development were observed by Jay *et al.* (1992) by using mixture of diluted DKW medium (containing gelrite) and vermiculite used in the proportions 250:200 (v/v) which strongly promoted root elongation (2- to 7-fold) and the development of secondary roots on induced shoots (using 24.6 micro M IBA) after 2 weeks of culture.

Maturation of somatic embryos of Persian walnut was evaluated on media containing various concentrations of Gelrite (0.2, 0.3, 0.4, and 0.5 %) and polythene glycol (0.0, 3.0, 5.0 and 7.5 %) in the presence of 7.56 μmol abscisic acid and 3.0 per cent sucrose. The highest rate of cotyledonary (70 %) and normal (50 %) somatic embryos was obtained using 0.3 per cent Gelrite without polythene glycol (Ali *et al.*, 2010).

Miao *et al.* (2010) studied the method of the tissue culture and rapid propagation of thin-shell walnut. The new axillary shoots segment of thin-shell walnut were used as explants which were cultivated in different DKW media with different types and concentrations of plant hormones. The results indicated that the optimum culture medium for shoot induction was DKW/MS+6-BA 1.0 mg/

L+IBA 0.01 mg/L; the medium for shoot propagation was DKW+6-BA 1.0 mg/L+IBA 0.05 mg/L; the suitable medium of the single plantlet growth was DKW+6-BA 0.5 mg/L+IBA 0.05 mg/L; the rooting medium was 1/2DKW+IBA 1.0 mg/L and rooting rate was up to 70%.

Micro-propagated plants need to be hardened off after planting in liner sized containers by gradually reducing humidity (from 100 per cent) over a period of weeks in a greenhouse. In the spring or fall, hardened plantlets can be transferred to a nursery field and then budded or grafted when they reach an appropriate size.

10. Variety

All the walnut trees in India are of seedling origin except few seedlings locally named as cultivars or some cultivars introduced from abroad which are grown at the Regional research stations. Due to its cross pollinated nature each plant is itself a genotype. During the last twenty years, field surveys were carried out in different walnut growing areas of North-eastern Himalayas (including Jammu and Kashmir, Himachal Pradesh and Uttarakhand) to select the superior selections from the seedling populations (Gupta, 1999; Sharma, 1999; Sharma, 2002; Mehta *et al.*, 2005; Pandey *et al.*, 2006; Sofi *et al.*, 2010, 2012). These surveys resulted in identification of potential seedlings were released as cultivars after through evaluation. Walnut germplasm of Jammu and Kashmir having high bearing potential with better nut and Kernel quality needs special attention for exploration, collection and evaluation. Besides the nut yield and quality, other tree characteristics, i.e., plant vigour, ecological tolerance and pest and disease resistance should also be assessed. Apart from the above selections, a quite number of commercial cultivars of walnut commonly grown in other countries having good qualitative traits have been introduced and planted in the field gene banks of different SAU's and ICAR institutes to see their performance. The salient features of the promising exotic cultivars are:

10.1. Amigo

A cross of Sharkey x Marchetti and a sibling of Chico; it was released in 1968. Tree productive with mostly four nuts per clusters, 80 per cent fruitful on lateral buds and leafing comes out two weeks after Payne. Protogynous in nature and a pollinizer for Midland, Pioneer, Tehama, and Pedro. Nuts of satisfactory quality with poor shell seal. Kernel having 56 per cent recovery with light colored and good quality. Susceptible to codling moth and walnut husk fly.

Table 4: Summary of *in vitro* studies with *Juglans* spp L. species

Species/cultivars	Explant	Study/results	Reference
J. nigra L.	Shoots, leaves	Micro propagation/shooted rooted, plants	Roschke and Pijut (2006)
J. regia L.	Mature embryo	Micro propagation/mediums comparison	Paghamzadeh and Kazemitabar (2008a)
J. regia L.	Immature cotyledons	Somatic embryogenesis	Paghamzadeh and Kazemitabar (2008d)
J. regia L.	Cotyledons	Adventitious regeneration/ Somatic embryogenesis	Long *et al.* (1995)
J. regia L. rootstock *cv. Perlata*	Mature fruit	Embryo culture/germinated embryo and developed plants	Sanchez-Zamora *et al.* (2006)
J. regia L. *cv. SU-2*	Somatic embryos	Germination of Somatic embryos/germinated embryos	Dumanogle (2000)
J. regia L.	Mature fruit	Micro propagation/ germinated embryos, fully development plant	Kaur *et al.* (2006)
J. cinerea	Immature cotyledons	Somatic embryogenesis; callus; roots; germinant	Pijut (1993a, b)
J.nigra L.	Immature cotyledons	Somatic embryogenesis;shoot organogenesis	Long *et al.* (1992)
J.nigra L.	Immature to mature cotyledons	Somatic embryogenesis; callus; roots	Neuman *et al.* (1993)
J.nigra, J. major	Immature cotyledons	Somatic embryogenesis; inter-specific hybrids	Cornu (1988)
J. nigra x *J. regia*	Embryonic axes	Germinant shoots	Cornu and Jay Allemand (1988)
J. nigra x *J. regia*	Shoot tips	Rooted plantlets	Meynier and Arnould(1989)
J. nigra x *J. regia*	Immature cotyledons	Maturation, Germination of somatic embryos	Deng and Cornu (1992)
Interspecific hybrids	Embryonic axes	Improved rooting of micro shoots; plants	Jay-Allemand *et al.* (1992)
J. regia	cotyledon	Somatic embryogenesis; plants	Tulecke and McGranahan (1985)

Contd.

Species/cultivars	Explant	Study/results	Reference
J. regia	Immature cotyledons	Histology of somatic embryo origin	Polito *et al.* (1989)
J.regia	Embryonic axes; nodal segments	Shoots; rooted plants	Revilla *et al.* (1989)
J. regia	Apical and lateral buds	Shoot formation	Felaliev (1990)
J. regia	Axillary buds	Shoot multiplication; rooted plants	Stephens *et al.* (1990)
J. regia	Ovules	Somatic embryo, origin determined by RFLP and isozyme analysis	Aly *et al.* (1992)
Clone TRS	Micro shoots	Improved acclimatization of plantlets	Voyiatzis and McGranahan (1994)
J. regia, *J. hindsiix**J. regia*	Nodal segments	Shoot multiplication; rooted shoots	Gruselle and Boxus (1990)
J. regia x *J. nigra*	Nodal segments	Pi nutrition in relation to callus and shootdevelopment	Barbas *et al.* (1993a)
J. regia x *J. nigra*	Embryonic axes	Gelling agent effects on shoot growth	Barbas *et al.* (1993b)
J. regia L.	Immature and matureembryos, immaturecotyledons	Micro propagation/callus induction, embryogermination	Payghamzadeh (2008)

Pi, Orthophosphate; RFLP, restriction fragment length polymorphism

Table 5: Commonly used methods for sterilization of explants

Species/Cultivar	Explants	Study	Disinfectants and dosage	Method	Reference
J. regia L. cultivars vis., ACO38853 NetarAkhort	Gobind, Solding Selection	Mature fruit	Embryo culture 0.1% sodium hypochlorite	Mature fruits were washed in runningtap water. Then epicarp was removed and the remaining part of the fruit (nut)was disinfected by treating with0.1% solution of sodium hypochlorite followedby washing twice with sterile distilled water for 5 min.	Kaur *et al.* (2006)
J. regia L. cvs vis., Yalova-1, Sebin, Bilecik, KR-1, KR-2, Sen-2, 07-KOR-1,Tokat-1, Kaman-1, Kaman-5	Immature fruit	Somat icembryogenesis	3.75% sodium hypochlorite	Fruits were sterilized by immersion for25 min in 3.75% (v/v) sodium hypochlorite, followed by three 5- min rinses in sterile distilled water.	San and Hatic (2006)
J. regia L.	Shoot and leaf	Adventitious shoot regeneration and micro propagation	For shoots: 70% ethanol, 15%bleach solution	For shoots: shoots washed under running tap water for 30 min. Shoots were cut into sections and washed for 30 s in 70% ethanol, disinfected for 20min in 15% bleach solution plus 0.01% Tween on an orbital shaker and rinsed in sterile water four times for 30 s.	Roschke and Paula, (2006)
			For leaves: 10% bleach solution	For leaves: explants were surface washed in running tap water for 5 min, sterilized for 10 min in 10% bleach solution and rinsed four times with sterile water.	
J. cinerea L.	Nodal segment	Axillary bud culture	0.8% (v/v) sodium hypochlorite (15%	Stems from actively growing seedlings were stripped of leaflets and surface	Pijut (1997)

Contd.

Species/Cultivar	Explants	Study	Disinfectants and dosage	Method	Reference
			clorox bleach)	disinfested in 0.8% (v/v) sodium hypochlorite (15% Clorox bleach) for 15-20 min, followed by four rinses in sterile, deionized water.	
J. regia L. cultivar vis., Plemiana 1 and Plemiana 2	Kernel	Embryo culture	NaOCl (1.0% w/v), 75% v/v EtOH	The nuts were cracked and the kernels were immersed in NaOCl (1.0% w/v) for 5 min followed by 75% v/v EtOH for another 5 min and were rinsed three times in sterile distilled water.	Scaltsoyiannes *et al.* (1997)
J. nigra L.	Immature Cotyledons	Adventitious regeneration	5.25% NaClO70% EtOH; HCL0.01 N	The immature fruit were surface disinfected immediately following collection for 1 min in 70% (v/v) ethanol/water solution, followed by a 5 min immersion in 5.25% NaClO. This was followed by a 1 min rinse in sterile deionized water, 1 min in 0.01 M HCl, and three 5-min rinses in sterile deionized water.	Long *et al.* (1995)
*J. regia*L. *J. regia* L. rootstock cv. Peralta	Fruit Nut	Micro propagation Embryo culture	0.5% NaClO, 75% EtOH	Seeds were previously immersed for 24h in water and disinfected for 5 min in 0.5% NaCIO solution followed by 5 min in 75% ethanol and 3 rinses in sterile distilled water.	Revilla *et al.* (1989) Sánchez-
			10% solution of Domestos®, 70%ethanol	The nuts were harvested in October 2002 and washed in a 10% solution of Domestos® for 2 h, rinsing and changing the solution every 30 min. The nuts	Zamora *et al*. (2006)

Contd.

Species/Cultivar	Explants	Study	Disinfectants and dosage	Method	Reference
				were cracked, the kernel removed and the lower half of the cotyledons cut off. These embryos were disinfested in a laminar flow cabinet, for 2 min in 70% ethanol and for 20 min in 20% Domestos®, finishing with three rinses of sterile water.	
J. regia L.	Embryonic axes	Morphogenic study and embryo culture	Alconox (0.1%), NaClO(5 glH1)	Embryonic axes from the seeds were isolated and then sterilized by soaking with alconox (0.1%) for 10 min, rinsed with distilled water for 10 min, soaked with NaClO (5 glH1) for 10 min and finally rinsed 3 times with distilled water.	Fernandez *et al.* (2000)
J. regia L.	Immature fruits	Callus induction from immatureem bryos	Hypochlorite sodium 3.75%; EtOH 50%	Immature fruit were selected at 6 weeks postanthesis and washed under running tap water for 30 min, then the fruits were treated for 30 s in 50% EtOH, disinfected for 20 min in 3.75% hypochlorite sodium solution made up with sterile water plus 0.01% Tween 20,after that samples were three 5-times rinsed with sterile distilled water.	Payghamzadeh (2008)

10.2. Ashley

A seedling or budsport released in 1945, similar to Payne in phenology, 90 per cent fruitful on lateral buds. Yields well, somewhat better than Payne, susceptible to blight. Shell seal adequate, kernel (52 per cent) having light color, good flavor and high quality. The tree is small and early bearing, requiring regular pruning.

10.3. Cascade

A seedling from a backcross to Manregian. Leafs out with Hartley and is late harvesting. Large nuts, dark shelled with a large kernel. Yield is moderate.

10.4. Chandler

A cross between Pedro and 56-224 released in 1979, highly fruitful on laterals, vigorous, needs pruning. Nuts are oval, smooth, with weak shell, good seal and light colored kernel. Nuts easily crack out in halves and stores well. It yields well if pollinizers are present.

10.5. Chico

A cross between Sharkey and Marchetti released in 1968, small, upright tree, vigorous, needs pruning, early leafing, fruitful on laterals, protogynous and high yielder. Nut small, round, with good seal and good quality kernel.

10.6. Cisco

A cross between Meylan and Pedro that released in 1990 as a pollinizer for midseason cultivars. Cisco is a late-leafing, late-maturing cultivar which is fruitful on laterals. It sheds pollen in moderate amounts and covers the peak pistillate bloom of Chandler and Howard. Nut has medium light color with good seal, kernel recovery 47 per cent with variable color.

10.7. Concha

Selected in 1979 by Forde. Leafs out with Payne, laterally fruitful and is harvested with Hartley. Nut seal is good, kernel (47 %) having variable kernel.

10.8. Early Ehrhardt

A chance seedling or possibly a budsport released in 1940. Soft-shelled require less chilling hours and leafs out almost two weeks earlier than Payne. Protogynous and bears only on the terminals. Nuts of lower quality, kernel (55 %) with poor color. It is subject to sunburn.

10.9. Eureka

A seedling tree originated in 1903 from a Persian or Kaghazi type of walnut from Iran. Leafing one week and harvesting one to two weeks after Payne, little fruitfulness on lateral buds. Yield moderate, elongated nuts with both ends rounded. Excellent shell quality, seal good, but kernel color poor. Less susceptible to sunburn, late maturing and susceptible to frost.

10.10. Franquette

Large terminal bearing cultivar, having late leafing and often escapes blight and codling moth. Yields low with good quality nuts. Lower yields than Hartley, both terminal bearers, because Hartley branches more and therefore has more terminals. Franquette also produces one or two nuts per terminal while Hartley produces two or three.

10.11. Scharsch Franquette

Good pollinizer for Hartley and bears catkins at a younger age than regular Franquette. Leafs out three to four weeks later than Payne with late harvesting. Nuts with medium thin shell, good seal, light colored kernels and good fill.

10.12. Forde

Protogynous in nature with 100 per cent laterally fruitful cultivar it is a cross of UC61-25 x Chico. Leaf comes out 5 days and harvesting done 5-10 days earlier than Chandler. Potential pollinizers are Payne, Vina, Serr and Sexton. Nuts are oval to round and medium textured with good seal and shell strength. Large kernels of weighing nine gram are light in colour having 54 per cent recovery.

10.13. Gillet

A cross between UC76-80 and Chico, it is vigorous, protogynous and laterally fruitful cultivar. Leaf comes out a week or more and harvesting 10-20 days before Chandler. Potential pollinizers are Payne, Vina, Serr and Sexton. Nuts are somewhat oblong with good seals and good strength, easy to remove light colored kernels with 50 per cent recovery.

10.14. Gustine

A cross of Waterloo and Payne released in 1968, highly fruitful on lateral buds, vigorous, productive, leafs out with Payne and a midseason variety. Nut elongated, pointed, with good shell seal and good fill, but kernel color poor. Susceptible to bark canker and have brittle wood.

10.15. Hartley

May be a seedling tree of cross between Franquette x Mayette, which released in 1925. Hartley has midseason phenology which is about two weeks after Payne and a week or more before Franquette. Nuts fairly large, broad based with pointed tip, shell light color and thin, with good seal, the kernel light colored but per cent kernel is low. Susceptible to deep bark canker, codling moth is minimal. Less susceptible to blight than Payne, since it leafs out two weeks later. It is a consistent producer, even though it is slightly fruitful on lateral buds.

10.16. Howard

A sibling of Chandler from the cross Pedro x 56-224 which was released in 1979. Laterally fruitful, midseason cultivar with heavy bearing potential. Nuts are large with strong shell of light colour.

10.17. Idaho

A protogynous selection released in 1947. Leaf out with Payne and bears mostly on terminals. Nuts are large having poor quality kernel (35 per cent). Yields poor and is cold hardy.

10.18. Lompoc

A cross between Waterloo x Payne released in 1968, it is moderately fruitful on lateral buds, productive, leafing date is close to Payne, but harvest is late. Large nuts, pointed with good seal, kernel recovery 54 per cent.

10.19. Marchetti

A chance seedling released in 1948 which is a natural cross of Eureka and Payne. Leaf out almost a week after Payne, mostly fruitful on lateral buds, bears heavily and requires a vigorous rootstock. Very good nut seal, kernel with variable color, quality satisfactory. Yields excellent and harvest a week after Payne. Trees tends overbear, so requires pruning every year. Adapts well to hot summers and is more susceptible to blight than Payne.

10.20. Mayette

Trees large, difficult to shake for harvest, late leafing date, nuts having flat shape on stem end like Hartley. Protogynous in nature and best pollinizer for Franquette. Nut seal excellent with shell strong, kernel small with low percentage, but color good.

10.21. Meylan

Terminal bearer, protogynous, late leafing and late in shedding pollen used as a pollinizer for Franquette. Poor nut seal, yields fair, kernel quality good with low percentage. Kernel lacks in astringency, so a good nut for people sensitive to walnuts.

10.22. Midland

A midseason variety, released in 1968, it is from a cross of Franquette x Payne. Moderately fruitful on lateral buds, more vigorous than Vina (a sibling) and leaf comes out eleven days after Payne. Shell seal good, kernel recovery 48 per cent of high quality.

10.23. Payne

A chance seedling with fruitful lateral buds, an early leafing, precocious and high yielding cultivar. Good nut seal and satisfactory kernel quality. Payne is susceptible to frost, blight, codling moth and sunburn. Serr and Forde used mostly in their walnut breeding program due to its lateral fruitfulness.

10.24. Pedro

Laterally fruitful, highly productive, midseason variety, released in 1968, it is cross between Conway Mayette and Payne. Leaf comes out two weeks after Payne and sheds abundant pollen for long period, covering cultivars as Ashley, Lompoc, Marchetti, Serr, Gustine, and Vina. Nuts are large with good shell and fair seal, kernel recovery 50 per cent with fair quality.

10.25. Pioneer

A cross between Franquette and Payne, released in 1968, moderately fruitful on lateral buds. Leafing date is one to two weeks after Payne and is harvested at midseason. Shell seal is fair to good, Nut suitable for in-shell than cracking, kernel recovery 45 per cent, light colored.

10.26. Placentia

Released in 1969, it is an earlier leafing variety than Payne and therefore susceptible to blight and frost damage. Protogynous in nature, terminal bearer having thin shell but strong and good seal. Kernel color is poor and is good source of genes for low chilling.

10.27. Serr

Vigorous, moderately lateral fruitful, it is a cross of Payne x PI 159568 that released in 1968. Leafing time equivalent to Payne, thin shelled nuts with good seal and well filled. Light colored kernel of excellent quality. Well adapted to hot climate and resistant to sunburn. It does not have a problem of pistillate flower abscission.

10.28. Sexton

A cross of Chandler x UC85-8, it is a high yielding variety, leafing out and harvested at least one week before Chandler, 100 per cent fruitful on laterals with abundant male and female flowers. Protandrous in nature with potential pollinizers are Howard, Tulare and Chandler. Nuts are smooth, round with good seals and good strength, kernels recovery >50 per cent, light colored, easy to remove from shell. Densely branching canopy and require pruning to prevent overbearing.

10.29. Sharkey

Early bearing and somewhat fruitful on lateral buds, it is protogynous in nature with pistillate bloom two weeks earlier than Payne and sheds pollen ten days later than Payne. Nuts are round, well filled with a fair seal.

10.30. Sinensis

It is a protogynous variety which bears nuts primarily on terminals. Leafing starts five days after Payne and pistillate bloom nine days earlier than Payne. Good nut seal with 50 per cent kernel recovery, but with poor quality.

10.31. Sunland

A cross between Lompoc and PI 159568, released in 1979, it is vigorous, early-leafing and late harvesting cultivar, with good bearing potential on lateral buds. Nuts are large, oval, with good shell and seal and have high percent kernel, of good quality, large and plump. Susceptible to blight, shallow bark canker and stress from rapid growth.

10.32. Tehama

A very productive (80 % fruitful on lateral buds) cultivar of cross between Waterloo and Payne that released in 1968. Leaf comes out two weeks after Payne and is harvested at midseason. Nut cracks easily, good seal, but the shell tends to be weak. Kernel recovery 54 per cent of good quality and light colored. Requires pruning to prevent overbearing.

10.33. Tulare

Released in 1993, it is a cross of Tehama x Serr, leafs out at midseason, laterally fruitful, and sheds pollen during most of its pistillate bloom period. Upright growth habit, vigour moderate, precocious with heavy nut production. Nut shell of medium texture and color, kernel recovery 54 per cent with mostly light colored. Suitable for hedgerow and other high density planting systems.

10.34. Vina

A cross between Franquette and Payne that released in 1968 is productive, consistent bearer, being 80 per cent fruitful on lateral buds and leafing out about a week after Payne. Good shell seal, kernel recovery 48 per cent of good quality. Mature early to midseason, tolerant to summer heat and resistance to blight. Due to vigorous growth in young trees, regular pruning is required to prevent overbearing and lower quality nuts.

10.35. Waterloo

Released in 1934, not laterally fruitful, but good producer. Leafing out with Hartley, but harvested after Hartley and before Franquette. Nuts are strong with good seal and poor kernel color.

10.36. Westside

A chance seedling, it is protogynous in nature, requires an early pollinizers and highly fruitful on lateral buds. Small nuts, thick shelled with low kernel percentage.

Selections released from India

10.37. CITH Walnut-1

A protogynous, terminal as well as lateral bearing and mid-season blooming variety. Nuts and kernels are very large, kernel recovery 47 per cent, yields high, light shell colour, soft shelled well filled kernel, plumy, very easy to remove full kernel, light in colour. Tolerant to biotic stress except white grub and susceptible to moisture stress (Sofi *et al.*, 2010).

10.38. CITH Walnut-2

A protogynous, predominantly terminal bearing and early season blooming variety. Nuts are large, good kernel recovery (51.27 %), higher yield, semi-soft, well filled kernel, plumy, easy to remove full kernel and light in colour. Tolerance to pests except white grub (Sofi *et al.*, 2010).

10.39. CITH Walnut-3

Protogynous in nature, terminal as well as lateral bearing and late maturing variety. Nuts are large, kernel recovery 54 per cent, higher yield, good kernel colour with medium size, well filled, plumy, moderate to remove kernel, light in colour. Tolerance to pests except white grub and susceptible to moisture stress (Sofi *et al.*, 2010).

10.40. CITH Walnut-4

Protandrous in nature, terminal as well as lateral and late season variety. Nuts are large, higher yields very high kernel recovery (59.39 %), light shell colour, thin shelled, good kernel flavour, well filled kernel, moderately plumy, very easy to remove kernel, light kernel colour (Sofi *et al.*, 2010).

10.41. CITH Walnut-5

A protandrous, terminal as well as lateral and late season maturing variety. Nuts are large, bold kernel, higher yields with high kernel recovery (52.42 %), smooth shell texture, light shell colour, good kernel flavour, well filled kernel, plumy, moderate to remove kernel and light kernel colour. Tolerance to diseases and pests except white grub (Sofi *et al.*, 2010).

10.42. CITH Walnut-6

A protandrous variety which is intermediate in vigour, semi-erect, intermediate branches, well apart, dense, stout, brown to grey and healthy. High yielding variety, nuts with superior quality having higher shelling percentage (51 %). Nuts are large weighing 23.99g, shape ovate, light shell and kernel colour, plumpy kernel, well filled, rough shell texture and intermediate shell seal. Bearing starts from third year of planting and yields 30 kg/tree at 15 years of age (Sofi *et al.*, 2012).

10.43. CITH Walnut-7

A high yielding chance seedling selection, with superior nut quality and higher shelling percentage (50 %). Vigorous growth habit, semi-erect, intermediate branches, well apart, dense, stout, brown to grey and healthy. Nut are large with elliptic nut shape, medium shell and light amber kernel colour, plumpy kernel, well filled, rough shell texture and strong shell seal. Protandrous in nature, starts yielding from third year of planting and yields 30 kg/tree at 15 years of age (Sofi *et al.*, 2012).

10.44. CITH Walnut-8

A high yielding chance seedling selection, protandrous in nature having superior nut quality and higher shelling percentage (54.10 %). Semi-vigorous, semi-erect, intermediate branches, well apart, dense, stout, brown to grey and healthy. Nut shape long trapezoid weighing 20.35 g, light shell and kernel colour, plumpy kernel, well filled, rough shell texture and intermediate shell seal. Bearing starts from third year of planting and 50 kg/tree at 20 years of age (Sofi *et al.*, 2012).

10.45. CITH Walnut-9

A protandrous, high yielding and chance seedling selection, having superior nut quality with higher shelling percentage (50.9 %). Vigorous growth, spreading type dense branches, well apart, dense, stout, brown to grey and healthy. Round nut shape and rough shell texture, with light shell and kernel colour, well filled, plumpy kernel, rough shell texture and intermediate shell seal. Bearing starts from third year of planting and 50 kg/tree at 20 years of age (Verma *et al.*, 2011).

10.46. CITH Walnut-10

A high yielding, protandrous and chance seedling selection of superior nut quality and higher shelling percentage (55.50 %). Vigorous growth, semi-erect, intermediate branches, well apart, dense, stout, brown to grey and healthy. Nut shape long trapezoid weighing 19.95 g, light shell and kernel colour, plumpy kernel, well filled, rough shell texture and intermediate shell seal. Bearing starts from third year of planting and yields 30 kg/tree at 20 years of age (Verma *et al.*, 2011).

10.47. Hamdan

Dwarf, spreading, precocious, starts bearing third years after planting, protandrous in nature. Nut large in size with shelling percentage (54 per cent) ovate in shape with > 80 per cent light coloured kernels. Tolerant to anthracnose and leaf spot disease (Bhat and Ahmad, 2003).

10.48. Sulaiman

Semi-dwarf, spreading, protogynous in nature and Hamdaan used as pollinizer. Highest bearing potential, productive upto 60 per cent on lateral. Nuts are large weighing 21.0 g with 52 per cent shelling capacity. Nuts are round in shape, with >50 per cent light colour kernels. Resistant to anthracnose leaf spot caused by *Gnomonia leptostyla* and *Marssonina juglandis*, respectively (Bhat and Ahmad, 2003).

10.49. Gobind

Ripens in the end of September under Kinnaur conditions. Kernel quality good, creamish in colour, well filled, thin shelled, nut size medium.

10.50. Pratap

Tree large and heavy bearer. Nuts are large, oblong and each nut weights about 24 g having smooth surface with light amber colour. Cracking and shelling quality excellent, semi thin shelled, light coloured kernel with excellent taste.

10.51. Solding Selection

Fruits medium sized (17.1 g) circular nuts, shell medium hard with extremely smooth surface, light coloured kernels with 40 per cent recovery, kernel fat 49 per cent, protein 21 per cent, quality and taste good with flavor, tree vigorous, regular bearer and moderate yielder.

10.52. Kotkhai Selection

Early maturing variety, nut size medium (15.4 g) with smooth shell surface, thin shelled (1.4 mm), excellent shelling quality, shell well sealed, more than 58 per cent kernel content, colour of the kernel is light with excellent taste.

Variety developed from China

10.53. Luzao

It is a new early ripening, precocious and thin shell walnut cultivar. It has some characteristics, such as early ripening (from the end of July to the beginning of August), full kernel, savory and crisp, the fresh kernel mass was 14.3 g, thin shell, big fruit, precocious, high and stable yields. It is suitable for planting to thickness layer and fertility soil (Li *et al*., 2008).

10.54. Luxiang

It is a new precocious walnut cultivar, which was selected from the walnut seedling resources in Shandong Province. The fresh fruit is ovoid in shape with a yellow-green skin. The average fruit weight is 54.53 g. Walnut green husk thickness is about 0.58 cm. The kernel total fat content is 62.96%, kernel total protein content is 17.27%, soluble protein content is 8.17%. It is easy to take out the whole kernel. The color of endotesta is milk white. The kernel is filled plumply and full of aroma. 'Luxiang' showed excellent characteristics such as strong growth, high yield, good quality. It is suitable for fresh-eating. (Hou *et al.,* 2010).

10.55. Jinxiang

A new walnut cultivar which selected from the seedlings of Xinjiang cultivar. The average mass of single nut is 11.5 g. The nut shell is smooth, which thickness is 0.75 mm. It is easy to take out the whole kernel because the crack seam is loose. Kernel weight accounts up 63.97% of nut. The color of 96% kernel is light. The kernel is filled plumply and full of aroma. It has high quality (Wang *et al.,* 2008).

10.56. Yuanlin

A new walnut cultivar derived from the cross 'Yuanfeng' x 'Chandler'. The average fruit weight is 16.84 g. The kernel rate is 55.42%, the total fat content is 63.6%, and total protein content is 18.25%. This early-maturing cultivar has high yield and strong fruiting ability. It can avoid the damage of late frost (Hou *et al*., 2008). :

10.57. Chuanzao 1

It is a new early-fruiting walnut cultivar derived from the cross 'Yunxin 7926' x 'Xiazao'. The nut is large and round, with the smooth surface. The shell thickness is 0.9 cm, and kernel percentage is 51.6%. The total fat content is 72.14%, and total protein content is 16.80%. The grafted seedlings could flower and fruit after planting half a year. The production of its 5-year-old crown projection area is 595 g . m-2. This early-maturing cultivar has high yield and strong fruiting ability. Meanwhile, it has good adaptability resistance to drought, wind, dust, disease and insects (Pu *et al.,* 2011).

10.58. Jinboxiang 7

A new walnut cultivar which bred from the Xinjiang walnut for over 20 years. The nut is round, smooth, full and plump. It is easy to get the kernel from the nut. The kernel is light yellow with milky white flesh and excellent quality. It has a good productivity, high resistant to cold, drought and disease, and late blossom (Tian *et al.,* 2009).

10.59. 'Qinquan 1'

It is a new cultivar of *Juglans regia* L., with the characteristics of apomixis, selected from the walnut seedling resources in Taihang Mountain area, the ratio of apomixis was steady in 17.6%. The size of the nut is large and round, the smooth surface and light color, shell is thin; the kernel was full and plump, with a membrane of diaphragm, a light yellow inner layer, yellow-white kernel, releasing fragrance, and excellent quality. In addition, it has good productivity, high resistant to cold and disease (Wu *et al.,* 2009).

10.60. Lvzao

It is an open-pollinated chance seedling of Xinjiang walnut cultivar, selected in 1999. Nuts mature in early August, around 30 days earlier than cv. Xiangling. Nuts weigh 11.8 g on average, and have a light yellow thin shell (0.73 mm). The nuts are easily shelled. The kernel rate is over 65%. The kernel contains 49.30% fat and 19.30% protein. The trees are moderately vigorous, precocious, and cropping is stable. A 5-year-old plantation can produce 3000 kg/ha (Li, 2008).

10.61. Daifeng

The new walnut [Juglans regia] cultivar Daifeng had bigger fruits, higher yield, thinner shells, better quality and was highly resistant to adverse conditions compared to other walnut cultivars. Fruit shape, thin shell and kernel percentage were similar to the international standard of excellence for walnut cultivars (Xu *et al.*, 2001, 2001).

10.62. NY83-05

A seedling selection of the Italian cv. Corne, is late flowering, allowing it to escape late frost damage. It is precocious, with over 90% of 5-year-old trees fruiting. Nuts weigh 13.7 g and have a thin shell (1.22 mm thick). The kernel is light yellow, lacks astringency and is of very good quality. Kernel percentage is 59.1% (Wu, 1997.

Variety developed in Turkey

10.63. Sebin

It is a Persian walnut cultivar that originated as a seedling in Sebinkarahisar, and has become a standard in many walnut producing areas of Turkey. The popularity of 'Sebin' is due to its high kernel ratio, nut quality, precocity and excellent yield. The mean nut weight varied from 9.95 to 12.40 g, kernel weight from 6.25 to 7.85 g, and the kernel ratio was between 61.0 and 64%. However, fruitlessness was a problem observed in some orchards established with this cultivar. Pistillate flower abortion (PFA) can reduce dry in-shell walnut yield 75-85% in Sebin (Akca, 2014).

Variety developed in Italy

10.64. Malizia

originated from the open pollination of a Sorrento seedling. It is resistant to spring frost and fruits ripen from late Sep. to early Oct., 4-6 days earlier than Sorrento. Yields are high and regular. The light coloured, sweet tasting kernel represents 46-49% of the nut weight. Nuts are large (36-45 mm in length and 29-33 mm in diameter) with a 100-nut weight of 1.1-1.2 kg. Malizia is recommended for growing in the Campania region of Italy (Limongelli, 1993).

Variety developed in Bulgaria

10.65. Vasden

This cultivar is a result of sexual hybridization through crossing the local walnut varieties Bachkovski x Plovdivski. The crown is medium in size and the tree growth is moderate. Fruiting is apical with a good yield potential. The cultivar is protogenous. The fruit size is large, oval in shape, with an average weight of 14.9 g. Fertility is very good and regular. Kernel percentage is high - 53-55%. Fruits mature at the beginning of September, before Sheynovo cultivar (Gandev, 2007).

10. 66. Cold hardy cultivars

Cold hardiness is important because freeze injury is a leading cause of walnut yield losses in many cold regions of Europe and Asia. Cold hardiness was determined on the basis of Proline content of fresh stem tissues using the Bates method (Aslamarz *et al.*, 2011). Cold hardiness of samples increased with accumulated seasonal chilling and decreased as they approached spring bud break. A 1.4 to 2.7 fold increase in proline content was observed in stems from November to midwinter. LT50s (temperature causing 50% lethal injury) ranged from -5.8 to -10 degrees C in November, -12.5 to -19.3 degrees C in midwinter and -5.8 to -9.2 degrees C in early spring. Using both the EL and TTC methods, cultivars and genotypes were classified into three groups based on their cold hardiness:

Sensitive: Z30 and 'Serr

Semi-hardy: Z53, Z67, and Hartley

Hardy: Lara, Z63 and Pedro

10.67. Development of resistant cultivar through bio-technological approach

Crown gall disease (*Agrobacterium tumefaciens*) can severely limit the growth and productivity of many tree crops and ornamental plants. Escobar *et al.* (2002) developed a novel crown gall disease resistance strategy that is based upon the silencing of two highly conserved *A. tumefaciens* oncogenes which are required for crown gall development. The tryptophan monooxygenase (iaaM) and isopentenyltransferase (ipt) genes are horizontally transferred from *A. tumefaciens* to the plant cell, mediating the de novo auxin and cytokinin production that initiates tumorigenesis. By expressing inducers of posttranscriptional gene silencing homologous to iaaM and ipt in plant cells, the corresponding oncogene mRNA transcripts are degraded in planta, providing functional resistance to crown gall disease. This strategy is highly efficacious in English (Persian) walnut (*Juglans regia*), as several transformed lines expressing self-complementary iaaM and ipt transgenes display a highly specific accumulation of small-interfering RNAs and a broad-spectrum suppression of tumorigenesis.

11. Cultivation

11.1. Planting

To ensure long productive life of walnut trees, planning with regards of selection of site, choice of location, preparation of the soils, planning the orchard, laying out the orchard, planting distance, time of planting, method of planting should be done in advance and in a proper way, because it is very difficult to rectify defects or make adjustments later on the establishment of trees (Wildman, 1985).

11.1.1. Selection of site

Walnut is a temperate fruit crop and mostly planted on the sloppy land and rarely on flat land. On flat land, square, quincunx or rectangular system and on sloping land, contour or terrace system of layout are done. In square system, the distance between plants and rows is the same while another tree is planted in the centre of four plants in the quincunx system. More distance is, however, kept between rows than plants in rectangular system. Square system is commonly adopted in walnut because it is easier to layout and convenient for cultural operations and tree thinning. In India, walnut is usually grown on hill slopes. When the gradient of slope is more than 10 per cent but gradual on a hillock, planting is done in contour design. All the trees in a row are at same elevation but the distance between rows varies, being closer where the slope is steep.

A number of factors need to be assessed before deciding the tree density. First, a walnut tree takes about 15-20 years to occupy the entire space allotted to it in conventional planting system leaving sufficient space unutilized in early stages. Secondly, in the early years of the orchard, production is directly related to the number of trees planted, i.e., more the tree number higher is the yield. Thirdly, a number of lateral bearing and precocious walnut cultivars are now available for closer planting. Fourthly, the processes of pruning and tree removal have made the high density planting possible. Finally, high density planting requires a higher level of orchard management, i.e., irrigation, fertilization, pruning and control of pests and diseases.

11.1.2. Spacing

Traditionally, walnuts are planted at a spacing of 8 to 10 m distance in square system of planting, but the growth habit and vigour of the cultivars to be grown play the pivotal role in spacing. Under high density plantation, plants are planted at a distance between 5 to 6 m, but which should be heavily pruned and gradually thinned on overcrowding. In certain areas, the distance from row to row and plant to plant is kept 12 metres, which accommodates 70 plants per hectare. Osgood (1985) have illustrated this view-point by citing three examples. Chico is a lateral bearing and early fruiting cultivar but is a relatively small and compact tree at maturity. Hence, a spacing of 6.6 x 6.6 m should be kept initially and should be adjusted to 9.3 x 9.3 m to final adjustment of 13.3 x 13.3 m is suggested for Vina, a highly fruitful lateral bearing but vigorous cultivar. Terminal bearing vigorous cultivars like Hartley is however, to be given 15 m to 20 m spacing, interspace being utilized by Chico fillers in early stages. Ramos *et al.* (2001) established hedgerows (3.6 x 7.3 m) of Chico cultivar to develop a high density system for walnuts. The trees were mechanically trained into fruiting walls. Hedgerow trees gave as much as double yield in early fruiting years than that of traditionally spaced and pruned trees. They were, however, in favour of further research to determine the best hedging procedures, closest spacing and suitable cultivars. When pollinizers are required, they should be placed in permanent position in order to avoid their elimination during temporary tree removal. Every eighth row of pollinizers be planted perpendicular to the usual direction of the wind.

11.1.3. Preparation of pit and planting

Pits of 1 x 1 x 1 m size should be dug one to two months before planting at marked positions. The pits should be filled up with orchard soil mixed with 50 kg well rotten FYM, 150 g Aldrin dust, 150 g urea, 500 g each of superphosphate and muriate of potash. All the extraneous materials, such as, weeds, stones and plant remnants need to be removed. If there is a hard layer on any side of the pit, it should be broken. The plants should be purchased well in advance of actual planting and be healed immediately on arrival after laying them in a trench and covering the root portion with moist soil which should never be allowed to go dry in the process. A plant is then placed in the pit and the roots are spread around in all sides and covered with soil. The soil should be filled in the pit by firming with hands and feet followed immediately by light irrigation. The plant should be well fixed in the soil and the adjoining portion of scion and stock union should be at least 15 cm above the ground surface. To protect the plant from collar rot it should be treated with Dithane Z-78 before planting. The planting is done in late dormant season or early spring season.

11.1.4. Care of young trees

The most important operation is irrigation during the first year of the planting in the orchard. Frequent watering is particularly required in the summer season, but waterlogging is to be avoided in the rainy season. The trunks should be painted white to prevent sunburn. The plant basin should be kept clean by removing weeds as soon as they arise. In windy situations, staking of plants becomes necessary. The young trees need to be protected from the attack of pests and diseases.

11.2. Canopy management

11.2.1. Training

Pruning of young plants is termed as training which is carried out with two primary objectives, i.e., developing a strong framework capable of supporting heavy crops and to encourage the shoot growth along with the early heavy fruiting. In order to achieve these objectives, it is imperative to know the bearing habit of the cultivars and their nature of growth. Walnut is both terminal and lateral bearer. It bears fruits on one year old wood which is produced either terminally or both terminally and laterally. Terminal bearing cultivars are thinned, but not headed back because vegetative growth keeps on increasing in absence of early fruiting. On the other hand, lateral bearing cultivars are both thinned and headed back in order to encourage shoot growth which gets suppressed due to early fruit production.

Modified leader system of training is followed in walnut (Serr, 1969). After the first year's growth, the plant is headed back above 2 m of its height. All lateral shoots on the leader are removed leaving one or two shoots at lower level on the trunk. To avoid narrow crotched, all primary buds 1.5 m above the ground should be removed forcing the secondary buds to grow (Freeman and Aldrich, 1985). After selecting the main scaffold limb 1.6 m above the ground, the other primary scaffold limbs are chosen in all directions on the trunk. These limbs should have wider angles and not have arisen just opposite to each other. They should also be more than 30 cm part vertically on the main trunk. Remove the rest of the vigorous branches leaving small branches undisturbed for fruiting. Head back all the selected limbs on terminal fruiting cultivars but not for lateral fruiting ones. In later years, secondary scaffold limbs on primary scaffold ones are also allowed to grow by removing the extra vigorous branches.

Terminal bearing cultivars (Hartley and Franquette) are thinned but not headed back because vegetative growth remains proliferating in absence of early fruiting. Conversely, lateral bearing cultivars (Vina and Chico) are both thinned and headed back heavily in order to encourage shoot growth which get suppressed due to early fruit production.

11.2.2. Pruning

Pruning of mature walnuts is done regularly to obtain regular and high production of quality nuts which is often hampered by overcrowding of branches, insufficient production of new wood and inadequate availability of sunlight. Selective thinning out of limbs in the top and sides of the tree should be done to avoid overcrowding. Pruning of young bearing trees is again determined by the bearing habit. The cultivars which are highly fruitful on lateral buds, the shoot growth is suppressed due to heavy crop load. In this case, a large number of new shoots should be headed back on the periphery to reduce fruiting and increase vigorous shoot growth throughout the tree's periphery. In contrast, on the terminal bearing cultivars, pruning consists of heading back of selected framework branches and thinning out of competing limbs.

Ramos *et al* (1990) studied the effect of effects of pruning on high density 'Hartley' walnuts. Limbs in the upper productive portion of the trees were thinned out when dormant in 5 successive years to allow rejuvenation of lower limbs by better light exposure. Photosynthetically active radiation beneath pruned trees was greater than that beneath unpruned trees but, in both, light flux densities were <3% of full sunlight. Yield was consistently lower in pruned trees, mainly because fruiting spurs were removed and fruiting was not promoted in lower limbs. It was concluded, however, that pruning is beneficial as it restricts the size and manageability of the trees, delays the further loss of fruiting wood on lower limbs, removes dead branches that would otherwise fall and impede harvest and results in larger, better quality nuts, the higher price of which helps to offset the lower yield.

Ryugo and Ramos (1979) reported the conversion of vegetative buds into reproductive ones within four weeks of formation in the cultivar Chico on the basis of a trial on heading back of second flush growth to basal buds done at different time of the year. Defoliation or defoliation with etiolation of seven terminal buds in July did not deter the initiation of female flowers for the next season. Thus fruit bud formations is not a problem in walnut. Shading becomes a problem in mature trees because it reduces photosynthesis thereby weakens spur growth. Hence selective thinning out of limbs the top and sides of the tree should be initiated well before overcrowding. Since the leaves are food manufacturing organs being essential for growth and fruiting, not more than 25 per cent of the branch structure is removed at any time (Stoke, 1969). On the basis of experimental evidence, Ramos *et al*. (2001) concluded that benefits of pruning are not immediate, but in time they can result in sustaining yield with increased nut quality and value.

The pruning operations are carried out in the dormant season but early spring is preferable (Maurer, 1976). Further delay causes excess bleeding. All the cut surfaces should be treated with a tree wound dressing. The best method of pruning young (16-17 year old) and middle aged (34-35 year old) walnut trees is to cut back to 3 year-old shoots; this method also provided 5.5 % and 6.5 %, more buds for budding, respectively. The best method for pruning mature trees (40 year old) is to cut back 4 to 5 year old shoots (Novak, 1974).

11.3. Intercropping

As most of the walnut plantation are of seedling origin and grown on undulated and non cultivable land, so mostly land between plants are used for pasture farming systems. In some areas legume fodders crops are grown between walnut plantations as intercropping. For other farming systems, low growing and non-competitive vegetables such as non-climbing beans and other legumes, tomato and onion and more competitive vegetables such as chilli and asparagus can be grown during first 4 to 6 years after planting only. If sustained intercropping is planned from the establishment of the orchard, wider spacing between rows of trees should be adopted.

11.4. Irrigation

Walnut trees are grown under rainfed conditions. However, whenever the rain is not sufficient and well distributed, irrigation is essential for walnut cultivation. Inadequate water supply to the trees results in slow growth, reduced yield and nut quality, but excess of it retards orchard development and causes root and crown diseases. At least 760 mm of rain is required and irrigation will benefit trees in drought time. Supplemental irrigation in dry spring and early summer before the start of monsoon period i.e. from April to June is beneficial. No irrigation is needed from June onwards until harvest in areas receiving monsoon. Wet weather at harvest cause shell and kernel stain.

Adequate supply of irrigations or water helps in better growth and early bearing. To the bearing trees, irrigation from the time of fruit set till its maturity proves helpful in reducing the fruit drop and better filling of nuts. Nuts grow most rapidly during the 5 to 6 weeks after the fruit set. Moisture shortage during nut development stage may result in large proportion of small sized nuts. Moisture stress during summer results in poor nut quality due to lack of plumpness of the kernels and the yield is reduced considerably. The water requirement decreases with the development of roots in adult trees.

Ramos *et al.* (1978) compared the unirrigated with irrigated walnut trees of the cultivar Serr, and found increased nut weight from 7.6 to 9.6 g and nut diameter form 40 to 42 mm due to irrigation. Kernel quality, as indicated by skin colour and percentage shrivel, was also much superior with watering and returns were 4.3 per cent higher. Withholding irrigation for one growing season significantly reduced trunk growth and kernel weight (Martin *et al.*, 1980). They further obtained heavier kernel weight on the resumption of irrigation in the following year. The nut size does not increase by mid-summer or late irrigation after the hardening of nutshell (Aldrich *et al.*, 1976).

11.5. Weed Control

Weeds can cause a multitude of problems in walnut orchards by reducing the growth of young trees because they compete for water, nutrients and space. Weeds also increase water use, cause vertebrate and invertebrate and other pest problems and may enhance the potential for disease. The increasing use of more efficient low-volume irrigation systems has increased the need for selective pre-emergence herbicide use in drip, microsprinkler and sprinkler irrigated orchards. Herbicides are generally used only in the tree row. This reduces the total amount of herbicides and prevents the surface roots in the tree row from being damaged by cultivation equipment. Weed species are controlled by pre-emergence, post-emergence or a combination of pre and post emergent herbicides.

The herbicides simazine and diuron, have been successfully used in weed control of young and established walnut groves for four months (Charlot, 1986). Kiryakov and Lukova (1974) found very good weed control in nurseries (second year) by using simazine at 3 kg/ha, which reduced the number of manual weeding from 6 to 3 and was found highly economical. Simazine and diurom at 2-4 lb/acre effectively controlled annual weeds throughout the season. Dichlobenil was found safe for trees on heavier soils but less so on lighter soils, particularly under sprinkler and flood irrigation (Lange *et al.*, 1969). In a 3 year herbicide trial in a young black walnut plantation the best result was obtained with simazine (6 lb/acre) plus pre-planting treatment with paraquat (0.5 pt/acre). This combination showed excellent weed control without causing visible injury to the trees (Roth, 1971).

Schlesselman *et al.* (1998) reported that the addition of thiazopyr gave excellent residual broad spectrum weed control and 75-90 per cent effective for 8 to 9 months after treatment in walnut orchards.

11.6. Mineral nutrition

Generally walnuts are not fertilized in India by the orchardists, but for achieving higher yields of quality crop, fertilization is very important (Ponder and Van-Sambeek, 1997). Walnut obtains mineral nutrients from the soil which normally meets all the requirement of the plant except in few cases when one or more elements may become deficient or toxic. Fertilizers are applied to rectify the deficient situations. For the accurate diagnosis of nutritional needs of the plant soil sampling or leaf tissue sampling should be done. Soil analysis is of limited utility in determining the nutritional requirement of walnuts because the plant deficiencies are not well correlated with soil analysis values (Proebsting and Davy, 1946). Moreover, walnut tree has a very extensive root system trapping nutrients from a large volume of the soil. The leaf analysis is accurate diagnosis of nutritional needs of the plant (Table 6) and leaf samples are taken in the month of July to August and then analyzed in laboratory for the formulation of nutrient guide. Abaev and Khvorova (1983) suggested the collection of leaves from the middle of the crown during active vegetative growth period in June for the most accurate diagnosis. The terminal leaflets of the compound leaf from spurs or middle of moderately growing shoots should be taken and only 20 to 30 leaflets are sufficient for a sample (Uriu, 1985).

Masarat and Tomar (2008) demonstrated that N, P and K contents declined with the advancement of season basipetally on the shoot, while Ca and Mg levels increased with age in basipetal direction. The least variation period of N, P, Ca and Mg contents accrued in median shoot leaflets from mid-July to mid-August and K contents showed the lowest least variation during mid-September to mid-October. The 3.5-to 4.5-month-old median shoot leaflets of walnut should be collected to determine nutrient status.

Table 6: Critical nutrient levels of walnut leaves

Element	Deficient	Adequate	Toxic	Element	Deficient	Adequate	Toxic
Nitrogen	< 2.1 %	2.2 to 3.2 %		Chlorine			0.3 %
Phosphorous		0.1 to 0.3 %		Boron	< 20 ppm	36-200 ppm	300 ppm
Potassium	< 0.9 %	1.2 %		Copper		4 ppm	
Calcium		1.0 %		Manganese		20 ppm	
Magnesium		0.3 %		Zinc	18 ppm		
Sodium			0.1 %				

Table 7: Fertilizers schedule of walnut

Age (in years)	Fertilizer/plant (gm)			Age (in years)	Fertilizer/plant (gm)		
	Urea	DAP	MOP		Urea	DAP	MOP
1	20	65	30	11	585	160	500
2	30	65	40	12	780	215	665
3	40	65	50	13	980	270	830
4	60	65	65	14	1175	325	995
5	85	65	80	15	1565	435	995
6	136	65	125	16	1960	540	1245
7	190	65	165	17	2460	650	1825
8	290	70	250	18	2950	760	1995
9	390	110	330	19	3415	975	2155
10	490	130	415	20 & onwards	3915	1085	2400

(*Source:* Anonymous, 2011)

Following fertilizer schedule (Table 7) shall help in obtaining better yields of quality nuts. Full dose of P and K and ½ dose of N should be applied two weeks before expected bloom. Rest ½ quantity of N should be split into two equal doses which should be applied three weeks after fruit set and during early July.

The deficiency of various nutrient elements may be responsible for poor growth, flowering, fruit-set and yields (Table 8).

Table 8: Nutrient deficiency symptoms of walnut

Element	Deficiency symptoms
Nitrogen	Nitrogen deficiencies are rarely found and are also difficult to diagnose. Initially, the symptoms are pale colour of foliage, small leaves, reduced shoot growth, low fruit set and yield, and formation of small and poorly filled nuts
Phosphorous	Weak growth, small leaves and sparse foliage are the symptoms of phosphorous deficiency. The leaves become yellow and develop necrosis in irregular patches
Potassium	Deficient leaves first become pale and thereafter margins of leaflets fold upward and curl in, and the underside turns grayish (bronzing). In severe cases, leaf size reduces and marginal and tip scorching occurs.
Zinc	Bud opening delayed and leaves small and chlorotic. The terms 'little leaf' and rosette' and yellows are used to describe Zn deficiency symptoms. Terminal die back of shoots is also observed in severe deficiency.
Boron	Deformed leaves, short internodes, chlorosis and die back of twigs and branches are the various symptoms exhibited on the extent of boron deficiency. Tip burn and marginal necrosis extending towards the midrib of leaflets appears
Manganese	In early to mid summer, interveinal chlorosis develops extending from midrib to margin and subsequently leaves fall early due to manganese deficiency
Iron	Leaves become yellow and in severe cases almost white due to iron chlorosis which begins early in the season, first appearing on terminal leaves.
Copper	Copper deficiency appears in the form of necrotic spots, yellowing and premature leaf fall. Small dark brown spots also appear on the shoots around the tip.
Magnesium	Distinct chlorotic region appears at apical and margins of leaflets in basal leaves.

Sources: Proebsting and Serr, 1966; Uriu, 1985,

Abaev *et al.*, (1976) obtained growth and development of walnuts by supplying N, P_2O_5 and K_2O at 180:180:60 kg/ha respectively. But there appears a general consensus on the application of 60 kg/ha each N, P and K for getting the highest yield and net profit (Kolov, 1977; Thagusev and Caroev, 1971). Kolov (1977) differed only for the dose of K and recommended 30 instead of 60 kg.

Liang *et al.* (2010) standardized the reasonable rate of fertilizer application for each walnut tree in southern Xinjiang (China). Each tree requires 3 kg urea, 3 kg Re-SSP and 1.5 kg potassium sulfate which resulted in yield increment up to 19.6-22.2 kg/tree. Compared with conventional fertilization, the complexation of N, P and K was more reasonable.

Beneficial effect of boron and zinc spray on walnut was observed by Keshavarz *et al.* (2011). The results showed that all B and Zn applications and combinations had a significant effect on all traits except nut and kernel diameter, shell percent, husk thickness, and pistillate flower abscission (PFA). Pollen germination, fruit set, vegetative growth, nut weight, kernel percent, nut and kernel length, and chlorophyll index were highest when B and Zn were applied simultaneously at 174 and 1050 mg L^{-1} concentrations, respectively.

11.6.1. Physiological aspect of mineral absorption

Absorption of zinc

The characteristics and mechanisms of foliar Zn uptake and translocation in pistachio (cv. Kerman) and walnut (cv. Gustine) were investigated by Zhang and Brown (1999) using 68Zn labelling in both intact and detached leaves from trees grown at Davis, California. Following washing, mature walnut and pistachio leaves retained 8% and 12% of the total Zn applied, respectively. About half of retained Zn (3.5% and 6.5% of total Zn, respectively) was absorbed into the leaf and translocated outside the treated area. Leaf age affected the Zn absorption capacity of pistachio but not walnut. The absorption of Zn by walnut leaves at high concentrations (7.5-15 mM Zn) was not significantly affected by the pH of the solution. The uptake process was not affected by light or addition of metabolic inhibitors. Foliar leaf absorption was only slightly affected by changes in temperature with an average Q10 of 1.2-1.4. These results suggest that foliar Zn uptake is dominated by an ion exchange and/or diffusion process rather than an active one.

11.7. Effect of plant growth regulators

11.7.1. Improvement of fruit setting and fruit retention

Pistillate flower abortion (PFA) in walnuts is defined as the drop of female flowers with or without pedicles within 10-15 days after flowering. This situation is commonly observed in some walnut cultivars causes dramatic yield losses. When a pistillate flower receives high pollen load, pollen tubes growing down the style induce high rate of ethylene biosynthesis activating pre-formed abscission zones and resulting in flower abscission (Beede and Polito, 2003; Polito *et al.*, 2005; Johnson *et al.*, 2006; Lemus *et al.*, 2007). PFA (Pistillate

Flower Abscission) is the loss of nut-producing pistillate (female) flowers due to an excessive pollen load on them. That pollen load stimulates production of ethylene, causing flower drop. With PFA, flowers stop growing at the 2-3 mm size and fall about 10 days after bloom, leading to reduced fruit set and low yields.

An ethylene inhibitor amino-ethoxy-vinyl-glycine (AVG - the active ingredient in ReTain® plant growth regulator) has been tested on walnuts to reduce PFA. The results showed increased yields especially on 'Serr' cultivar (Beede and Polito, 2003, Lemus *et al.* 2007). The use of ReTain on walnut orchards with PFA problems offers significant return on investment because it delivers reduced PFA pressure, increased fruit set, increased yield, improved balance between nut production and tree canopy management and increased net returns.

Akca *et al.* (2012) investigated the effects of AVG (Aminoethoxyvinylglycine) and GA3 treatments on pistillate (female) flower abortion in "Sebin" walnut cultivar. The treatments were carried out just before the receptivity period of female flowers of cv. 'Sebin'. The highest fruit set was obtained from 125 ppm AVG combined 270 ppm GA3 treatment. It was found that AVG alone or in combination with GA3 is effective treatments in reducing female flower abortion for 'Sebin' walnut cultivar.

Emasculated and bagged female flowers (5 %, 50 % and 100 %) were pollinated at receptivity and whole trees were sprayed with 125 ppm ethylene inhibitor AVG (amino-ethoxy-vinyl-glycine) when anthesis reached to 5-30 per cent. Application of AVG reduced PFA from 82.4 to 43.6 per cent in average and increased fruit set significantly in all cultivars except 'Sen-1' (Gun *et al.*, 2010).

San and Dumanoglu (2007) obtained highest percentage of germination (69.1 %) was determined with desiccated embryos originating from open-pollinated seeds of Bilecik genotype on Driver and Kuniyuki walnut basal medium containing 8.6 μm GA_3.

11.7.2. Effect on seed germination and seedling growth

Gao *et al.* (2003) studied the effect of plant growth regulators on seed germination and seedling growth of walnut. Walnuts cv. Jianshi were soaked in water (control), IBA at 80 mg/kg, IAA at 100 mg/kg, NAA at 80 mg/kg, ABT root-growing powder at 1 g/kg, or 6-BA at 5 mg/kg. The 6-BA showed the highest germination percentage (44.5%), which was 10.8% higher than that of the control. The effect of other plant growth regulators was not very different from the control. IAA affected seedling growth the most, producing seedlings 37.5% taller than the control. IBA resulted in seedlings 30.7% taller than the control. No significant difference was observed in seedling growth among 6-BA, ABA, IBA, NAA and the control.

11.7.3. Increase in efficiency in mechanical harvesting

It is necessary to benefit from mechanization applications to decrease costs in walnut growing. Mechanical harvest especially plays a very important role in decreasing cost of harvesting in the growing period. The kernel of walnut ripens approximately three weeks before the husk (outer green skin) cracks. The husk not ripening at the same time with the kernel makes both traditional and mechanical harvests difficult. Yalcin *et al.* (2014) found that ethephon at 1000 ppm dose, applied with a turbo pulverizer which was P.t.o. driven until the run-off of the whole trees in the early morning hours, and concluded that applying ethephone increased the efficiency of the machine. It was investigated that facilitating of disjunction of the fruit by ethephon application increased the success of the mechanical harvest. It was also determined that positive contribution of ethephon was apparent in the hulling of the husk.

11.8. Insect pests and their management

Pest and disease management is a problem in walnut because of giant trees. In controlling, pest and disease, the sanitary measures can play an effective role but are often overlooked. Sanitary measures involve the planting of healthy trees, precautions against the transfer of disease and insects through tools used in pruning and cultivation, removed of diseased branches and leaves by pruning and destroying these by pruning.

11.8.1. Codling moth (*Cydia pomonella*)

It is the key pest of walnut and is the most economically important insect pest and must be controlled in many early blooming varieties. Three generations of moth damage the fruits. Its first generation larvae enter the growing fruit through the blossom and cause dropping of immature fruits. The second and third generations of pest enter the nut through the side of the husk. Its larvae are whitish to pink in color with mature larvae being about ¾ inch long.

It can be checked with pheromone traps which are utilized to assess population levels and calculate degree days for timing treatments. Synthetic pheromone traps should be installed in May. Insecticides like diazinon or carbaryl (0.2 %) or chloropyrifos (1.75 to 2.0 lb. ai./acre) may be sprayed for its control.

11.8.2. Walnut weevil (*Alcidodes porrectirostris*)

The affected plant parts are fruits and leaves. Its female lays eggs on the fruits. On hatching it feed by boring hole on kernels causing pre mature fruit drop. The adult weevils emerge from fruits and feed on petioles and tender shoots. The grubs are more destructive because it kills new shoots upto the old wood. The grubs are 12-16 mm long creamy white in colour while the adults are 9-12 mm long, black brown in colour. It has two generations, one comes in April-July and second comes in June- September. First generation is more damaging as it feeds on developing kernels while second generation often feed on rind. Kernels turn into black mass.

No cheap and satisfactory control is available so far that the fallen fruits be collected and destroyed at every 10 days interval to kill the grubs. This operation is useful and successful with the combined efforts of all the owners in an area. The pest can be checked by spraying twice carbaryl (0.2 %) at 7 days interval. If feasible, spray the trees into quinalphos (0.04 %) during the month of May.

11.8.3. Walnut green aphid (*Chromaphis juglandicola*)

Walnut aphid was historically a major pest in walnuts until the successful introduction of a wasp parasite Walnut aphids are greenish, much smaller than the dusky veined aphid and typically found scattered on the lower side of the leaf, whereas the dusky-veined aphid feeds in rows along the mid-vein. Both species overwinter in the egg stage on the tree. After overwintered eggs hatch in early spring and they suck sap from the underside of leaves and tender shoot during April to September. Due to severe infestation, leaves fall, kernel shrivels and sooty mould develops on honeydew.

The granules of disulfotan (5 %) should be mixed thoroughly in the soil around the tree trunk to check the aphid attack. Spray the infested tree with chloropyriphos (0.02 %) or dimethoate (0.03 %) when damage is noticed.

11.8.4. Leopard moth (*Zeuzera* spp.)

The caterpillars bore into shoots and branches and pupate inside them. The attacked leaves and shoots wilt and die. The moth completes a single life cycle in about 2 years. To control the insect, insert cotton wool soaked in carbon bisulphide, paradichlorobenzene or ethyl acetate in the hole. Seal the holes then with mud.

11.8.5. Indian gypsy moth (*Lymantria obfuscata*)

It hatched in the month of March. The caterpillars feed on buds and leaves. The pest hides during day under rocks, stones, soil and attacks the foliage after sunset which results into defoliation. The pest can be effectively checked by applying Endosulfan (0.5 %) and fenvalerate (0.03 %).

11.8.6. Mites

Two spotted mite (*Tetranychus urticae*), Pacific mite (*Tetranychus pacificus*) and European red mite (*Panonychus ulmi*)

Although European red mite can build up to high numbers, it is seldom considered a serious pest. However, both two spotted and Pacific mites can cause almost complete defoliation which exposes trees and nuts to sunburn, reduces nut size, and can interfere with harvest. Pacific and two spotted mites overwinter as adult females in the trees or on the orchard floor. Both species are favored by hot, dry conditions and as the weather becomes warmer, they increase in numbers and move throughout the tree. The leaf size is reduced and turns yellowish which fall prematurely. Severe defoliation early in the season can cause a 25 per cent reduction in yield.

Mites are monitored by counting the number of infested leaf clusters. Minimize dust by oiling orchard roads and maintaining a well-managed ground cover. Well irrigated, vigorous trees are less susceptible to mite damage. Spray the infested tree with Karathene (0.05 %) or chloropyrifos (0.02 %) or dimethoate (0.02 %) at bud bursting stage. Repeat the spray if needed with alternate miticide [propargite (0.25 %) or Dicofol (1.0 %)] during middle of June.

11.8.7. Hairy Caterpiller (*Schizura concinna*)

Larvae of caterpillar are yellow with longitudinal reddish and white stripes. The head and fourth abdominal segment are red. They pass the winter as pupae in the soil and there are three generations per year. Larvae feed on leaves in heavy infestation, trees may be completely defoliated. Repeated defoliation may weaken and kill most of the trees. Control involves collection and destruction of egg masses, pupae, larvae and even female adults. Use of burlap skirts around the tree trunk during May-June and Spray of chlorpyriphos (0.02 %) or endosulfan (0.05 %).

11.8.8. Other minor pests

These pests are not encountered frequently, but it might be useful to know them. *Lecanium coni* – European fruit Lecanium, *Pratylenchus vulnus*- root lesion nematode. The following pests are (usually) found in North America, *Conotrachelus juglandis*-walnut weevil, *Amyelous transitella*- Navel orange worm, *Datuna integerrima*- walnut caterpillar, husk flies are found in USA and Canada: they may spoil the flavor of the kernel and feed in the husk. They may also stain the fruits.

11.8.9. Nematodes

Lesion Nematode (*Pratylenchus vulnus*), Ring Nematode (*Criconemella xenoplar*) and Root Knot Nematode (*Meloidogyne* spp.)

The extent of damage caused by nematodes in walnuts depends largely on the density of the nematode population, soil conditions and rootstock selection. The root lesion nematode builds to its highest population levels on walnut roots and a hypersensitive response by the plant results in a distinct decayed lesion on invaded roots. Symptoms of a nematode infestation include lack of vigor, small leaves, dieback of twigs and a sparse root system, particularly the lack of small feeder roots. Root galls are an indication of root knot nematode. Black walnut rootstocks are a good source of resistance to the root knot nematodes, whereas English walnut parentage can carry a high degree of susceptibility to this nematode.

Paradox hybrid rootstock exhibits greater tolerance to the root lesion nematode than do the black seedling rootstocks. However, it will also exhibit dramatic damage by high populations of root lesion nematode. Cut off trees at their trunks and painting the cambial region with Triclopyr systemic herbicide, to kill the roots completely and after 18 months one can replant without experiencing the replant problem. Methyl bromide is used as a pre-plant treatment when replanting into soils previously in orchard crops. It is applied one to two feet deep, usually with a plastic tarpaulin stretched over the field surface. Once trees are established, there are no effective methods for nematode control except for methods such as cover cropping to improve soil structure, thereby reducing tree stress.

11.9. Diseases

11.9.1. Walnut Anthracnose

This disease also goes by the name of walnut leaf blotch which was caused by fungus *Gnomonia leptostyla*, *Marssonina juglandis* or *Marssoniella juglandis*. This widespread fungus is prominent in Kashmir valley (Parihar *et al.*, 1980) and it causes nuts to turn black and then fall as well as brown blotches on leaves, young shoots and fruits. Initially small, circular light brown spots develop into large blotches involving a major part of lamina. It also causes a great number of leaves to fall. This disease spreads well in wet weather, usually during May and early June and will over winter on leaf litter.

Collect and burn the fallen leaves. Spray Bordeaux mixture (4:4:50) or copper oxychloride (250-300 g in 100 litres of water) or captan (0.3 %) at the leaf unfolding stage, then after two weeks and at full leaf stage.

The effects of 22 fungicides on the mycelial growth and ascospore germination of G. leptostyla in vitro were studied by Nakova and Dimova (2003) and found that the most effective were Anvil (hexaconazole) and Fundasol (benomyl) which control of more than 90%, followed by Topsin M (control of more than 80%).

11.9.2. Walnut blight

Its causal organism is *Xanthomonas arboricola* pv. *juglandis*. It is the major disease affecting leaves and nutlets of walnuts. Walnut blight is spread by wind-blown rain drops, sprinkler water or pollen. The severity of blight each season depends primarily on overwintering populations of the walnut blight pathogen in dormant walnut buds and the presence of free moisture. The disease enters into the tree through stomata, wounds and insect puncture in the presence of free water. The bacteria attack all tissues of leaves, catkins and nuts. Brown and black spots with yellowish green perimeter develop on the leaves. Individual florets die in the distorted infected catkins. Nuts are dropped due to early infection but shriveled and become discoloured due to late infection. As nuts mature, lesions can develop anywhere on the husk surface causing kernels to shrivel or darken which renders them unmarketable. These later infected nuts often harbor navel orange worm. All walnut cultivars are affected, but it is most serious on early blooming cultivars.

Moragrega and Ozaktan (2010) stated that Characteristic disease symptoms consist of an apical necrosis originating at the stigmatic end of the hull. In early stages of fruit development, differentiation of "apical necrosis" from "blight", another walnut disease, requires a detailed observation of externally and internally affected tissues. Recent studies indicate that the bacterium Xanthomonas arboricola pv. juglandis is the microorganism most frequently associated with apical necrosis and may produce the initial infections in young nuts. Fusarium spp. and Alternaria spp. seem also to be involved in the induction of apical necrosis, causing secondary infections, or growing as saprophytes on bacterial-infected tissues, thus enhancing disease symptoms and severity. Nutritional problems and soil characteristics could be partially related to apical necrosis, in addition to microbial infection.

It is controlled by cutting out by damaged parts and burning them. The first spray of Bordeaux mixture (2:2:250) is applied on the opening of first female flower. It should be followed by subsequent sprays of one to two weeks interval. Ginibre and Prunet (2001) stated that Copper hydroxide, even with restricted dose, was more effective. In fact, copper compounds plus EBDC material were the most effective.

Prevention may be attained by using a soil with a pH above 6, avoiding excessive nitrogen, excessive wetness and allowing for good aeration by pruning (Teviotdale *et al.*, 1985).

11.9.3. Black line disease

It is reported in different walnut growing regions in India. This is a dangerous disease caused by a virus (a strain of the cherry leaf roll virus – *CLRV-W*), and it only affects non-regia rootstocks. Plants affected by blackline suffer yellow drooping leaves; stunted growth and early leaf fall. The name of the disease comes from the typical dark line between the stock and grafted tissue. The tree may suffer dieback and death. The disease can be transmitted through infected pollens, seeds and scion wood. For its control virus free scion wood should be used for grafting and budding.

11.9.4. Phytophthora Root and Crown Rot (*Phytophthora* spp.)

About ten different *Phytophthora* species attack walnut trees. The pathogen enters the tree either at the crown near the soil from the major roots or at the feeder roots, depending on the species. Trees affected with *Phytophthora* first show small leaves, sparse foliage and lack of terminal growth. Infected trees may decline for several years or die within the same growing season in which the foliage symptoms first appear. *Phytophthora* can survive in the soil for many years and spreads and infects the trees during moist cool to moderate temperatures in spring and fall, and some infection may occur in the summer depending on species.

To control this disease plant resistant rootstock however, Paradox rootstock is more tolerant than northern California black walnut. Plant on soil with good surface and internal drainage. Plant on ridges to keep standing water from around the base of the trees and manage irrigation to minimize soil saturation. Apply Metalaxyl (1 %) to the soil as a drench.

11.9.5. Die-back or Canker

The disease is caused by *Glomerella cingulata.* The disease is prevalent in Jammu and Kashmir and Uttaranchal. Irregular dark brown spots appear on leaves which involve the entire leaf causing defoliation. Blighting of twigs is also common and young plants may die.To control this disease, spray Mancozeb or copper oxychloride (300 g in 100 litres of water) before rainy season. Repeat the spray if the disease is severe.

11.9.6. Downy leaf spot

It is caused by *Microstroma juglandis*. The disease is serious in Kashmir. The white mould appears in geometric patches on ventral side of leaves which may coalesce turning into large irregular shapes. It can be controlled by spraying twice with bavistin 50 WP (0.03 %) in June at an interval of 15 days.

11.9.7. Powdery mildew

The causal organism of this disease is *Phyllactinia ruboris*. Grayish to white powdery spore mass appears on upper and lower surface of leaves. Spraying the trees with wettable sulphur (300 g/100 litres of water) or carbendazim fungicide (50 g/100 litres of water) was found effective. The spraying should be done (i) during the late dormancy (ii) at bud swelling stage (iii) at petal fall and (iv) two weeks later.

12. Physiological Disorders

The most serious postharvest physiological disorder that affects walnut quality is oil rancidity. The problem appears to be caused by poor seed storage conditions, elevated temperature and relative humidty and failure to use CA with reduced O_2 concentration.

13. Harvesting

13.1. Maturity

Walnuts are harvested when 10-15 per cent of the fruits have dropped naturally, however, maturity and harvesting of fruits starts from August to October depending on the elevation and cultivars. For production of quality walnuts, harvesting at proper maturity and subsequent handling of the product are of prime importance. Any delay in picking after maturity of kernels, deteriorates the quality and increases the incidence of mould and pests. Two maturity indices are commonly used for determining the harvesting date of the walnut. A nut is rightly covered with a thick hull which splits on maturity causing the fall of nut on the ground. When about eighty percent of the hulls have cracked exposing the nuts, it is the time for walnut picking. But the synchrony between hull and kernel maturity is not necessary. In higher altitudes, kernel matures earlier to hull dehiscence while in the lower altitude hull dehiscence takes place prior to kernel maturity. Proper maturity is therefore, assessed by observing the packing tissue between and around the kernel halves which turns brown on maturity. Depending upon the climate, the harvesting is usually done after about two weeks of browning. The kernel becomes darker if the harvesting is further delayed and deteriorates the quality.

13.2. Manual Harvesting

Walnut is mostly harvested in the months of September and October, earlier at lower altitudes and later at higher altitudes. The preparation of orchard floor is important for facilitating the harvest. The ground should be leveled and cleared of all external material, such as, dried leaves, twigs, pebbles, etc. Fruits are harvested by thrashing the nuts from the branch bearing fruits with bamboo pole. The fallen nuts are collected in huge heaps in the orchard and covered with leaves and grass for hull fermentation taking about 9-15 days period. Dehulling, washing and cleaning are carried out manually causing staining of hand, clothes and nuts. Sometimes in orders to improve the appearance of nuts, these are bleached with either alkali or acid solution. After cleaning, the nuts are dried under sun for many days by spreading them on a canvass sheet or hard floor. Occasionally the nut drying is not complete due to lesser number of sunny hours and lack of warmth in autumn season.

13.3. Machine Harvesting

All operation such as harvesting, dehulling, washing, cleaning, drying, bleaching, grading and packing can also be performed by machines which are desirable for standardizing the product for export. The use of ethephon (2-chloroethylphosphonic acid) has greatly facilitated the walnut picking (Martin, 1971). The spray of 200-500 ppm ethephon on kernel maturity advances harvesting by about ten days, makes harvest of entire crop possible at a time and promotes hull splitting (Sibbet *et al.*, 1978; Woodroof, 1979). Ethephon is useful for both mechanical harvesting and dehulling. Dipping of the nuts, while the hulls are still on, in a 1500 ppm ethephon solution loosens the hulls in three or four days (Woodroof, 1979). The hulls are removed by motor driven abrasion hullers; the bleaching is done by dipping the dried nuts in 26 to 28 per cent calcium hypochlorite followed by a thorough rinsing in one per cent sulfurous acid for redrying, warm air at a temperature range of 32-38°C is circulated around the individual nut in a drier till the moisture content is 8 per cent. If the temperature in the drier goes more than 43°C, the kernels will develop rancidity becoming unfit for consumption. After sorting out the small, misshapen, dark coloured, deeply furrowed and off types, the nuts are size graded. As there is an increasing demand for the kernels, these can be graded for colour and quality with an 'electronic eye' before packing in wooden boxes.

14. Yield

Seedling trees start bearing nuts at the age of 10-12 years and full commercial production 18-20 years from planting, while grafted walnut starts production from 4-5 years after planting and it takes 8-10 years for commercial production. Because of differences in tree age, size, soil, fertilization and irrigation practices, a great variation is found in the yield of walnut. A fully grown big size tree having 40 year old tree can produce as much as 100-150 kg nuts while an average yield is only around 40 kg (Metzer, 1977). The nut yield of about two and half tons per hectare is obtained from 8 or 9 year-old trees in the USA under quadruple planting system (Serr, 1969).

15. Drying, Storage and Marketing

The dried nuts are graded according to cultivar, size and colour. Ungraded product fetches lower price in the market. Delay in drying causes rapid loss in nut quality and makes it susceptible to mould. The nut quality deteriorates even within nine hours of harvest if walnut remains on orchard floor at a temperature of 32.2°C or more. Walnuts are dried to remove excess moisture from the shell and kernel. Drying stabilizes the product weight, prevents deterioration in kernel quality makes bleaching efficient and prolongs storage life (Olson *et al*., 1998). The nuts are dried up to 8 per cent moisture level. In the past, sun drying was prevalent which was time consuming, labour intensive and often resulted in poor nut quality. Three types of drier, viz, stationery bin, forklift pallet bin and trailer, are now popular for nut drying which is achieved within 24 to 48 hours (Thompson *et al*., 1985). Drying air temperature should not exceed 43.3°C because kernels become rancid and inedible at higher temperature.

The nuts are disposed of in the ensuing winter season, the problem of their shelf life is seldom felt. The shelf life of nut meat is examined on the basis of fatty acids, colour and taste. The quality of nut meat, however, deteriorates due to darkening and rancidity which are affected by air, moisture, heat and light. With an antioxidant, kernels at 2.8-4 per cent moisture can be kept for 18 months at 4.4° C with about 3.1 per cent humidity (Woodroof, 1979).

Walnuts are ordinarily stored in gunny bags in small ventilated rooms. Packing cases are used to send produce to distant markets. The boxes are lined with paper on all sides to avoid shaking and breaking of walnuts. The nuts which fall on tile ground with their husk intact should be stored and marketed separately as otherwise they will lower the price of the whole lot.

16. Special problems

16.1. Pistillate flower abscission

Pistillate flower abscission has emerged as a very serious non-pathogenic problem of some walnut cultivars. It has been shown to cause up to 90 per cent abscission of female flowers followed by a similar reduction in yield (Catlin *et al*., 1987) under normal circumstances in the field. PFA is characterized by cessation of pistillate flower growth when the ovary is about 3-4 mm in diameter followed by flower abscission 1-2 weeks later (Catlin and Polito, 1989, Polito *et al*, 2005, Rovira and Aleta, 2006). Near the time of abscission, PFA type flowers show signs of necrosis at the tips of the stigma, in the integument, and throughout the placental invaginations. Flower abscission due to PFA is clearly distinguishable from abscission that results from lack of pollination. Unfertilized flowers usually expand to approximately 7 mm in diameter and remain attached for 3 weeks or longer.

Pistllate flower abortion (PFA) is the loss of nut producing pistillate flowers early in the season, typically 2 to 3 weeks after bloom. PFA is sometimes taken to mean pistillate flower abscission, because the pistillate flowers ultimately abscise. PFA flowers emerge and grow normally to an early receptive stage, stigma expands partially and the pistil grows to a diameter of about 1/16 inch. At this stage differences between PFA flowers and normal flowers become apparent. PFA flowers stop soon afterward, they become necrotic and drop. Losses to PFA can be very high, especially in 'Serr' cultivar, which appears to be the most susceptible. The result of recent research about PFA, provides orchard manages with information to improve their understanding of individual situations.

High pollen loads can also cause pistillate flower abortion (McGranahan *et al.*, 1994). According to Polito *et al.* (1998) PFA loss is stimulated by an excess of pollen in the stigmas, causing decomposition and the production of ethylene, which spreads to the rest of the flower, generating high levels of this hormone and producing detritions and harm to vegetal tissue. Rovira and Aleta (1997) and Polito (1998) observed in field conditions losses was up to 90 per cent by abscission owing to excess pollen, noting that this situation varies widely among varieties, sites and years.

The formerly popular cultivar 'Serr' is thought to be most susceptible but most of the commercial cultivars are susceptible to some extent. Catlin and Olsson (1990) measured PFA in four cultivars for seven consecutive years and reported high (54 per cent) PFA percentage in Serr cultivar. Researchers found that response of the cultivars to PFA varies with cultivars. 'Chico' showed the lowest levels while 'Chandler' and 'Vina' were intermediate (Polito *et al.*, 1996). Similarly, Spanish selections were the most affected by PFA showing 73.4 per cent of flower loss while Chilean selections had only 6.8 per cent flower drop due to PFA. Californian and French cultivars showed medium values but closer to Chilean selections (Rovira and Aleta, 2006). PFA was found between 11 to 92 per cent among Iranian genotypes (Hassani *et al.*, 2006).

PFA can be minimized in existing orchards by either mechanically shaking catkins from pollinizers, from both pollinizers and main cultivar or removal of pollinizer rows altogether (Sibbett *et al.*, 1995).

16.2. Dichogamy

Dichogamy is another major problem of walnut crop. Earlier a great diversity was reported in the opinion of dichogamy in *Juglans* as to its extent and type. Many growers believed that light crops were due (in some cases) to sterility or incompatibility of pollen and stigma or that pollination under natural conditions was inadequate or somewhat dichogmay is responsible. Dichogamy is often the cause of low yields in orchard planted. Dichogamy is mainly attributed to the maturation of the male and female flowers at different time or when blooming period of each variety differs from that of the others in the orchard. Regarding its extent (varies from zero to 100 per cent) and type (protandrous, protogynous and homogamous) there is great variation. Most of the varieties tend to be protandrous in nature, few are protogynous and rare homogamous.

Dichogamy varies with the age of the tree, it is more pronounced in young than in old trees. Variation in dichogamy with the age of the tree is especially pronounced in some of the protandrous varieties. More specifically, dichogamy is affected by temperature, amount and intensity of sunlight, winds, air drainage, humidity of the atmosphere, soil moisture and perhaps such factors as type of soil and rootstock. Dichogamy varies not only according to the season, but according to the geographic location. In general, lower provinces emphasize protogyny and decrease protandry, while in the interior areas protandry is more pronounced and protogyny diminished. During the same season a variety having a protogynous tendency may be protogynous in low valleys areas and protandrous in the northern part.

Because no variety or locality is always free from dichogamy, it would appear that any grouping is better than none at all, but to secure the best results, careful grouping based on data collected for an area is necessary. Inter-planting early and late varieties is not advisable because in some seasons their blooming periods don't overlap. Probably the best combinations are those which bring together early and intermediate bloomers or intermediate and late bloomers or all three.

17. Post-harvest management

Post-harvest management of walnut includes dehulling, drying, sorting, grading and packaging.

17.1. Dehulling

After harvesting, walnuts are collected on a flat ground and put in a heap for few days (8-10 days) and then washed under the running water for the removal of hull. Sometimes light hammer is also used to remove the hull or nuts can also be driven over with an automobile to crush the soft hulls. After washing, the dehulled nuts are dried under direct sun or dried in gas dryers for 24 hours or less.

17.2. Drying

After dehulling, the nuts are dried. Delay in drying causes rapid loss in nut quality and makes it susceptible to mould. Once the hulls have been removed from either type of walnut, the nuts must be dried. The nut quality deteriorates even within nine hours of harvest if walnut remains on orchard floor at a temperature of 32.2°C or more. Walnuts are dried to remove excess moisture from the shell and kernel. Drying stabilizes the product weight, prevents deterioration in kernel quality makes bleaching efficient and prolongs storage life (Olson *et al.*, 1998). The nuts are dried up to 8 per cent moisture level. In the past, sun drying was prevalent which was time consuming, labour intensive and often resulted in poor nut quality. Three types of drier, viz, stationery bin, forklift pallet bin and trailer, are now popular for nut drying which is achieved within 24 to 48 hours (Thompson *et al.*, 1985). Drying air temperature should not exceed 43.3°C because kernels become rancid and inedible at higher temperature.

The nuts are disposed of in the ensuing winter season, the problem of their shelf life is seldom felt. The shelf life of nut meat is examined on the basis of fatty acids, colour and taste. The quality of nut meat, however, deteriorates due to darkening and rancidity which are affected by air, moisture, heat and light. With an antioxidant, kernels at 2.8-4 per cent moisture can be kept for 18 months at 4.4° C with about 3.1 per cent humidity (Woodroof, 1979).

17.3. Sorting and Grading

Walnuts are usually sorted by size and colour, which can be done manuallyor through a conveyor belt as too dark and too green by use of an electronic eye. A blower will remove any worm infested or empty walnuts as their lack of weight lets them be blown away. Then the walnuts are sorted by handand then stored in large cargo containers. These containers have large grates in the bottom that allow air to flow through. A hot air blower is attached to the container and the walnuts are then dried. This is to reduce the acidity that naturally occurs in a too-green walnut.

Walnuts are transported to a packing plant where they are graded based on usage, in-shell or shelled. Shelled walnuts are further graded by color. Walnut kernels are screened and separated into different sizes.

17.3.1. In-shell walnuts-Following drying, sizing of the in-shell nut occurs. In-shell walnuts are sized as jumbo, large, medium or baby according to USDA standards.

17.3.2. Shelled walnuts-Walnuts are mechanically cracked as needed. The shelled material is air-separated and screened for size and sent to electronic laser-sorting units for kernel color and shell removal. The walnuts are certified to meet USDA grade standards and customer specifications. Product is then packed for shipment to the market place.

The Indian walnuts are categorized in to 4 categories viz, paper shelled, thin shelled, medium shelled and hard shelled. Walnuts produced in India have different sizes and shapes and are categorised into paper-shelled, thin shelled, medium-shelled and hard shelled. These are classified on the basis of their size and shell cracking rate also. The main grades are:

1. Indian special light half, 2. Indian light broken, 3.Indian light pieces, 4. Indian light crumbs, 5. Indian light my-fire, 6. Indian light amber halves,7.Indian light amber broken, 8.Indian light amber pieces.

17.4. Packaging

Walnut kernels must be packed in such a way as to protect the produce properly.Correct selection, design and construction of the packaging is just as important for loss-free transport as are the requirements placed on the product itself.Walnuts are the packaged by weight in giant sacks for shipping. Generally walnuts are packaged in, net bags (1.5 - 2.5 kg capacity), poly sacks (5 and 10 k capacity g), cartons (10 kg capacity) and flat jute fabric bags (55 kg capacity).For export purpose, these are packed in double gunny bags. Walnuts are consumed in the winter season, so the problem of their shelf life is seldom felt. The quality of nut meat, however, deteriorates due to darkening and rancidity which are affected by air, moisture, heat and light.

If wooden packaging is used, the produce must be separated from the bottom, sides and lid by paper or suitable protective material.The materials used inside the package must be new, clean and of a quality such as to avoid causing any external or internal damage to the produce. The use of materials and particularly of paper or stamps bearing trade specifications is allowed provided that the printing or labelling has been done with non-toxic ink or glue.Kernels may be packed in airtight sealed containers, in a vacuum or in an inert gas.

Walnuts are ordinarily stored in gunny bags in small ventilated rooms. Packing cases are used to send produce to distant markets. The boxes are lined with paper on all sides to avoid shaking and breaking of walnuts. The nuts which fall on tile ground with their husk intact should be stored and marketed separately as otherwise they will lower the price of the whole lot.

Products that contain walnuts, especially roasted walnuts, are subject to oxidation which can reduce shelf life and cause rancidity. To prevent rancidity and extend finished product shelf life, the product should be packaged in materials that do not allow light or air to come into contact with the product. Packages that can be resealed are ideal for snack mixes which may be opened and closed several times by the consumer. Nitrogen flushing (replacing the oxygen with nitrogen) can also extend the shelf life of unopened sealed product. Walnuts that are formulated with coatings like chocolate, starches, gums, egg whites and sugar will have a longer shelf life than plain or roasted walnuts because they are protected by the coated exterior.Fresh taste of walnuts can be maintained by keeping them cold. Heat causes the fat in walnuts to change structure, which creates off odors and flavours. Fresh walnuts smell mildly nutty and taste sweet.

17.5. Storage

Low water content and high fat content of the kernel make it relatively metabolically stable and able to tolerate low temperatures. The primary objectives of storage are to maintain the low water content attained after preliminary drying (for enzyme activity suppression, retention of texture and reduction of microbial activity) and limit exposure to O_2 to minimize rancidity. The optimum temperature range for storage is 0 to 10°C, with the lower temperature being better. Within this temperature range, a 50 to 65 per cent relative humidity will maintain walnuts at 4 per cent moisture (Beuchat, 1978). Lopez *et al.,* (1998) modeled sorption isotherms for walnut kernels and concluded that optimum stability and texture were retained in a storage environment of 10°C and 60 per cent relative humidity in air. The best place to store walnuts is in refrigerator or freezer, depending on when to use them. If walnuts are going to use right away, place them in your refrigerator. If walnuts to be store for a month or longer, store in freezer.

18. Standards for export

Generally the export specification of walnut depends upon quality, size, tolerance and presentation of the shell or kernels.

A. Quality

The purpose of the standard is to define the quality requirements for walnut kernels at the export control stage, after preparation and packaging.

Minimum requirements

In all classes, subject to the special provisions for each class and the tolerances allowed, walnut kernels must be sufficiently dry to ensure keeping quality, sound, firm, sufficiently developed; shriveled kernels are to be excluded, clean, free from insects or mites, free from damage caused by pests, free of rancidity or oily appearance, free from mould, free of abnormal external moisture. The walnut kernels shall have a moisture content of not greater than 5 per cent.

Classification

Walnut kernels are classified in the three classes as defined below according to their quality and colour.

Extra Class	Class I	Class II
Kernels with superior quality, uniformly light-coloured with practically no dark straw and/or lemon-yellow colour and with no dark brown. They must have characteristic features of variety. No defects allowed in the kernels. Scuffing is allowed on: - quarters and all pieces, - halves, provided it covers no more than 10% of the surface area of the skin.	Kernels with good quality, light brown or lemon yellow in colour. Slight defects may be allowed provided that these do not affect the general appearance. Scuffing is allowed on: - quarters and all pieces, - halves, provided it covers no more than 20% of the surface area of the skin.	Kernels which do not qualify for inclusion in the higher classes, but satisfy the minimum requirements. Light dark brown in colour. Defects may be allowed, provided that the walnut kernels retain their essential characteristics as regards general appearance, quality, keeping quality and presentation. Scuffing is not considered as a defect.

Source: www.nutfruit.org/stand

B. Sizing (styles)

Walnut kernels are classified by size as follows:

halves	kernels separated into two more or less equal and intact parts
quarters	kernels separated lengthways into four more or less equal pieces
large pieces	portions smaller than a "chipped kernel" but larger than a "broken piece"
broken pieces	portions of kernels which can pass through a 8mm sizing screen but not through a 3mm sizing screen
large pieces and halves	a mixture of kernels corresponding to the styles large pieces (iii) and halves (i) and of which the proportion of halves may be specified in the marking

Source: www.nutfruit.org/stand

C. Tolerances

Tolerances in respect of quality, colour and type shall be allowed in each package for produce not satisfyingthe requirements of the class indicated.

Quality and Colour Tolerances

Defects allowed	Tolerances allowed (% by weight of kernels)		
	Extra	Class I	Class II
(1) Kernels not satisfying the minimum requirements, which include not more than:	4	6	8
- Rotten kernels 0.5 1 b 2 b	0.5	1 [b]	2 [b]
- Mouldy kernels 0.5 1 b 2 b	0.5	1 [b]	2 [b]
- Shell fragments or foreign matter	0.1	0.1	0.1
(2) Kernels darker in colour,	8	9	10
(3) Scuffing (halves only)	10	10	-

Source: www.nutfruit.org/stand

Size tolerance (styles)

For all sizes, a minimum percentage of kernels corresponding to the style indicated in the marking are required and a maximum percentage by weight of kernels different from the style indicated is tolerated:

D. Presentation

Uniformity

The contents of each package may be uniform and contain only kernels of the same origin, crop year, quality, size and when applicable of the same variety and commercial style. Uniformity of colour is compulsory for Extra Class and Class I. However, with regard to shape, 'halves' which pass through a 15 mm mesh and 'chipped kernels' may be included without limitation in consignments of 'large pieces'. The visible part of the contents of the package must be representative of the entire contents.

Kernels must be presented in small unit packages of uniform weight intended for sale directly to the consumer or packaged in bulk.Each package must bear the following particulars in letters grouped on the same side, legibly and indelibly marked and visible from the outside:

A. Identification

B. Nature of produce

C. Origin of produce

D. Commercial specifications

19. Processing

After drying properly (moisture content upto 8 per cent), walnuts are then returned to the processing center and put into a sorter that again sorts them by size. After they are sorted, they drop down into a giant cracking machine. This machine is conical in shape with a solid rotating center. As the walnuts fall into a more narrow space, they are cracked against the side of the rotator's bin and fall below to a secondary rotator which captures the walnuts that have not quite been cracked. From the cracked kernels following products are prepared

Walnut Halves: Walnut halves in products demonstrate its natural shape and visual appearance. Applications include trail mixes, confectionary, garnishing and decorating baked goods, and whole nut snack mixes.

Large Pieces: Typical applications include trail mixes, energy bars, frozen dairy inclusions and baked good toppings.

Medium Pieces: Use medium walnut pieces in products where equal flavor and texture is needed throughout the product. Baking mixes, pizza toppings, chocolate based candy bars.

Small Pieces: Use small pieces in seasoning blends, baking mixes, breading mixes and frozen dairy toppings. Small pieces work especially well in premixed bakery blends because the small size allows for even distribution throughout the entire mix allowing the nut flavor and texture to come through with every bite.

Roasted Walnuts: Roasting walnuts heightens the flavor and aroma while increasing the crunchy mouth-feel. Roasted walnuts work well in applications that will not be further baked such as no-bake energy bars and ready-to-eat snack mixes. For items that will be subjected to further heating (like cakes and muffins) roasted nuts can be used but should be limited to the interior of the baked item and not part of the surface area that will be directly exposed to the heat..

Walnut Meal and Gluten Formation: The fat in walnut meal shortens or "interrupts" the gluten strands preventing them from getting longer and will yield tender products. A combination of gluten free flour and walnut meal may result in a product without enough structure, thus R&D work is always necessary to achieve final desired texture.

Blanched Walnuts: Walnuts cannot be easily blanched because their general shape and contour makes it difficult to remove the skin. The best way to blanch walnuts is to either first warm them in the oven or boil them in water and then rub the skins off. Blanched walnuts can be used to help developers create refined baked good products that will not have any discoloration from the walnut skins. Sometimes lighter colored walnuts are available and those can be used as an alternative to blanching.

20. Effect of climate change

Climate change and increased climate variability in terms of rise in temperature, prolong winter season, heavy snow fall, shifting rainfall pattern and incidences of extreme weather events like severe drought, devastating flood, etc. have now started posing a real threat to the production walnut crop. There is few published data on the effect of climate change on the walnut trees. From the last one decade early flowering was noticed in few areas of Kashmir valley. In a study in California, Luedeling *et al.* (2011) reported that in the coming five decades the pest pressure like codling moth(*Cydiapomonella* L.), navel orangeworm (*Amyeloistransitella* Walker), two-spotted spider mite (*Tetranychusurticae*Koch), and European red mite (*Panonychusulmi*Koch)) can be expected to increase substantially.

21. Economic

Expenditure and income from walnut cultivation has been presented in Table 9. An estimation cost of a newly established orchard was calculated when orchard start full bearing. Grafted walnut trees starts bearing at the age of 3-4 year and full bearing after six years. In one kanal of land (1 Kanal=1.5 bigha or 20 Kanal= 1 hectare), fourteen plants were accommodated in a square system of planting at a spacing of 6 m x 6 m. Different inputs alongwith the purchasing of 14 plants (14 x @ Rs. 250 = 3500/-) including measuring tape (@ Rs.150/-), leveling and ploughing of land (@ Rs. 300/-), transportation cost (@ Rs. 50/-), two labours/year (@ Rs. 300/-) required for digging of pits, planting, hoeing, weeding etc., (6 x 2 x @ Rs. 300 = 3600/-), organic manure (FYM) @ 10 kg/ plant in 1st year and 10 kg/plant increased every alternate year and it costs Rs. 3360/-, chemical fertilizers viz. Urea (@ Rs. 150/-), DAP (@ Rs. 180/-), MOP (@ Rs. 180/-), weedicides (6 x @ Rs. 100 = 600/-), fungicides/ pesticides (6 x @ Rs. 200 = 1200/-). The total investment for six years in a kanal was Rs. 13,270/-. As bearing starts in the 4th year, fourteen plants produced 280, 1120 and 2100 nuts in the 4th, 5th and 6th year, respectively. The total weight of nuts produced was 4.2 kg, 16.8 kg and 31.5 kg which generated Rs 1260/-, Rs. 5040/- and Rs. 9450/-, respectively. The total gross return was Rs 15,750/- with a cost of cultivation Rs 13,270/- thus the net profit during the period estimated to Rs. 2480/-.

Table 9: An estimation cost for growing of walnut plants per kanal for six year

S.No.	Components	Quantity	Cost of item	1st year	2nd year	3rd year	4th year	5th year	6th year	Total cost
1.	Grafted Plants	14	Rs. 250	3500/-	-	-	-	-	-	3500
2.	Measuring Tape	01	Rs. 150	-	-	-	-	-	-	150
3.	Ploughing and Levelling	Rs. 300/kanal		300	-	-	-	-	-	300
4.	Transportation cost	Rs. 50		-	-	-	-	-	-	50
5.	Number of labors (hoeing & weeding)	02	Rs. 300	600	600	600	600	600	600	3600
6.	Chemical fertilizer	Urea	Rs. 15/kg	15	15	15	30	30	45	150
		DAP	Rs. 30/kg	30	30	30	30	30	30	180
		MOP	Rs. 20/kg	20	20	20	40	40	40	180
7.	Organic manure	FYM*	Rs. 2/kg	280	280	560	560	840	840	3360
8.	Weedicides		100	100	100	100	100	100	100	600
9.	Fungicides & Pesticides		200	200	200	200	200	200	200	1200
							TOTAL INVESTMENT			13,270
10.	Number of walnut produced			-	-	-	20	80	150	15,750
11.	Total number of nuts			-	-	-	280	1120	2100	
12.	Total weight of nuts			-	-	-	4.2 kg	16.8 kg	31.5 kg	
13.	Total money generated			-	-	-	1260.00	5040.00	9450.00	
	Total money generated	-		**Total investment**				=	**Total income**	
	Rs. 15,270.00	-		**Rs. 13,270.00**				=	**Rs. 2,480.00**	

* FYM was applied @ 10 kg/plant in the first year and @ 10 kg increased every alternate year

*Average size of nut produced=15 g and rate of one kg of walnut = 300/-

22. Future thrust

Walnut is a major dry fruit (nut) crop of the country which has much potential and is successfully grown in all the temperate states. The walnut industry in India is definitely going to grow rapidly. Because the fresh domestic market is limited due to the lack of proper marketing facilities as well as great demand for inshell walnuts of high quality with cream colored kernels in the international market. At present majority of the produce come from the plantation of seedling origin, however in the last two decades few varieties have been developed from selection procedure having good nut and kernel quality traits. Introduction of few new varieties are also helpful directly in the production of the quality nuts and also in the breeding programs. A strategic research and development program and ensured implementation are needed to popularize walnut cultivation in India in order to compete within the World Trade Organization regime.

1. Identification and testing of superior genotypes based on inshell nut weight, creamish kernel color and identification of superior trees from naturalized populations from the walnut growing areas need to be initiated to extend the new plantations.
2. Collection, characterization, conservation, evaluation and utilization of germplasm.
3. Development of precocious, lateral bearing, self-fruitful and cross compatible varieties having better in shell nut as well as kernel characteristics.
4. Development of varieties resistant biotic and abiotic stresses
5. Clonal rootstocks and dwarf cultivars of walnut need to be introduced for adaptive research and high-density plantation culture.
6. Sound research programs on cultural techniques, propagation, pollination, nutrition and postharvest handling and storage of walnuts need to be strengthened and initiated for commercialization
7. Developing of methodologies to overcome negative impacting factors of climate variables on productivity.
8. Establishing/strengthening the mother blocks.
9. Assessment of cost effective water and nutrient management technologies under different agro-climatic conditions.
10. Production of healthy quality planting material.

11. Expansion of walnut growing in areas of northeastern hill region of India especially in the states of Meghalaya, Arunachal Pradesh, Sikkim and Darjeeling district of West Bengal need to be commercialized as scattered plantings of walnuts in farmers' backyards were already established.

References

Abaev, M.N. and Khvorova, N.E. 1983. Khimiya V sef, skom Khozyalstve No. 4: 31-33.

Abaev, M.N., Lukin, N.F. and Khvorova, N.E. 1976. *Agrokhakhamiya,* No. 12 pp. 113-116.

Achim, G. and Botu, I. 2001. Results in walnut propagation by using different methods. *Acta Hort.*, **544**: 504-510.

Ahmad, M.F., Iqbal, U. and Khan, A.A. 2007. Response of different environments and dates of patch budding on success in walnut. *Indian Journal of Horticulture,* **64(3)**: 286-289.

Ahmad, N., Singh, S.R., Srivastava, K.K., Shagoo, P.A. and Hayat, S. 2012. Effect of different environments, grafting methods and times on sprouting, graft success and plant growth of walnut (*Juglans regia*). *Indian Journal of Agricultural Sciences,* **82(12)**: 1022-26.

Akca, Y.2014. 'Sebin' Persian walnut. *Acta Horticulturae,* 1050: 139-142.

Akca, Y; Ozgen, M; Erturk, U; Ercisli, S. 2012. The effects of AVG and GA3 treatments on pistillate (female) flower abortion in "Sebin" walnut cultivar. *Acta Scientiarum Polonorum Hortorum Cultus,* **11**(4): 179-185

Akopov, I. E. 1981. Hemostatic plants. Meditsina. Tashkent. (in Russian).

Alasalvar, C. and Shahidi, F. 2009. Tree Nuts: Composition, Phytochemicals and Health Effects. CRC Press. Boca Raton, FL, USA.

Aldrich, T. M. *et al.*, 1976. Irrigation. Diamond walnut. *News,* **58**(2): 33-34.

Ali, S.B.G.M., Vahdati, K., Hassan, B.S., Kalantari, S. and Leslie, C. 2010. Enhancement of maturation and germination of somatic embryos in Persian walnut (*Juglans regia* L.) using osmolites, hormones and cold treatments. *African Journal of Food Science,* **4**(12): 735-743.

Aly, M.A.M., Fjellstrom, R.G., McGranahan, G.H. and Parfitt, D.E. 1992.Origin of walnut somatic embryos determined by RFLP and isozyme analysis. *Hort. Science,* **27**(1): 61-63.

Amiri, M.E. and Gharati, S. 2012. Influence of medium composition on multiplication of walnut (*Juglans regia* L.) *Journal of Medicinal Plants Research,* **6(8)**: 1482-1485.

Anonymous, 2011. Walnut. *In:* Temperate Fruits Package of Practices. Directorate of Extension Education, SKUAST-Kashmir. Pp. 56-62.

Anonymous, 2013.http://fruitandnuteducation.ucdavis.edu

Aslamarz, A. A; Vahdati, K; Hassani, D; Rahemi, M; Mohammadi, N. and Leslie, C. 2011. Cold hardiness and its relationship with proline content in Persian walnut. *European Journal of Horticultural Science,* **76**(3): 84-90.

Atefi, J. 1990. Preliminary research of Persian walnut and correlation between pair characters. *Acta Hort.*, **284**: 97–104.

Atefi, J. 1997. Comparison of hypocotyls and hot callus cable graft with traditional grafting method. *Acta Hort.*, **442**: 306-312.

Avanzato D. and Tamponi, G. 1988. The effect of heating of walnut graft unions on grafting success. *Acta Hort.*, **227**: 79-83.

Avanzato, D., Ducci, F., Gorian, F., Gui, L. Major, A., Malvolti, E., Mezzalira, G., Pollegioni, P. and Proietti, R. 2006. Propagation ability of selected walnut hybrids (*Juglans regia* L. x *J. nigra* L.). *Acta Hort.*, **705**: 359-364.

Avanzato, D.2010. Traditional and modern uses of walnut. *Acta Horticulturae*, **861**: 89-96.

Barbas, E., Chaillou, S., Cornu, D., Doumas, P., Jay-Allemand, C. and Lamaze, T. 1993a.Orthophosphate nutrition of *in vitro* propagated hybrid walnut (*Juglans nigra* x *J. regia*) trees: Pi (32Pi) uptake and transport in relation to callus and shoot.

Barbas, E., Jay-Allemand, C., Doumas, P., Chaillou, S. andCornu, D. 1993b. Effects of gelling agents on growth, mineral composition and naphthoquinone content of in vitro explants of hybrid walnut tree *Juglans regia-J. nigra* L. *Ann. Sci. For.,* **50:** 177-186.

Bayazit, S., Imrak, B. and Kuden, A. 2005. Determination of grafting times and methods of walnut under Adana ecological conditions. *Bance Ceviz.,* **34**(1): 231-234.

Beede, R.H. 1985. Origin of walnut. *In*: Ramos, D. (ed.). Walnut production manual. Univ. California Pub. 33373: 3-7.

Beede, R.H. and Polito, V.S. 2003. Effect of retain on reducing pistillate flower abortion in 'Serr' walnut. University of California. Fruit and Nut Research Info Center walnut research reports, p.197.

Bemmann, A. 1998. Characteristics and use of walnut timber. *In*: Blaser, J., Carter, J. and Gilmour, D. (eds.) Biodiversity and Sustainable use of Kyrgyzstan's Walnut-Fruit Forests. IUCN, Gland, Switzerland. pp. 107–109.

Beuchat, L.R. 1978. Relationship of water activity to moisture content in tree nuts. Jr. Food Sci. 43: 754-755.

Bhat, A.R. and Ahmad, M.F. 2003. Hamdaan and Sulaiman: New walnut varieties for north-western Himalayas. *Indian Horticulture,* **47**: 33-34.

Brinkman, K.A. 1965. Black walnut (*Juglans nigra* L.). *In*: Silvics of forest trees of the United States. (ed.). Fowells, H.A. USDA, Agriculture Handbook 762 p.

Burmistrov, L. A. 1996. The nut industries of Russia (*Part I*). Yearbook. *West Australian Nut and Tree Crops Association,* **20**:69–80.

Catlin, P.B. and Olsson, E.A. 1990. Pistillate flower abscission of walnut – 'Serr', 'Sunland', 'Howard' and 'Chandler'. *Hort. Science,* **25**(11): 1391-1392.

Catlin, P.B. and Polito, V.S. 1989. Cell and tissue damage associated with pistillate flower abscission of Persian walnut. *Hort. Science,* **24**: 1003–1005.

Catlin, P.B., Ramos, D.E., Sibbett, G.S., Olson, W.H. and Olsson, E.A. 1987. Pistillate flower abscission of the Persian walnut. *Hort. Science,* **22**:201-205.

Chandel, J.S., Gautam, D.R. and Sharma, N.C. 2006. Chip budding: an excellent method of propagation of walnut (*Juglans regia* L.) *Acta Hort.,* **705:** 335-339.

Charlot, G. 1986. Performance of French and foreign varieties in the Isere Department. Compte Rendu d' Experimentation, CTIFL (86012): 9.

Chauhan, J.S. and Sharma, S.D. 1980. Veneer grafting: A new technique for walnut propagation. *Fruit Science Reports,* **7**(2): 45- 48.

Chebotar, E.I. 1974. The time for bench grafting walnuts. *Sadovodstvo Vinogradarstvo I Vinodelie Moldavii,* **10**: 32-34.

Chen QuanMing; Gao XingZhong 2004. Raising walnut planting material by tender rootstock grafting. *China Fruits,* **6:** 36-37.

Cornu, D. 1988.Somatic Embryogenesis in tissue cultures of walnut (*Juglans nigra, J. major* and hybrids *J. nigra* × *J. regia*).*In:* Somatic Cell Genetics of Woody Plants, (ed.) Ahuja, M.R., Kluwer Academic Pub., Dordrecht, The Netherlands, pp. 45-49.

Cornu, D. and Jay-Allemand, C. 1988.Micropropagation of hybrid walnut trees (*Juglans nigra* × *J. regia*) through culture and multiplication of embryos. *Ann. Sci. For.*, **46:** 113-116.

Cornu, D. and Jay-Allemand, C. 1989. Micropropagation of walnut hybrid trees (*J. nigra* x *J. regia*) through culture and multiplication of embryos. *Ann. Sci. (Suppl.)*: 113-116.

Dehghan, B; Vahdati, K; Rezaee, R; Hassani, D. and Papachatzis, A. 2009. Topworking of mature walnut trees as affected by different bleeding control methods and scion cultivars. *Analele Universitatii din Craiova Biologie, Horticultura, Tehnologia Prelucrarii Produselor Agricole, Ingineria Mediului*, **14**: 5-12.

Delbos, R., Kerlan, C, Dunez, J., Lansac, M., Dosba, F. andGermain, E. 1983. Virus infection of walnuts in France. *Acta Hortic.* **130**:123–131

Deng, M.D. and Cornu, D. 1992. Maturation and germination of walnut somatic embryos.*Plant Cell, Tissue Organ Culture,* **28**: 195-202.

Driver, J.A. and Kuniyuki, A.H. 1984. *In-vitro* propagation of Paradox walnut rootstock. *Hort., Science,***19**: 507–509.

Dumanoglu, H. 2000. Dessiccation using saturated salt solutions and improvement germination rate of walnut (*Juglansregia* L.) somatic embryos. *Turk. J. Agric. For.* 24: 491-498.

Erdogan, V. 2006. Use of hot callusing cable in walnut propagation. *Acta Hort.*, **705**: 313-317

Escobar, M.A; Leslie,.C.A; McGranahan,.G.H. and Dandekar, A. M. 2002. Silencing crown gall disease in walnut (*Juglans regia* L.). *Plant Science,* **163**(3): 591-597.

Fan ZhiYuan and Xi-XueLiang. 2004. The new method of raising planting material for walnut. *South China Fruits*, **6**: 93.

FAO, 2015. www.faostat.org.

Felaliev, A.S. 1990. Morphogenesis of *Juglans regia* L. *in vitro*.*Ukr J. Bot.,* **47**(3): 85-87.

Fernandez, H., Perez, C. and Sanchez-Tames, R. 2000. Modulation of the morphogenic potential of the embryonic axis of *Juglans regia* by cultural conditions. *Plant Growth Regul.,* **30**: 125-131.

Forde, H.I. 1975. Walnuts: p. 84–97. *In*: J. Janick and J.N. Moore (eds.). Advances in fruit breeding. Purdue Univ. Press, West Lafayette, Ind.

Freeman, M.W. and Aldrich, T.M. 1985. *In*: Walnut orchard management, Coop. Ext. Serv., University of California. 57-61.

Gandev, S. 2007. Propagation of walnut under controlled temperature by the method of omega bench grafting, hot callus and epicotyl grafting. *Propagation of Ornamental Plants,* **41**: 105-111.

Gandev, S. 2007. Vasden - a new walnut cultivar. *Rasteniev'dni Nauki,* **44**(4): 383-384.

Gandev, S. and Arnaudov, V. 2011. Propagation method of epicotyl grafting in walnut (*Juglans regia* L.) under production condition. *Bulg. J. Agric. Sci.,* **17**: 173-176.

Gandev, S. and Dzhuvinov, V. 2006. Performance of hypocotyls grafting of walnut under uncontrolled temperature conditions. *Acta Hort.*, **705:** 351-353.

Gao-HuanZhang; Bao-XinMei; Hu-XiaoXiang. 2003. Effect of plant growth regulators on seed germination and seedling growth of walnut. *Journal of Hubei Agricultural College,* **23**(3): 161-163

Gautam, D.R. 1990. Studies on the winter and summer vegetative propagation techniques of walnut (*J. regia* L.). *Acta Hort.,* **284**: 27-31.

Germain, E., Jalmat, J. and Marcheu, M. 1975. Divers aspects de la biologieflorale du noyer.*In*: CTIFEL, INVUFLEC (eds.). Noyer, Paris, pp 13-27

Germain, E., Prunet, J.P. and Garcin, A. 1999. Le Noyer. Centre Technique Interprofessionnel des Fruits et Legumes Publication, Paris.

Ginibre, T. and Prunet, J. P. 2001. Chemical control of bacterial blight of walnut, three years of results. : *Acta Horticulturae*, **544** : 409-412.

Glonti, I.S., Tutberidze, B.D. and Kalandarishvili, T.L. 1987.Selection of rootstocks and development of methods for walnut propagation by grafting in western Georgia.*SubtropicheskieKultury,* **2**: 102-110.

Gonzalez R, C; Lemus S, G; Reginato, G. 2008. Pistillate flower abscission symptoms of 'Serr' walnut (*Juglans regia* L. *Chilean Journal of Agricultural Research,* **68**(2): 183-191.

Gruselle, R. and Boxus, P. 1990. Walnut micropropagation. *Acta Hort.,* **284**: 45-52.

Gruselle, R., Badia, N. and Boxus, P. 1987. Walnut micropropagation: first results. *Acta Horti.*, **212**: 511-515.

Gun, A., Erdogan, V., Akcay, M.E., Fddanci, A. and Tosun, D. 2010. Pistillate flower abscission in Turkish walnut cultivars and its reduction by AVG.

Gupta, S. K. 1999. Characterization and evaluation of walnut (*J. regia* L.) germplasm accessions. M.Sc. Thesis, Dr. YSP UHF, Nauni-Solan, India.

Gyambo, T. 2000.Causes of graft failure in walnut.M. Phil Thesis, Reading University, England.

Hassani, D., Eskandari, S. and Jarrahi, K. 2006. Pistillate flower abscission of walnut genotypes. *Acta Hort*., **705**: 257-260.

Heile-Sudholt, C., Huetteman, C.A., Preece, J.E., van Sambeek, J.W. and Gaffoney, G.R. 1986. *In-vitro* embryonic axis and seedling shoot tip culture of *Juglans nigra* L. *Plant Cell Tissue Organ Cult*., **6**: 189-197.

Hou LiQun; Zhao-DengChao; Wang-JunYi; Yang-KeQiang; Han-ChuanMing; Cui-ShuYing 2010. A new fresh-eating walnut cultivar 'Luxiang'. *Acta Horticulturae Sinica,* **37**(7):1193-1194.

Jay Allemand, C; Capelli, P. and Cornu, D. 1992. Root development of in vitro hybrid walnut micro-cuttings in a vermiculite-containing gelrite medium. *Scientia Horticulturae*, **51**(3-4): 335-342.

Jay-Allemand, C. and Cornu, D. 1986. Culture in-vitro of embryos isolated of noyer commun (*Juglans regia* L.). *Ann. Sci. For.,* **43**: 189-192.

Jay-Allemand, C.,Capelli, P. andCornu, D. 1992. Root development of *in vitro* hybrid walnut microcuttings in a vermiculite-containing gelrite medium. *Sci. Hortic.,* **51**: 335-342.

Ji, B. 1980. Walnut cultivation and Utilization in China. *Fruit Vars. Jr.* 34: 98-99.

Johnson, H., Grant, J., and Polito, V. 2006. Pistillate flower abortion and ethylene production in walnut. University of California Fruit and Nut Research Info Center. Walnut research reports, p.101.

Kaur, R., Sharma, N., Kumar, K., Sharma, D.R. and Sharma, S.D. 2006. *in-vitro* germination of walnut (*Juglans regia* L.) embryos. *Sci. Horticulturae,* **109**: 385-388.

Keshavarz, K; Vahdati, K; Samar, M; Azadegan, B. and Brown, P. H. 2011. Foliar application of zinc and boron improves walnut vegetative and reproductive growth. *Hort. Technology,* **21** (2): 181-186.

Kiryakov, D. and Lukova, V. 1974. Chemical weed control in fruit tree nurseries in vegetable transplants. *Rastitelna Zashchita,* **22**(12): 6-7.

Kolov, O.V. 1977. Effect of mineral fertilizers on the growth of walnut seedlings. *Agris,* **1**: 61- 62.

Kumar, A., Kumar, K. and Sharma, S.D. 2005. Extent of fruit set and retention under different modes of pollination in Persian walnut (*Juglans regia* L.). *Acta Horticulturae,* **696**: 327-330.

Kuniyuki, A.H. and Forde, H.I. 1985. Walnut propagation. *In:* Walnut orchard Management. Ramos, D.E. (ed.). Division of Agriculture and Natural Resources, University of California, Davis, CA, USA. 38-45.

Lal, H., Nautiyal, M.C. and Sharma, R.M. 1984. Walnut seed germination- Effect of planting depth and seed position in soil. *Prog. Hort.,* **16**(1): 6- 8.

Land, S.B. and Cunningham, M. 1994. Rooted cutting macro propagation of hardwoods. *In*: Applications of vegetative propagation in forestry. Proc. of the Southern Regional Information exchange Group Biennial Symposium on Forest Genetics. Foster, G.S. and Diner, A.M. (eds.). Published by Southern Forest Experiment Station. New Orleans, Louisiana. Pp. 75-96.

Lange, A.H., Crane, J.C., Fischer, B.B., Roberts, K.O. and Elmore, C.L. 1969. Pre-emergence weed control in young deciduous fruit trees. *Jr. Amer. Soc. Horti. Sci.* 94: 57-60.

Lemus, G., González, C. and Retamales, J., 2007. Control of pistillate flower abortion in 'Serr' walnut in Chile by inhibiting ethylene biosynthesis with AVG. *Adv. Plant Ethylene Res.,* **48:** 305–307.

Li BaoGuo. 2008. Lvzao, an early season walnut cultivar with thin shell. *China Fruits,* **6:** 76.

Li BaoGuo; Qi GuoHui; Guo SuPing and Chen LiYing 2008. A new early ripening, precocious and thin shell walnut cultivar 'Luzao'. *Acta,Horticulturae Sinica,* **35**(7): 1088.

Li, YongTao; Ma, SanMei; Wang, YongFei; Xuan, XiaoLi; Hou, LiQun; Sun, QingRong and Yang-KeQiang. 2012. The dynamics of fat, protein and sugar metabolism during walnut (*Juglans regia* L.) fruit development. *African Journal of Biotechnology,* **11**(5): 1267-1276.

Liang Zhi; Zou YaoXiang and Zhang JiFeng. 2010. Effect of N, P and K fertilizers on walnut tree in southern Xinjiang. *Xinjiang Agricultural Sciences,* **47**(5): 958-963.

Limongelli, F. 1993. A new walnut cultivar: "Malizia". *Acta Horticulturae,* **311** : 46-48.

Liu Hao; Gao Ying; Dong NingGuang; Xu HuZhi and Pei-Dong. 2011. Study on pollination activity of pistillate and staminate flowers in walnut excellent varieties. *Journal of Beijing Forestry University,* **33**(6): 119-123.

Long, L.M., Preece, J.E. and Van Sambeek, J.W. 1995.Adventitious regeneration of *Juglans nigra* L. (eastern black walnut). *Plant Cell Rep.,* 14: 799-803.

Long, L.M., Preece, J.H., Gaffney, G.R. and Van Sambeek, J.W. 1992.Somatic embryogenesis and organogenesis of eastern black walnut (*J. nigra*). *Hort. Sci.,* **27**(6): p. 584.

Lopez, A., Pique, M.T., Romero, A. and Aleta, N. 1998. Modelling of walnut sorption isotherms. *Italian Journal of Food Science,* **10**: 67-74.

Losche, C.K. 1973. Black walnut growth better on deep well drained bottomland soils. North Central Forest Experimental Station, Forest Service-USDA, Research Note NC-154.

Luedeling, E., Girvetz, E.H., Semenov, M.A. and Brown, P.H. 2011. Climate Change affects winter chill for temperate fruit and nut trees. *PLoS one,* **6**(5): e20155.

Mamadjanov, D.K. 2006b. Walnut fruit forests and diversity of walnut tree (*Juglans regia* L.) in Kyrgyzstan. *Acta Horticulturae,* **705**: 173–176.

Martin, G.C. 1971. 2-chloroethylphosphonic acid as an aid to mechanical harvesting of English walnuts. *Jr. Amer. Soc. Hort. Sci.,* **96**: 434-436.

Martin, G.C., Uriu, K. and Nishijima, C. 1980. The effect of drastic reduction of water input on mature walnut trees. *Hort. Science,* **15**: 157–158.

Masarat Jabeen and Tomar, C. S. 2008. Standardization of leaf position and sampling time for macro-nutrient estimation in walnut (*Juglans regia*). *Indian Journal of Agricultural Sciences,* **78**(6): 531-533.

Matheron, M.E. and Mircetich, S.M. 1985. Control of *Phytophthora* root and crown rot and trunk canker in walnut with metalaxyl and fosetyl Al. *Plant Disease,* **69**:1042-1043.

Maurer, K.J. 1976. The pruning of walnut trees finishes in April. *Baumschulpraxis,* **6**(4): 110-112.

McGranahan, G., Voyiatzis, D., Catlin, P. and Polito, V. 1994. High pollen loads can cause pistillate flower abscission in walnut. *J. Amer. Soc. Hort. Sci.,* **119**: 505-509.

McGranahan, G.H. and Catlin, P.B. 1987. *Juglans* Rootstocks. *In*: Rom, R.C. and Carlson, R.F. (eds.). Rootstocks for Fruit Crops. John Wiley and Sons. New York, pp. 411–450.

McGranahan, G.H., Driver, J.A. and Tulecke, W. 1987. Tissue culture of *Juglans*, pp. 261–271. *In*: Bonga, J.M. and D.J. Durzan (eds.). Cell and Tissue Culture in Forestry. V3 Martinus Nijhoff Publishers Dordrecht.

McGranahan, G.H., Leslie, C.A., Uratsu, S.L., Martin, L.A. and Dandekar, A.M. 1988. Agrobacterium-mediated transformation of walnut somatic embryos and regeneration of transgenic plants. *Biotechnology,* **6**: 800–804.

Mehta, K., Sharma, O.C., Kashyap, A.S. and Thakur, B.S. 2005. Variability for various nut and kernel characters in some promising seedling walnuts (*Juglans regia* L.) *Acta Horticulturae,* **696**: 91-95.

Metzger, F.T. 1977. Sugar maple and yellow-birch seedling growth after simulated browsing. NC-140, USDA, Forest Service, North Central Research Station. 11 p.

Meynier, V. and Arnould, M.F. 1989.Compared effectiveness of antibiotic treatments and shoot tip culture on bacterial decontamination of an *in vitro*-propagated clone of hybrid walnut (*Juglans nigra* × *J. regia*). *Biol. Plant,* **31**(4): 269-275.

Miao YuQing; Li Guan; Wu SongLin; Song XiangChun and Zhang TieGang. 2010. Tissue culture and rapid propagation of thin-shell walnut. *Xinjiang Agricultural Sciences*, **47**(3): 503-507.

Mir, M. and Kumar, A. 2011. Effect of different methods, time and environmental conditions o grafting in walnut. *International Journal of Farm Sciences,* **1(2):** 17-22.

Moragrega, C. and Ozaktan, H. 2010. Apical necrosis of Persian (English) walnut (Juglans regia): an update. Journal of Plant Pathology, **92** (Supplement 1): S1.67-S1.71.

Nakova, M. and Dimova, M. 2003. Anthracnose disease (Gnomonia leptostyla (Fr.) Ces et de Not) on walnuts - chemicals for control. *Rasteniev' dni Nauki,* **40**(4): 366-369.

Navacchi, O; Zuccherelli, G. and Cozzolino, E. 2009. Walnut (*Juglans regia* L.) micropropagation: technical problems and agronomic implications. *Italus Hortus,* **16**(2): 67-70.

Nedev, N. Vasilev, V., Kavardzhikov, L. and Zdravkov, K. 1976. Propagation of walnut plant material. *In*: Nedev, N. (ed.), Nut Fruits. Plovdiv, pp 80-133.

Nedev, N. 1975. Inter-relations in the pollination and fertilization of some Bulgarian walnut cultivars. *Ovoshcharstvo,* **54**: 25-29.

Nedev, N.V. 1973. Effect of scion in rootstocks of common, eastern black and Manchurian walnuts during first aging period and at beginning of second aging period. II. Reproductive Behavior of Trees. *Grad.Lozar.Nauk,* **10**(1): 3-8.

Neuman, M.C., Preece, J.E., Van Sambeek, J.W. and Gaffney, G.R. 1993. Somatic embryogenesis and callus production from cotyledon explants of Eastern black walnut. *Plant Cell Tissue Organ Culture,* **32**: 9-18.

NHB. 2014. www.nhb.gov.in

Novak, Y.V. 1974. Stimulating shoots formation in walnut trees by grown prinning. *Lesov. I Agrolesomelior. Resp. Mezhved. Temat. Nauch. Sbornik,* **38**: 115-119.

O'Rourke, F.L.S. 1969. *In*: Handbook of North America Nut Trees. Jaynes, R.A. (ed.). The W.F. Humphery Press. Inc. Geneva, New York., Pp. 232-239.

Olson, W.H., Labavitch, J.M., Martin, G.C. and Beede, R.H. 1998. Maturation, harvesting and nut quality. *In*: D.E. Ramos, (ed.). Walnut Production Manual, pp. 273-276.

Osgood, J.W. 1985. Transplanting walnut trees. Walnut Research Reports Database. UC, Davis.

Ozkan, Y. and Gumus, A. 2001. Effect of different applications on grafting under controlled conditions of walnut (*Juglans regia* L.). *Acta Hort.*, **544**: 515-520.

Pandey, C. and Tomar, C S. 2012. Growth, flowering fruit set and yield in some cultivars/ selections of walnut (*Juglans regia*). *Indian Journal of Agricultural Sciences,* **82(5)**: 402–404.

Pandey, G., Tripathi, A.N. and Mandal, S. 2006. Promising genotypes of Persian walnut (*Juglans regia* L.) from Uttaranchal. *Indian Journal of Horticulture,* **63**: 256-259.

Parihar, M.C. and Bajpai, P.N. 1985. Pollination studies in walnut (*Juglans regia* L.). *Haryana J. Horti. Sci.,* **14**: 48-51.

Parihar, M.C., Bajpai, P.N. and Pathak, V.N. 1980. Diseases of fruit crops. Oxford and IBH Publishing Co. New Delhi, Pp 267.

Park, K.S. 1984. Studies on the heteroplastic grafting of Carpathian walnut. *Agris,* **66**: 95-108.

Pathak, R.K. and Srivastava, R.P. 1975. Walnut propagation by vegetative method. *Indian Horticulture,* **19**(4): 13-16.

Paunovic, S.M., Miletic, R., Mitrovic, M. and Jankovic, D. 2010. Walnut grafting success as affected by stratification. *Acta Agriculturae Serbica,* **30**: 165-172.

Payghamzadeh, K. 2008. Somatic embryogenesis from immature cotyledons and meristemic culture of walnut (*Juglans regia* L.).The M.Sc. thesis.College of Agriculture, Deptt.of

Plant breeding and Biotechnology, University of Agricultural and Natural Resources of Sari, Iran. pp. 48-77.

Payghamzadeh, K. and Kazemitabar, S.K. 2008a. Comparison effects of MT novel medium with modified DKW basal medium on walnut micropropagation. *Proceedings of the 1st National Conference of Student Biology and Modern World,* p. 204.

Payghamzadeh, K. and Kazemitabar, S.K. 2008d. Somatic embryogenesis from immature cotyledons of a local cultivar of walnut. *Proceedings of the 1st National Conference of Student Biology and Modern World*, p. 25.

Pijut, P.M. 1993a. Somatic embryogenesis in butternut, *Juglans cinerea*. *Can. J. For. Res.,* **23**: 835-838.

Pijut, P.M. 1993b. Regeneration of *Juglans cinrea* through somatic embryogenesis. Part II, *In Vitro Cell. Dev. Biol.,* **29**(3): 69.

Pijut, P.M. 1997. Micro-propagation of *J.cinerea* L. (Butternut).High-Tech and Micro-propagation V. (ed.) Y.P.S. Bajaj. Springer- Verlag, Berlin, *Biotechnol. Agric. For.,* **39**: 345-357.

Polito, V. 1998. Floral biology: flower structure, development and pollination. p. 127-132. *In*: Ramos, D.E. (ed.). Walnut production manual. Univ. Calif. Div. Agr. Natural Resources. Publ 3373.

Polito, V., Coates, B., Grant, J., Hasey, J., Micke, W., Olson, B., and Pinney, K. 1996. Pollen, Pistillate flower abortion/abscission. University of California, Fruit and Nut Research Info Center, Walnut Research Reports. p.77.

Polito, V., Grant. J. and Johnson, H. 2005. Walnut pollination and pistillate flower abortion. University of California Fruit and Nut Research Info Center. p.133.

Polito, V., Pinney, K. and Labavitch, J. 1998. Fruit growth and development. p. 139-143. *In:* Ramos, D.E. (ed.). Walnut production manual. Univ. Calif. Div. Agr. Natural Resources. Publ 3373.

Polito, V.S., McGranahan, G., Pinney, K. and Leslie, C. 1989. Origin of somatic embryos from repetitively embryogenic cultures of walnut (*Juglansregia* L.): implications for Agrobacterium-mediated transformation. *Plant Cell Rep.,* **8**: 219-221.

Ponder, F. Jr. and Van-Sambeek, J.W. 1997. Walnut fertilization and recommendations for wood and nut production. *In*: Proceedings of fifth Black Walnut symposium, Stockton, Missouri, USDA, 191: 128-137.

Porebski, S., Rzeznicka, B. and Poniedzialck, W. 2002. Comparison of two methods of walnut grafting. *Journal of Fruit and Ornamental Plant Research,* **10**: 55-62.

Proebsting, E.L. and Davy, S. 1946. In: Fruit Nutrition (eds. Childers, N.F.) Rutgers State Univ. New Brunswick, N.J. p. 262.

Proebsting, E.L. and Serr, E.F. 1966. Edible nuts. *In:* N.F. Childers (ed.). Nutrition of Fruit Crops. Hortic. Publs., New Brunswick, N. J.

Pu GuangLan; Xiao QianWen; Zhou LanYing. 2011. A new early-fruiting walnut cultivar 'Chuanzao 1'. *Acta Horticulturae Sinica,* **38**(10): 2025-2026.

Qureshi, A.S. and Dalal, M.A. 1985. Status of nut crops in Jammu and Kashmir state. *Progressive Horticulture,* **17**: 197-205.

Qureshi, S. N; Wani, M. S; Javaid, K; Raja, R. H. S; Basu, Y. A. 2014. Epicotyl grafting: a new vegetative propagation method in walnut under field conditions of Kashmir valley. *Indian Journal of Horticulture*, **71**(3): 306-310.

Ramos, D. E; Olson, W. H; Snyder, R. G. 1990. Effects of pruning on high density 'Hartley' walnuts (*Juglans regia* L.). *Acta Horticulturae,* **284:** 329-335.

Ramos, D., Kelley, K., Reil, W., Sibbett, G.S. and Synder, R. 2001. Establishment and management considerations for walnut hedgerow orchards. *Acta Horticulturae,* **544**: 427-435.

Ramos, D., McGranahan, G. and Hendricks, L. 1984. Walnuts. *Fruit Var. J.,* **38**: 112-120.

Ramos, D.E., Brown, L.C., Uriu, K. and Marangoni, B. 1978. Water stress affects size and quality of walnuts. *California Agriculture,* **10**:29-30.

Rashid, A. 1978. Effect of plant growth regulators on stooling of walnut. *Indian J. Hort.,* **35**(3): 202- 206.

Revilla, M.A., Majada, J. and Rodriguez, R. 1989. Walnut (*Juglans regia* L.) micro-propagation. *For. Tree Physiol.,* **46**: 149-151.

Rezaee, R., Vahdati, K., Grigoorian, V. and Valizadeh, M. 2008. Walnut grafting success and bleeding rate as affected by different grafting methods and seedling vigour. *Journal of Horticultural Science and Biotechnology,* **83**(1): 94-99.

Rezaee, R; Hassani, D. and Vahdati, K. 2014. Long term trials on top-working of walnut trees in Iran. *Acta Horticulturae,* **1050**: 197-201.

Rodriguez, R. 1982a. Callus initiation and root formation from *in-vitro* culture of walnut cotyledons. *Hort. Science,* **17**(2): 195-196.

Rodriguez, R. 1982b. Stimulation of multiple shoot bud in walnut seeds. *Hort Science,* **17**(4): 592.

Rodriguez, R. and Sanchez-Tames, R. 1981. Cultivo de tejidos y differenciacionen nogal. Rev Fac Cienc Univ Oviedo, **22**: 21-28.

Roschke, C. and Paula, M.P. 2006. http://ncrs.fs.fed.us/4157/local resources/downloads/ posters/ 2006/Roschke.pdf

Roth, P.L. 1971. 62nd Ann. Rept. Northern Nut Grower's Association USA, pp 45-47.

Rovira, M. and Aleta, N. 2006. Flower drop by precocious abortion in walnuts. *Nucis Newsletter,* **13**: 27-30.

Rovira, M. and Aleta, N., 1997. Pistillate flower abscission on four walnut cultivars. *Acta Horticulturae,* **442**: 231–234.

Ryugo, K.M. and Ramos, D.E. 1979. Tree training and pruning and management of close planted orchards. Walnut Research reports. 4.

Saida Haider; Zehra-Batool; Saiqa Tabassum; Tahira Perveen; Sadia Saleem; Fizza Naqvi; Huma Javed and Haleem, D. J. 2011. Effects of walnuts (*Juglans regia*) on learning and memory functions. *Plant Foods for Human Nutrition,* **66** (4): 335-340.

San, B. and Dumanoglu, H. 2007. Effect of desiccation, cold storage and gibberellic acid on germination of somatic embryos in walnut (*Juglans regia* L.). *New Zealand Journal of Crop and Horticultural Science,* **35:** 73-78.

San, B. and Hatic, D. 2006. Somatic embryogenesis from immature cotyledons of apomictic and non-apomictic seeds in walnut (*Juglans regia* L.).*Turk. J. Agric. For.,* **30**: 111-117.

Sanchez-Zamora, M.A., Tomas, J.C.T. and Garcia-Lopez, R. 2006.Embryo germination and proliferation *in vitro* of *Juglans regia* L. *Sci. Horti.,* **108**(3): 317-321.

Sartorius, R. and Stosser, R. 1992. Fruit development in walnut (*Juglans regia*). *Erwerbsobstbau,* **34**(4): 98-106.

Scaltsoyiannes, A., Tsoulpha, P., Panetsos, K.P. and Moulalis, D. 1997. Effect of genotype on micropropagation of walnut trees (*Juglans regia* L.). *Sil. Gen.,* **46**(6): 326-332.

Schlesselman, J.T., Smith, R.L., West, D. and Yoshida, H.A. 1998. The pre-emergence activity of thiazopyr under field conditions in California. *Proceedings of the Annual Meeting of the Western Society of Weed Science,* **51**: 81-85.

Serr, E.F. 1968. Dwarfing interstocks for Persian walnuts. *Plant Propagator,* **14**(1): 10-13.

Serr, E.F. 1969. *In*: Handbook of North America Nut Trees. Jaynes, R.A. (ed.). The W.F. Humphery Press. Inc. Geneva, New York., Pp. 240-263.

Sharma, A.K. and Dar, M.Y. 2006. Walnut grafting as influenced by environment conditions and rootstocks. *Indian Journal of Horticulture* **63**(3): 264-266.

Sharma, O.C. 1999. Studies on variability and selection of superior persian walnut (*Juglans regia* L.) seedling trees in Himachal Pradesh. Ph.D. Thesis, Dr. YSP UHF, Nauni-Solan, India.

Sharma, R. 2002. Selecting walnuts from seedling tree populations in some villages of district Sirmour (H.P.). M.Sc. Thesis, Dr. YSP UHF, Nauni-Solan, India.

Sharma, S.D., Dhuria, H.S. and Chauhan, J.S. 1982. Standardization of suitable time for grafting walnut under controlled conditions. *Indian Journal of Horticulture,* **39**: 201-204.

Sibbett, G.S., Martin, G.C. and Draper, T.M. 1978. Effects of prolonged drying and harvest delay following ethephon on walnut kernel quality. *California Agriculture,* **32**: 12-13.

Sibbett, G.S., Polito, V., McGranahan, G., Kelley, K. Olson, W., Hendricks, L., Catlin, P., and Grant, J. 1995. Minimizing PFA in walnuts. *Sun Diamond Grower,* **14**(1): 16-18.

Simovski, K. and Kuzmanov, I. 1972. *Jugoslovensko Vocarstvo,* **6**: 699-704.

Sofi, A.A., Ahmed, N., Singh, S.R., Verma, M.K., Pandey, G., Sharma, O.C., Shagoo, P.A., Pandit, A.H., Mir, H., Rashid, M., Dar, S.A. and Shah, S. 2012. New varieties released - Walnut. *Indian J. Hort*., **69**(2): 4-6.

Sofi, A.A., Verma, M.K., Ahmad, N., Verma, R.K., Mir, J.I., Singh, D.B., Kumar, D., Lal, S., Singh, S.R., Srivastava, K.K., Bhat, H.A., Pandit, A.H., Sheikh, M.A. and Shah, S. 2010. New Walnut varieties released. *Indian J. Hort.,* **67**(1): 1-2.

Solar, A., Stampar, F., Trost, M., Barbo, J. and Avsec, S. 2001. Comparison of different propagation methods in walnut (*Juglans regia* L.) made in Slovenia. *Acta Hort*., **544**: 527-530.

Srivastava, K. K; Banday, F. A; Singh, S. R; Sundouri, A. S; Sharma, M. K. 2008. Regeneration of walnut (Juglans regia L.) plants through single bud grafting and in situ grafting. *Indian Journal of Ecology,* **35**(2): 112-114.

Stephans, L.C., Krell, S.L. and Domoto, P.A. 1990. *In-vitro* propagation of *Juglansregia*.81 sf. Hamden, Connecticut USA. *Ann. Rep. Northern Nut Growers,* **9**: 122-126.

Stoke, H.F.1969. In: Handbook of North American Nut trees. (ed. Jaynes, R.A.) The W.F. Hamphery Press Inc. Geneva, NY, pp. 39-41.

Taha, N. A and Al-Wadaan M. A. 2011. Utility and importance of walnut, *Juglans regia* Linn: a review. African Journal of Microbiology Research, **5**(32): 5796-5805.

Teviotdale, B.L., Schroth, M.N. and Mulrean, E.N. 1985. *In:* walnut orchard management (ed. D. E. Ramos). Divi. Agri. Nat Res, Univ. of Calif. Publ. 21410, Pp. 153-157.

Thagusev, N.A. and Caroev, B.J. 1971. Manuring walnut orchards. *Sadovodstvo,* **4**: 30-31.

Thompson,J.F., Rumsey, T.R. and Grant, J.A. 1985. Walnut dehydration. *In*: Walnut Production Manual. 277-284. Davis, California, University of California.

Tian JianBao; Wang Yong; Wu YanXia; Cheng EnMing; Shao JiaMing; Liu ZhaoHong. 2009. A new walnut cultivar 'Jinboxiang 7'.. *Acta Horticulturae Sinica,* **36**(7): 1086.

Tulecke, W. and McGranahan, G. 1985. Somatic embryogenesis and plant regeneration from cotyledons of walnut, *Juglans regia* L. *Plant Sci.,* **40**: 57-63.

Unal, A. 1995. Studies on the effect of different practices on grafting success in bark, whip and top cleft grafting in walnuts. E.U.Z.F. *Dergisi Izmir, Turkey* **32**(1): 85-90.

Uriu, K. 1985. *In*: Walnut Orchard Management (ed. Ramos, D.E.) Divi. Agri. Nat. Res. Univ. of Calif. Publi. 21410 pp. 115-121.

Vadathi, K.; Razaee, R. and Mirmasoomi, M. 2009. Micropropagation of some dwarf and early mature walnut genotypes. *Biotechnology,* **8**(1): 171-175.

Vahdati, K. and Khalighi, A. 2001. Persian walnut stooling in Iran. *Acta Horti.,* **544**: 527-530.

Vahdati, K. and Zareie, N. 2006. Evaluation of side stub and hypocotyls grafting efficiency for walnut propagation in Iran. *Acta Hort*., **705**: 347-350.

Vahdati, K., Leslie, C., Zamani, Z. and McGranahan, G. 2004. Rooting and acclimatization of *in-vitro* grown shoots from mature trees of three Persian walnut cultivars. *Hort. Science,* **39**: 324–327.

Vahdati, K., Rezaee, R., Grigoorian, V., Valizadeh, M. and Azar, A.M. 2008. Rooting ability of Persian walnut as affected by seedling vigour in response to stool layering. *Journal of Horticultural Science and Biotechnology,* **83**(3) 334–338.

Vahdati, K; Ashrafi, E N; Ebrahimzadeh, H; Mirmasoumi, M. 2009. Improved micropropagation of walnut (*Juglans regia* L.) on media optimized for growth based upon mineral content of walnut seed. *Acta Horticulturae,* **839**: 117-124.

Verma, M.K., Ahmad, N., Sofi, A.A., Verma, R.K., Mir, J.I., Singh, D.B., Lal, S., Singh, D.K., Srivastava, S.R., Bhat, K.K., Pandit, H.A., Sheikh, A.H. and Shah, S. 2011. High yielding of better quality walnut released 2011. www.scribd.com

Voyiatzis, D.G. and McGranahan, G.H. 1994. An improved method for acclimatizing tissue-cultured walnut plantlets using an antitranspirant. *Hort. Sci.,* **29**(1): 4-2.

Wang Gui; Chang YueMei and Wu-Jing. 2008. A new walnut cultivar 'Jinxiang'. *Acta Horticulturae Sinica,* **35**(9): 1397.

Weber, G.C. and MacDaniels, L.H. 1969. Propagation. p. 29. *In*: Jaynes, R.A. (ed.). Handbook of North American nut trees. Northern Nut Growers Assn., Knoxville, TN.

Wilbur, O.R., Leslie, C.A., Forde, H.I. and Mckenna, J.R. 1998. Propagation. In: Ramos, D.E. (ed.). Walnut production manual. University of California, Division of Agriculture and Natural Resources, *Publication* 3373.

Wildman, W. E. 1985. Site preparation and correction of soil problems. pp 27-35 *In*: Ramos, D. E. (ed.) Walnut Orchard Management. Univ. Calif., Div. Agri. & Nat. Res., Pub. 21410.

Woodroof, J.G. 1979. *In*: Tree Nuts Production Processing Products, 2nd ed. pp. 604-55.

Wu GuoLiang. 1997. A promising walnut selection - NY83-05. *China Fruits*, **3:** 55.

WunGuoLiang; LiunHe; Liu QunLong; Xu YanGang; Wang Yong; Zhang PengFei and Shi XiangYu.2009. New late fruiting walnut cultivar 'Qinquan 1'. *Acta Horticulturae Sinica.* **36**(10): 1549-1550.

Xin, G; Zhang, .J D; Wang, H. M; Wang, H; Hu, Y.Q; Zhang, H.Y; Mao, H. H. and Jia, F. T . 2014. *A rapid grafting technique using direct sowing of fresh walnut seeds. Acta Horticulturae,* **1050** : 177-180.

Xu Ying; Zhang MeiYong and An-GuoNing, 2001. A new early fruit walnut variety - Daifeng. *Acta Horticulturae Sinica*, **28**(1): 87.

Yakabe, L.E., Parker, S.R. and Kluepfel, D.A. 2010. Incidence of *Agrobacterium tumefaciens* on walnut seeds used for rootstock production: Implications for crown gall management strategies. *Phytopathology,* **100**: 141.

Yalcin, M; Acican, T; Alibas, K; Erturk, U; Akca, Y; Polat, R 2014. Effects of ethephone application on different harvesting methods and hulling of walnut husk. *Acta Horticulturae,* **1050**: 323-329.

Yaskina, L. V. 1980. Dendrology. Ukituvchi. Tashkent. (in Russian).

Zachej, S. 1974. Pomological analysis of selected types of walnut. Vedecke Prace Vyskumneho Ustavu Rastlinnej Vyroby v Piest anoch, **12**: 257-265.

Zhang QingLong and Brown, P.H. 1999. The mechanism of foliar zinc absorption in pistachio and walnut. *Journal of the American Society for Horticultural Science,* **124**(3): 312-317.

Colour Plates

Chapter 14: Strawberry (*Fragaria x ananassa Duch.)*

Chandler strawberries

Runner production in strawberry

A view of a strawberry field in Meghalaya

Black polythene mulch in strawberry

A view of strawberry harvesting in India

Subject Index

Almond

Apple

Apricot

Cherry

Hazelnut

Kiwifruit

Olive

Peach

Pear

Pecan nut

Pistachio

Plum

Quince

Strawberry

Walnut